$GABA_B$ Receptors in Mammalian Function

GABA$_B$ Receptors in Mammalian Function

Edited by

N. G. BOWERY
The School of Pharmacy, London, UK

H. BITTIGER
CIBA-Geigy AG, Switzerland

and

H.-R. OLPE
CIBA-Geigy AG, Switzerland

JOHN WILEY & SONS
Chichester • New York • Brisbane • Toronto • Singapore

Other Wiley Editorial Offices

John Wiley & Sons. Inc. 605 Third Avenue,
New York, NY 10158-0012, USA

Jacaranda Wiley Ltd. G.P.O. Box 859, Brisbane,
Queensland 4001, Australia

John Wiley & Sons (Canada) Ltd, 22 Worcester Road,
Rexdale, Ontario M9W 1L1, Canada

John Wiley & Sons (SEA) Pte Ltd, 37 Jalan Pemimpin 05-04,
Block B, Union Industrial Building,
Singapore 2057

Library of Congress Cataloging-in-Publication Data:

GABA B receptors in mammalian function/edited by N. G. Bowery,
H. Bittiger, H.-R. Olpe.
p. cm.
On t.p. 'B' is subscript.
Includes bibliographical references and index.
ISBN 0 471 92461 X
1. GABA—Receptors—Physiological effect. I. Bowery, N. G. II. Bittiger, H. III. Olpe, H.-R. IV. Title: GABA B receptors in mammalian function.
QP364.7.G34 1990
599'.0188—dc20

90-39778
CIP

British Library Cataloguing in Publication Data:

GABA receptors in mammalian function.
1. Mammals. Brain. Receptors
I. Bowery, N. G. II. Bittinger, H. III. Olpe, H.-R.
599.0188

ISBN 0 471 92461 X

Typeset by APS Ltd., Salisbury, Wiltshire
Printed in Great Britain by Biddles Ltd, Guildford

Contents

Contributors ix

Introduction xv

I GENERAL

1 GABA$_B$ Receptors: Past, Present and Future 3
N. G. Bowery, G. D. Pratt and C. Knott

2 GABA$_B$ Receptor Agonists and Antagonists 29
D. I. B. Kerr, J. Ong and R. H. Prager

3 Biochemistry, Electrophysiology and Pharmacology of a New GABA$_B$ Antagonist: CGP 35348 47
H. Bittiger, W. Froestl, R. Hall, G. Karlsson, K. Klebs, H.-R. Olpe, M. F. Pozza, M. W. Steinmann and H. van Riezen

II CONTROL OF TRANSMITTER RELEASE

4 GABA$_B$ Receptors and Transmitter Release 63
P. C. Waldmeier and P. A. Baumann

5 Release-regulating GABA Autoreceptors in Human and Rat Central Nervous System 81
M. Raiteri, M. T. Giralt, G. Bonanno, A. Pittaluga, E. Fedele and G. Fontana

III PERIPHERAL ASPECTS

6 Review of Peripheral GABA$_B$ Effects 101
A. Giotti, A. Bartolini, P. Failli, G. Gentilini, M. Malcangio and L. Zilletti

IV BIOCHEMICAL ASPECTS

7 Evidence for the Existence of GABA$_B$ Receptor Subpopulations in Mammalian Central Nervous System 127
E. W. Karbon, S. H. Zorn, R. J. Newland and S. J. Enna

8 GABA$_B$ Receptors and Inhibition of Cyclic AMP Formation 141
W. J. Wojcik, M. Bertolino, R. A. Travagli, S. Vicini and M. Ulivi

9 Modulation by GABA of Agonist-induced Inositol Phospholipid Metabolism in Guinea-pig and Rat Brain 161
M. L. A. Crawford and J. M. Young

10 Solubilization and Partial Purification of Cerebral $GABA_B$ Receptors 183
K. Kuriyama and Y. Ohmori

V ELECTROPHYSIOLOGICAL ASPECTS

11 Pre- and Postsynaptic $GABA_B$ Responses in the Hippocampus 197
R. A. Nicoll and P. Dutar

12 Presynaptic $GABA_B$ Receptors on Rat Hippocampal Neurons 207
N. L. Harrison, N. A. Lambert and D. M. Lovinger

13 $GABA_B$ Receptor Function in the Hippocampus: Development and Control of Excitability 223
P. A. Schwartzkroin

14 $GABA_B$ Receptor-mediated Frequency Dependence of Inhibitory Postsynaptic Potentials of Neocortical Neurons In Vitro 239
R. A. Deisz and W. Zieglgänsberger

15 $GABA_B$-mediated Inhibition of Calcium Currents: A Possible Role in Presynaptic Inhibition 259
A. C. Dolphin, E. Huston and R. H. Scott

16 Physiological and Pharmacological Characterization of $GABA_B$ Receptor-mediated Potassium Conductance 273
N. Ogata

VI APPLIED ASPECTS

Antidepressants

17 Antidepressants and $GABA_B$ Site Upregulation 297
K. G. Lloyd

18 $GABA_B$ Binding Sites in Depression and Antidepressant Drug Action 309
J. A. Cross, S. C. Cheetham, M. R. Crompton, C. L. E. Katona and R. W. Horton

19 Autoradiographic Analysis of $GABA_B$ Receptors in Rat Frontal Cortex Following Chronic Antidepressant Administration 319
G. D. Pratt and N. G. Bowery

20 Neurochemical Effects of Prenatal Antidepressant Administration on Cortical $GABA_B$ Receptor Binding and Striatal [^{3}H]Dopamine Release 335
C. Knott, N. G. Bowery, C. DeFilipe, D. Montero and J. Del Rio

Epilepsy

21 $GABA_B$ Receptors and Experimental Models of Epilepsy 349
G. Karlsson, M. Schmutz, C. Kolb, H. Bittiger and H.-R. Olpe

Analgesia

22 $GABA_B$ Receptors and Analgesia 369
J. Sawynok

Closing Remarks 387
B. Gähwiler

Abstracts 391

Index 433

Contributors

A. Bartolini
Department of Preclinical and Clinical Pharmacology 'Mario Aiazzi Mancini', University of Florence, Viale G. B. Morgagni 65, 50134 Florence, Italy

P. A. Baumann
Research and Development Department, Pharmaceuticals Division, CIBA-Geigy AG, CH-4002 Basel, Switzerland

M. Bertolino
Fidia Georgetown Institute for the Neurosciences, School of Medicine, Georgetown University, 37th and 0 Sts, NW, Washington DC, WA 20057, USA

H. Bittiger
Research and Development Department, Pharmaceuticals Division, CIBA-Geigy AG, CH-4002 Basel, Switzerland

G. Bonanno
Istituto di Farmacologia e Farmacognosia, Università degli Studi di Genova, Viale Cembrano 4, 16148 Genoa, Italy

N. G. Bowery
Department of Pharmacology, School of Pharmacy, University of London, 29–39 Brunswick Square, London WC1N 1AX

S. C. Cheetham
Department of Pharmacology and Clinical Pharmacology, St George's Hospital Medical School, Cranmer Terrace, London SW17 0RE

M. L. A. Crawford
Department of Pharmacology, University of Cambridge, Tennis Court Road, Cambridge CB2 1QJ

M. R. Crompton
Department of Forensic Medicine, St George's Hospital Medical School, Cranmer Terrace, London SW17 0RE

J. A. Cross
Department of Pharmacology and Clinical Pharmacology, St George's Hospital Medical School, Cranmer Terrace, London SW17 0RE

C. DeFilipe
Department of Neuropharmacology, Cajal Institute, Madrid, Spain

R. A. Deisz
Department of Clinical Neuropharmacology, Max Planck Institute for Psychiatry, 8000 Munich 40, FRG

J. Del Rio
Department of Neuropharmacology, Cajal Institute, Madrid, Spain

A. C. Dolphin
Department of Pharmacology and Clinical Pharmacology, St George's Hospital Medical School, Cranmer Terrace, London SW17 0RE

P. Dutar
Department of Physiology, University of California at San Francisco, S-1210, San Francisco, CA 94143–0450, USA

S. J. Enna
Nova Pharmaceutical Corporation, 6200 Freeport Center, Baltimore, MD 21224–2788, USA

P. Failli
Department of Preclinical and Clinical Pharmacology 'Mario Aiazzi Mancini', University of Florence, Viale G. B. Morgagni 65, 50134 Florence, Italy

E. Fedele
Istituto di Farmacologia e Farmacognosia, Università degli Studi di Genova, Viale Cembrano 4, 16148 Genoa, Italy

G. Fontana
Istituto di Farmacologia e Farmacognosia, Università degli Studi di Genova, Viale Cembrano 4, 16148 Genoa, Italy

W. Froestl
Research and Development Department, Pharmaceuticals Division, CIBA-Geigy AG, CH-4002 Basel, Switzerland

G. Gentilini
Department of Preclinical and Clinical Pharmacology 'Mario Aiazzi Mancini', University of Florence, Viale G. B. Morgagni 65, 50134 Florence, Italy

A. Giotti
Department of Preclinical and Clinical Pharmacology 'Mario Aiazzi Mancini', University of Florence, Viale G. B. Morgagni 65, 50134 Florence, Italy

M. T. Giralt
Departamento de Farmacologia y Terapeutica, Facultad de Medicina y Odontologia, Universidad del Pais Vasco, 48940 Leioa (Vizcaya), Spain

R. Hall
Central Research Laboratories, CIBA-Geigy Ltd, Tennax Road, Trafford Park, Manchester M17 1WT

N. L. Harrison
Department of Anesthesia and Critical Care, University of Chicago, Chicago, IL 60637, USA

R. W. Horton
Department of Pharmacology and Clinical Pharmacology, St George's Hospital Medical School, Cranmer Terrace, London SW17 0RE

E. Huston
Department of Pharmacology and Clinical Pharmacology, St George's Hospital Medical School, Cranmer Terrace, London SW17 0RE

E. W. Karbon
Nova Pharmaceutical Corporation, 6200 Freeport Center, Baltimore, MD 21224–2788, USA

G. Karlsson
Research and Development Department, Pharmaceuticals Division, CIBA-Geigy AG, CH-4002 Basel, Switzerland

C. L. E. Katona
Department of Psychiatry, University College and Middlesex School of Medicine, Gower Street, London

D. I. B. Kerr
Department of Anaesthesia and Intensive Care, University of Adelaide, South Australia 5000, Australia

K. Klebs
Research and Development Department, Pharmaceuticals Division, CIBA-Geigy AG, CH-4002 Basel, Switzerland

C. Knott
Department of Pharmacology, School of Pharmacy, University of London, 29–39 Brunswick Square, London WC1N 1AX

C. Kolb
Research and Development Department, Pharmaceuticals Division, CIBA-Geigy AG, CH-4002 Basel, Switzerland

K. Kuriyama
Department of Pharmacology, Kyoto Prefectural University of Medicine, Kawaramachi-Hirokoji, Kamikyo-Ku, Kyoto 602, Japan

N. A. Lambert
Department of Neurobiology, Northeastern Ohio Universities College of Medicine, Rootstown, OH 44272, USA

K. G. Lloyd
Wyeth Research (UK) Ltd, Huntercombe Lane South, Taplow, Maidenhead, Berks SL6 0PH

D. M. Lovinger
Laboratory of Physiological and Pharmacological Sciences, NIAAA, Room 2, 12501 Washington Ave, Rockville, MD 20852, USA

M. Malcangio
Department of Preclinical and Clinical Pharmacology 'Mario Aiazzi Mancini', University of Florence, Viale G. B. Morgagni 65, 50134 Florence, Italy

D. Montero
Department of Neuropharmacology, Cajal Institute, Madrid, Spain

R. J. Newland
Glassboro State College, Glassboro, NJ 08028, USA

R. A. Nicoll
Department of Pharmacology, University of California at San Francisco, S-1210, San Francisco, CA 94143–0450, USA

N. Ogata
Department of Pharmacology, Faculty of Medicine, Kyushu University, Fukuoka 812, Japan

Y. Ohmori
Department of Pharmacology, Kyoto Prefectural University of Medicine, Kawaramachi-Hirokoji, Kamikyo-Ku, Kyoto 602, Japan

H.-R. Olpe
Research and Development Department, Pharmaceuticals Division, CIBA-Geigy AG, CH-4002 Basel, Switzerland

J. Ong
Department of Anaesthesia and Intensive Care, University of Adelaide, South Australia 5000, Australia

A. Pittaluga
Istituto di Farmacologia e Farmacognosia, Università degli Studi di Genova, Viale Cembrano 4, 16148 Genoa, Italy

M. F. Pozza
Research and Development Department, Pharmaceuticals Division, CIBA-Geigy AG, CH-4002 Basel, Switzerland

R. H. Prager
School of Physical Sciences, Flinders University, Bedford Park, South Australia 5042, Australia

G. D. Pratt
Department of Pharmacology, School of Pharmacy, University of London, 29–39 Brunswick Square, London WC1N 1AX

M. Raiteri
Istituto di Farmacologia e Farmacognosia, Università degli Studi di Genova, Viale Cembrano 4, 16148 Genoa, Italy

H. van Riezen
Research and Development Department, Pharmaceuticals Division, CIBA-Geigy AG, CH-4002 Basel, Switzerland

J. Sawynok
Department of Pharmacology, Dalhousie University, Halifax, Nova Scotia, Canada B3H 4H7

M. Schmutz
Research and Development Department, Pharmaceuticals Division, Ciba-Geigy AG, CH-4002 Basel, Switzerland

P. A. Schwartzkroin
Department of Neurological Surgery and Department of Physiology and Biophysics, University of Washington, Seattle, WA 98195, USA

R. H. Scott
Department of Pharmacology, St George's Hospital Medical School, Cranmer Terrace, London SW17 0RE

M. W. Steinmann
Research and Development Department, Pharmaceuticals Division, CIBA-Geigy AG, CH-4002 Basel, Switzerland

R. A. Travagli
Fidia Georgetown Institute for the Neurosciences, School of Medicine, Georgetown University, 37th and 0 Sts, NW, Washington DC, WA 20057, USA

M. Ulivi
Fidia Georgetown Institute for the Neurosciences, School of Medicine, Georgetown University, 37th and 0 Sts, NW, Washington DC, WA 20057, USA

S. Vicini
Fidia Georgetown Institute for the Neurosciences, School of Medicine, Georgetown University, 37th and 0 Sts, NW, Washington DC, WA 20057, USA

P. C. Waldmeier
Research and Development Department, Pharmaceuticals Division, CIBA-Geigy AG, CH-4002 Basel, Switzerland

W. J. Wojcik
Fidia Georgetown Institute for the Neurosciences, School of Medicine, Georgetown University, 37th and 0 Sts, NW, Washington DC, WA 20057, USA

J. M. Young
Department of Pharmacology, University of Cambridge, Tennis Court Road, Cambridge CB2 1QJ

W. Zieglgänsberger
Department of Clinical Neuropharmacology, Max Planck Institute for Psychiatry, 8000 Munich 40, FRG

L. Zilletti
Department of Preclinical and Clinical Pharmacology 'Mario Aiazzi Mancini', University of Florence, Viale G. B. Morgagni 65, 50134 Florence, Italy

S. H. Zorn
Pfizer Central Research, Eastern Point Road, Groton, CT 06340, USA

Introduction

It never ceases to amaze me that γ-aminobutyric acid (GABA), which has such a simple structure, should play such an important neurotransmitter role within the mammalian brain. But there is no doubt that it is one of the most important inhibitory mediators within higher centres. Present estimates indicate that more than 40% of all inhibitory synaptic activity is mediated through this neutral amino acid. Despite this importance it is only relatively recently that its role has been established.

The presence of GABA in plants has been known for more than half a century but it was not until 1950 that Roberts and Frankel and Awapara *et al.* first described its existence in brain tissue. Many studies ensued in attempts to establish its significance such that by the end of the decade it had been both proposed and rejected as an inhibitory neurotransmitter (see Table 1). However, in 1966 Krnjevic and Schwartz obtained firm electrophysiological evidence to implicate GABA as an inhibitory transmitter in the cerebral cortex. This provided a major impetus for further studies in neurochemistry, pharmacology, medicinal chemistry as well as electrophysiology, which together have now substantiated its physiological role.

A brief chronological record of the events associated with the history of GABA research is shown in Table 1. One of the major advances occurred in 1970, when the selective competitive antagonist bicuculline was first described by Curtis *et al.* and it is the receptor at which bicuculline acts, rather than other elements of the synaptic process, which has probably attracted the majority of research studies. The recognition site for GABA now designated $GABA_A$ appears to form part of a receptor complex which comprises sites for the benzodiazepines, barbiturates and the sedative steroids. Drugs with affinity for these sites modulate the GABA recognition site and it is probably these loci which are implicated in the therapeutic action of such agents. Recent molecular biology studies have provided valuable information about the sequence(s) of the receptor complex and indicate that recognition sites for the modulators are present within the subunit sequences.

This $GABA_A$ receptor is not the only locus of GABA action. In line with most other receptors for neurotransmitters, those for GABA are not homogeneous. In 1981 a second population of receptors for GABA, designated $GABA_B$, was recognized which differed from $GABA_A$ receptors in nearly all respects. The pharmacological profile for activation as well as the associated secondary

Table 1. Brief history of research into GABA in mammalian brain

Event	Year	Reference
Presence in brain	1950	Roberts and Frankel *J. Biol. Chem.*, **187**, 55–63.
	1950	Awapara, Landau, Fuerst and Seale *J. Biol. Chem.*, **187**, 35–39.
Cerebral metabolic pathway	1956	Roberts In: S.R. Korey and J.I. Nurnberger (eds) *Progress in neurobiology*, vol. 1. London, Cassel, pp. 11–25.
Action on brain neurons	1956	Hayashi *Chemical physiology of excitation in muscle and nerve.* Tokyo: Nakayama-Shoten.
Proposed as inhibitory neurotransmitter	1957	Bazemore, Elliott and Florey *J. Neurochem.*, **1**, 334–339.
Rejected as inhibitory neurotransmitter	1959	McLennan *J. Physiol.*, **146**, 358–368.
	1959	Curtis, Phillis and Watkins *J. Physiol.*, **146**, 185–203.
Proposed as non-specific depressant	1962	Bindman, Lippold and Redfearn *J. Physiol.*, **162**, 105–120.
	1964	Crawford and Curtis *Br. J. Pharmacol.*, **23**, 313–329.
Evidence for role as inhibitory neurotransmitter	1966	Krnjevic and Schwartz *Nature*, **211**, 1372–1374.
Selective receptor antagonism	1970	Curtis, Duggan, Felix and Johnston *Nature*, **226**, 1222–1224.
Evoked release in vivo	1971	Iversen, Mitchell and Srinivasan *J. Physiol.*, **212**, 519–534.
Benzodiazepine facilitation of GABA synaptic activity	1975	Haefely, Kulcsár, Möhler, Pieri, Polc and Schaffer *Adv. Biochem. Psychopharmacol.*, **14**, 131–151.
Receptor subtype, $GABA_A$ and $GABA_B$	1981	Hill and Bowery *Nature*, **290**, 149–152.
First sequence of $GABA_A$ receptor	1987	Schofield, Darlison, Fujita, Burt, Stephenson, Rodriguez, Rhee *et al.* *Nature*, **328**, 221–227.

mechanisms were (and are) quite distinct. Fortuitously, a selective agonist for the new site was already available. Baclofen, *β-p*-chlorophenyl-GABA, was first introduced 9 years earlier as a muscle relaxant and still remains as the drug of choice in spasticity.

The compound arose from an original idea of the CIBA chemist, Keberle, who set out to design a GABA mimetic which would cross the blood–brain barrier with greater ease than GABA itself. Apparently Keberle's idea arose whilst listening to a seminar given by the eminent physiologist Feldberg at the Royal Society in the mid-1960s. Having missed his plane connection to Switzerland, Keberle took the opportunity of attending the seminar in which Feldberg spoke, among other items, about GABA, which had just emerged as a neurotransmitter. One wonders if and when baclofen would have appeared had Keberle missed the lecture instead of his plane.

The recognition of a second functional receptor for GABA has kindled a research interest which has increased steadily since 1981 in much the same way, though to a lesser extent, as general GABA research. Last year $GABA_B$ receptor research reports formed about 13 % of all original articles on GABA (Figure 1). It is worth noting, in Figure 1, the occasional sudden increase in the number of published papers on GABA. These may correspond to the introduction of a new compound or a new concept. For example, the $GABA_A$ antagonist bicuculline became available from 1970 and, following a brief lag phase, there appears to have been a significant increase in the research output (Figure 1). Of course, there may be other reasons for the increase but only an examination of each of

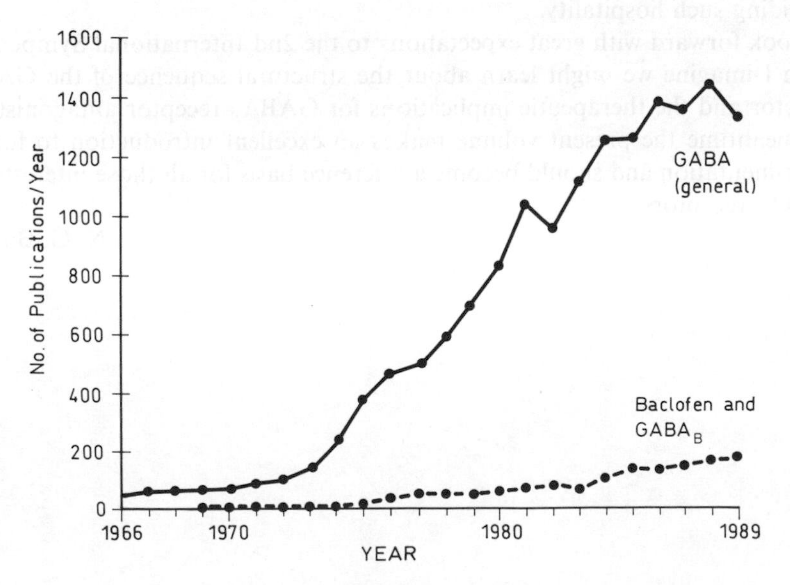

Figure 1. Time-course of original publications on GABA and on baclofen and $GABA_B$ since 1966. (Source: Medline. Data compiled by: L. Lisgarten and K. Henderson, School of Pharmacy, London.)

the published papers would provide the exact answer. Interestingly, another increase in the publication rate appears post-1975, which corresponds to the origin of the concept of benzodiazepines facilitating the action of GABA. However, this also coincides with the availability of a radiolabelled binding method for detecting GABA receptors in synaptic membranes. Perhaps a similar sharp increase in $GABA_B$ research will occur as a result of the recent availability of selective antagonists.

There is now a firm basis for further research in the $GABA_B$ field and the present volume provides the background for this. The contents derive from the proceedings of the 1st International $GABA_B$ Symposium, which took place at King's College, Cambridge, England in September 1989. The meeting provided a forum for much discussion and it is hoped that the chapters of the volume have captured some of the enthusiasm of the presenters as well as documenting their data.

Symposia do not happen spontaneously. They require much planning, coordination, financial support and the help of many individuals. I would therefore like to take this opportunity, on behalf of my fellow editors, to express my sincere thanks to the pharmaceutical companies - CIBA-Geigy (the major sponsor); Smith, Kline & French; Glaxo; Wyeth; Merck, Sharp & Dohme; Pfizer; Nova; and Beecham - for their financial support, and to all the individuals, in particular Yvonne Haseldine, Derek King and colleagues from the School of Pharmacy, who helped make the meeting run so smoothly. My additional thanks go to David Smyth and his colleagues at King's College for providing such hospitality.

I look forward with great expectations to the 2nd International Symposium, when I imagine we might learn about the structural sequence of the $GABA_B$ receptor and the therapeutic implications for $GABA_B$ receptor antagonists. In the meantime the present volume makes an excellent introduction to further experimentation and should become a reference basis for all those interested in $GABA_B$ receptors.

N. G. Bowery

Part I

GENERAL

1 $GABA_B$ Receptors: Past, Present and Future

N. G. BOWERY, G. D. PRATT and C. KNOTT

INTRODUCTION

The term $GABA_B$ originated in 1981 (Hill and Bowery, 1981), when it became necessary to differentiate a novel receptor for the amino acid, GABA, from the 'classical' receptor which had been delineated earlier in the 1960s (see Krnjevic, 1974 for review). Hill and Bowery (1981) designated the classical site, $GABA_A$, to encompass all GABA receptors which are associated with chloride ion channels and blocked by bicuculline, thereby distinguishing them from $GABA_B$ sites which are neither coupled to chloride ion channels nor blocked by bicuculline. Much research has focused on $GABA_B$ sites since 1981, enabling a functional as well as a pharmacological separation of the two receptors.

HISTORICAL ASPECTS

It often happens in scientific studies that the eventual outcome of experimental endeavours is not as predicted. Such was the situation with the advent of $GABA_B$ receptors. Experiments were designed to obtain a model for GABA-mediated presynaptic inhibition in the spinal cord using peripheral nervous tissue. For this purpose rat postganglionic sympathetic nervous tissue was chosen because previous experiments had shown that neurons of the superior cervical ganglion, which innervate the heart, possess bicuculline-sensitive GABA receptors (Adams and Brown, 1975; Bowery and Brown, 1974). When these receptors were activated, an increase in membrane chloride ion conductance, occurred. It was reasoned that, if the receptors were present on the cell soma, they might also be present on the cell membrane at the nerve terminal (Figure 1.1). Since there was no electrophysiological means of measuring GABA-mediated responses in nerve terminals, it seemed that a possible way forward might be to measure the evoked release of transmitter from the terminals in the absence and presence of GABA. This should, of course, be the ultimate effect of any presynaptic modulator. Fortunately the postganglionic processes of the superior cervical ganglion had been studied extensively and many neurochemical experiments had been performed on the sympathetic innervation of the heart (see Iversen, 1967). In particular, Iversen (1967) had shown that

$GABA_B$ Receptors in Mammalian Function. Edited by N.G. Bowery, H. Bittiger and H.-R. Olpe

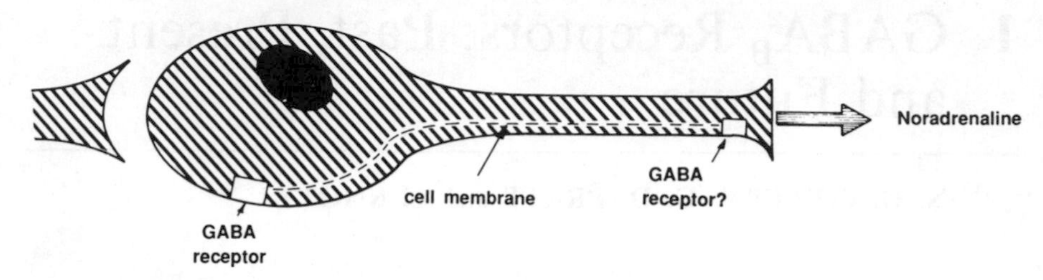

Figure 1.1. Basis of the original hypothesis for obtaining a model of spinal presynaptic GABA receptors using a peripheral sympathetic neuron. Bicuculline-sensitive GABA receptors defined on the ganglion cell body might also have been present on the nerve terminal. Activation was expected to produce an increase in Cl^- conductance with a resulting decrease in evoked transmitter (noradrenaline) release. Whilst inhibition of release was subsequently demonstrated, the receptor on the terminal was different from that on the cell body and Cl^- conductance was not implicated

[^{3}H]noradrenaline is accumulated by sympathetic nerve terminals in isolated atria and could be subsequently released by transmural stimulation in a manner consistent with a neurotransmitter function. This provided the basis for our experiments in which GABA was applied just prior to and during transmural stimulation of superfused isolated atria. The α_2-adrenoceptor antagonist, yohimbine, was present in the superfusion solution to increase the overflow of [^{3}H]noradrenaline (Langer, 1977).

As predicted, GABA inhibited the evoked release of [^{3}H]noradrenaline in a dose-dependent manner (Bowery and Hudson, 1979) (Figure 1.2), although the maximum inhibition achieved was never greater than 50–60%. But what we did not predict were the pharmacological characteristics of this inhibitory response. The response to GABA was not blocked by any recognized GABA antagonist and was not mimicked by GABA receptor agonists such as isoguvacine, THIP (4,5,6,7-tetrahydroisooxazolo[5,4-*c*]pyridin-3-ol) and piperidine-4-sulphonic acid. Even muscimol was only a weak mimetic (Bowery *et al.*, 1981). The most striking observation was that the GABA analogue, β-*p*-chlorophenyl-GABA (baclofen), was active. This compound, which was designed as a GABA mimetic, had previously been shown to be ineffective at bicuculline-sensitive GABA receptors (Curtis *et al.*, 1974; Davies and Watkins, 1974; Olpe *et al.*, 1977). In the atrial model it not only mimicked GABA but did so in a stereospecific manner. Activity resided in the (−)-isomer and this difference between the isomers has subsequently become one of the requirements for substantiating the presence of $GABA_B$ receptors.

The consequence of these findings was that the original hypothesis had to be revised. Whilst GABA inhibited transmitter release, the effect was not mediated through the classical receptor and subsequent experiments showed that the inhibition was not chloride-dependent (Bowery *et al.*, 1981). We concluded

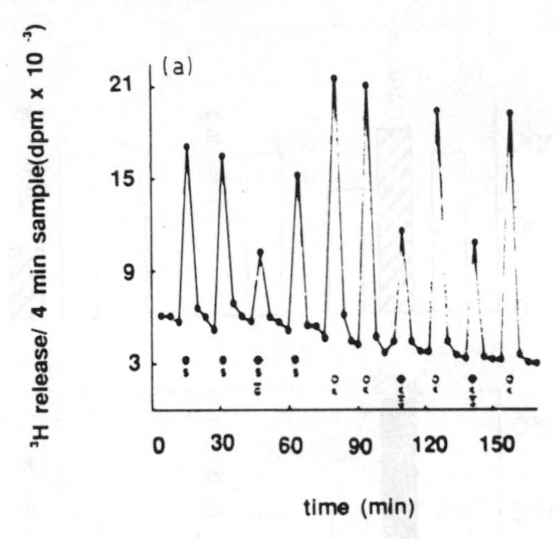

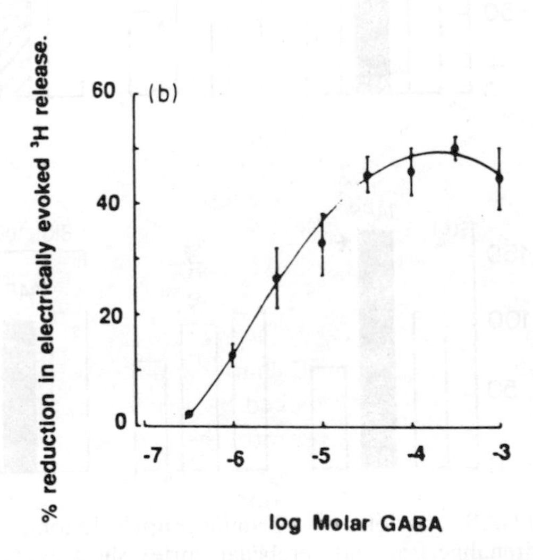

Figure 1.2. GABA-induced depression of the K^+-evoked and transmurally-stimulated release of [^{3}H]noradrenaline from rat isolated atria. (a) Two atria from a single rat were incubated in 0.4 μM [^{3}H]noradrenaline for 40 min followed by incubation in fresh Krebs–Henseleit solution for a further 45 min. The tissue was then superfused at 32 °C and the effluent collected at consecutive periods of 4 min. The tritium content of each sample (●) is represented on the ordinate (d.p.m./1.8 ml sample) and plotted against time (abscissa). The tissue was stimulated transmurally (3 Hz, 0.5 ms at 10 V) for 1-min periods (●, S) or by superfusion with high (60 mM) KCl-containing solution for 2-min periods (○, K). GABA (G, 100 μM) was added 30 s before S or K and left in contact during the stimulation periods. (b) shows the log dose–response curve for the inhibitory action of GABA on transmural stimulation ($n = 4$ determinations for each point). Percentage inhibition (ordinate) was determined by reference to the stimulated release immediately before and after the addition of GABA. Vertical bars indicate SEM

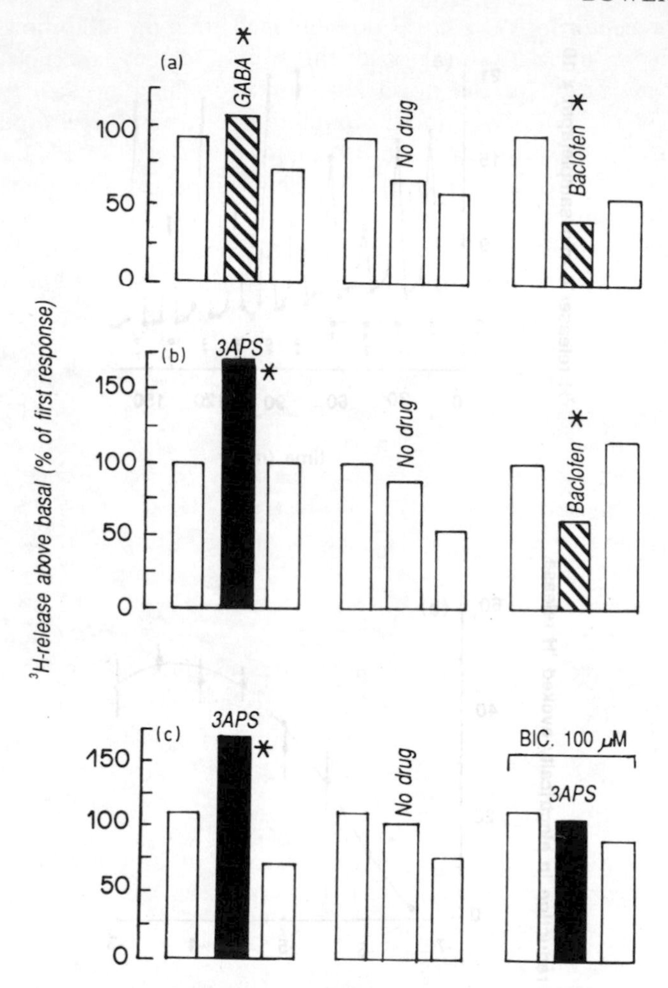

Figure 1.3. Effect of GABA, baclofen and 3-aminopropylsulphonic acid (3-APS) on the efflux of [^{3}H]noradrenaline from rat cerebellar cortex slices evoked by 15 mM KCl. Three or four slices of cortex (0.25 mm) were incubated for 20 min at 32 °C in Krebs solution containing 0.8 μM [^{3}H]noradrenaline, 0.5 mM iproniazid and 0.1 mM ascorbic acid. The slices were then incubated for a further 20 min in fresh solution at 32 °C before transferring to a superfusion chamber. Superfusion fluid was collected every 4 min (2 ml volume). The tritium content of each sample was measured by liquid scintillation spectrometry. KCl was added to the superfusion fluid for 2-min periods at intervals of 16 min. The data derive from three experiments (a, b and c). In each experiment, control slices were perfused with KCl alone for three periods (middle three histogram bars in each experiment). In experiment (a), GABA (100 μM, first set) or (±)-baclofen (100 μM, third set) was added during the second period of KCl addition. 3-APS (100 μM) and baclofen were similarly applied in experiment (b) and 3-APS was also applied in both the first and third sets of experiment (c). Bicuculline (BIC, 100 μM) was also present in the superfusion fluid of the third set of experiment (c). Asterisks indicate significant difference from control ($P < 0.05$ Student's t-test)

that a novel receptor for GABA was possibly mediating the inhibition and that (−)-baclofen was a selective agonist for this site. In further experiments it was shown that activation of the novel receptor in isolated organs diminished contractions evoked by transmural stimulation but not those produced by applied agonists (Anwar and Mason, 1982; Bowery *et al.*, 1981; Kleinrok and Kilbinger, 1983; Ong and Kerr, 1983). This supported the concept of a presynaptic location.

These studies prompted us to examine whether similar data could be obtained on the control of neurotransmitter release from central nervous tissue. We chose to study the release of three radiolabelled transmitters, [^{3}H]noradrenaline, [^{3}H]dopamine and [^{3}H]5-hydroxytryptamine (5-HT, serotonin), from rat cerebellar, striatal and frontal cortical slices respectively. In all cases we observed that (−)- but not (+)-baclofen (1–100 μM) inhibited the release of the radiolabelled amines evoked by 25 mM K^+ (Bowery *et al.*, 1980). A detailed study of this effect was performed in cerebellar slices. In the presence of 25 mM K^+, GABA mimicked the effect of (−)-baclofen. However, in 15 mM K^+, GABA enhanced the release of [^{3}H]noradrenaline (Figure 1.3). If bicuculline (100 μM) was present, then GABA, like baclofen, reduced the evoked release (data not shown). Baclofen never enhanced the evoked release. 3-Aminopropylsulphonic acid (3-APS) enhanced the release in the absence of bicuculline but, unlike GABA, never inhibited release in the presence of this antagonist (Figure 1.3). We concluded that two types of GABA receptor were present in the cerebellar slice but both were only manifest at the lower potassium concentration.

Final confirmation for the existence of two GABA receptor types came when we obtained evidence for distinct binding sites for [^{3}H]GABA in brain synaptic membranes. The novel site was detected when either Ca^{2+} or Mg^{2+} was added to the incubation buffer. This increased the total amount bound without altering the non-displaceable portion (Table 1). The additional binding could not be

Table 1. Influence of [Ca^{2+}] on [^{3}H]GABA binding to rat brain synaptic membranes in Tris–HCl buffer solution

	[^{3}H]GABA bound (c.p.m.)	
	$CaCl_2$ absent	$CaCl_2$(2.5 mM) present
Alone	2170 ± 20	2900 ± 20
+GABA	920 ± 5	850 ± 9
+Isoguvacine	900 ± 6	1550 ± 18
+(±)-baclofen	1800 ± 22	1870 ± 4
+Isoguvacine and baclofen	900 ± 3	900 ± 5

Data show amount of [^{3}H]GABA (10 nM) bound in the absence or presence of excess GABA (100 μM), isoguvacine (100 μM), (±)-baclofen (100 μM) or isoguvacine plus baclofen in the presence and absence of 2.5 mM $CaCl_2$.

prevented by isoguvacine but was completely suppressed by unlabelled GABA or (−)-baclofen. Baclofen had no effect on the binding in the absence of Ca^{2+}. It was at this point that we were convinced that it was necessary to introduce the terms $GABA_A$ and $GABA_B$ to describe these separate sites (Bowery *et al.*, 1983; Hill and Bowery, 1981).

CHARACTERISTICS OF $GABA_B$ RECEPTORS AND THEIR ACTIVATION

After the initial observations on the pharmacological separation of $GABA_B$ and $GABA_A$ sites, numerous distinct characteristics became apparent (Table 2).

Unlike $GABA_A$ receptors, the function of $GABA_B$ sites does not appear to be modulated by benzodiazepines or barbiturates, and low concentrations of the detergent, Triton X-100, decrease the binding capacity of $GABA_B$ sites, whereas they increase that of $GABA_A$ sites in synaptic membranes (Bowery *et al.*, 1983; Enna and Snyder, 1977). Guanyl nucleotides reduce the binding affinity of $GABA_B$ sites, whereas $GABA_A$ sites are unaffected (Hill *et al.*, 1984). This influence of guanyl nucleotides provided the initial clue that $GABA_B$ sites, like other receptors similarly affected by guanyl nucleotides (Rodbell, 1980), may be coupled to one or more G-proteins. This was borne out by the studies of Asano *et al.* (1985), who showed that synaptic membranes incubated in pertussis toxin were able to bind less [^{3}H]GABA. This was reversed if a source of G-protein

Table 2. Contrasting aspects of $GABA_A$ and $GABA_B$ receptors

	$GABA_A$	$GABA_B$
Consequences of receptor activation		
channels	Cl^- conductance	K^+conductance ↑ Ca^{2+}conductance ↓
adenylyl cyclase	—	activation or inhibition
Synaptic activation	fast IPSPs	slow IPSPs
Influence of ions or chemicals		
Ca^{2+}/Mg^{2+} dependence	no significant effect	absolute
guanyl nucleotide	no effect	binding affinity ↓
benzodiazepines	ligand binding and functional response ↑	no effect
barbiturates	functional response ↑	no effect
Triton detergent	binding capacity ↑	binding capacity ↓
Selective agonist	isoguvacine	(−)-baclofen
Selective antagonist	bicuculline	saclofen

IPSP, inhibitory postsynaptic potential.

was added to the incubation mixture containing the pertussis toxin-treated membranes. Pertussis toxin ADP-ribosylates the G-protein to prevent its coupling to the receptor and facilitation of high-affinity binding.

Further studies showed that the functional response to GABA$_B$ receptor activation was also impaired by pertussis toxin treatment. In particular the electrophysiological responses and the inhibition of neurotransmitter release provoked by GABA$_B$ site activation were prevented by pertussis toxin treatment (Andrade *et al.*, 1986; Harrison, 1989; Thalmann, 1988; see also Wojcik *et al.*, Chapter 8; Nicoll and Dutar, Chapter 11; and Dolphin *et al.*, Chapter 15).

The coupling through G-protein(s) mediates the inhibition of adenylyl cyclase, with a resulting decrease in intracellular cyclic AMP. This will reduce phosphorylation within the cell. Coupling with G-protein also appears to mediate the electrophysiological responses to GABA$_B$ site activation, namely an increase in membrane K^+ conductance and a decrease in membrane Ca^{2+} conductance, since both are reduced by pertussis toxin pretreatment (Andrade *et al.*, 1986; Holz *et al.*, 1986). These electrophysiological responses probably do not result from any alteration in adenylyl cyclase activity (Dolphin *et al.*, 1989).

Surprisingly, GABA$_B$ site activation can also produce an increase in cyclic AMP accumulation (Hill, 1985; Hill *et al.*, 1984; Karbon and Enna, 1985; Watling and Bristow, 1986). However, this only occurs in brain slices and not membranes and the increase is only manifest when adenylyl cyclase has been activated by other receptor agonists such as isoprenaline, noradrenaline, vasoactive intestinal peptide (VIP) or histamine. This enhancement by GABA$_B$ receptor activation does not appear to be mediated through a G-protein since treatment with pertussis toxin did not prevent the enhancement (Bowery *et al.*, 1989; but see Wojcik *et al.*, Chapter 8). The mechanism of this enhancement has been postulated by Karbon and Enna (1989) to be mediated through a product of phospholipase A_2 metabolism, such as arachidonic acid, since inhibitors of this enzyme decreased the effect. However, these inhibitors, for example quinacrine and glucocorticoids, are not selective and therefore the significance of these findings remains to be confirmed. An important aspect of this dual effect on cyclic AMP levels is whether the decrease and increase represent separate processing systems mediated through distinct receptor subtypes. This is a point we will come to later.

ELECTROPHYSIOLOGICAL ASPECTS

It is often suggested that receptor binding studies, and perhaps even neurochemical data, do not necessarily reflect functional receptor-mediated events and that binding sites may be little more than artefacts. We would be the first to agree that data obtained from radiolabelled binding studies do not stand alone. They require the support of more functional responses. Whilst the results obtained from the transmitter-release experiments provide some of this information, perhaps the most critical evidence emanates from electrophysiological studies. The original electrophysiological evidence was obtained by three groups working on isolated dorsal root ganglia and hippocampal slices. The studies by Desarmenien *et al.* (1984) and Dunlap (1981) on dorsal root ganglion demonstrated that $GABA_B$ site activation decreases the duration of action potentials in C and Aδ neurons. This they attributed to an inhibition of Ca^{2+} conductance, allowing faster repolarization. The reduction in duration could have been due to an increase in K^+ conductance, but subsequent experiments in dorsal root ganglion have shown that the former argument is correct (Dolphin and Scott, 1986; Feltz *et al.*, 1987). Nevertheless an increase in K^+ conductance seemed a likely explanation at the time since intracellular recording in CA1 cells of hippocampal slices performed by Newberry and Nicoll (1984, 1985) clearly indicated that an increase in K^+ conductance occurs on stimulating $GABA_B$ receptors.

Pharmacological separation of these two systems has so far not been demonstrated and we are left with $GABA_B$ receptors affecting two ion channels apparently in an independent manner. This is in marked contrast to $GABA_A$ sites, which are linked solely to chloride ion channels in mammalian neurons (see Bormann, 1988). Details of the electrophysiological events associated with $GABA_B$ site activation are covered in the chapters by Nicoll and Dutar (Chapter 11), Harrison *et al.* (Chapter 12), Schwartzkroin (Chapter 13), Dolphin *et al.* (Chapter 15), Ogata (Chapter 16) and Karlsson *et al.* (Chapter 21).

RECEPTOR AUTORADIOGRAPHY

One of the distinctions between $GABA_B$ and $GABA_A$ sites is their different distributions within mammalian brain tissue (Bowery *et al.*, 1987; Price *et al.*, 1987; Wilkin *et al.*, 1981). This has been demonstrated using receptor autoradiography, which allows their discrete regional locations in thin brain sections to be ascertained. The technique developed for $GABA_B$ autoradiography is an extension of that employed in membrane binding studies (Wilkin *et al.*, 1981). The important requirements are that Ca^{2+} (2.5 mM) is present in the incubation medium and excess isoguvacine, or another selective $GABA_A$ ligand, is there to prevent binding to $GABA_A$ sites. Under these conditions [^{3}H]GABA can be used as the label so that alteration of the incubation conditions, i.e. removal of Ca^{2+} and isoguvacine with the addition of excess baclofen, enables $GABA_A$ sites to be labelled in adjacent sections.

Information obtained from autoradiographic studies has been invaluable in providing evidence for pre- and postsynaptic locations. Evidence for a presynaptic location for $GABA_B$ sites has been obtained in the dorsal horn of the spinal cord (Price *et al.*, 1987), the interpeduncular nucleus (Price *et al.*, 1984) and striatum (Moratalla and Bowery, 1988). In all cases the data derive from lesioned brain tissue in which nerve terminals have been allowed to degenerate. This approach has been necessary because the resolution of autoradiography at the level of the light microscope is too low to distinguish between pre- and postsynaptic elements, and electron micrography is not possible with present ligands for $GABA_B$ sites due to their rapid dissociation rate.

Lesioning presynaptic pathways as a procedure for determining receptor locations can be criticized on the basis that any chronic changes in the presynaptic element may influence the postsynaptic site. Neurotrophic factors released from the presynaptic nerves may modulate the postsynaptic response and/or the binding of the transmitter ligand. In an attempt to allow for this possibility in the study of the rat spinal cord we compared the autoradiographic profiles of $GABA_B$ ($[^3H]$GABA) and neurotensin ($[^3H]$neurotensin) receptor binding sites (Price *et al.*, 1987). Binding sites for $[^3H]$neurotensin have been reported to be restricted to postsynaptic elements in the dorsal horn, and following rhizotomy of afferent fibres we observed no change in the quantitative distribution of $[^3H]$neurotensin binding sites (15 days after rhizotomy). By contrast the number of $GABA_B$ binding sites was reduced by up to 50% in laminae I–IV following rhizotomy or neonatal capsaicin treatment. Little difference was observed between the rhizotomy and capsaicin treatment and we therefore concluded that up to half of the $GABA_B$ sites in the dorsal horn are on presynaptic terminals of small-diameter fibres rather than larger 1a afferents since capsaicin is known to affect only the small-diameter fibres. The results do not exclude the possibility that more than 50% are located presynaptically since the terminals of descending fibres would still be intact. It may be inferred that $GABA_B$ sites have a significant role in the modulation of transmitter release from afferent terminals within the spinal cord. This is borne out by studies showing that the evoked release of substance P from isolated rat spinal cord can be reduced by $GABA_B$ receptor activation (Del Rio *et al.*, 1984; Ray *et al.* Abstract 6). It might also provide the basis for the analgesic action of baclofen (Cutting and Jordan, 1975; Levy and Proudfit, 1977), although this role for $GABA_B$ sites within the cord has been questioned (see Sawynok, Chapter 22). Similarly, presynaptic $GABA_B$ sites might also be implicated in the muscle-relaxant effect of baclofen (Fox *et al.*, 1978), although it seems unlikely that the site of action is in laminae I–IV since the primary effect of baclofen appears to be on monosynaptic pathways controlling motoneuron output (Davies, 1981).

The interpeduncular nucleus is a region containing a high density of $GABA_B$ but not $GABA_A$ sites (Bowery *et al.*, 1987). Autoradiographic data indicate that all of these $GABA_B$ sites appear to be located presynaptically since electrolytic lesion of the main afferent fibres, the retroflexus of Meynert, reduced their density by more than 90% after 12 days. Kainic acid or ibotenic acid injected directly into this region failed to alter the binding to $GABA_B$ sites after the same period (Price *et al.*, 1984). The physiological and pharmacological significance of these findings has yet to be determined since evidence for any GABA-mediated innervation of afferent terminals, outside the spinal cord, has not been demonstrated and any behavioural effects which might result from the direct injection of baclofen into the interpeduncular nucleus remain to be studied. However $GABA_B$-mediated electrophysiological events have been recorded in vitro from the interpeduncular nucleus, indicating the presence of functional receptors in this brain region (Docherty and Halliwell, 1984).

A presynaptic location for $GABA_B$ sites has also been demonstrated autoradiographically in the caudate putamen. The density of $GABA_B$ sites diminished by about 25% 14–60 days after injection of 6-hydroxydopamine into the substantia nigra, to destroy dopaminergic afferents (Moratalla and Bowery, 1988). A similar decrease in $GABA_B$ binding also occurred in the same structure after chronic ablation of the frontal cerebral cortex (14–60 days earlier). By contrast the density of $GABA_A$ and [^{3}H]flunitrazepam binding sites in the same animals increased after cortical ablation (Moratalla *et al.*, 1989). The explanation for this apparent increase is unknown although one possibility is that it may be due to sprouting of nerve processes; however, we have no evidence for this at present. The decrease in $GABA_B$ binding can be attributed to a loss of terminals of either nigrostriatal or corticostriatal afferents. Presumably combined nigral and cortical lesions would produce a greater additive reduction in $GABA_B$ binding but this has yet to be examined.

RECEPTOR HETEROGENEITY

Recent evidence indicates that $GABA_B$ receptors are probably not homogeneous and that receptor subtypes exist. A neurochemical examination of this is provided by Karbon *et al.* in Chapter 7. In brief, two neurochemical pathways have been associated with $GABA_B$ site activation, namely inhibition of basal or forskolin-stimulated cyclic AMP formation (Wojcik and Neff, 1984; Xu and Wojcik, 1986) and potentiation of β-adrenoceptors, as in VIP- and histamine-induced cyclic AMP accumulation (Hill, 1985; Hill *et al.*, 1984; Karbon and Enna, 1985; Watling and Bristow, 1986). The latter effect only occurs in brain slices, whereas the inhibitory effect occurs in synaptic membranes as well. This would suggest that synaptic integrity is required for the potentiation to be produced but details of the mechanism(s) are unknown.

Studies with certain $GABA_B$ agonists have revealed that these two responses are not produced by all $GABA_B$ mimetics (Bowery *et al.*, in press; Pratt *et al.*, 1989; Scherer *et al.*, 1988). 3-Aminopropylphosphinic acid (3-APA) (Figure 1.4) exhibits high affinity for $GABA_B$ binding sites (Bittiger *et al.*, 1988; Dingwall *et al.*, 1987; Pratt *et al.*, 1989), but it is only equipotent with baclofen in inhibiting forskolin-stimulated cyclic AMP formation. However, it produces the same maximum effect as baclofen. By contrast the maximum potentiation of β-adrenoceptor-mediated cyclic AMP accumulation by 3-APA is much less than that produced by baclofen, suggesting a partial agonist effect. In line with this

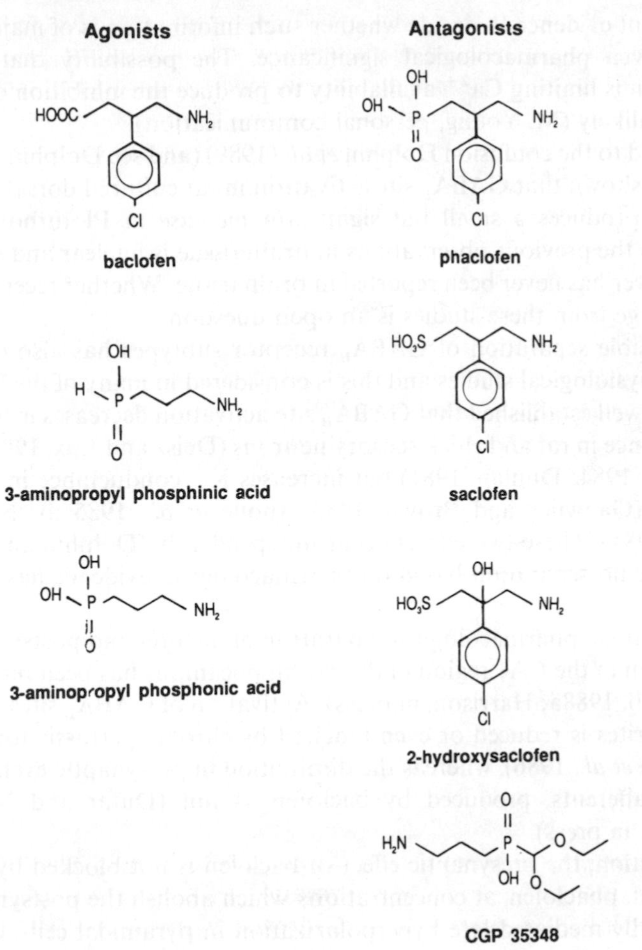

Figure 1.4. Structures of the most potent and selective agonists and antagonists of $GABA_B$ receptors so far described

proposal a high concentration of 3-APA antagonized the response to baclofen (Pratt *et al.*, 1989). This result could be explained by different receptor populations associated with each of the two neurochemical effects. The data presented by Karbon *et al.* in Chapter 7 concur with this possibility.

A third neurochemical process has been linked with $GABA_B$ site activation–phosphatidyl inositol (PI) turnover. Two research groups have demonstrated an inhibition of 5-HT- and histamine-induced PI turnover in brain tissue (Crawford and Young, 1988; Godfrey *et al.*, 1987). Crawford and Young (Chapter 9) discuss the possible significance of this interaction, but because of the wide species variation and the lack of detailed pharmacology there is at present insufficient evidence to decide whether such information is of major physiological or even pharmacological significance. The possibility that $GABA_B$ site activation is limiting Ca^{2+} availability to produce the inhibition of PI turnover seems unlikely (M. Young, personal communication).

To add to the confusion Dolphin *et al.* (1989) (and see Dolphin *et al.*, Chapter 15) have shown that $GABA_B$ site activation in rat cultured dorsal root ganglion neurons produces a small but significant *increase* in PI turnover. How this relates to the previous observations in brain tissue is unclear and stimulation of PI turnover has never been reported in brain tissue. Whether receptor subtyping will emerge from these studies is an open question.

A possible separation of $GABA_B$ receptor subtypes has also emerged from electrophysiological studies and this is considered in many of the later chapters. It is now well established that $GABA_B$ site activation decreases membrane Ca^{2+} conductance in rat and chick sensory neurons (Deisz and Lux, 1985; Desarmenien *et al.*, 1984; Dunlap, 1981) but increases K^+ conductance in hippocampal neurons (Gähwiler and Brown, 1985; Inoue *et al.*, 1985a,b; Newberry and Nicoll, 1984). These two effects occur independently (Dolphin and Scott, 1986) but so far no separation based on pharmacological evidence has been demonstrated.

By contrast, pharmacological separation of the pre- and postsynaptic actions of baclofen in the CA1 region of the rat hippocampus has been reported (Dutar and Nicoll, 1988a; Harrison, in press). Activation of $GABA_B$ sites on pyramidal cell dendrites is reduced or even blocked by chronic pertussis toxin treatment (Andrade *et al.*, 1986), whereas the diminution in presynaptic events in pyramidal cell afferents, produced by baclofen, is not (Dutar and Nicoll, 1988a; Harrison, in press).

In addition, the presynaptic effect of baclofen is not blocked by the $GABA_B$ antagonist, phaclofen, at concentrations which abolish the postsynaptic action. Synaptically mediated late hyperpolarization in pyramidal cells, equivalent to the postsynaptic action of baclofen, is also blocked by this weak, although selective, antagonist (Dutar and Nicoll, 1988b; Soltesz *et al.*, 1988, 1989). However, the low potency of phaclofen might account for this difference as observed in the cat spinal cord by Kerr *et al.* (1987). In this preparation, baclofen elicits in the dorsal horn pre- and postsynaptic effects which can be separated. Initial experiments with phaclofen showed that the presynaptic effects could be blocked, whereas the postsynaptic actions were unaffected by the antagonist (Kerr *et al.*, 1987). In subsequent experiments with the relatively more potent antagonist 2-hydroxysaclofen (Figure 1.4), both pre- and postsynaptic effects were blocked (Curtis *et al.*, 1988). The authors suggested that a lack of potency of phaclofen was the reason for this apparent discrepancy between the two antagonists rather than different receptors. At present no systematic pharmacological studies have been performed on pre- and postsynaptic sites within the hippocampus or other brain region. Thus to conclude that separate receptor types exist at these sites is still premature but the possibility is attractive.

AUTORADIOGRAPHY AND G-PROTEINS

Receptor autoradiography has been used to great effect in studies on the association of GABA$_B$ sites with G-proteins. The results of Dutar and Nicoll (1988a) referred to above prompted us to examine whether the apparent separation between pre- and postsynaptic sites and their pertussis toxin sensitivities could be demonstrated using receptor autoradiography. It had previously been established by Asano *et al.* (1985) that binding to GABA$_B$ sites in synaptic membranes from rat brain is inhibited by in-vitro incubation in pertussis toxin. We have examined the influence of both in-vivo and in-vitro pertussis toxin administration on the distribution of GABA$_B$ sites in rat brain sections using receptor autoradiography (Bowery *et al.*, 1989, in press). In the in-vivo studies pertussis toxin (4 μg) was injected unilaterally into the hippocampus (right side) and 4–8 days later 10 μm cryostat sections of rat brain were prepared and the binding to GABA$_A$ and GABA$_B$ sites was performed as described previously (Bowery *et al.*, 1987).

GABA$_B$ but not GABA$_A$ binding was significantly reduced throughout the hippocampus although the decrease was not homogeneous (Table 3). In particular, binding in the CA1 region was reduced by 43%, whereas in the CA3 region it was only reduced by 21% even though ventral and dorsal CA1 regions were equally affected (Moratalla, Knott and Bowery, in preparation). This suggested that limited diffusion might not be the reason for the difference in

Table 3. Effect of pertussis toxin (PTX) in vivo on GABA receptor binding in rat hippocampal sections

Hippocampal region	Normal (left side)	PTX treated (right side)	Difference	% decrease
	GABA$_B$ binding (nCi/g; n = 22)			
CA1	1007 ± 109	579 ± 74 **	428 ± 49	43
CA3	980 ± 79	775 ± 60 *	205 ± 31†	21
Dentate gyrus	1552 ± 187	862 ± 118**	690	45
	GABA$_A$ binding (nCi/g)			
CA1	2141 ± 139	1903 ± 129	238 ± 20	11
CA3	1393 ± 94	1225 ± 98	177 ± 18	12
Dentate gyrus	3178 ± 213	2724 ± 186	454	14

*$P < 0.05$, **$P < 0.005$, compared to untreated side. $P < 0.005$, compared with difference in CA1. Student's t-test. Data from Moratalla, Knott and Bowery (in preparation).

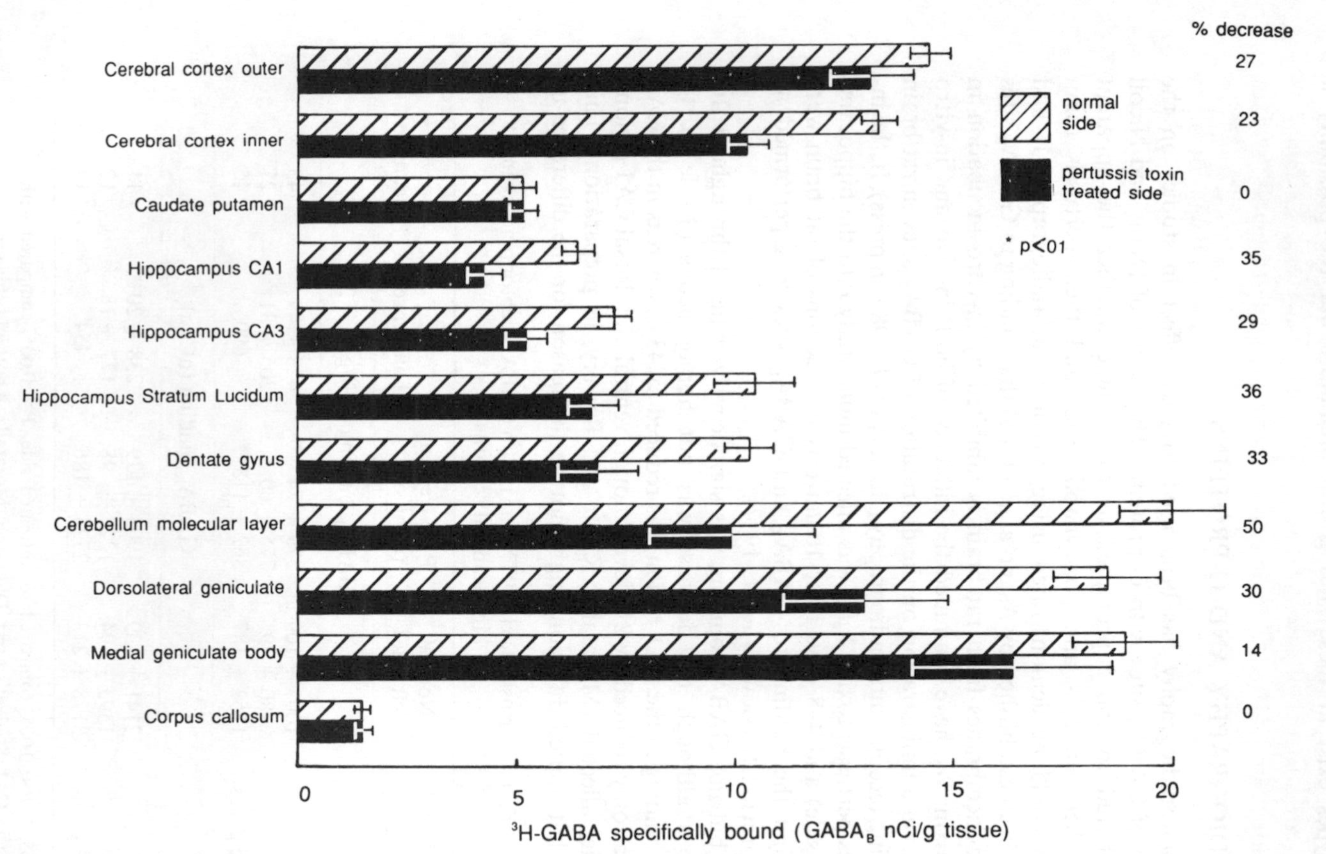

Figure 1.5. Effect of pertussis toxin in vitro on the regional binding of [^{3}H]GABA to $GABA_B$ sites in rat brain slices. Horizontal bars indicate the mean regional densities ($\pm$ SEM nCi/g tissue) of $GABA_B$ binding sites determined from 27 sections (three rat brains). The control (hatched bars) and pertussis toxin-treated (solid bars) values derive from the same brain sections, one half incubated for 24 h in Krebs–bicarbonate buffer alone, the other half in buffer containing pertussis toxin (10 μM) for the same period

binding between CA1 and CA3 but instead this may reflect a difference in pertussis toxin sensitivity between these regions. To assess this possibility we have developed an in-vitro technique to examine the effect of pertussis toxin, avoiding possible differences in regional diffusion.

Coronal blocks of fresh brain, 250–400 μm thick, were incubated for up to 24 h at 25 °C in Krebs-bicarbonate buffer containing 10 μM pertussis toxin (Porton Products Ltd) and NAD, ATP, thymidine, EDTA, dithiothreitol and $MgCl_2$. To compare untreated and toxin-treated tissue, one half of each block was incubated in pertussis toxin-free medium whilst the other half was incubated for the same period in toxin-containing medium. After incubation the half-blocks were mounted together on cork chucks and rapidly frozen for the preparation of cryostat sections (10 μm). GABA$_B$ (and GABA$_A$) receptor autoradiography was then performed.

Whilst pertussis toxin does not readily penetrate cells in vitro, our preliminary experiments indicated a 25–30% reduction in the density of GABA$_B$ binding sites within the hippocampus after 24 h incubation in 10 μM toxin (Figure 1.5). However, the reduction in binding in the CA3 region was the same as within the CA1 region, which would suggest that limited diffusion could have accounted for the apparent difference observed in CA1 and CA3 after in-vivo administration (Table 3).

EFFECTS OF GABA$_B$ RECEPTOR ACTIVATION

Numerous effects have been implicated in the response to GABA$_B$ receptor activation both within and outside the central nervous system (CNS) (Table 4; see Bowery, 1989). The question is which, if any, are of physiological significance and which might be pursued as targets for therapeutic manipulation.

Table 4. Actions associated with GABA$_B$ site activation

Inhibition of evoked neurotransmitter release: heteroreceptors and autoreceptors
Analgesia
Hypothermia
Catatonia
Muscle relaxation
Inhibition of 5-hydroxytryptophan-induced head twitch
Stimulation of brown-fat thermogenesis
Hypotension
Mediation of slow inhibitory postsynaptic potentials
Inhibition of voltage-dependent synaptosomal Ca^{2+} uptake
Increased gastric motility
Reduction of memory consolidation and retention
Suppression of hippocampal epileptiform activity
Epileptogenic stimulation of gastric acid secretion
Inhibition of corticotropin-releasing hormone release

The criteria required to be fulfilled to establish a neurotransmitter–receptor system as physiological are rigorous and in the majority of cases are never met completely. In particular, demonstrating the physiological release of a putative transmitter is often difficult to perform. Studies on GABA have enabled many of these criteria to be fulfilled, including release, and its physiological role in mediating inhibition through $GABA_A$ receptor-controlled chloride conductance mechanisms is now well established. Recent studies by Dutar and Nicoll (1988b) and Soltesz *et al.* (1988) in the hippocampus and by other groups in the neocortex, lateral geniculate and septum (Hasuo and Gallagher, 1988; Karlsson *et al.*, 1988; Soltesz *et al.*, 1988) have shown that $GABA_B$ receptor-mediated transmission is also of physiological importance.

It seems likely therefore that a number of the central effects of $GABA_B$ receptor stimulation are of physiological relevance but the question is which ones?

Baclofen has been used for more than 15 years in the treatment of spasticity by decreasing motoneuron output within the spinal cord. This probably results from a depression in mono- and maybe even polysynaptic reflex activity produced by a reduction in excitatory transmitter release (Fox *et al.*, 1978). The site of this action is likely to be within the ventral horn, particularly if monosynaptic mechanisms are implicated.

Receptor autoradiography studies (Price *et al.*, 1987) showing a high density of $GABA_B$ sites in the dorsal rather than ventral horn may cast doubt on whether $GABA_B$ sites are responsible for mediating the antispastic action of baclofen. However, results obtained by F. Brugger (personal communication) indicate that the inhibitory action of baclofen on spinal reflexes can be blocked by the new $GABA_B$ antagonist CGP 35348, and previous data by Schwarz *et al.* (1988) and Wüllner *et al.* (1989) have shown that the weak antagonists δ-aminovaleric acid and phaclofen can reduce the decrease in muscle tone of spastic mutant rats and depression of spinal mono- and polysynaptic reflexes produced by baclofen. It would seem therefore that the control of motoneuron output may be an important role for $GABA_B$ sites. Similarly, it is possible that the modulation of small-diameter afferent output within the dorsal horn is also an important physiological mechanism. $GABA_B$ sites are clearly present on these terminals (Price *et al.*, 1987) and presynaptic GABAergic innervation has been described (Barber *et al.*, 1978; McLaughlin *et al.*, 1975). In addition, activation of these receptors reduces the evoked release of substance P (Del Rio *et al.*, 1984; Ray *et al.*, Abstract 6), which may be one of the small-fibre transmitters in the cord. The obvious significance of this effect is the production of analgesia. Baclofen is an analgesic and part of this effect is mediated within the spinal cord (see Sawynok, 1987; Terrence *et al.*, 1985). However, the role of $GABA_B$ sites in this action has been questioned by Sawynok (see Chapter 22). D-Baclofen, which is a very weak $GABA_B$ agonist, appears to be an antagonist of the spinal analgesic action of L-baclofen (Sawynok and Dickson, 1985), although this may stem from partial agonist activity. In fact there have been no other reports of L-baclofen or GABA$_B$ receptor antagonism by D-baclofen in any other system (Haas *et al.*, 1985; Howe and Zieglgänsberger, 1986).

Despite numerous actions having been reported for baclofen in vivo as well as in vitro in laboratory animals (see Bowery, 1989), surprisingly few centrally mediated effects, other than muscle relaxation and analgesia, have been reported in man. Even effects generated in the periphery are relatively few. Why should this be? Two reasons come to mind: firstly, desensitization or tolerance and secondly, in the case of CNS activity, poor brain access. There is no doubt that, whilst baclofen can cross the blood-brain barrier, it does so rather poorly. Only the exquisite sensitivity of spinal reflexes to the action of baclofen allows any effects to be seen. Tolerance to agonists is a general phenomenon and it would not be surprising if the same occurred with baclofen. Indeed the analgesic actions of baclofen can decline with time (Fromm *et al.*, 1984). Perhaps desensitization prevents any manifestation of more subtle responses such as effects on hormone release. Certainly, in view of the marked neuronal hyperpolarization produced by baclofen in many neuronal populations in vitro (see e.g. Newberry and Nicoll, 1984), it seems surprising that baclofen does not produce more gross effects in vivo. A more potent agonist which readily penetrates the brain is required to answer this question. 3-APA is the only agonist at present which appears to exhibit greater affinity for GABA$_B$ binding sites in vitro. However, its efficacy in functional receptor systems appears to be little better than that of baclofen.

Antagonists

Predicting the in-vivo effects of a selective GABA$_B$ antagonist within the CNS is not easy. Even in the periphery one cannot be sure that any actions will result from its in-vivo administration. Although GABA$_B$ receptors have been demonstrated in the mammalian periphery (for review see Erdö and Bowery, 1986), the equivalent information about neural processes or humoral release processes which might be responsible for their activation is not available. It is possible that GABA neurons which form part of the enteric nervous system (Jessen *et al.*, 1979, 1983) may provide the input to GABA$_B$ receptors on autonomic fibres innervating the intestine (Giotti *et al.*, 1983; Ong and Kerr, 1983, 1984). Certainly GABA and its forming enzyme, glutamic acid decarboxylase, have been demonstrated in many peripheral organs such as the intestine, bladder and pancreas (Kusunoki *et al.*, 1984; Tanaka 1985; see Erdö and Bowery, 1986).

For any pure antagonist to elicit effects in vivo, the presence of physiological 'tone' is required. Without the presence of any agonist activation it might be expected that an antagonist would appear to be inert. In fact initial indications from the studies with the GABA$_B$ antagonist, CGP 35348 (Figure 1.4), reported by Bittiger *et al.* (Chapter 3) and Karlsson *et al.* (Chapter 21), suggest that it produces few behavioural changes in normal animals at doses which block applied baclofen. What will be most important to examine are its effects under abnormal conditions where $GABA_B$ sites are possibly being activated. But what abnormal conditions are likely to be associated with an increase in $GABA_B$ function, and what therapeutic indications might there be for a $GABA_B$ antagonist? A number of possibilities arise, such as depression, epilepsy and cognitive impairment, but these are speculative and derive from our knowledge about the actions of baclofen in animal models. Baclofen decreases the release of neurotransmitter amines (Bowery *et al.*, 1980; Gray and Green, 1987) and, since the amine theory of depression (Schildkraut, 1965) suggests that an increase in amine transmitter function may reduce depression, the expected increase in synaptic levels of amines produced by a $GABA_B$ antagonist may be advantageous.

Data obtained with baclofen on neuronal firing activity in brain slices are conflicting. Both suppression and enhancement of seizure activity have been reported (Ault *et al.*, 1986; Cottrell and Robertson, 1987; Ogata *et al.*, 1986; Swartzwelder *et al.*, 1986, 1987a). In the majority of studies baclofen has been described as a potential anti-convulsant since it reduces neuronal firing activity. As a consequence different groups have suggested that baclofen may have therapeutic potential in the treatment of seizures (Ault *et al.*, 1986; Jones, 1989). However, in man baclofen appears to exhibit little or no anticonvulsant activity. Contrasting animal data, obtained by Swartzwelder *et al.* (1978a), suggest that baclofen may even be epileptogenic in vivo. This group have shown that baclofen decreases interictal activity in hippocampal slices and this would tend to increase ictal (seizure) activity rather than decrease it. Thus an increase in $GABA_B$ receptor activation may precipitate or facilitate epileptogenic activity and therefore an antagonist might be expected to suppress this. Unfortunately, preliminary reports by Karlsson *et al.* (Chapter 21) with the new $GABA_B$ antagonist CGP 35348 do not support this view. CGP 35348 was neither a strong anticonvulsant nor convulsant in whole-animal models. Thus the role of $GABA_B$ sites in seizure activity has not been substantiated . However, it does not indicate that at least one pharmacological action of an antagonist of $GABA_B$ receptors is distinct from that of a $GABA_A$ antagonist. All $GABA_A$ antagonists including bicuculline and picrotoxin produce convulsions in vivo and seizure activity in brain slice preparations. This may in part be due to the production of an imbalance with excitatory processing mechanisms since excitatory amino-acid (*N*-methyl-D-aspartate, NMDA) receptor antagonists can reduce the seizure activity associated with $GABA_A$ receptor antagonism.

A third potential therapeutic role for a $GABA_B$ antagonist may be in the improvement of cognition. Modulation of hippocampal activity is evident with $GABA_B$ receptor agonists and $GABA_B$ antagonists might be expected to increase pyramidal cell activation by preventing the late inhibitory postsynaptic potential (IPSP). In-vivo data from Swartzwelder *et al.* (1987b) indicate that baclofen decreases the retention and consolidation of memory; thus an antagonist may produce the reverse effect, particularly under abnormal conditions of pronounced memory loss.

THE FUTURE

The advent of a GABA$_B$ antagonist, CGP 35348, which can cross the blood-brain barrier is, presumably, only the beginning of a series of even more active compounds capable of blocking the effects of GABA$_B$ site activation. Hopefully a full analysis of their actions will reveal the physiological and pathological significance of GABA$_B$ receptors. At present we can only guess.

One pathologically induced change in GABA$_B$ sites appears to occur in the brains of bulbectomized rats and in the learned helplessness rat model. Both of these models are considered to reflect depression in man, antidepressants being able to reverse the abnormal behaviour of these animals. Interestingly the density of GABA$_B$ binding sites in frontal cortex membranes from these rats is reduced and treatment with various antidepressants reverses the deficit as well as the behavioural pattern (Joly *et al.*, 1987; Lloyd, 1989). Even in normal rats treatment with antidepressants can raise the level of GABA$_B$ binding (Lloyd *et al.*, 1985) and other groups have also reported increases in functional GABA$_B$ receptor activity following antidepressant treatment (Suzdak and Gianutsos, 1986). However, these findings have not been confirmed by others (Cross and Horton, 1986, 1988; H. Bittiger, personal communication), and even where an increase has been obtained using receptor autoradiography this did not correlate with changes in another marker of antidepressant treatment, i.e. downregulation of β-adrenoceptor binding (Pratt and Bowery, Chapter 19).

The molecular aspects of GABA$_B$ receptors have so far received little attention although some studies on purification and isolation are in progress (see Kuriyama *et al.*, Chapter 10). This must be a focus of attention in the future but the topic is not without its problems. To define the correct receptor sequence it is essential that it is expressed ultimately in a suitable vector. Whilst this has been possible for receptors linked to fast ion channels (e.g. GABA$_A$ and glycine; Grenningloh *et al.*, 1987; Schofield *et al.*, 1987), the expression of receptors functionally coupled to G-proteins appears to be much more difficult. Measurement of functional activity is essential for the assessment of complete receptor expression. Undoubtedly this problem will be overcome in the near future and we suspect that the GABA$_B$ sequence will be obtained by searching existing cDNA libraries rather than through the receptor purification procedure used to great effect in understanding the molecular biology of the GABA$_A$ benzodiazepine receptor (see Schofield, 1989). The question of multiple $GABA_B$ receptors will surely then be resolved.

FINAL COMMENT

The $GABA_B$ receptor has come a long way since its discovery 10 years ago. In this time it has passed from a mere pharmacological 'surprise' to a more 'respectable' physiological function. Above all, its discovery in simple mammalian peripheral systems prior to obtaining evidence for any physiological significance in the brain shows the value of pharmacological models using the 'lock–key' analogy for drug-receptor classification.

Confirmation of the existence of $GABA_B$ sites within the brain came from radiolabelled receptor binding studies which, while they did not stand alone, have provided a firm basis for the receptor classification and have enabled the identification of substances for more functional analyses. Subsequent electrophysiological procedures have proved invaluable in our understanding of the role of $GABA_B$ sites. One wonders if molecular biology will provide the next detail.

We began this chapter by referring to the superior cervical ganglion, which provided the basis for the initial studies on $GABA_B$ receptors. After discovering the presence of $GABA_B$ sites on the nerve terminals of ganglion cell axons, we never considered that $GABA_B$-mediated responses might even be detectable in the ganglion itself. It was therefore very interesting to read the recent article by Newberry and Gilbert (1989) which demonstrates a pertussis toxin-sensitive ganglion hyperpolarization in response to $GABA_B$ site activation (Figure 1.6). It seems to have taken 10 years to come back to the future.

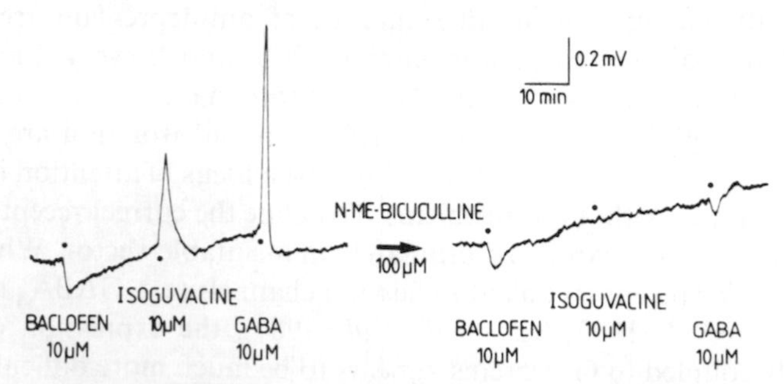

Figure 1.6. Bicuculline-sensitive and -insensitive actions of GABA recorded extracellularly in a rat isolated superior cervical ganglion. In the absence of bicuculline, GABA depolarized the ganglion, whereas in the presence of the antagonist GABA produced a hyperpolarizing response. Only this latter effect was mimicked by baclofen, whereas the former response was mimicked by isoguvacine. Each agonist was applied for 1 min at the points indicated by the filled circles. (From Newberry and Gilbert, 1989, with permission.)

ACKNOWLEDGEMENTS

Financial support from the Medical Research Council is gratefully acknowledged. We are also grateful to Angela Marshall for typing the manuscript.

REFERENCES

Adams, P.R. and Brown, D.A. (1975) Actions of γ-aminobutyric acid on sympathetic ganglion cells. *J. Physiol. Lond.*, **250**, 85–120.

Andrade, R., Malenka, R.C. and Nicoll, R.A. (1986) A G-protein couples serotonin and GABA$_B$ receptors to the same channels in hippocampus. *Science*, **234**, 1261–1265.

Anwar, N. and Mason, D.F.J. (1982) Two actions of γ-aminobutyric acid on the responses of the isolated basilar artery from the rabbit. *Br. J. Pharmacol*, **75**, 177–181.

Asano, T., Ui, M. and Ogasawara, N. (1985) Prevention of the agonist binding to γ-aminobutyric acid B receptors by guanine nucleotides and islet-activating protein, pertussis toxin, in bovine cerebral cortex. *J. Biol. Chem.*, **260**, 12653–12658.

Ault, B., Gruenthal, M., Armstrong, D.R., Nadler, J.V. and Wang, C.M. (1986) Baclofen suppresses bursting activity induced in hippocampal slices by differing convulsant treatments. *Eur. J. Pharmacol.*, **126**, 289–292.

Barber, R.P., Vaughn, J.E., Saito, K., McLaughlin, B.J. and Roberts, E. (1978) GABAergic terminals are presynaptic to primary afferent terminals in the substantia gelatinosa of the rat spinal cord. *Brain Res.*, **141**, 35–55.

Bittiger, H., Reymann, N., Hall, R. and Kane, P. (1988) CGP 27492, a new, potent and selective radioligand for GABA$_B$ receptors. In: *Proceedings of the 11th Annual Meeting of the European Neuroscience Association, Zurich, September 1988. Eur. J. Neurosci.* (Suppl.), Abstr. 16.10.

Bormann, J. (1988) Electrophysiology of GABA$_A$ and GABA$_B$ receptor subtypes. *Trends Neurosci.*, **11**, 112–116.

Bowery, N.G. (1989) GABA$_B$ receptors and their significance in mammalian pharmacology. *Trends Pharmacol. Sci.*, **10**, 401–407.

Bowery, N.G. and Brown, D.A. (1974) Depolarizing actions of γ-aminobutyric acid and related compounds on rat superior cervical ganglia in vitro. *Br. J. Pharmacol.*, **50**, 205–218.

Bowery, N.G. and Hudson, A.L. (1979) γ-Aminobutyric acid reduces the evoked release of ^{3}H-noradrenaline from sympathetic nerve terminals. *Br. J. Pharmacol.*, **66**, 108P.

Bowery, N.G., Hill, D.R., Hudson, A.L., Doble, A., Middlemiss, D.N., Shaw, J. and Turnbull, M.J. (1980) Baclofen decreases neurotransmitter release in the mammalian CNS by an action at a novel GABA receptor. *Nature*, **283**, 92–94.

Bowery, N.G., Doble, A., Hill, D.R., Hudson, A.L., Shaw, J.S., Turnbull, M.J. and Warrington, R. (1981) Bicuculline-insensitive GABA-receptors on peripheral autonomic nerve terminals. *Eur. J. Pharmacol.*, **71**, 53–70.

Bowery, N.G., Hill, D.R. and Hudson, A.L. (1983) Characteristics of GABA$_B$ receptor binding sites on rat whole brain synaptic membranes. *Br. J. Pharmacol.*, **78**, 191–206.

Bowery, N.G., Hudson, A.L. and Price, G.W. (1987) GABA$_A$ and GABA$_B$ receptor site distribution in the rat central nervous system. *Neuroscience*, **20**, 365–383.

Bowery, N.G., Hill, D.R. and Moratalla, R. (1989) Neurochemistry and autoradiography of GABA$_B$ receptors in mammalian brain: second messenger system(s). In: E.A. Barnard and E. Costa (eds) *Allosteric modulation of amino acid receptors: therapeutic implications.* New York: Raven Press, pp. 159–172.

Bowery, N.G., Knott, C., Moratalla, R. and Pratt, G.D. (in press) $GABA_B$ receptors and their heterogeneity. In: G. Biggio and E. Costa (eds) *GABA and benzodiazepine receptor subtypes: from molecular biology to clinical practice*. New York: Raven Press.

Cottrell, G.A. and Robertson, H.A. (1987) Baclofen exacerbates epileptic myoclonus in kindled rats. *Neuropharmacology*, **26**, 645–648.

Crawford, M.L.A. and Young, J.M. (1988) $GABA_B$ receptor-mediated inhibition of histamine H_1-receptor-induced inositol phosphate formation in slices of rat cerebral cortex. *J. Neurochem.*, **51**, 1441–1447.

Cross, J.A. and Horton, R.W. (1986) Cortical $GABA_B$ binding is unaltered following chronic oral administration of desmethylimipramine and zimelidine in the rat. *Br. J. Pharmacol.*, **89**, 521P.

Cross, J.A. and Horton, R.W. (1988) Effects of chronic oral administration of the antidepressants desmethylimipramine and zimelidine on rat cortical $GABA_B$ binding sites: a comparison with $5HT_2$ binding site changes. *Br. J. Pharmacol.*, **93**, 331–336.

Curtis, D.R., Game, C.J.A., Johnston, G.A.R. and McCulloch, R.M. (1974) Central effects of β-(*p*-chlorophenyl)-γ-aminobutyric acid. *Brain Res.*, **70**, 493–499.

Curtis, D.R., Gynther, B.D., Beattie, D.T., Kerr, D.I.B. and Prager, R.H. (1988) Baclofen antagonism by 2-hydroxy-saclofen in the cat spinal cord. *Neurosci. Lett.*, **92**, 97–101.

Cutting, D.A. and Jordan, C.C. (1975) Alternative approaches to analgesia: baclofen as a model compound. *Br. J. Pharmacol.*, **54**, 171–179.

Davies, J. (1981) Selective depression of synaptic excitation in cat spinal neurones by baclofen: an iontophoretic study. *Br. J. Pharmacol.*, **72**, 373–384.

Davies, J. and Watkins, J.C. (1974) The action of β-phenyl-GABA derivatives on neurones of the cat cerebral cortex. *Brain Res.*, **70**, 501–505.

Deisz, R.A. and Lux, H.D. (1985) γ-Aminobutyric acid-induced depression of calcium currents of chick sensory neurons. *Neurosci. Lett.*, **56**, 205–210.

Del Rio, J., Arnedo, A. and Naranjo, J.R. (1984) GABA inhibits the potassium-induced release of substance P from slices of rat spinal cord. In: *Proceedings of the 9th International Congress of Pharmacology, London*, Abstr. 203.

Desarmenien, M., Feltz, P., Occhipinti, G., Santangelo, F. and Schlichter, R. (1984) Coexistence of $GABA_A$ and $GABA_B$ receptors on AS and C primary afferents. *Br. J. Pharmacol.*, **81**, 327–333.

Dingwall, J.G., Ehrenfreund, J., Hall, R.G. and Jack, J. (1987) Synthesis of γ-aminopropylphosphonous acids using hypophosphorous acid synthons. *Phosph. Sulf.*, **30**, 571–574.

Docherty, R.J. and Halliwell, J.V. (1984) 4-Aminopyridine reveals a novel response to baclofen in the rat interpeduncular nucleus in vitro. *J. Physiol.*, **357**, 18P.

Dolphin, A.C. and Scott, R.H. (1986) Inhibition of calcium currents in cultured rat dorsal root ganglion neurones by (−)baclofen. *Br. J. Pharmacol.*, **88**, 213–220.

Dolphin, A.C., McGuirk, S.M. and Scott, R.H. (1989) An investigation into the mechanisms of inhibition of calcium channel currents in cultured sensory neurones of the rat by guanine nucleotide analogues and (−)baclofen. *Br. J. Pharmacol.*, **97**, 263–273.

Dunlap, K. (1981) Two types of γ-aminobutyric acid receptor on embryonic sensory neurones. *Br. J. Pharmacol.*, **74**, 579–585.

Dutar, P. and Nicoll, R.A. (1988a) Pre- and postsynaptic $GABA_B$ receptors in the hippocampus have different pharmacological properties. *Neuron*, **1**, 585–598.

Dutar, P. and Nicoll, R.A. (1988b) A physiological role for $GABA_B$ receptors in the central nervous system. *Nature*, **332**, 156–158.

Enna, S.J. and Snyder, S.H. (1977) Influence of ions, enzymes and detergents on GABA receptor binding in synaptic membranes of rat brain. *Mol. Pharmacol.*, **13**, 442–453.

Erdö, S.L. and Bowery, N.G. (1986) *GABAergic mechanisms in mammalian periphery.* New York: Raven Press.

Feltz, A., Demeneix, B., Feltz, P., Taleb, O., Trouslard, J., Bossu, J.-L. and Dupont, J.-L. (1987) Intracellular effectors and modulators of GABA-A and GABA-B receptors: a commentary. *Biochimie*, **69**, 395–406.

Fox, S., Krnjevic, K., Morris, M.E., Puil, E. and Werman, R. (1978) Action of baclofen on mammalian synaptic transmission. *Neuroscience*, **3**, 495–515.

Fromm, G.H., Terrence, C.F. and Chattha, A.S. (1984) Baclofen in the treatment of trigeminal neuralgia: double-blind study and long-term follow-up. *Ann. Neurol.*, **15**, 240–244.

Gähwiler, B.H. and Brown, D.A. (1985) GABA$_B$-receptor-activated K^+ current in voltage-clamped CA3 pyramidal cells in hippocampal cultures. *Proc. Natl Acad. Sci. USA*, **82**, 1558–1562.

Giotti, A., Luzzi, S., Spagnesi, S. and Zilletti, L. (1983) GABA$_A$ and GABA$_B$ receptor-mediated effects in guinea-pig ileum. *Br. J. Pharmacol.*, **78**, 469–478.

Godfrey, P.E., Grahame-Smith, D.G., Gray, J.A. and McClue, S.J. (1987) GABA$_B$ receptor-mediated inhibition of 5-HT stimulated phosphatidylinositol turnover in mouse cerebral cortex. *Br. J. Pharmacol.*, **90**, 251P.

Gray, J. and Green, A.R. (1987) GABA$_B$-receptor mediated inhibition of potassium-evoked release of endogenous 5-hydroxytryptamine from mouse frontal cortex. *Br. J. Pharmacol.*, **91**, 517–522.

Grenningloh, G., Rienitz, A., Schmitt, B., Methfessel, C., Zensen, M., Beyreuther, K., Gundelfinger, E.U. and Betz, H. (1987) The strychnine-binding subunit of the glycine receptor shows homology with nicotinic acetylcholine receptors. *Nature*, **328**, 215–220.

Haas, H.L., Greene, R.W. and Olpe, H.-R. (1985) Stereoselectivity of L-baclofen in hippocampal slices of the rat. *Neurosci. Lett.*, **55**, 1–4.

Harrison, N.L. (1990) On the presynaptic action of baclofen at inhibitory synapses between cultured rat hippocampal neurones. *J. Physiol.*, **422**, 433–446.

Hasuo, H. and Gallagher, J.P. (1988) Comparison of antagonism by phaclofen of baclofen induced hyperpolarizations and synaptically mediated late hyperpolarizing potentials recorded intracellularly from rat dorsolateral septal neurons. *Neurosci. Lett.*, **86**, 77–81.

Hill, D.R. (1985) GABA$_B$ receptor modulation of adenylate cyclase activity in rat brain slices. *Br. J. Pharmacol.*, **84**, 249–257.

Hill, D.R. and Bowery, N.G. (1981) ^{3}H-Baclofen and ^{3}H-GABA bind to bicuculline-insensitive GABA$_B$ sites in rat brain. *Nature*, **290**, 149–152.

Hill, D.R., Bowery, N.G. and Hudson, A.L. (1984) Inhibition of GABA$_B$ receptor binding by guanyl nucleotides. *J. Neurochem.*, **42**, 652–657.

Holz, G.G., Rane, S.G. and Dunlap, K. (1986) GTP-binding proteins mediate transmitter inhibition of voltage-dependent calcium channels. *Nature*, **319**, 670–672.

Howe, J.R. and Zieglgänsberger, D. (1986) Baclofen does not antagonize the actions of L-baclofen on rat neocortical neurons in vitro. *Neurosci. Lett.*, **72**, 99–104.

Inoue, M., Matsuo, T. and Ogata, N. (1985a) Baclofen activates voltage-dependent and 4-aminopyridine sensitive K^+ conductance in guinea-pig hippocampal pyramidal cells maintained in vitro. *Br. J. Pharmacol.*, **84**, 833–841.

Inoue, M., Matsuo, T. and Ogata, N. (1985b) Characterization of pre- and postsynaptic actions of (–)baclofen in the guinea-pig hippocampus in vitro. *Br. J. Pharmacol.*, **84**, 843–851.

Iversen, L.L. (1967) *The uptake and storage of noradrenaline in sympathetic nerves.* Cambridge: Cambridge University Press.

Jessen, K.R., Mirsky, R., Dennison, M.E. and Burnstock, G. (1979) GABA may be a neurotransmitter in the vertebrate peripheral nervous system. *Nature*, **281**, 71–74.

Jessen, K.R., Hills, J.M., Dennison, M.E. and Mirsky, R. (1983) γ-Aminobutyric acid as an autonomic neurotransmitter: release and uptake of ^{3}H-γ-aminobutyric acid in guinea-pig large intestine and cultured enteric neurons using physiological methods and electron microscopic autoradiography. *Neuroscience*, **10**, 1427–1442.

Joly, D., Lloyd, K.G., Pichat, P. and Sanger, D.J. (1987) Correlation between the behavioral effect of desipramine and $GABA_B$ receptor regulation in the olfactory bulbectomized rat. *Br. J. Pharmacol.*, **90**, 125P.

Jones, R.S.G. (1989) Ictal epileptiform events induced by removal of extracellular magnesium in slices of entorhinal cortex are blocked by baclofen. *Exp. Neurol.*, **104**, 155–161.

Karbon, E.W. and Enna, S.J. (1985) Characterization of the relationship between γ-aminobutyric acid B agonists and transmitter-coupled cyclic nucleotide-generating systems in rat brain. *Mol. Pharmacol.*, **27**, 53–59.

Karbon, E.W. and Enna, S.J. (1989) In: N.G. Bowery and G. Nistico (eds) *GABA: basic research and clinical applications.* Rome: Pythagora Press, pp. 205–223.

Karlsson, G., Pozza, M. and Olpe, H.-R. (1988) Phaclofen: a $GABA_B$ blocker reduces long-duration inhibition in the neocortex. *Eur. J. Pharmacol.*, **148**, 485–486.

Kerr, D.I.B., Ong, J., Prager, R.H., Gynther, B.D. and Curtis, D.R. (1987) Phaclofen: a peripheral and central baclofen antagonist. *Brain Res.*, **405**, 150–154.

Kleinrok, A. and Kilbinger, H. (1983) γ-Aminobutyric acid and cholinergic transmission in the guinea-pig ileum. *Naunyn-Schmiedeberg's Arch. Pharmacol.*, **322**, 216–220.

Krnjevic, K. (1974) Chemical nature of synaptic transmission in vertebrates. *Physiol. Rev.*, **54**, 418–540.

Kusunoki, M., Taniyama, K. and Tanaka, C. (1984) Neuronal GABA release and GABA inhibition of acetycholine release in guinea-pig urinary bladder. *Am. J. Physiol.*, **246**, 502–509.

Langer, S.Z. (1977) Presynaptic receptors and their role in the regulation of transmitter release. *Br. J. Pharmacol.*, **60**, 481–497.

Levy, R.A. and Proudfit, H.K. (1977) The analgesic action of baclofen [β-(4-chlorophenyl)-γ-aminobutyric acid]. *J. Pharmacol. Exp. Ther.*, **202**, 437–445.

Lloyd, K.G. (1989) GABA and depression. In N.G. Bowery and G. Nistico (eds) *GABA: basic research and clinical applications.* Rome: Pythagora Press, pp. 301–343.

Lloyd, K.G., Thuret, F. and Pilc, A. (1985) Upregulation of γ-aminobutyric acid ($GABA_B$) binding sites in rat frontal cortex: a common action of repeated administration of different classes of antidepressants and electroshock. *J. Pharmacol. Exp. Ther.*, **235**, 191–199.

McLaughlin, B.J., Barber, R., Saito, K., Roberts, E. and Wu, J.-Y. (1975) Immunocytochemical localization of glutamic decarboxylase in rat spinal cord. *J. Comp. Neurol.*, **164**, 305–322.

Moratalla, R. and Bowery, N.G. (1988) Autoradiographic measurement of $GABA_A$ and $GABA_B$ binding sites in rat caudate putamen after denervation of neuronal inputs. *Br. J. Pharmacol.*, **95**, 476P.

Moratalla, R., Barth, T. and Bowery, N.G. (1989) Benzodiazepine receptor autoradiography in corpus striatum of rat after large frontal cortex lesions and chronic treatment with diazepam. *Neuropharmacology*, **28**, 893–900.

Newberry, N.R. and Gilbert, M.J. (1989) Pertussis toxin sensitivity of drug-induced potentials on the rat superior cervical ganglion. *Eur. J. Pharmacol.*, **163**, 245–252.

Newberry, N.R. and Nicoll, R.A. (1984) Direct hyperpolarizing action of baclofen on hippocampal pyramidal cells. *Nature*, **308**, 450–452.

Newberry, N.R. and Nicoll, R.A. (1985) Comparison of the action of baclofen with γ-aminobutyric acid on rat hippocampal pyramidal cells in vitro. *J. Physiol.*, **360**, 161–185.

Ogata, N., Matsuo, T. and Inoue, M. (1986) Potent depressant action of baclofen on hippocampal epileptiform activity in vitro: possible use in the treatment of epilepsy. *Brain Res.*, **377**, 362–367.

Olpe, H.R., Koella, W.P., Wolf, P. and Haas, H.L. (1977) The action of baclofen on neurons of the substantia nigra and of the ventral tegmental area. *Brain Res.*, **134**, 577–580.

Ong, J. and Kerr, D.I.B. (1983) $GABA_A$ and $GABA_B$ receptor mediated modification of intestinal motility. *Eur. J. Pharmacol.*, **86**, 9–17.

Ong, J. and Kerr, D.I.B. (1984) Evidence for a physiological role of GABA in the control of guinea-pig intestinal motility. *Neurosci. Lett.*, **50**, 339–343.

Pratt, G.D., Knott, C., Davey, R. and Bowery, N.G. (1989) Characterisation of 3-aminopropyl phosphinic acid as a $GABA_B$ agonist in rat brain tissue. *Br. J. Pharmacol.*, **96**, 141P.

Price, G.W., Blackburn, T.P., Hudson, A.L. and Bowery, N.G. (1984) Presynaptic $GABA_B$ sites in the interpeduncular nucleus. *Neuropharmacology*, **23**, 861–862.

Price, G.W., Kelly, J.S. and Bowery, N.G. (1987) The location of $GABA_B$ receptor binding sites in mammalian spinal cord. *Synapse*, **1**, 530–538.

Rodbell, M. (1980) The role of hormone receptors and GTP-regulatory proteins in membrane transduction. *Nature*, **264**, 17–22.

Sawynok, J. (1987) GABAergic mechanisms of analgesia: an update. *Pharmacol. Biochem. Behav.*, **26**, 463–474.

Sawynok, J. and Dickson, C. (1985) D-Baclofen is an antagonist at baclofen receptors mediating antinociception in the spinal cord. *Pharmacology*, **31**, 248–259.

Scherer, R.W., Ferkary, J.W. and Enna, S.J. (1988) Evidence for pharmacologically distinct subsets of $GABA_B$ receptors. *Brain Res. Bull.*, **21**, 439–443.

Schildkraut, J.J. (1965) The catecholamine hypothesis of affective disorders: a review of supporting evidence. *Am. J. Psychiatr.*, **122**, 509–522.

Schofield, P.R. (1989) The $GABA_A$ receptor: molecular biology reveals a complex picture. *Trends Pharmacol. Sci.*, **10**, 476–478.

Schofield, P.R., Darlison, M.G., Fujita, N., Burt, D.R., Stephenson, F.A., Rodriguez, H., Rhee, L.M., Ramachandra, J., Reale, V., Glencorse, T.A., Seeburg, P.H. and Barnard, E.A. (1987) Sequence and functional expression of the $GABA_A$ receptor shows a ligand-gated receptor super family. *Nature*, **328**, 221–227.

Schwarz, M., Klockgether, T., Wüllner, U., Turski, L. and Sontag, K.-H. (1988) δ-Aminovaleric acid antagonizes the pharmacological actions of baclofen in the central nervous system. *Exp. Brain Res.*, **70**, 618–626.

Soltesz, I., Haby, M., Leresche, N. and Crunelli, V. (1988) The $GABA_B$ antagonist phaclofen inhibits the late K^+-dependent IPSP in cat and rat thalamic and hippocampal neurones. *Brain Res.*, **448**, 351–354.

Soltesz, I., Lightowler, S., Leresche, N. and Crunelli, V. (1989) On the properties and origin of the $GABA_B$ inhibitory postsynaptic potential recorded in morphologically identified projection cells of the cat dorsal lateral geniculate nucleus. *Neuroscience*, **33**, 23–34.

Suzdak, P.D. and Gianutsos, G. (1986) Effects of chronic imipramine or baclofen on GABA-B binding and cyclic AMP production in cerebral cortex. *Eur. J. Pharmacol.*, **131**, 129–133.

Swartzwelder, H.S., Bragdor, A.C., Sutch, C.P., Ault, B. and Wilson, W.A. (1986) Baclofen suppresses hippocampal epileptiform activity at low concentrations without suppressing synaptic transmission. *J. Pharmacol. Exp. Ther.*, **237**, 881–887.

Swartzwelder, H.S., Lewis, D.V., Anderson, W.W. and Wilson, W.A. (1987a) Seizure-like events in brain slices: suppression by interictal activity. *Brain Res.*, **410**, 362–366.

Swartzwelder, H.S., Tilson, H.A., McLamb, R.L. and Wilson, W.A. (1987b) Baclofen disrupts passive avoidance retention in rats. *Psychopharmacology*, **92**, 398–401.

Tanaka, C. (1985) γ-Aminobutyric acid in peripheral tissues. *Life Sci.*, **37**, 2221–2235.

Terrence, C.F., Fromm, G.H. and Tenicela, R. (1985) Baclofen as an analgesic in chronic peripheral nerve disease. *Eur. Neurol.*, **24**, 380–385.

Thalmann, R.H. (1988) Evidence that guanosine triphosphate (GTP)-binding proteins control a synaptic response in brain: effect of pertussis toxin and GTPγS on the late inhibitory postsynaptic potential of hippocampal CA3 neurons. *J. Neurosci.*, **8**, 4589–4602.

Watling, K.J. and Bristow, D.R. (1986) $GABA_B$ receptor-mediated enhancement of vasoactive intestinal peptide-stimulated cyclic AMP production in slices of rat cerebral cortex. *J. Neurochem.*, **46**, 1756–1762.

Wilkin, G.P., Hudson, A.L., Hill, D.R. and Bowery, N.G. (1981) Autoradiographic localization of $GABA_B$ receptors in rat cerebellum. *Nature*, **294**, 584–587.

Wojcik, W.J. and Neff, N.H. (1984) γ-Aminobutyric acid B receptors are negatively coupled to adenylate cyclase in brain, and in the cerebellum these receptors may be associated with granule cells. *Mol. Pharmacol.*, **25**, 24–28.

Wüllner, U., Klockgether, T. and Sontag, K.-H. (1989) Phaclofen antagonizes the depressant effect of baclofen on spinal reflex transmission in rats. *Brain Res.*, **496**, 341–344.

Xu, J. and Wojcik, W.J. (1986) γ-Aminobutyric acid B receptor-mediated inhibition of adenylate cyclase in cultured cerebellar granule cells: blockade by islet-activating protein. *J. Pharmacol. Exp. Ther.*, **239**, 568–573.

2 GABA$_B$ Receptor Agonists and Antagonists

D. I. B. KERR, J. ONG and R. H. PRAGER

INTRODUCTION

Until recently, the greatest emphasis has been placed on GABA actions mediated through the classical GABA$_A$ receptors, which have been well characterized and had their functions explored in a great variety of preparations. Largely this has been due to the introduction nearly 20 years ago of bicuculline as a selective competitive antagonist of these GABA-mediated actions (Curtis *et al.*, 1971), whilst the importance of GABA receptors in controlling normal brain function has been emphasized by the realization that these receptors are subject to modulation by a range of agents of therapeutic interest. By contrast, GABA$_B$ receptors, so defined by Hill and Bowery (1981) on the basis of bicuculline insensitivity and response to baclofen as a specific agonist, are only now becoming understood, and any likely physiological or clinical significance is just beginning to emerge. Again, this has very much been due to our recent development of antagonists for these receptors.

The notion of a second GABA-receptor type arose from observations on isolated tissues, where GABA exerts a presynaptic depressant action on transmitter release, which is insensitive to bicuculline (Bowery *et al.*, 1981) and hence evidently not mediated through the classical GABA receptor; furthermore, baclofen is a specific agonist for this action. Indeed, synaptically mediated twitch responses in isolated segments of ileum, vas deferens or anococcygeus muscle have become routinely used for screening agonists or antagonists at GABA$_B$ receptors (Bowery *et al.*, 1981; Luzzi *et al.*, 1985; Muhyaddin *et al.*, 1982; Ong and Kerr, 1983a,b; Ong *et al.*, 1987a,b), complemented by calcium-dependent binding assays using [^{3}H]($-$)-baclofen in brain synaptosomal membranes (Hill and Bowery, 1981). In addition we have found a very useful preparation for surveying GABA$_B$-receptor ligands in the central nervous system (CNS) to be the baclofen-induced depression of spontaneous ictaform activity, recorded from rat cortical slices in magnesium-free medium (Kerr *et al.*, 1988, 1989b,c). So far, using any of these, there is no evidence for modulatory sites on these receptors, in contrast to GABA$_A$ receptors, although GABA$_B$ receptors are linked to second messengers through G_i/G_O-proteins and require divalent cations for binding and activity (Ong *et al.*, 1987a). Furthermore, there is increasing evidence that GABA$_B$ receptors may in fact be heterogeneous.

GABA$_B$ Receptors in Mammalian Function. Edited by N.G. Bowery, H. Bittiger and H.-R. Olpe

Whilst a considerable range of GABA analogues have been examined for $GABA_B$ agonist or antagonist actions, surprisingly few specific agents have to date been found, and the structural requirements for ligands at $GABA_B$ receptors are evidently quite stringent. Indeed, the only new specific $GABA_B$-receptor agonist so far reported is the phosphinic analogue of GABA (3-aminopropylphosphinic acid, 3-APA) (Dingwall *et al.*, 1987), which is extremely potent (10–100 times more than baclofen) in binding studies and on some isolated preparations (Hills *et al.*, 1989; Pratt *et al.*, 1989). In general, the greatest problem has been that many promising compounds also exhibit significant activity at $GABA_A$ receptors. This is particularly true of δ-aminovaleric acid (DAVA), introduced as an antagonist at $GABA_B$ receptors by Muhyaddin *et al.* (1982), and of 3-aminopropylsulphonic acid (3-APS; Giotti *et al.*, 1983), which is a potent $GABA_A$-receptor agonist but only a weak $GABA_B$-receptor antagonist. Unfortunately, mixed enantiomers of substituted GABA analogues have perforce generally been used, which presents a very real difficulty as will be seen with the hydroxy-substituted GABA analogues.

In contrast to the $GABA_A$-receptor complex, where bicuculline and picrotoxin are available, no naturally occurring specific antagonist for $GABA_B$ receptors has so far emerged; this is not to say that one may not yet be found, although it is not clear what central effects a $GABA_B$-receptor antagonist might show. With the $GABA_A$ receptor, the ionophore is a chloride channel, and the well-known antagonist picrotoxin binds at some region close to the barbiturate/GABA site(s) to act as a non-competitive inhibitor, probably by blocking the chloride channel; bicuculline on the other hand is a competitive antagonist of $GABA_A$-receptor-mediated responses. In the case of the $GABA_B$ receptor, there does not seem to be any direct coupling to ion channels, and instead its actions appear to be expressed indirectly through G_i/G_o-proteins (guanine nucleotide binding proteins), with at least two different intracellular mechanisms involved. In some preparations, activation of $GABA_B$ receptors leads to an increase in potassium conductance, particularly postsynaptically, whereas in others there is a decrease in calcium conductance, leading to decreased transmitter release at presynaptic terminals. These two actions can be discriminated on the basis of sensitivity to ribosylation of the G_i/G_o-proteins by pertussis toxin (Dutar and Nicoll, 1988b), but the G-protein target is indiscriminate and the toxin can in no way be considered a specific antagonist for $GABA_B$-receptor-mediated effects.

GABA ANALOGUES AND $GABA_B$ RECEPTORS (Figure 2.1)

GABA, which exhibits $GABA_B$-agonist potency close to that of baclofen in many preparations, is a highly flexible molecule capable of assuming a number of stable conformations, with extremes representing those fitting either to $GABA_A$ or to $GABA_B$ receptors. The preferred GABA conformation for attachment at $GABA_A$ receptors has been inferred from the configuration of a number of rigid GABA analogues, including bicuculline (Aprison and Lipkowitz, 1989). However, no such rigid analogues of restricted conformation are known for $GABA_B$ receptors, other than muscimol and dihydromuscimol, neither of which is specific or at all potent (Falch *et al.*, 1986). From this it is evident that any GABA conformation congruent with muscimol in its 'bicuculline conformation' cannot provide an effective fit at $GABA_B$ receptors, there being only one carbon with rotational flexibility that would allow muscimol to attach at these receptors.

Neither *trans*- nor *cis*-4-aminocrotonic acid (TACA, CACA) is a $GABA_B$-receptor agonist or antagonist (Ong, unpublished results), although TACA does displace [^{3}H]baclofen (Falch *et al.*, 1986). CACA has been described as exerting bicuculline-insensitive depression in the spinal cord ($GABA_C$ receptors; Johnston and Allan, 1984; Johnston *et al.*, 1975), but no $GABA_B$-receptor-mediated actions have so far been found in a variety of preparations (Ong, unpublished observations). These two partially restricted conformations can thus be eliminated as having any relevance for the preferred GABA conformation(s) at $GABA_B$ receptors. More interesting is the related *Z*-5-aminopent-2-enoic acid, which combines $GABA_A$-agonist properties with highly potent $GABA_B$-antagonism (Dickenson *et al.*, 1988), but is for that reason virtually useless as an antagonist in most preparations. Nevertheless, together with the considerable body of work on $GABA_A$ agonists, all of this tells us something of what not to look for in the structure of $GABA_B$ agonists or antagonists. However, attempts to design $GABA_B$-receptor agonists or antagonists of restricted conformation, as with $GABA_A$ and glutamate ligands (Allan and Johnston, 1983; Watkins, 1981) have as yet failed. In part, this reflects the extraordinary specificity of agonist structure for $GABA_B$ receptors, exemplified in the baclofen series, where the 4-chlorophenyl β-substituent of baclofen itself is by far the most potent (Bowery, 1982). For the main part, then, we are left with analogues of GABA or baclofen based on modification of the acidic or basic function as potential antagonists, and of these the latter have proven less profitable.

One compound with a modified basic function, through conversion of GABA to an imino-derivative, has been described as active at $GABA_B$ receptors (Bowery *et al.*, 1982). This is SL 75102, a metabolic product from progabide, and is curious in that it retains agonist activity at both $GABA_A$ and $GABA_B$ receptors, particularly as the latter are so sensitive to modification of the GABA basic function. One possibility would appear to be that SL 75102 is in reality a pro-drug, since such Schiff's bases are inherently unstable; hydrolytic cleavage of the imino group would then yield GABA itself to act at the $GABA_B$ receptors, although this has previously been discounted.

With $GABA_A$ receptors, there has been considerable success using modification of the basic function of GABA to yield antagonists. In particular, SR 95531, developed by Wermuth *et al.* (1987), is a highly potent $GABA_A$-receptor

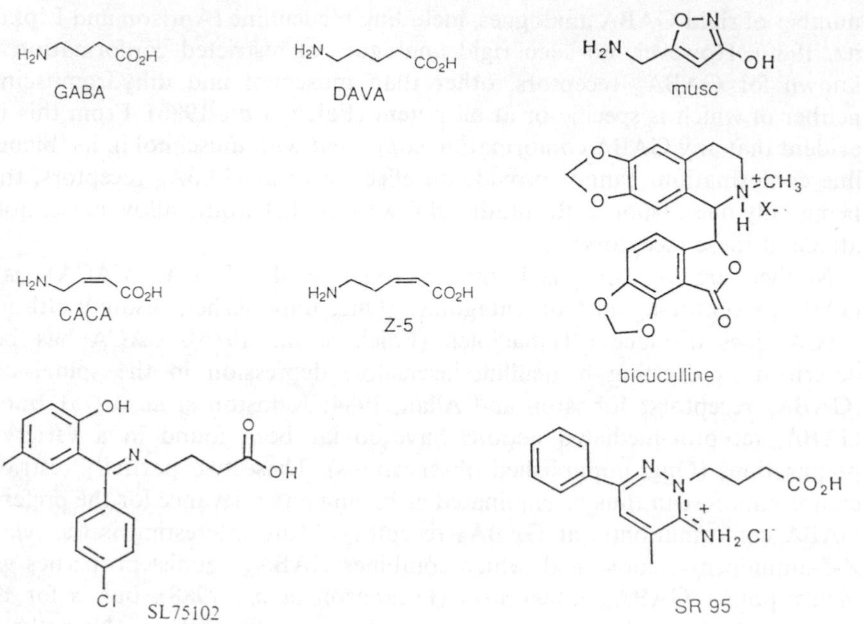

Figure 2.1. GABA analogues and derivatives active at GABA receptors: GABA, γ-aminobutyric acid; DAVA, δ-aminovaleric acid; musc, muscimol; CACA, *cis*-aminocrotonic acid; Z-5, *Z*-5-aminopent-2-enoic acid; bicuculline; SL 75102, [α-(4-chlorophenyl)-5-fluoro-2-hydroxybenzilidene-amino]-4-butanoic acid; SR 95, the basic pyridazinyl-GABA derivative with charge delocalization over the pyridazinyl moiety (Wermuth *et al.*, 1987). GABA, DAVA, muscimol, Z-5 and SL 75102 are active at both $GABA_A$ and $GABA_B$ receptors, where DAVA and the unsaturated analogue Z-5 are $GABA_B$-receptor antagonists. CACA is a putative agonist for $GABA_C$ receptors (Johnston and Allan, 1984). Bicuculline and the SR 95 series are competitive $GABA_A$ antagonists, of which SR 95531 is the most potent known, but they are totally inactive at $GABA_B$ receptors

antagonist, bearing a charge-delocalized amino pyridazinyl heterocyclic basic function; yet this compound shows absolutely no activity at $GABA_B$ receptors despite structurally being an *N*-substituted GABA retaining considerable flexibility of the GABA backbone. Evidently such charge delocalization over a heterocycle is not compatible with attachment at $GABA_B$-receptor sites. Somewhat similar compounds were amongst the earliest that we prepared and tested for $GABA_B$-receptor antagonism. Since guanidino-baclofen is a moderately potent $GABA_B$ agonist, we incorporated this structure into 2-imidazoloyl-substituted ω-amino acids. However, none were at all active as agonists or antagonists, and other analogues with modification or substitution of the basic function of GABA were weaker or totally inactive. For this reason such compounds have not been pursued further.

PHOSPHONO-ANALOGUES OF GABA AND BACLOFEN AS ANTAGONISTS (Figure 2.2)

Following the definition of GABA$_B$ receptors as a distinct group, it was immediately obvious that antagonists for these receptors were required in order that their physiological functions might be explored. Our own efforts in this regard began in 1983, when we first considered the possibility of synthesizing phosphonic and sulphonic analogues of baclofen as GABA$_B$-receptor antagonists. The primary lead for this came from the demonstration by Watkins (1981) that ω-phosphono-analogues of glutamate with various chain lengths were NMDA antagonists. In addition, 3-aminopropylphosphonic acid (3-APPA) had already been described as a central depressant (Curtis and Watkins, 1965), but unfortunately had been examined before bicuculline was available to eliminate the possibility of any GABA$_A$-receptor involvement in this action.

3-APPA was of interest as the first naturally occurring substance with a C–P bond, but apart from the work of Curtis and Watkins it had not been explored for pharmacological actions. That 3-APPA and higher analogues might be of interest as ligands for GABA$_B$ receptors was reinforced by two further pieces of work; Bioulac *et al.* (1979) pointed out that the depressant action of 3-APPA was bicuculline-insensitive, again suggesting a GABA$_B$-receptor-mediated action. Also, by 1984, 3-APPA had been shown to displace baclofen under conditions for binding at GABA$_B$ receptors (Cates *et al.*, 1984). Added to this, Muhyaddin *et al.* (1982) showed that DAVA, a weak GABA$_A$ agonist, would antagonize baclofen in the anococcygeus muscle, which lacks GABA$_A$ receptors. (Incidentally, at that time they tested 3-APPA on this preparation, but only at 10 μM, which was too low to show any GABA$_B$-receptor-mediated effects.) We quickly confirmed that DAVA would antagonize peripheral actions of baclofen (Ong and Kerr, 1983b), as did Luzzi *et al.* (1985). This activity of DAVA, together with the action of 3-APPA, suggested that the corresponding 4-aminobutylphosphonic acid (4-ABPA) might be a better GABA$_B$-receptor antagonist devoid of interfering GABA$_A$-receptor-mediated actions such as with DAVA. 4-ABPA was therefore prepared by R.H. Prager and, when tested in the guinea-pig ileum and vas deferens, proved to be a competitive, albeit weak, antagonist of GABA and baclofen at GABA$_B$ receptors. By contrast, we found 3-APPA to be a partial agonist/antagonist at these receptors, with a potency comparable to that of 4-ABPA.

When the synthesis of phosphono-baclofen (phaclofen, 3-amino-2-(4-chlorophenyl)-propylphosphonic acid) was achieved (Chiefari *et al.*, 1987), it too proved to be a GABA$_B$-receptor antagonist in the periphery, with a potency close to that of 4-ABPA ($pA_2 = 4$). Furthermore, phaclofen was effective against baclofen-induced depression of synaptic transmission in the cat spinal cord (Kerr *et al.*, 1987). Contemporaneously with this account, Luzzi *et al.* (1986) also described the partial agonist/antagonist activity of 3-APPA. In the course of developing phaclofen, a substantial number of phosphono-analogues and their esters were prepared, but apart from 4-ABPA and phaclofen, the only other phosphono-analogue that we have found to exhibit $GABA_B$-receptor antagonism has been 3-amino-2-cyclohexylpropylphosphonic acid, which occurs as a dechlorinated side-product of the reduction of the acrylic precursor of phaclofen (K.N. Mewett, Department of Pharmacology, University of Sydney, personal communication). Evidently $GABA_B$-receptor antagonism is sensitive to β-substitution in the propyl backbone of the phosphonates, in the same way as is found for agonist activity in baclofen carboxylic analogues. In particular, β-phenyl-3-APPA is inactive (up to 1 mM), whereas β-phenyl-GABA is a partial agonist/antagonist in the isolated ileum (Ong *et al.*, 1987b) and cortical slice (Kerr *et al.*, 1989c).

Phaclofen has subsequently become more widely available, and has been used to demonstrate $GABA_B$-receptor-mediated effects in a variety of isolated preparations from the CNS. It is now apparent that the late, slow inhibitory postsynaptic potential (IPSP) in many of these is due to phaclofen-sensitive GABA activation of $GABA_B$ receptors, leading to a hyperpolarizing increase in potassium conductance, which is also induced by baclofen. This has been seen in the hippocampus (Dutar and Nicoll, 1988a,b; Karlsson and Olpe, 1989; Malouf *et al.*, 1988; Soltesz *et al.*, 1988) and neocortex (Karlsson *et al.*, 1988), as well as in dorsolateral septal neurons (Hasuo and Gallagher, 1988), the locus coeruleus (Olpe *et al.*, 1988) and the dorsal lateral geniculate nucleus (Soltesz *et al.*, 1988). Phaclofen and baclofen have also been used to differentiate between inhibitory inputs to ventral or dorsal lateral geniculate neurons, only the latter receiving a GABAergic input that activates $GABA_B$ receptors (Soltesz *et al.*, 1989). In striate cortex, the suppression of visual evoked responses by baclofen is phaclofen-sensitive, whereas the suppression by GABA is not affected, whilst in some preparations phaclofen itself suppressed these responses (Baumfalk and Albus, 1988), possibly by blocking autoreceptors and releasing bicuculline-sensitive inhibition.

Phaclofen does not penetrate the blood–brain barrier, but its direct intracerebroventricular administration antagonizes baclofen antinociceptive activity (Giuliani *et al.*, 1988) and reveals $GABA_B$-receptor involvement in tonic and phasic modulation of central respiratory activity (Schmid *et al.*, 1989). There is, in addition, more direct evidence for $GABA_B$ autoreceptors that are sensitive to phaclofen (Bonanno *et al.*, 1988, 1989). This is related to $GABA_B$-receptor-mediated presynaptic actions which in one model were shown to be sensitive to phaclofen (Ticku and Delgado, 1989), but in another, were insensitive (Stirling *et al.*, 1989). However, with the latter study, phaclofen appears to have behaved as a partial agonist, in that phaclofen itself partially inhibited potassium-stimulated calcium influx but prevented any additive effect of subsequently co-applied baclofen.

Using neurochemical methods, phaclofen also completely prevents the inhibitory effects of GABA and baclofen on forskolin-stimulated adenylyl cyclase activity in cortical membrane preparations (Nishikawa and Kuriyama, 1989). However, Scherer *et al.* (1988) have found that 3-APPA and 4-ABPA displace baclofen binding with comparable potency, yet 4-ABPA is totally inactive in influencing either isoproterenol- or foskolin-stimulated cyclic AMP production, whilst 3-APPA only inhibits the forskolin-mediated response without affecting isoproterenol-stimulated second messenger accumulation. Curiously, in their study, phaclofen did not inhibit $GABA_B$-receptor binding, neither did it affect isoproterenol- or forskolin-stimulated cyclic AMP accumulation in brain slices. Nevertheless, in the spontaneously discharging neocortical slice maintained in magnesium-free medium, baclofen attenuates or suppresses the discharges, and this action is antagonized by both phaclofen and 4-ABPA, as well as by 3-APPA, which is evidently a partial agonist in this preparation (Kerr *et al.*, 1989c).

Clearly, much more functional work must be done before sensitivity, or lack of it, to these phosphono-compounds can be taken as a basis for defining $GABA_B$-receptor subtypes. Nevertheless, there is a strong temptation to recognize at least $GABA_{B1}$ and $GABA_{B2}$ receptors. Differences in phaclofen sensitivity between pre- and postsynaptic receptors (Dutar and Nicoll, 1988a,b) further suggest a heterogeneity of $GABA_B$ receptors.

At first sight, it might be thought that these phosphono-analogues of GABA are $GABA_B$-receptor antagonists, because they are more highly acidic than GABA, and the phosphono-head is bulkier than the carboxylate function. Against this, 3-APA (Dingwall *et al.*, 1987), which more nearly resembles GABA (Cates *et al.*, 1984), is actually more acidic than 3-APPA, whilst the phosphinic group has nearly the same bulk as that of the phosphonate. This argues against the importance of bulk or strength of the acidic function in imparting antagonism in these phosphonates. More likely, it is some mismatch at the anionic attachment site on the receptor that leads to partial agonist/antagonist properties in these analogues. Surprisingly, 3-APA is a very potent $GABA_B$-receptor agonist in depressing transmitter release in the periphery (Hills *et al.*, 1989), and in displacing baclofen binding (Pratt *et al.*, 1989).

Our own observations confirm that 3-APA is indeed very potent in depressing transmitter release in the periphery and sensitive to $GABA_B$ antagonists. 3-APA also effectively activates central $GABA_B$ autoreceptors, where it is again sensitive to these antagonists (Ong *et al.*, 1990). However, it is relatively weak at postsynaptic sites, and behaves as a partial agonist in spontaneously discharging neocortical slice preparations, where it weakly slows the discharge rate and attenuates the amplitude but prevents the depressant action of co-superfused baclofen. Curiously, the corresponding phosphinic analogue of baclofen is a weak agonist in the rat vas deferens, and behaves as a partial agonist/antagonist here and in the cortical slice preparation. As already mentioned, *Z*-5-aminopent-2-enoic acid is a potent $GABA_B$-receptor antagonist, but phosphonic or

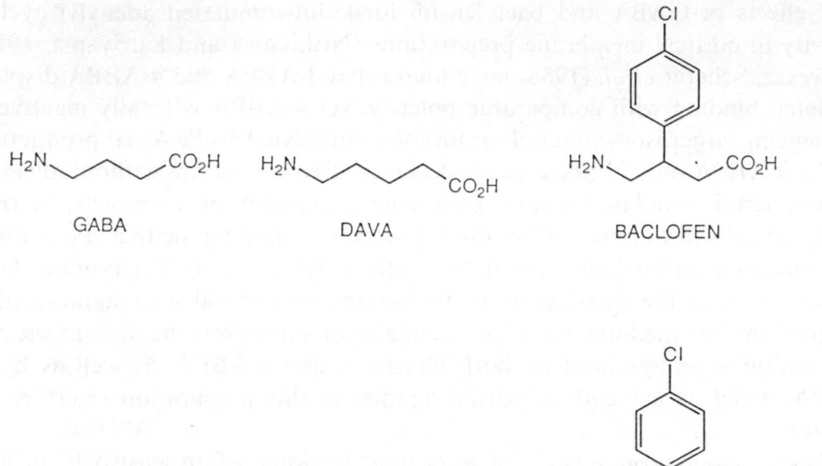

Figure 2.2. Structures of phosphonic and phosphinic analogues of GABA and baclofen investigated for activity at $GABA_B$ receptors: GABA, γ-aminobutyric acid (agonist); DAVA, δ-aminovaleric acid (antagonist); baclofen, 4-amino-3-(*p*-chlorophenyl)-butyric acid (agonist). Phosphonic analogues: 3-APPA, 3-aminopropylphosphonic acid (partial agonist); 4-ABPA, 4-aminobutylphosphonic acid (antagonist); phaclofen, 3-amino-2-(*p*-chlorophenyl)-propylphosphonic acid (antagonist). Phosphinic analogues: 3-APA, 3-aminopropylphosphinic acid (agonist/partial agonist); 4-ABA, 4-aminobutylphosphinic acid (weak antagonist); ACPPA, 4-amino-2-(*p*-chlorophenyl)-propylphosphinic acid (partial agonist). All the dibasic phosphonic compounds except 3-APPA are competitive antagonists at $GABA_B$ receptors with pA_2 values close to 4, whilst the monobasic phosphinic compounds, which more closely resemble the parent carboxylic acids, are agonists/partial agonists. Of these, 3-APA is a potent agonist at peripheral autonomic presynaptic $GABA_B$ receptors and autoreceptors in the hippocampus, whilst the baclofen analogue ACPPA is a partial agonist that can antagonize baclofen in the vas deferens and rat isolated neocortical slice

phosphinic variants of this DAVA analogue did not yield improved $GABA_B$-receptor antagonists, although the interfering $GABA_A$-receptor-mediated actions of the parent compound were abolished; neither was the direct phosphinic analogue of DAVA any more active. However, a disubstituted phosphinic derivative based on 3-APA where an alkoxy replaces the hydrogen attached to the phosphorus of the acid moiety, as in CGP 35348 (Bittiger *et al.*, Chapter 3), is a $GABA_B$ antagonist of medium potency, devoid of agonist actions.

ANTAGONISM BY SULPHONIC ANALOGUES OF GABA AND BACLOFEN (Figure 2.3)

Earlier, the account by Giotti *et al.* (1983) appeared, describing $GABA_B$-receptor antagonism with 3-APS, which displaces both $GABA_A$- and $GABA_B$-receptor binding (Bowery, 1982). The latter action is relatively weak with 3-APS, so that $GABA_B$-receptor antagonism is greatly complicated by potent $GABA_A$-agonist actions; consequently, 3-APS is of little use as an antagonist in many preparations. It was anticipated that insertion of the 4-chlorophenyl group at the β-position of 3-APS would eliminate this undesirable $GABA_A$-agonist property, by analogy with baclofen, and leave a possibly enhanced $GABA_B$-receptor antagonism. In practice, the β-hydroxy-substituted variant of this analogue was easier to prepare, giving 3-amino-2-(4-chlorophenyl)-2-hydroxypropylsulphonic acid (2-hydroxysaclofen, 2-OH-saclofen), which exhibited some 10-fold improvement over phaclofen as a $GABA_B$-receptor antagonist (Curtis *et al.*, 1988; Kerr *et al.*, 1988).

Subsequently, saclofen itself (3-amino-2-(4-chlorophenyl)-propylsulphonic acid) was prepared, along with the unsaturated precursor 3-amino-2-(4-chlorophenyl)-prop-l-enylsulphonic acid, as well as the sulphonamide of saclofen. It was anticipated that, being a weaker acid, the sulphonamide might be an agonist, but it exhibited only weak antagonism. In the ileum, the propenyl analogue was, in turn, a weak antagonist, which is surprising since the corresponding propenyl-phosphono or carboxylic analogues are totally inactive. Saclofen proved to be the most potent specific $GABA_B$-receptor antagonist yet found ($pA_2 = 5.3$), devoid of $GABA_A$-agonist side-actions and some twofold more potent than 2-OH-saclofen ($pA_2 = 5$). Both saclofen and 2-OH-saclofen also antagonize baclofen in the neocortical slice preparation (Kerr *et al.*, 1989a), and 2-OH-saclofen reversibly blocks the late IPSP, as well as the baclofen-induced hyperpolarization in CA1 neurons (Lambert *et al.*, 1989). In addition, we have evidence for 2-OH-saclofen antagonism at hippocampal presynaptic receptors. If saclofen were resolved, its pA_2 value of 5.6 would then approach that for bicuculline ($pA_2 = 5.9$–6.0) at $GABA_A$ receptors. In this regard, there is a real need for the resolution of the enantiomers of all the β-substituted phosphonic and sulphonic antagonists so far prepared.

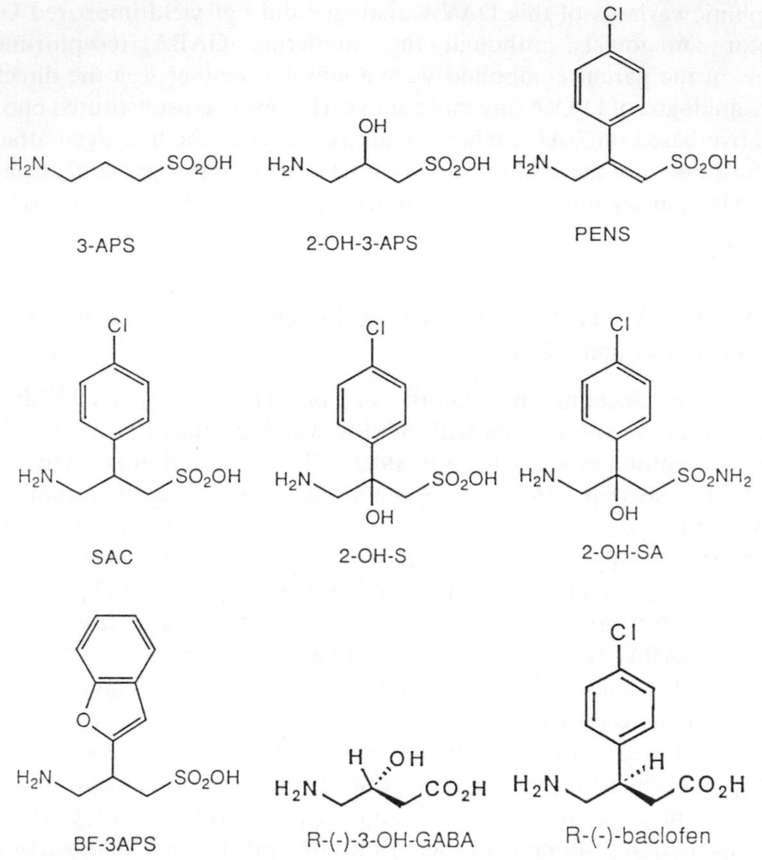

Figure 2.3. Structures of sulphonic analogues of GABA and baclofen that are antagonists at $GABA_B$ receptors: 3-APS, 3-aminopropylsulphonic acid; 2-OH-3-APS, 3-amino-2-hydroxypropylsulphonic acid; PENS, 3-amino-2-(*p*-chlorophenyl)-prop-1-enylsulphonic acid; SAC, saclofen, 3-amino-2-(*p*-chlorophenyl)-propylsulphonic acid; 2-OH-S, 2-hydroxysaclofen, 3-amino-2-(*p*-chlorophenyl)-2-hydroxypropylsulphonic acid; 2-OH-SA, the sulphonamide of 2-OH-S; BF-3-APS, 3-amino-2-benzo[*b*]furanyl-butanoic acid; (*R*)-(−)-3-OH-GABA, (*R*)-GABOB; (*R*)-(−)-baclofen. The absolute configurations of (*R*)-GABOB and (*R*)-baclofen are compared to show the opposite orientation of the *β*-substituent, presumably both the chlorophenyl and hydroxy substituents can be present in the correct orientation for antagonism with 2-OH-saclofen

ARYL AND HETEROCYCLIC SUBSTITUTION AT THE *β*-CARBON (Figure 2.4)

With baclofen analogues, there is an absolute preference for *β*-aryl substitution on the GABA backbone, and the 4-chlorophenyl derivative is by far the most active (Bowery, 1982). Other substituents give a substantial loss of activity, as does substitution at either the α- or γ-carbon of GABA. However, β-phenyl-GABA (phenibut) is of interest in that it has been used clinically in the USSR, not as a muscle relaxant like baclofen but as a mood elevator. In all our tests, β-phenyl-GABA is a partial agonist/antagonist at GABA$_B$ receptors (Kerr *et al.*, 1989c; Ong *et al.*, 1987b), and the (*R*)-isomer is the more active, as with baclofen (Ong, unpublished observations on (*R*)- and (*S*)-β-phenyl-GABA resolved by Dr R. Duke, Department of Pharmacology, University of Sydney). Incidentally β-*p*-isopropylphenyl-GABA is evidently more active as an agonist on the vas deferens than on [^{3}H]noradrenaline release from atria, and may also prove to be a partial agonist; this difference also further suggests possible heterogeneity amongst GABA$_B$ receptors.

In addition, two β-disubstituted baclofen analogues have been prepared, prompted by the suggestion that (*S*)-baclofen may antagonize (*R*)-baclofen (Sawynok, 1986). Of these, the β-(*p*-*OMe*phenyl)-β-(*p*-chlorophenyl) compound when tested as the racemate was the most potent as a baclofen antagonist, although at that only weak ($pA_2 < 4.0$). Most recently, two further baclofen analogues have been found to be antagonists at GABA$_B$ receptors; these are the

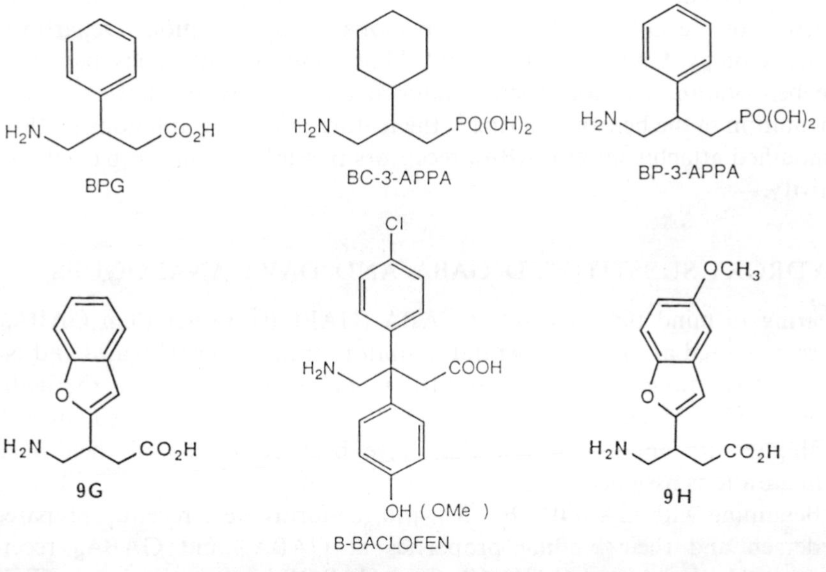

Figure 2.4. Structures of baclofen analogues with modified substituents at the β-carbon: BPG, β-phenyl-GABA (partial agonist/antagonist); BC-3-APPA, 3-amino-2-cyclohexyl-propylphosphonic acid (antagonist); BP-3-APPA, 3-amino-2-phenylpropylphosphonic acid (extremely weak or inactive as antagonist); 9G, 4-amino-3-benzo[*b*]furan-2yl-butanoic acid (antagonist); B-baclofen, 4-amino-3-(*p*-chlorophenyl)-3-(*p*-hydroxy-phenyl)-butanoic acid (OMe, the *O*-methylated derivative; each of these are antagonists, the last being more potent); 9H, 4-amino-3-(5-methoxybenzo[*b*]furan-2yl)-butanoic acid (antagonist). Both 9G and 9H were synthesized by Berthelot *et al.* (1987)

β-benzo[*b*]furans prepared by Berthelot *et al.* (1987). In the ileum, the unsubstituted β-benzo[*b*]furan GABA analogue (9G) is almost as potent as the 5-methoxy analogue (9H), despite the latter showing a substantially more potent baclofen displacement in binding assays; both analogues also antagonize baclofen in the isolated neocortical slice (Kerr *et al.*, 1989b), and the 5-methoxy compound antagonizes baclofen in the cat spinal cord (Beattie *et al.*, 1989). The success of these benzofurans as $GABA_B$-receptor antagonists suggested to us that the equivalent sulphonic analogue, with a β-benzo[*b*]furan on 3-APS, might be an even better antagonist. But when prepared, this analogue, combining the features of the Berthelot compounds and of saclofen, proved to be less active than either alone, and was little more active than phaclofen at $GABA_B$ receptors.

In attempting to explain the $GABA_B$ receptor antagonism by such benzofuran analogues, comparisons have been made by modelling of baclofen and the benzofurans. It is known from X-ray crystallography that the *p*-chlorophenyl substituent in (*R*)-baclofen is preferentially oriented across the plane of the GABA backbone (Chang *et al.*, 1982). Minimal energy calculations confirm this, whereas the benzofuran ring system shows a minimum when oriented in the plane of the backbone with the oxygen of the furan ring approaching the amine function of the GABA (H. Capper, personal communication; Department of Pharmacology, University of Sydney). Thus these two analogues, baclofen and the benzofuran, are not strictly matched and it may be that this differing orientation of the benzofuran ring is the feature imparting antagonism, through a modified attachment at $GABA_B$ receptors possibly leading to partial agonist activity.

HYDROXY-SUBSTITUTED GABA AND DAVA ANALOGUES

Bearing in mind that β-hydroxy-GABA (GABOB), rather than GABA, was once proposed as the major central inhibitory transmitter (Hayashi and Nagai, 1956), it is strange that relatively little work has been done on the hydroxy-substituted analogues of GABA or DAVA. This is the more so, since these analogues are among the few that have been resolved and their absolute configurations assigned.

Beginning with GABOB, the (*R*)- and (*S*)-forms were recently prepared by Pedersen and their binding properties at $GABA_A$ and $GABA_B$ receptors examined by Falch *et al.* (1986). (*R*)-(−)-GABOB binds with a lower affinity at $GABA_A$ receptors than at $GABA_B$ receptors, whereas (*S*)-(+)-GABOB shows the reverse, binding with a higher affinity at $GABA_A$ than at $GABA_B$ receptors. In the transmurally stimulated ileum, (*R*, *S*)-GABOB slightly potentiated the twitch (possibly related to $GABA_A$-receptor stimulatory actions) and antagonized the baclofen-induced depression of the twitch. This use of the racemate is deceptive since the pure (*R*)-GABOB turned out to be a weak partial agonist with good antagonist properties, whilst the (*S*)-enantiomer showed the twitch potentiation but lesser baclofen antagonism. The (*S*)-form also exhibited $GABA_A$-stimulatory actions. The lesser $GABA_B$-receptor antagonism correlates with the lower affinity of (*S*)-GABOB at these receptors. Thus one enantiomer is a weak antagonist, whereas the other is a partial agonist/antagonist at $GABA_B$ receptors. This is in keeping with the curious fact that (*R*)-(−)-GABOB has the opposite configuration to that of (*R*)-baclofen about the β-carbon (Falch *et al.*, 1986). The corresponding sulphonic analogue (*R*, *S*)-2-OH-3-APS in turn showed reduced $GABA_A$ agonist activity, and partial agonist/antagonist properties at $GABA_B$ receptors in the ileum, but has not been properly quantified. This gives some insight into the actions of 2-OH-saclofen, where it is evidently possible for both the hydroxy and 4-chlorophenyl substituents to be in the (*R*)-configuration at the β-position with retention of antagonist activity due to the sulphonic head. Although 2-OH-phaclofen has not yet been prepared, the corresponding 2-OH-baclofen showed a five-fold loss of activity in depressing the ileal twitch, but was used as a racemate, which again may be misleading.

Hydroxy-derivatives of DAVA represent a further group of compounds with mixed activity at $GABA_B$ receptors, depending on the carbon of the hydroxy substituent. From binding studies using the racemates, at $GABA_B$ receptors, the order of potency in displacing baclofen is highest for 2-OH-DAVA and 4-OH-DAVA, and lowest for 3-OH-DAVA; 2-OH-DAVA is also substantially more potent than 2-OH-GABA (Falch *et al.*, 1986). Using the ileal twitch, the OH-DAVA compounds were all antagonists, with potencies (*S*)-2-OH-DAVA > (*R*)-2-OH-DAVA > (*R*)-4-OH-DAVA $\gg$ (*S*)-4-OH-DAVA. Here, (*S*)-2-OH-DAVA is approximately equipotent with DAVA itself.

PARTIAL AGONISTS OR ANTAGONISTS

Structurally, there is no very obvious reason why any of these phosphonic or sulphonic analogues should be $GABA_B$-receptor antagonists, rather than agonists, although it is true that antagonism of NMDA is also found with similar glutamate analogues bearing an altered acidic function. However, one possibility that must be seriously entertained is that all these acidic analogues are in fact partial agonists rather than true antagonists. Certainly the phosphinic analogues appear to be partial agonists/antagonists, despite the potent [^{3}H]baclofen-binding displacement by 3-APA (Pratt *et al.*, 1989). Indeed, phaclofen has recently been described as a (partial) agonist (Stirling *et al.*, 1989). If so, then we have never achieved sufficient bath-concentration of any of these antagonists to demonstrate such partial agonist activity, other than with 3-APPA or 3-APA, or else their efficacies are very low. In keeping with this, β-phenyl-GABA is a partial agonist/antagonist (Ong *et al.*, 1987b), and thus for example the benzofuran analogues may also act in a similar fashion.

All the compounds so far identified as partial agonists or antagonists are uniformly active in displacing [^{3}H]baclofen under conditions for $GABA_B$ binding in the presence of calcium. In general, their half-maximal inhibitory concentration (IC_{50}) values reflect their pA_2 values, with the exception of the benzofuran-baclofen analogues, which in comparison to their IC_{50} values are anomalously poor antagonists. Of the acidic analogues, phaclofen is relatively weak (IC_{50} approximately 100 μM; pA_2 = 4.0), whilst saclofen is the most potent (IC_{50} approximately 8 μM; pA_2 = 5.3). None, however, approach the activity of bicuculline or SR 95531 at $GABA_A$ receptors.

Because the $GABA_B$-receptor antagonists so far developed are unable to penetrate the blood–brain barrier, there is little evidence that they might show any clinical significance. However, one can speculate that a good antagonist might elicit interesting clinical actions by preventing autoreceptor-mediated depression of GABA release at $GABA_A$-receptor sites, for which there is accumulating evidence. The resultant increased GABAergic inhibition might than attenuate seizure activity, or contribute antidepressive and anxiolytic actions. Alternatively, the antagonist might exhibit antidepressive activity by directly preventing $GABA_B$-receptor-mediated suppression of central amines. At any event, we look upon the progress so far as only the beginning, and the challenge remains to find improved, potent agonists and antagonists capable of true selectivity between the suggested $GABA_B$-receptor subtypes, and of penetrating the blood–brain barrier.

ACKNOWLEDGEMENTS

We thank the Australian National Research Fellowship Advisory Committee for the award of a Queen Elizabeth II Fellowship to Jennifer Ong, and Professors Graham Johnston and Povl Krogsgaard-Larsen for help and encouragement. We are grateful to Ciba-Geigy (Australia) for financial support to Jennifer Ong.

REFERENCES

Allan, R.D. and Johnston, G.A.R. (1983) Synthetic analogs for the study of GABA as a neurotransmitter. *Med. Res. Rev.*, **3**, 91–118.

Aprison, M.H. and Lipkowitz, K.B. (1989) On the $GABA_A$ receptor: a molecular modeling approach. *J. Neurosci. Res.*, **23**, 129–135.

Baumfalk, U. and Albus, K. (1988) Phaclofen antagonizes baclofen-induced suppression of visually evoked responses in the cat's striate cortex. *Brain Res.*, **463**, 398–402.

Beattie, D.T., Curtis, D.R., Debaert, M., Vaccher, C. and Berthelot, P. (1989) Baclofen antagonism by 4-amino-3-(5-methoxybenzo(*b*)furan-2-yl)butanoic acid in the cat spinal cord. *Neurosci. Lett.*, **100**, 292–294.

Berthelot, P., Vaccher, C., Musadad, N., Flouquet, M., Debaert, M. and Luyckx, M. (1987) Synthesis and pharmacological evaluation of gamma-aminobutyric acid analogues. *J. Med. Chem.*, **30**, 743–746.

Bioulac, B., deTinguy-Moreaud, E., Vincent, J.-D. and Neuzil, E. (1979) Neuroactive properties of phosphonic amino acids. *Gen. Pharmacol.*, **10**, 121–125.

Bonanno, G., Fontana, G. and Raiteri, M. (1988) Phaclofen antagonises GABA at autoreceptors regulating release in rat cerebral cortex. *Eur. J. Pharmacol.*, **154**, 223–224.

Bonanno, G., Pellegrini, G., Asaro, D., Fontano, G. and Raiteri, M. (1989) GABA$_B$ autoreceptors in rat cortex synaptosomes: response under different depolarizing and ionic conditions. *Eur. J. Pharmacol.* (*Mol. Pharmacol. Section*), **172**, 41–49.

Bowery, N.G. (1982) Baclofen: 10 years on. *Trends Pharmacol. Sci.*, **3**, 400–403.

Bowery, N.G., Doble, A., Hill, D.R., Hudson, A.L., Shaw, J.S., Turnbull, M.J. and Warrington, R. (1981) Bicuculline-insensitive GABA receptors on peripheral autonomic nerve terminals. *Eur. J. Pharmacol.*, **71**, 53–70.

Bowery, N.G., Hill, D.R. and Hudson, A.L. (1982) Evidence that SL75102 is an agonist at GABA$_B$ as well as GABA$_A$ receptors. *Neuropharmacology*, **21**, 391–395.

Cates, L.A., Li, V.-S., Yakshe, C.C., Fadeyi, M.O., Andree, T.H., Karbon, E.W. and Enna, S.J. (1984) Phosphorous analogues of gamma-aminobutyric acid: a new class of anticonvulsants. *J. Med. Chem.*, **27**, 654–659.

Chang, C.-H., Yang, D.S.C., Yoo, C.S., Wang, B.-C., Pletcher, J., Sax, M. and Terrence, C.F. (1982) Structure and absolute configuration of (*R*)-baclofen monohydrochloride. *Acta Cryst.*, **B38**, 2065–2067.

Chiefari, J., Galanopoulos, S., Janowski, W.K., Kerr, D.I.B. and Prager, R.H. (1987) The synthesis of phosphonobaclofen, an antagonist of baclofen. *Aust. J. Chem.*, **40**, 1511–1518.

Curtis, D.R. and Watkins, J.C. (1965) The pharmacology of amino acids related to gamma-aminobutyric acid. *Pharmacol. Rev.*, **17**, 347–392.

Curtis, D.R., Duggan, A.W., Felix, D. and Johnston, G.A.R. (1971) Bicuculline, an antagonist of GABA and synaptic inhibition in the spinal cord. *Brain Res.*, **32**, 69–96.

Curtis, D.R., Gynther, B.D., Beattie, D.T., Kerr, D.I.B. and Prager, R.H. (1988) Baclofen antagonism by 2-hydroxy-saclofen in the cat spinal cord. *Neurosci. Lett.*, **92**, 97–101.

Dickenson, H.W., Allan, R.D., Ong, J. and Johnston, G.A.R. (1988) GABA$_B$ receptor antagonist and GABA$_A$ receptor agonist properties of a δ-aminovaleric acid derivative, *Z*-5-aminopent-2-enoic acid. *Neurosci. Lett.*, **86**, 351–355.

Dingwall, J.-G., Ehrenfreund, J., Hall, R.G. and Jack, J. (1987) Synthesis of gamma-aminopropylphosphonous acid using hypophosphorous acid synthons. *Phosph. Sulf.*, **30**, 571–574.

Dutar, P. and Nicoll, R.A. (1988a) A physiological role for GABA$_B$ receptors in the central nervous system. *Nature*, **332**, 156–158.

Dutar, P. and Nicoll, R.A. (1988b) Pre- and postsynaptic GABA$_B$ receptors in the hippocampus have different pharmacological properties. *Neuron*, **1**, 585–598.

Falch, E., Hedegaard, A., Nielsen, L., Jensen, B.R., Hjeds, H. and Krogsgaard-Larsen, P. (1986) Comparative stereostructure–activity studies on GABA$_A$ and GABA$_B$ receptor sites and GABA uptake using rat brain membrane preparations. *J. Neurochem.*, **47**, 898–903.

Giotti, A., Luzzi, S., Spagnesi, S. and Zilletti, L. (1983) Homotaurine: a GABA$_B$ antagonist in guinea-pig ileum. *Br. J. Pharmacol.*, **79**, 855–862.

Giuliani, S., Evangelista, S., Borsini, F. and Meli, A. (1988) Intracerebroventricular phaclofen antagonises baclofen antinociceptive activity in hot plate test with mice. *Eur. J. Pharmacol.*, **154**, 225–226.

Hasuo, H. and Gallagher, J.P. (1988) Comparison of antagonism by phaclofen of baclofen induced hyperpolarizations and synaptically mediated late hyperpolarizing potentials recorded intracellularly from rat dorsolateral septal neurons. *Neurosci. Lett.*, **86**, 77–81.

Hayashi, T. and Nagai, K. (1956) Action of ω-amino acids on the motor cortex of higher animals, especially gamma-amino-β-oxybutyric acid as the real inhibitory principle in brain. *Abstr. XX. Int. Physiol. Congr.*, 410.

Hill, D.R. and Bowery, N.G. (1981) ^{3}H-Baclofen and ^{3}H-GABA bind to bicuculline-insensitive $GABA_B$ sites in rat brain. *Nature*, **290**, 149–152.

Hills, J.M., Dingsdale, R.A. and Howson, E. (1989) 3-Aminopropylphosphinic acid inhibits the cholinergic twitch in the guinea-pig ileum through $GABA_B$ receptor agonist activity. *Br. J. Pharmacol.*, **96**, 51P.

Johnston, G.A.R. and Allan, R.D. (1984) GABA agonists. *Neuropharmacology*, **23**, 831–832.

Johnston, G.A.R., Curtis, D.R., Beart, P.M., Game, C.J.A., McCulloch, R.M. and Twitchin, B. (1975) *Cis* and *trans*-4-aminocrotonic acid as GABA analogues of restricted conformation. *J. Neurochem.*, **24**, 157–160.

Karlsson, G. and Olpe, H.-R. (1989) Inhibitory processes in normal and epileptic-like rat hippocampal slices: the role of $GABA_B$ receptors. *Eur. J. Pharmacol.*, **163**, 285–290.

Karlsson, G., Pozza, M. and Olpe, H.-R. (1988) Phaclofen: a $GABA_B$ blocker reduces long-duration inhibition in the neocortex. *Eur. J. Pharmacol.*, **148**, 485–486.

Kerr, D.I.B., Ong, J., Prager, R.H., Gynther, B.D. and Curtis, D.R. (1987) Phaclofen: a peripheral and central baclofen antagonist. *Brain Res.*, **405**, 150–154.

Kerr, D.I.B., Ong, J., Johnston, G.A.R., Abbenante, J. and Prager, R.H. (1988) 2-Hydroxy-saclofen: an improved antagonist at central and peripheral $GABA_B$ receptors. *Neurosci. Lett.*, **92**, 92–96.

Kerr, D.I.B., Ong, J., Johnston, G.A.R., Abbenante, J. and Prager, R.H. (1989a) Antagonism at $GABA_B$ receptors by saclofen and related sulphonic analogues of baclofen and GABA. *Neurosci. Lett.*, **107**, 239–244.

Kerr, D.I.B., Ong, J., Johnston, G.A.R., Berthelot, P., Debaert, M. and Vaccher, C. (1989b) Benzofuran analogues of baclofen: a new class of central and peripheral $GABA_B$-receptor antagonists. *Eur. J. Pharmacol.*, **164**, 361–364.

Kerr, D.I.B., Ong, J., Johnston, G.A.R. and Prager, R.H. (1989c) $GABA_B$-receptor-mediated actions of baclofen in the rat isolated cortical slice preparation: antagonism by phosphono-analogues of GABA. *Brain Res.*, **480**, 312–316.

Lambert, N.A., Harrison, N.L., Kerr, D.I.B., Ong, J., Prager, R.H. and Teyler, T.J. (1989) Blockade of the late IPSP in rat CA1 hippocampal neurons by 2-hydroxysaclofen. *Neurosci. Lett.*, **107**, 125–128.

Luzzi, S., Maggi, C.A., Spagnesi, S., Santicioli, P. and Zilletti, L. (1985) 5-Aminovaleric acid interactions with $GABA_A$ and $GABA_B$ receptors in guinea-pig ileum. *J. Auton. Pharmacol.*, **5**, 65–69.

Luzzi, S., Franchi-Micheli, S., Ciuffi, M., Pajoni, A. and Zilletti, L. (1986) GABA-related activities of amino phosphonic acids on guinea-pig ileum longitudinal muscle. *J. Auton. Pharmacol.*, **6**, 163–169.

Malouf, A.T., Robbins, C.A. and Schwartzkroin, P.A. (1988) Effects of bicuculline methiodide and phaclofen on spontaneous and evoked PSP's in rat hippocampal rollertube cultures. *Soc. Neurosci. Abstr.*, **14**, 814.

Muhyaddin, M., Roberts, P.J. and Woodruff, G.N. (1982) Presynaptic gamma-aminobutyric acid receptors in the rat anococcygeus muscle and their antagonism by 5-aminovaleric acid. *Br. J. Pharmacol.*, **77**, 163–168.

Nishikawa, M. and Kuriyama, K. (1989) Functional coupling of cerebral gamma-aminobutyric acid ($GABA_B$) receptor with adenylate cyclase system: effect of phaclofen. *Neurochem. Int.*, **14**, 85–90.

Olpe, H.-R., Steinmann, M.E., Hall, R.G., Brugger, F. and Pozza, M.F. (1988) $GABA_A$

and $GABA_B$ receptors in locus coeruleus: effects of blockers. *Eur. J. Pharmacol.*, **149**, 183-185.

Ong, J. and Kerr, D.I.B. (1983a) $GABA_A$- and $GABA_B$-receptor-mediated modification of intestinal motility. *Eur. J. Pharmacol.*, **86**, 9-17.

Ong, J. and Kerr, D.I.B. (1983b) Interactions between GABA and 5-hydroxytryptamine in the guinea-pig ileum. *Eur. J. Pharmacol.*, **94**, 305-312.

Ong, J., Kerr, D.I.B. and Johnston, G.A.R. (1987a) Calcium dependence of baclofen- and GABA-induced depression of responses to transmural stimulation in the guinea-pig isolated ileum. *Eur. J. Pharmacol.*, **134**, 369-372.

Ong, J., Kerr, D.I.B. and Johnston, G.A.R. (1987b) Differing actions of β-phenyl-GABA and baclofen in the guinea-pig isolated ileum. *Neurosci. Lett.*, **77**, 109-112.

Ong, J., Harrison, N.L., Hall, R.G., Barker, J.L., Johnston, G.A.R. and Kerr, D.I.B. (1990) 3-Aminopropanephosphinic acid is a potent agonist at peripheral and presynaptic $GABA_B$ receptors. *Brain Res.*, in press.

Pratt, G.D., Knott, C., Davey, R. and Bowery, N.G. (1989) Characterisation of 3-aminopropylphosphinic acid as a $GABA_B$ agonist in rat brain tissue. *Br. J. Pharmacol.*, **96**, 141P.

Sawynok, J. (1986) Baclofen activates two distinct receptors in the rat spinal cord and guinea-pig ileum. *Neuropharmacology*, **25**, 795-798.

Scherer, R.W., Ferkany, J.W. and Enna, S.J. (1988) Evidence for pharmacologically distinct subsets of $GABA_B$ receptors. *Brain Res. Bull.*, **21**, 439-443.

Schmid, K., Bohmer, G. and Gebauer, K. (1989) $GABA_B$ receptor mediated effects on central respiratory system and their antagonism by phaclofen. *Neurosci. Lett.*, **99**, 305-310.

Soltesz, I., Haby, M., Leresche, N. and Crunelli, V. (1988) The $GABA_B$ antagonist phaclofen inhibits the late K^+-dependent IPSP in cat and rat thalamic and hippocampal neurones. *Brain Res.*, **448**, 351-354.

Soltesz, I., Lightowler, S., Leresche, N. and Crunelli, V. (1989) Optic tract stimulation evokes $GABA_A$ but not $GABA_B$ IPSPs in the rat ventral lateral geniculate nucleus. *Brain Res.*, **479**, 49-55.

Stirling, J.M., Cross, A.J., Robinson, T.N. and Green, A.R. (1989) The effects of $GABA_B$ receptor agonists and antagonists on potassium-stimulated $[Ca^{2+}]_i$ in rat brain synaptosomes. *Neuropharmacology*, **28**, 699-704.

Ticku, M.K. and Delgado, A. (1989) $GABA_B$ receptor activation inhibits Ca^{2+}-activated potassium channels in synaptosomes: involvement of G-proteins. *Life Sci.*, **44**, 1271-1276.

Watkins, J.C. (1981) Pharmacology of excitatory amino acid receptors. In: P.J. Roberts, J. Storm-Mathisen and G.A.R. Johnston (eds) *Glutamate: transmitter in the central nervous system.* Wiley: Chichester, pp. 1-24.

Wermuth, C.G., Bourguignon, J.J., Schlewer, G., Gies, J.P., Schoenfelder, A., Melikian, A., Bouchet, M.J., Chantreux, D., Molimard, J.C., Heulme, M., Chambon, J.P. and Biziere, K. (1987) Synthesis and structure-activity relationships of a series of aminopyridazine derivatives of gamma-aminobutyric acid acting as selective GABA-A antagonists. *J. Med. Chem.*, **30**, 239-249.

3 Biochemistry, Electrophysiology and Pharmacology of a New $GABA_B$ Antagonist: CGP 35348

H. BITTIGER, W. FROESTL, R. HALL, G. KARLSSON, K. KLEBS, H.-R. OLPE, M. F. POZZA, M. W. STEINMANN and H. VAN RIEZEN

INTRODUCTION

Research on the physiological role of $GABA_B$ receptors in the nervous system is presently in a critical phase. In large part, this is due to the recent availability of $GABA_B$ antagonists (see Kerr *et al.*, Chapter 2). To date, however, these antagonists have been of low potency and inactive in vivo. As a consequence, current ideas about the behavioural effects of $GABA_B$ receptor blockade and the possible therapeutic applications of $GABA_B$ antagonists derive almost exclusively from in-vitro data.

Recent studies in our laboratories have identified a novel, centrally active $GABA_B$ antagonist, CGP 35348, which enters the brain after systemic administration. Here we describe the properties and characteristics of this compound in vitro and in vivo, and attempt to answer the following questions:

(1) The $GABA_B$ agonist baclofen acts as a muscle relaxant and antispastic agent. Does a $GABA_B$ antagonist have spastic or any other effects on the motor system?

(2) Baclofen induces sedation in man and animals. Neuronal activity is suppressed and the release of several neurotransmitters reduced (see Bowery *et al.*, Chapter 1). Does an antagonist elicit opposite effects, i.e. does it act as a stimulator and increase vigilance, stimulate neurons and increase neurotransmitter release?

(3) Blockade of $GABA_A$ receptors for example by bicuculline, results in major convulsions. Are $GABA_B$ antagonists also convulsive agents?

PROPERTIES AND CHARACTERISTICS OF CGP 35348

Chemistry of CGP 35348

CGP 35348 is a substituted linear phosphoamino acid (*P*-[3-aminopropyl]-*P*-diethoxymethylphosphinic acid) and not a baclofen analogue (see Figure 3.1). The compound is highly water-soluble (700 g/l).

$GABA_B$ Receptors in Mammalian Function. Edited by N.G. Bowery, H. Bittiger and H.-R. Olpe

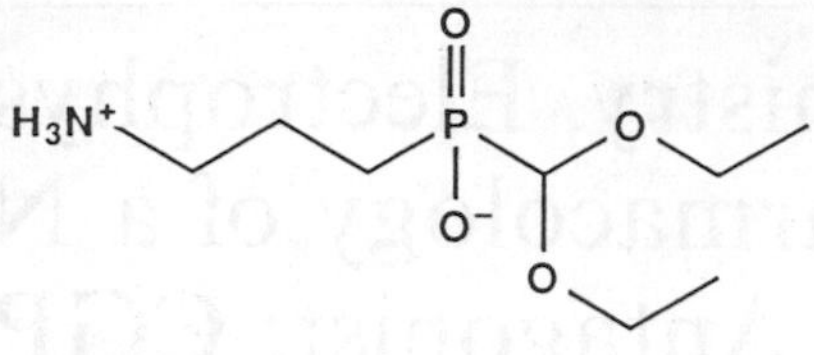

Figure 3.1. Chemical structure of CGP 35348

CGP 35348 Interacts Selectively with $GABA_B$ Receptors

For the measurement of the interaction of CGP 35348 with $GABA_B$ receptors, the potent $GABA_B$ ligand 3-aminopropylphosphinic acid, synthesized (Dingwall *et al.*, 1987) and tritiated in CIBA-Geigy laboratories, was used. This radioligand is selective for $GABA_B$ receptors (Bittiger *et al.*, 1988). In this assay CGP 35348 shows a half-maximal inhibitory concentration (IC_{50}) of 34 ± 5 μM (Table 1). Its interaction with $GABA_A$ receptors (Table 1) and 10 other receptors (including *N*-methyl-D-aspartate (NMDA), quisqualate and kainate

Table 1. Inhibition of binding by CGP 35348 in $GABA_B$ and $GABA_A$ receptor assays

Type of receptor	Radioligand	Inhibition of binding	Method
$GABA_B$	[^{3}H]CGP 27492	$IC_{50} = 34 \pm 5$ μM	Bittiger *et al.* (1988)
$GABA_A$	[^{3}H]muscimol	15% inhibition at 1 mM	Beaumont *et al.* (1978)

The IC_{50} value was determined in three independent experiments.

Method for the $GABA_B$ Receptor Assay Using [^{3}H]CGP 27492 as radioligand

Preparation of membranes Twenty male rats (Tif:RAlf(SPF)) of about 200 g body weight were used. The animals were decapitated, the brains removed, and the cerebral cortices dissected and homogenized in 10 volumes of ice-cold 0.32 M sucrose, containing $MgCl_2$ (1 mM) and K_2HPO_4 (1 mM), with a glass/Teflon homogenizer. The membranes were centrifuged at 750 × ***g***, the pellet was resuspended and the centrifugation repeated. The supernatants were pooled and centrifuged at 18 000 × ***g*** for 15 min. The pellet was osmotically shocked in 5 ml H_2O and kept on ice for 30 min. The suspension was centrifuged at 39 000 × ***g***, resuspended in Krebs–Henseleit buffer pH 7.4 containing 20 mM Tris, and kept for 2 days at −20 °C. The membranes were thawed at room temperature, washed three times with Krebs–Henseleit buffer by centrifugation at 18 000 × ***g*** for 15 min, left overnight at 4 °C and washed again three times. The final pellet was resuspended with a glass/Teflon homogenizer in 20 ml of the same buffer. Two-ml aliquots were frozen and stored in liquid nitrogen. Just before use, membranes were thawed quickly in a water-bath at 37 °C and again washed by centrifugation at 18 000 × ***g*** for 15 min with the same buffer three times.

Radioreceptor assay The radioreceptor assay was performed in 2 ml Krebs–Henseleit buffer pH 7.4, containing 20 mM Tris, 200–300 μg membrane protein, 2 nM [^{3}H]CGP 27492 (15 Ci/mmol, CIBA-Geigy, Horsham, UK) and the compound to be tested. The incubation was performed at 20 °C for 40 min and terminated by rapid filtration on Whatman GF/B glass-fibre filters, which were washed twice with 5 ml ice-cold buffer. Filter-bound radioactivity was counted in Irgascint A300 (CIBA-Geigy). Incubations were performed in triplicate and non-specific binding was determined in the presence of 10 μM L-baclofen. IC_{50} values were obtained by computer-aided curve fitting, according to a single-site model.

Table 2. Inhibition of binding by CGP 35348 in different receptor assays

Type of receptor	Radioligand	Inhibition of binding (% at 1 mM)	Method
Benzodiazepine	[^{3}H]flunitrazepam	0	Speth *et al.* (1978)
NMDA	[^{3}H]L-glutamate	9	Foster and Fagg (1987)
Quisqualate	[^{3}H]AMPA	10	Honoré and Nielsen (1985)
Kainate	[^{3}H]kainate	32	Foster *et al.* (1981)
			Simon *et al.* (1976)
Cholinergic Muscarinic	[^{3}H]*cis*-methyldioxolane	0	Closse *et al.* (1987)
α_1-Adrenergic	[^{3}H]prazosin	0	Greengrass and Bremner (1979)
α_2-Adrenergic	[^{3}H]clonidine	10	Tanaka and Starke (1980)
5-HT_1	[^{3}H]5-HT	0	Nelson *et al.* (1978)
$Histamine_1$	[^{3}H]doxepine	7	Tran *et al.* (1981)
Adenosine	[^{3}H]CH-adenosine	0	Patel *et al.* (1982)

NMDA, *N*-methyl-D-aspartate; AMPA, amino-3-hydroxy-5-methylisoxazole-4-propionic acid; 5-HT, 5-hydroxytryptamine (serotonin).

receptors) was negligible (Table 2). Thus CGP 35348 appears to be highly selective for $GABA_B$ receptors. The IC_{50} value of 34 μM is quite high (cf. L-baclofen, 30 nM; i.e. 1000 times lower). However, it should be remembered that the radioligand used here is an agonist and therefore that the IC_{50} values for antagonists may be overestimated. For example, the $GABA_A$ antagonist bicuculline shows an IC_{50} value of 2 μM in the [^{3}H]muscimol agonist binding assay (Beaumont *et al.*, 1978) and 68 nM in the [^{3}H] bicuculline methiodide antagonist assay, respectively (Möhler and Okada, 1977).

CGP 35348 Antagonizes the Effects of Baclofen In Vitro and In Vivo

Baclofen-induced potentiation of noradrenaline-stimulated adenylyl cyclase in rat cortical slices

In rat cortical slices, adenylyl cyclase is stimulated by noradrenaline (NA) via β-adrenergic receptors. This effect can be potentiated by baclofen (Karbon and Enna, 1985). In Figure 3.2 the percentage conversion of ATP into cyclic AMP induced by NA alone and by NA combined with L-baclofen is shown. The potentiation induced by 30 μM L-baclofen was dose-dependently antagonized by CGP 35348 with an IC_{50} of about 300 μM. CGP 35348 alone did not have any effect on the stimulation of adenylyl cyclase by NA.

Hyperpolarization of CA1 neurons in rat hippocampal slices

Bath-applied baclofen hyperpolarizes hippocampal CA1 pyramidal neurons in slice preparations by opening $GABA_B$ receptor-coupled K^+ channels (New-

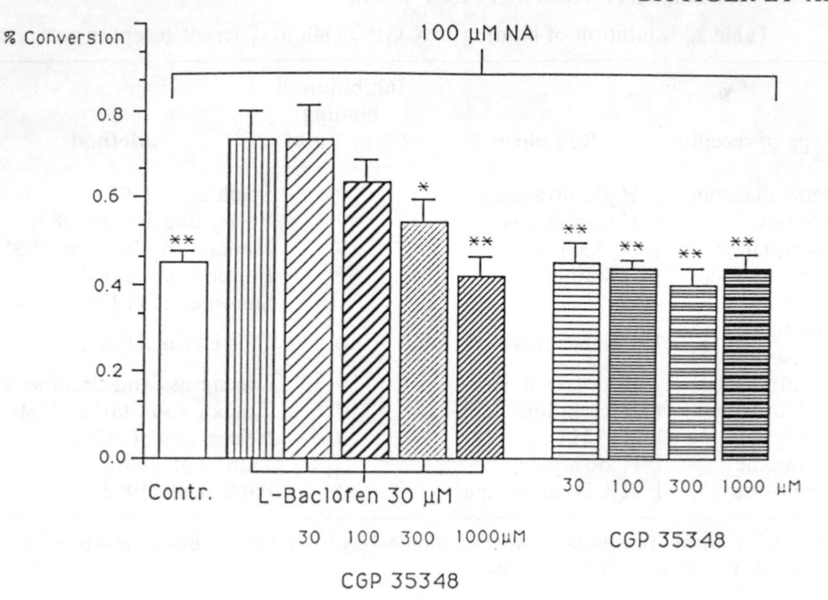

Figure 3.2. The potentiation by baclofen of noradrenaline (NA) stimulated adenylyl cyclase and its antagonism by CGP 35348 is depicted. Rat cortical slices were exposed to 100 μM NA, 30 μM L-baclofen and different concentrations of CGP 35348. Each value is the mean percentage conversion $\pm$ SEM of ATP into cyclic AMP of determinations each performed in triplicate in three independent experiments. *$P < 0.05$, **$P < 0.01$: significantly different from percentage conversion of cyclic AMP induced by 100 μM NA and 30 μM L-baclofen (second bar from left) (Dunnett's test)

Method

The prelabelling technique of Schimizu *et al.* (1969) was used. In this test, rat cortex slices are incubated with [^{3}H]adenine, which is taken up into the slices and transformed intracellularly into [^{3}H]ATP. [^{3}H]Cyclic AMP is formed from [^{3}H]ATP by the stimulation of adenylyl cyclase.

Briefly, male rats (Tif:RAlf(SPF), 150–200 g) were sacrificed by decapitation and the cortices dissected and minced (250 × 250 μm) using a McIllwain tissue chopper. The cortex slices were transferred to Krebs–Ringer–bicarbonate buffer (composition in mM: NaCl, 122; KCl, 3; KH_2PO_4, 0.4; $MgSO_4$, 1.2; $CaCl_2$, 1.5; $NaHCO_3$, 25; D-glucose, 10.0) at a concentration 25 mg/ml and incubated under constant gassing (95 % O_2/5 % CO_2, 0.2 bar) at 37 °C for 10 min. Then the slices were washed by sedimentation at 350 × ***g*** for 2–3 min, aspiration of the supernatant and resuspension of the slices in fresh buffer. The tissue was taken up in buffer (about 25 mg tissue/ml buffer), and [^{3}H]adenine (5.2 mCi/ml final concentration, 24 Ci/mmol; Amersham) and cold adenine (120 nM final concentration) were added and incubated at 27 °C for 40 min under constant gassing. After this incubation the suspension was washed three times at 350 × ***g*** and resuspended in buffer. Then 200 μl aliquots of the slice suspension were distributed in vials containing buffer at 37 °C and the suspension was preincubated for 10 min. Compounds (NA, L-baclofen, test compounds) were added every 10 s for a 12 min reaction. The total reaction volume was 5 ml. Aliquots of 1 ml were taken out and added to 0.25 ml of 25% trichloric acetic acid (TCA). After addition of 1 ml of 5% TCA the suspension was sonicated. The suspension was sedimented at 1800 × ***g*** for 20 min. One-ml aliquots were taken and analysed for [^{3}H]cyclic AMP using the double-column method according to Salomon *et al.* (1974).

The results were expressed as percentages of the total tritium pool represented by [^{3}H]cyclic AMP.

Figure 3.3. Intracellular recording from a CA1 hippocampal neuron (slice preparation). The neuron was spontaneously firing action potentials. Throughout the recordings, hyperpolarizing current pulses were passed through the recording electrode in order to discover any changes in membrane conductance. Bath application of D/L-baclofen (B, 10 μM) inhibited cell firing and induced hyperpolarizations of the membrane potential. The effect was attenuated by 30 μM CGP 35348 in a reversible manner. The resting membrane potential was -63 mV

Method

Hippocampal slices were used to study the action of CGP 35348 on the membrane hyperpolarization induced by D/L-baclofen. Rats were decapitated under halothane anaesthesia. The brains were removed and immersed prior to slicing in artificial cerebrospinal fluid (ACSF) at room temperature. The hippocampi were carefully dissected free. Using a tissue chopper, transverse slices 400 μm thick were cut and transferred to an interface chamber and perfused with hyperosmolar ACSF at room temperature for 30 min. At the end of this period the temperature was raised to 32 °C. The ACSF level was adjusted to just below the upper surface of the slices and the chamber was continuously supplied with a warm, humidified 95% O_2/5% CO_2 mixture. The slices were left to equilibrate for a minimum of 1 h before starting recording. The ACSF had the following composition in mM: NaCl, 120; KCl, 2.5; KH_2PO_4, 1.2; $MgSO_4$, 2.0; $NaHCO_3$, 30.0; $CaCl_2$, 2.5; glucose, 10.0.

Recordings were obtained with electrodes filled with potassium methylsulphate (2 M) having resistances ranging from 50 to 90 MΩ. Electrical signals were studied using an Axoclamp-2A amplifier. CGP 35348 and D/L-baclofen were dissolved in ACSF, either alone or together. For drug exposure, the infusion line was switched from drug-free to drug-containing solutions. D/L-Baclofen was administered for periods lasting 1 min. CGP 35348 was applied for 7–10 min.

berry and Nicoll, 1984). A concentration of 10 μM D/L-baclofen hyperpolarized CA1 neurons and inhibited cell firing in those neurons that were spontaneously active (Figure 3.3). CGP 35348 at a concentration of 30 μM completely and reversibly antagonized the effects of baclofen (4/4 cells). The threshold concentration for antagonizing the hyperpolarizing effect of bacolfen was 10 μM. In the presence of CGP 35348 ($\geq$ 30 μM) alone, the firing frequency of spontaneously active neurons was occasionally increased.

Systematically applied CGP 35348 antagonizes ionophoretically applied baclofen in the rat cerebral cortex

In these experiments we examined the effects of systematically applied CGP 35348 on the inhibition of cortical neuronal firing induced by D/L-baclofen

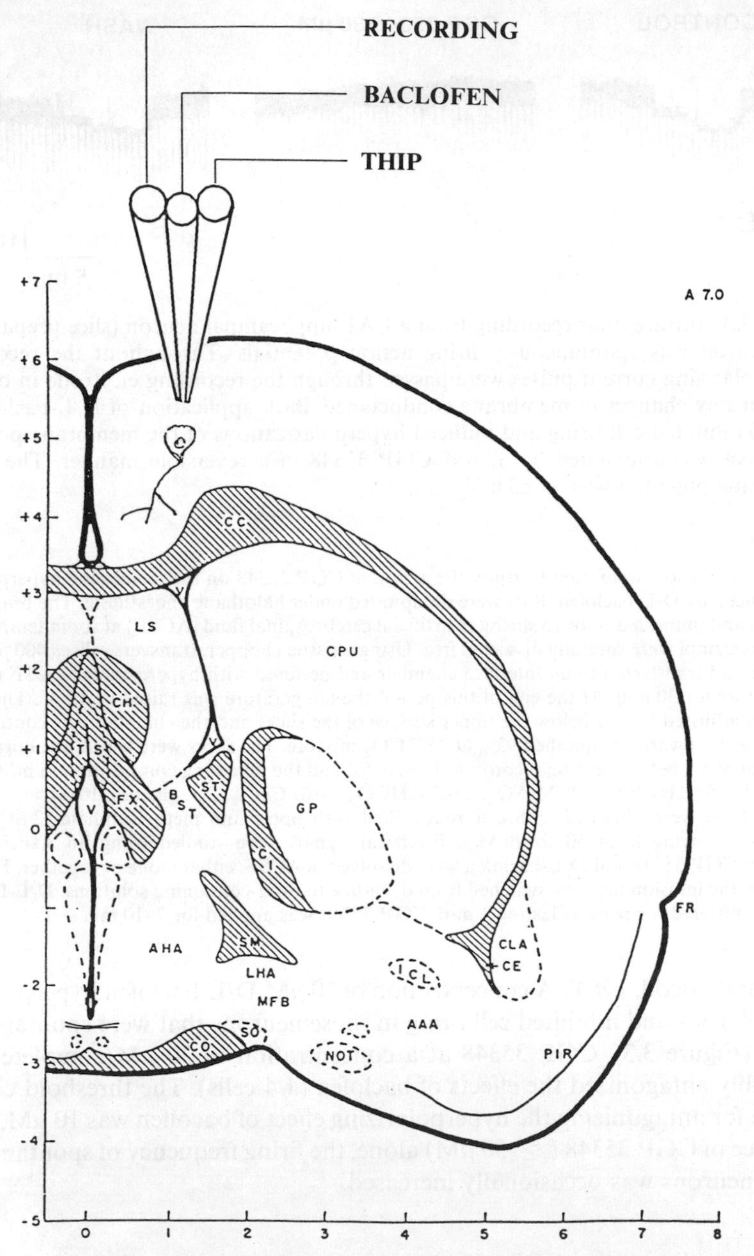

Figure 3.4. Placement of the electrode for ionophoretic administration of D/L-baclofen and THIP. (Coronal section adapted from the atlas of J. De Groot, 1972.) D/L-Baclofen and THIP were ionophoretically administered via a three-barrelled micropipette. *Abbreviations*: CC, corpus callosum; CPU, nucleus caudatus, putamen; LS, nucleus lateralis septi; GP, globus pallidus (for further details see atlas of J. De Groot)

ionophoretically applied near to rat cortical neurons via micropipettes (Figure 3.4). As shown in Figure 3.5, the spontaneous electrical activity of cortical neurons was inhibited by D/L-baclofen and the GABA$_A$ agonist THIP (4,5,6,7-tetrahydroisooxazolo[5,4-*c*]pyridin-3-ol; Krogsgaard-Larsen *et al.*, 1978). The inhibitory effect of D/L-baclofen was reduced by 10 mg/kg i.v. (4 rats) and strongly suppressed by 30 mg/kg i.v. (6 rats) CGP 35348. The antagonism was clearly visable 5–10 min after i.v. injection. The effects of THIP were not affected, indicating a selectivity for GABA$_B$ receptors. An almost complete antagonism against baclofen was obtained with 100 mg/kg i.p. CGP 35348 (6 rats). At 30 mg/kg i.p. the effect was moderate (6 rats). The effect of i.p. CGP 35348 appeared after 10 min. These experiments document the in-vivo activity of CGP 35348 and show that the compound is able to penetrate the blood–brain barrier. There was pratically no antagonism after oral applications up to doses of 1000 mg/kg.

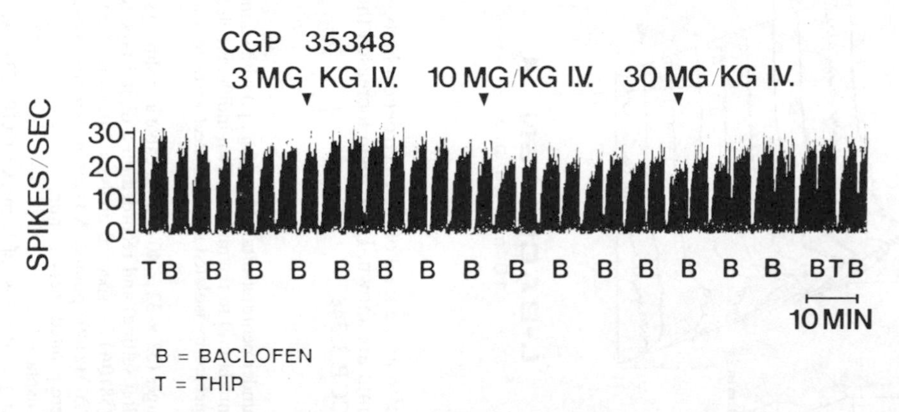

Figure 3.5. Intravenously injected CGP 35348 blocked the depressant responses of cortical neurons to ionophoretically administered D/L-baclofen (B) without affecting the depressant responses evoked by the GABA$_A$ agonist THIP (T)

Method

Male rats (Tif: RAlf(SPF)) weighing 280–320 g were anaesthetized with chloral hydrate (400 mg/kg i.p., supplemented approximately every hour) and mounted in a stereotaxic frame with the incisor bar located 5 mm above the ear bar. D/L-Baclofen (10 μM, pH 3) and THIP (10 μM, pH 3) were administered ionophoretically by means of a three-barrelled mircopipette positioned near spontaneously active neurons in the rostromedial cortex using the following coordinates: anterior 8.0–9.0 mm (zero at ear bar); 0.5–2 mm lateral; 0.8–1.5 mm ventral (from the cortical surface). Ejection currents applied to the drug barrels were adjusted to produce equipotent inhibitory responses with the two drugs. D/L-Baclofen was applied with currents of 50–150 nA for periods of 20–30 s. THIP was applied with similar currents for periods of 20–30 s. Retain currents of 20–50 nA were used routinely between applications. The activity of individual neurons was recorded by means of the third electrode filled with 4 M NaCl. Action potentials were separated from the background noise by means of a window discriminator and integrated over periods of 3–10 s. D/L-Baclofen and THIP were administered alternately at intervals of 3–5 min. CGP 35348 was dissolved in physiological saline solution and administered i.v. or i.p.

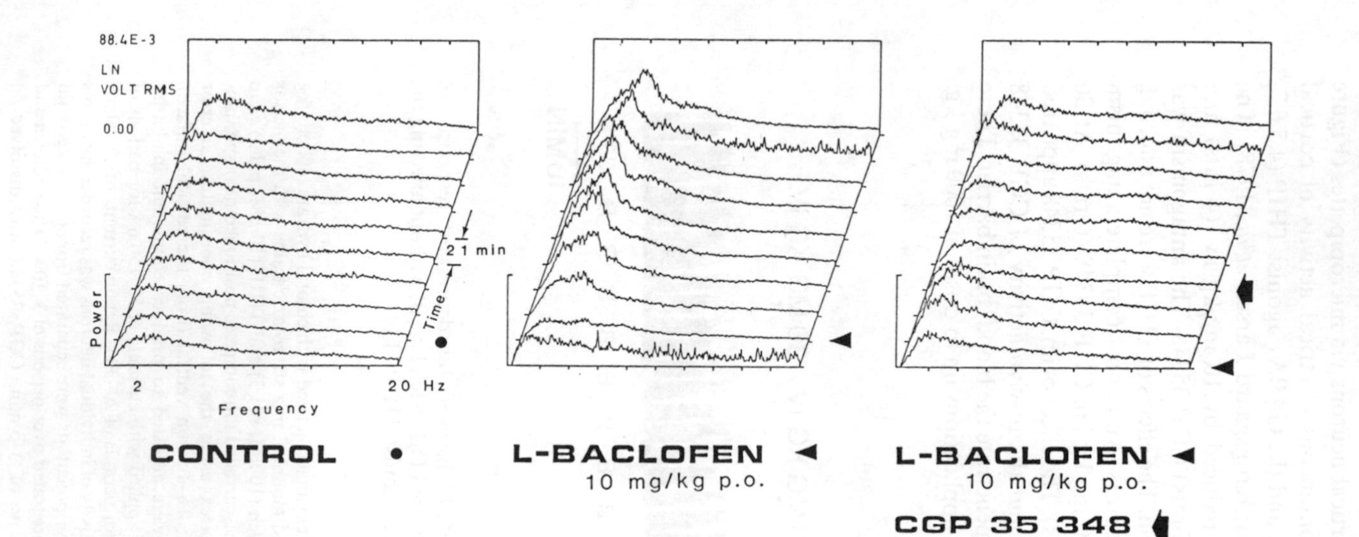

Figure 3.6. Interaction of CGP 35348 with baclofen at the level of the electroencephalogram (EEG) in the freely moving rat. The power/frequency spectra integrated over periods of 21 min are shown. L-Baclofen increased the power at low frequencies (2–4 Hz). This effect was completely antagonized by 100 mg/kg i.p. CGP 35348

Method (*Developed by A. Glatt, CIBA-Geigy*)

Male rats (Tif:RAlf(SPF), 280–320 g) were implanted under pentobarbital (54 mg/kg i.p.) with stainless steel screw electrodes located in the skull over the occipital cortex 4 mm posterior to the bregma and 4 mm lateral to the midline. Two indifferent electrodes were placed 6 mm anterior to the bregma and 1 mm lateral to the midline. The electrodes were connected by isolated stainless steel wires to a four-pole socket, fixed to the skull with dental acrylic cement.

Monopolar recordings were performed in wooden cages (32 × 32 × 40 cm) via cable and sliping system. The EEG signals were amplified by Grass P7B Polygraph amplifiers (low 1 Hz, high 35 Hz cut-off filter setting) and recorded on a digital tape recorder (EDR 8000, Earth Data Ltd). The amplified signals were processed in a Wavetek Mini Analyser (5810A) for spectral analysis. Intervals of 20 s were analysed. Sample frequency was 51.2 Hz and voltage was calculated for 0.05 Hz bands from 0.05 to 20 Hz (400 points). A Hanning window was applied, but no smoothing was performed. Frequency spectra at 3.5 h were analysed and average spectra were printed every 21 min.

On the ordinate, volts RMS (rate meter scale) are displayed.

L-Baclofen was dissolved in distilled water and given p.o. at a dose of 10 mg/kg. CGP 35348 was dissolved in saline and administered i.p. at a dose of 100 mg/kg.

Electroencephalogram of the freely moving rat

Baclofen has a pronounced effect on the electroencephalogram (EEG) of the freely moving rat (chronically implanted electrodes). As shown in Figure 3.6 the power spectra of the cortical EEG exhibit a strong increase of activity (power) at low frequencies (2–3 Hz) after administration of 10 mg/kg p.o. L-baclofen. This change in the EEG was completely normalized by 100 mg/kg i.p. CGP 35348.

Muscle relaxation

Muscle relaxation and effects on the motor system in general can be evaluated in the rotarod model. In the tests, rats or mice are placed on a rotating cylinder and, in the absence of pharmacologically induced motor impairment, the animals run on the cylinder without falling off. After 15 mg/kg p.o. D/L-baclofen the endurance of rats was reduced from more than 300 s to approximately 20 s (Table 3). This effect was dose-dependently antagonized by CGP 35348 after i.p. application. In mice similar effects were observed after doses of 60 mg/kg p.o. D/L-baclofen. Interestingly, no motor impairment was observed when the antagonist was given alone, even at the highest tested dose of 300 mg/kg i.p.

Table 3. Antagonism of CGP 35348 against the effects of D/L-baclofen in the rotarod test

Dose CGP 35348 (mg/kg i.p.)	Rats		Mice	
	CGP 35348 only	CGP 35348 with D/L-baclofen	CGP 35348 only	CGP 35348 with D/L-baclofen
0	>300 (0/8)	13 (8/8)	>300 (0/10)	52 (10/10)
30	>300 (0/8)	113* (7/8)	>300 (0/10)	87* (8/10)
100	>300 (0/8)	>300* (1/8)	>300 (0/10)	255* (5/10)
300	>300 (0/8)	>300* (0/8)	>300 (0/10)	>300* (0/10)

*$P < 0.05$ (Dunnett's test); comparison to D/L-baclofen-treated controls (zero-dose group).

Data are median endurance times in seconds; numbers in parentheses show numbers of animals failing to stay on the rotarod for 300 s. The test was arbitrarily stopped when the animals walked on the cylinder for more than 300 s. Rats were treated with 15 mg/kg p.o. D/L-baclofen dissolved in water, CGP 35348 i.p. (in water) or placebo (water) (8 animals/group). D/L-Baclofen and CGP 35348 were given simultaneously and the tests were performed 1 h after drug application. Mice were treated with 60 mg/kg p.o. D/L-baclofen (10 animals/group).

Method

Male rats (Tif:RAlf(SPF), 125–150 g) were trained on the rotating rod apparatus (UGO Basile, Milano) 1 day before the experiment. Only animals which walked for at least 300 s on the rotating rod were used. The same procedure was applied for mice (Tif:MAGF (SPF), 15–20 g).

Effects of CGP 35348 Alone

Release of GABA from electrically stimulated cortical slices

L-Baclofen (10 μM) decreased the release of GABA evoked by electrical field stimulation in rat cortical slices. This effect was antagonized by 100 μM CGP 35348. The compound not only antagonized the effect of L-baclofen, but also increased the release of GABA at 30–100 μM in the absence of baclofen. These effects are described in detail by Waldmeier and Baumann in Chapter 4.

Reduction of the late inhibitory postsynaptic potential in hippocampal slices

Electrical stimulation of Schaffer collateral/commissural fibres in rat hippocampal slices evokes inhibitory postsynaptic potentials (IPSPs) in CA1 neurons (see Figures 3.7 and 3.8). These potentials recorded intracellularly show two hyperpolarizing components: an early major component mediated by $GABA_A$ receptors and a late component assigned to $GABA_B$ receptors (Dutar and Nicoll, 1988; Karlsson *et al.*, 1988). The late IPSP was completely blocked by 100 μM CGP 35348. The refractory period of the neuron was shortened by CGP 35348 as shown by the appearance of spontaneous action potentials (Figure 3.8, middle panel). The blockade of the late IPSP was observed approximately 5 min after the superfusion was switched to artificial cerebrospinal fluid (ACSF)

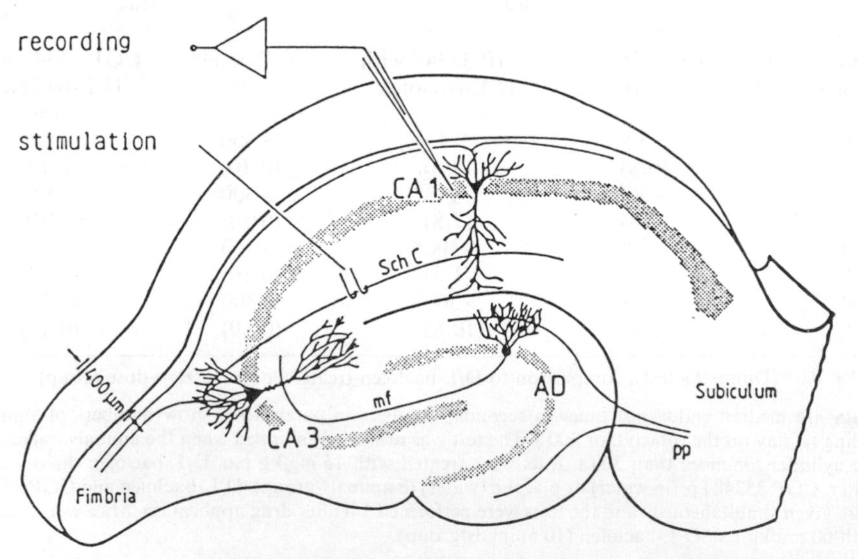

Figure 3.7. Scheme of the experimental design for measuring late inhibitory postsynaptic potentials (IPSPs). In hippocampal slices Schaffer collaterals/commissural fibres (SchC) were stimulated and intracellular recording was performed from a CA1 neuron. *Abbreviations:* PP, perforant path; mf, mossy fibres; AD, area dentata

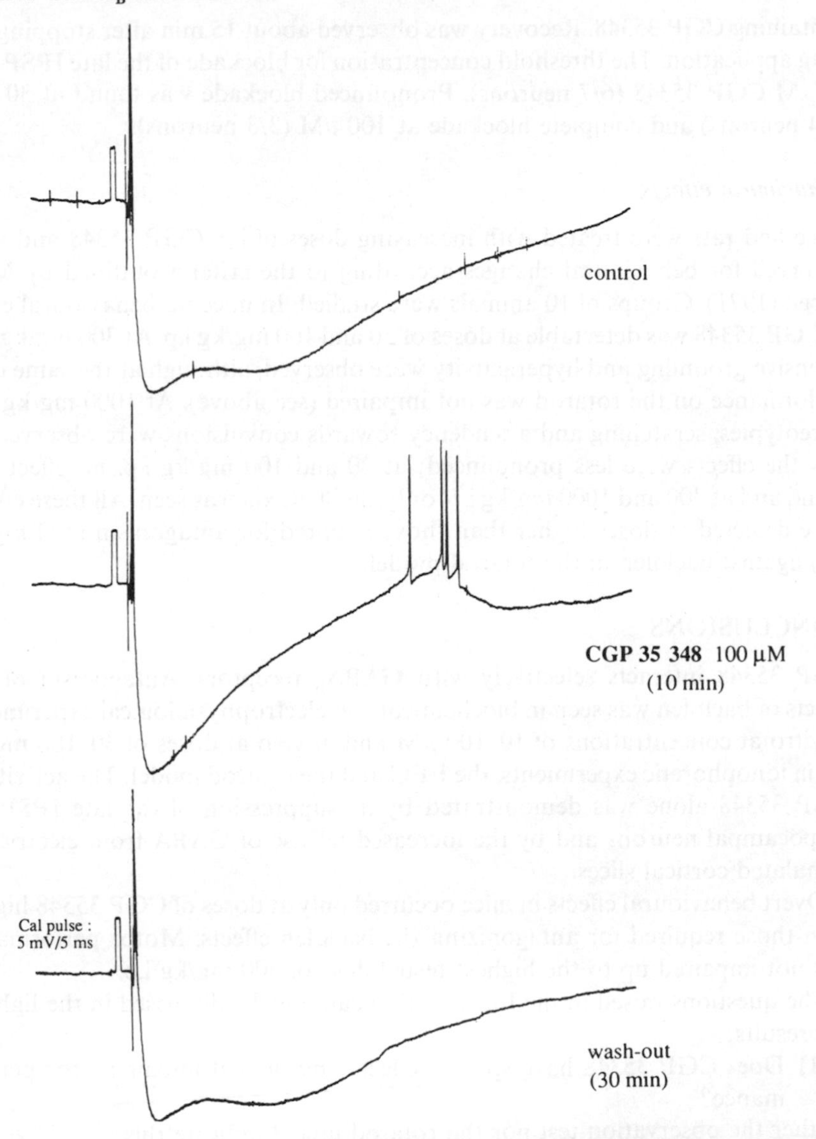

Figure 3.8. Reduction of the late inhibitory postsynaptic potential (IPSP) by CGP 35348 in a hippocampal pyramidal neuron. The late IPSP was evoked synaptically by applying brief stimuli to the Schaffer collateral/commissural fibres. Bipolar 50 μm platinum/iridium electrodes (monophasic wave pulse, 0.1 ms, 15–100 μA) were used for the stimulation. Each trace is the average of four successive sweeps. Preparation of hippocampal slices, the composition of the medium and recording were as described in Figure 3. Bath-applied CGP 35348 reduced only the late IPSP, leaving the early IPSP unaffected. In the middle panel it is shown that the neuron fired spontaneous action potentials during the exposure to CGP 35348, indicating that the refractory period was reduced as a consequence of blocking the late IPSP

containing CGP 35348. Recovery was observed about 15 min after stopping the drug application. The threshold concentration for blockade of the late IPSP was 10 μM CGP 35348 (6/7 neurons). Pronounced blockade was found at 30 μM (4/4 neurons) and complete blockade at 100 μM (3/3 neurons).

Behavioural effects

Mice and rats were treated with increasing doses of i.p. CGP 35348 and were observed for behavioural changes according to the criteria outlined by Morpurgo (1971). Groups of 10 animals were studied. In mice no behavioural effect of CGP 35348 was detectable at doses of 30 and 100 mg/kg i.p. At 300 mg/kg i.p., intensive grooming and hyperactivity were observed, although at the same dose performance on the rotarod was not impaired (see above). At 1000 mg/kg i.p., stereotypies, scratching and a tendency towards convulsions were observed. In rats the effects were less pronounced; at 30 and 100 mg/kg i.p. no effect was found and at 300 and 1000 mg/kg i.p. only slight ataxia was seen. All these effects were detected at doses higher than those required for antagonism (100 mg/kg i.p.) against baclofen in the rotarod model.

CONCLUSIONS

CGP 35348 interacts selectively with $GABA_B$ receptors. Antagonism of the effects of baclofen was seen in biochemical and electrophysiological experiments in vitro at concentrations of 10–100 μM and in vivo at doses of 30–100 mg/kg i.p. in ionophoretic experiments, the EEG and the rotarod model. The activity of CGP 35348 alone was demonstrated by its suppression of the late IPSPs in hippocampal neurons and by the increased release of GABA from electrically stimulated cortical slices.

Overt behavioural effects in mice occurred only at doses of CGP 35348 higher than those required for antagonizing the baclofen effects. Motor performance was not impaired up to the highest tested dose of 300 mg/kg i.p.

The questions raised in the Introduction can now be discussed in the light of our results:

(1) Does CGP 35348 have specific effects and does it impair motor performance?

Neither the observation test nor the rotarod model indicate this.

(2) Is CGP 35348 a stimulating agent?

There are several hints of an activation of neurons by CGP 35348:

- The increase in firing frequency of spontaneously active neurons in hippocampal slices.
- The suppression of the late IPSPs which resulted in higher excitability of neurons.
- The behavioural stimulation of mice at higher doses.

(3) Is CGP 35348 a convulsant?

Proconvulsive effects were seen in mice only at very high doses (10 times the ID_{50} of antagonism against baclofen in the rotarod model). This is in strong contrast to the marked convulsant activity of the $GABA_A$ antagonist bicuculline. The role of the $GABA_B$ system in epilepsy is discussed in detail by Karlsson *et al.* in Chapter 21.

In conclusion, CGP 35348 appears to be a 'soft' positive modulator of neurotransmission via the blockade of $GABA_B$ receptors. Possible therapeutic effects of $GABA_B$ antagonists, as discussed by Bowery *et al* in Chapter 1, may include improvement of cognitive processes and antidepressant activity. However, studies in a variety of complex animal models of human behaviour will be necessary to substantiate such speculations.

ACKNOWLEDGEMENT

We thank Dr G.E. Fagg, CIBA-Geigy, for testing CGP 35348 in NMDA, kainate and quisqualate receptor assays.

REFERENCES

Beaumont, K., Chilton, W.S., Yamamura, H.I. and Enna, S.J. (1978) Muscimol binding in rat brain: association with synaptic GABA receptors. *Brain Res.*, **148**, 153–162.

Bittiger, H., Reymann, N., Hall, R. and Kane, P. (1988) CGP 27492, a new, potent and selective radioligand for GABA-B receptors. In: *Proceedings of the 11th Annual Meeting of the European Neuroscience Association, Zurich, September 1988. Eur. J. Neurosci.* (Suppl.), Abstr. 16.10.

Closse, A., Bittiger, H., Langenegger, D. and Wanner, A. (1987) Binding studies with [^{3}H]*cis*-methyldioxolane in different tissues. *Arch. Pharmacol.*, **335**, 372–377.

de Groot, J. (1972) *The rat forebrain in stereotaxic coordinates*, 4th edition. Amsterdam: North-Holland Publishing Co.

Dingwall, J.G., Ehrenfreund, J., Hall, R.G. and Jack, J. (1987) Synthesis of gamma-aminopropylphosphonous acids using hypophosphorous acid synthesis. *Phosph. Sulf.*, **30**, 571–574.

Dutar, P. and Nicoll, R.A. (1988) A physiological role for $GABA_B$ receptors in the central nervous system. *Nature*, **332**, 156–158.

Foster, A.C. and Fagg, G.E. (1987) Comparison of L-[^{3}H]glutamate, D-[^{3}H]aspartate, DL-[^{3}H]AP5 and [^{3}H]NMDA as ligands for NMDA receptors in crude postsynaptic densities from rat brain. *Eur. J. Pharmacol.*, **133**, 291–300.

Foster, A.C., Mena, E.E., Monaghan, D.T. and Cotman, C.W. (1981) Synaptic localization of kainic acid binding sites. *Nature*, **289**, 73–75.

Greengrass, P. and Bremner, R. (1979) Binding characteristics of ^{3}H-prazosin to rat brain alpha-adrenergic receptors. *Eur. J. Pharmacol.*, **55**, 323–326.

Honoré, T. and Nielsen, M. (1985) Complex structure of quisqualate-sensitive glutamate receptors in rat cortex. *Neurosci. Lett.*, **54**, 27–33.

Karbon, E.W. and Enna, S.I. (1985) Characterization of the relationship between gamma-aminobutyric acid, beta-agonists and transmitter-coupled cyclic nucleotide generating systems in rat brain. *Mol. Pharmacol.* **27**, 53–59.

Karlsson, G., Pozza, M.F. and Olpe, H.-R. (1988) Phaclofen: a $GABA_B$ blocker reduces long-duration inhibition in the neo-cortex. *Eur. J. Pharmacol.*, **148**, 485–486.

Krogsgaard-Larsen, P., Hejds, H., Curtis, D.R., Lodge, D. and Johnston, G.A.R. (1978) Dihydromuscimol, thiomuscimol and related heterocyclic compounds as GABA analogues. *Neurochem.*, **32**, 1717–1724.

Möhler, H. and Okada, T. (1977) GABA receptor binding with ^{3}H(+)bicucul line-methiodide in rat CNS. *Nature*, **267**, 65–67.

Morpurgo, C. (1971) A new design for the screening of CNS-active drugs in mice. *Arzneim.-Forsch.*, **21**, 1727–1734.

Newberry, N.R. and Nicoll, R.A. (1984) A bicuculline-resistant inhibitory post-synaptic potential in rat hippocampal pyramidal cells in vitro. *J. Physiol.*, **348**, 239–254.

Nelson, P.L., Herbet, A., Bourgoin, S., Glowinski, J. and Hamon, M. (1978) Characteristics of central 5-HT receptors and their adaptive changes following intracerebral 5,7-dihydroxytryptamine administration in the rat. *Mol. Pharmacol.*, **14**, 983–995.

Patel, J., Marangos, P.J., Stivers, J. and Goodwin, F.K. (1982) Characterization of adenosine receptors in brain using N6 cyclohexyl[^{3}H]adenosine. *Brain Res.*, **237**, 203–214.

Salomon, Y., Londos, C. and Rodbell, M. (1974) A highly sensitive adenylate cyclase assay. *Anal. Biochem.*, **58**, 541–548.

Schimizu, H., Daly, J.W. and Creveling, C.R. (1969) A radioisotopic method for measuring adenosine 3′,5′-cyclic monophosphate in incubated slices of brain. *J. Neurochem.*, **16**, 1609–1619.

Simon, J.R., Contrera, J.F. and Kuhar, M. (1976) Binding of [^{3}H]kainic acid, an analogue of L-glutamate, to brain membranes. *J. Neurochem.*, **26**, 141–147.

Speth, R.C., Wastek, G.J., Johnson, P.C. and Yamamura, H.I. (1978) Benzodiazepine binding in human brain; characterization using ^{3}H-flunitrazepam. *Life Sci.*, **22**, 859–866.

Tanaka, T. and Starke, K. (1980) Antagonist/agonist preferring alpha-adrenoceptors or alpha$_1$/alpha$_2$-adrenoceptors? *Eur. J. Pharmacol.*, **63**, 191–194.

Tran, V.T., Lebovitz, R., Toll, L. and Snyder, S.H. (1981) [^{3}H]Doxepine interactions with histamine H_1 receptors and other sites in guinea pig and rat brain homogenates. *Eur. J. Pharmacol.*, **70**, 501–509.

Part II

CONTROL OF TRANSMITTER RELEASE

4 $GABA_B$ Receptors and Transmitter Release

P. C. WALDMEIER and P. A. BAUMANN

INVOLVEMENT OF $GABA_B$ RECEPTORS IN THE RELEASE OF OTHER TRANSMITTERS

In investigations into the mechanism of the antispastic effect of the GABA analogue, baclofen, this agent was found to depress mono- and polysynaptic reflex transmission in the spinal cord (Bein, 1972), and transmission in the cuneate nucleus (Polc and Haefely, 1976). It caused presynaptic inhibition of primary afferent terminals (Davidoff and Sears, 1974; Pierau *et al.*, 1975), which was suggested to be due to a reduction of transmitter release (Davidoff and Sears, 1974; Fox *et al.*, 1976; Morris *et al.*, 1977). Glutamate being considered as the most likely transmitter candidate in these terminals, Potashner (1978, 1979) investigated the effects of baclofen on the release of this and other amino acids. He indeed found the compound to inhibit the electrically evoked release of glutamate and aspartate, but not of GABA itself. At that time, however, it had not been considered that this effect might be mediated by a GABA receptor which was different from the classical, bicuculline-sensitive type activated by GABA analogues such as muscimol and THIP (4,5,6,7-tetrahydroisooxazolo[5,4-*c*]pyridin-3-ol). It was, however, only a little later that the existence of a different type of GABA receptor was proposed, on the basis of data obtained from peripheral tissues. GABA and baclofen, the latter in a stereospecific manner, inhibited noradrenaline release in atria (Bowery and Hudson, 1979; Bowery *et al.*, 1981) and acetylcholine release in superior cervical ganglia (Brown and Higgins, 1979). This type of receptor was subsequently termed $GABA_B$, in contrast to the classical GABA receptor, which was designated $GABA_A$ (Hill and Bowery, 1981). The existence of $GABA_B$ receptors regulating the release of noradrenaline (Bowery *et al.*, 1980), dopamine (Bowery *et al.*, 1980; Reimann *et al.*, 1982), serotonin (5-hydroxytryptamine, 5-HT; Bowery *et al.*, 1980; Gray and Green, 1987; Schlicker *et al*, 1984) and glutamate (Potashner, 1979) was also reported in the central nervous system (CNS).

While there is little doubt that, in peripheral tissues, baclofen depresses [^{3}H]noradrenaline release via $GABA_B$ receptors, the idea that this is also the explanation for the observed inhibition of the release of this amine in the CNS (Bowery *et al.*, 1980) has been contested (Fung *et al.*, 1985). The effect of

$GABA_B$ Receptors in Mammalian Function. Edited by N.G. Bowery, H. Bittiger and H.-R. Olpe

baclofen on the electrically stimulated release of [^{3}H]noradrenaline from rat atria was not antagonized by the α_2-noradrenoceptor antagonist, yohimbine (Bowery *et al.*, 1981), whereas a corresponding effect on [^{3}H]noradrenaline release from hippocampal and cerebellar synaptosomes was attenuated by yohimbine and other α_2-noradrenoceptor antagonists (Fung *et al.*, 1985). These authors argued that, therefore, baclofen must cause the decrease in [^{3}H]noradrenaline release in the CNS by an albeit weak agonistic effect on α_2-noradrenoceptors. For a full clarification of this issue, a study of the effects of a $GABA_B$ antagonist seems to be necessary.

Also, [^{3}H]dopamine release evoked by 25 mmol/ l K^+ from rat striatal slices was originally reported to be inhibited by 100 μmol/l baclofen (Bowery *et al.*, 1980), and this was subsequently confirmed in rabbit caudate slices (Reimann *et al.*, 1982). In slices of the neurointermediate lobe and the median eminence, however, such an effect of baclofen could not be demonstrated (Anderson and Mitchell, 1985a).

In the experiments of Bowery *et al.* (1980) and Schlicker *et al.* (1984), concentrations of 100 μmol/l baclofen were needed to cause an appreciable inhibition of [^{3}H]5-HT release elicited by 25 mmol/l K^+ or electrical stimulation at 3 Hz from rat cortical slices. Moreover, in the experiments of Schlicker *et al.* (1984), the maximal inhibition of [^{3}H]5-HT release by GABA, progabide or baclofen was about 30%. In contrast, Gray and Green (1987), who measured the release of endogenous 5-HT evoked by 35 mmol/l K^+ from mouse cortical slices, found about 70% of the evoked release to be baclofen-sensitive, and these authors reported a half-maximal inhibitory concentration (IC_{50}) of this drug of about 0.1 μmol/l. As possible explanations of this difference, Gray and Green (1987) cited measurement of endogenous instead of [^{3}H]5-HT release (exogenous 5-HT may not evenly mix with the endogenous pool(s)), species differences, and the fact that they used a batch rather than a superfusion procedure. The issue awaits clarification.

Little is known of a possible role of $GABA_B$ receptors in the regulation of acetylcholine release in the CNS. Wichmann *et al.* (1987) found an inhibitory effect of baclofen on electrically stimulated release of [^{3}H]acetylcholine in slices of rabbit superior colliculus; however, the agent was much more potent than GABA itself, even in the presence of a GABA uptake inhibitor, whereas the two substances are normally about equipotent at typical $GABA_B$ receptors. This makes an involvement of $GABA_B$ receptors doubtful, but does not, however, exclude such a role in other areas of the brain.

Baclofen was reported to inhibit the electrically stimulated (100 Hz) release of radiolabelled glutamate and aspartate, newly synthesized from glucose, from guinea-pig cortical slices at 4 μmol/l (Potashner, 1979). A similar effect with similar potency on [^{3}H]D-aspartate release induced by high K^+ or protoveratrine from rat cortical slices was found by Johnston *et al.* (1980). Also Spencer *et al.* (1986) reported an inhibition of K^+-evoked release of [^{14}C]glutamate from slices of rat dentate gyrus. Zhu and Chuang (1987) found baclofen to inhibit K^+-evoked [^{3}H]D-aspartate release from cultured cerebellar granule cells. On the other hand, baclofen at up to 100 μmol/l and the GABA$_B$ antagonist, phaclofen, at 250 μmol/l were not found to affect the release of endogenous glutamate or aspartate evoked by 50 mmol/l K^+ from rat hippocampal CA1 slices (Burke and Nadler, 1988).

INVOLVEMENT OF GABA$_B$ AUTORECEPTORS IN THE RELEASE OF GABA

The search for a GABA autoreceptor which regulates the release of this amino acid from nerve endings, in a manner analogous to what was known of aminergic and cholinergic neuronal systems, was begun more than 10 years ago. The early reports appeared to suggest the existence of autoreceptors of the GABA$_A$ type. The release of radiolabelled GABA from rat cortical (Mitchell and Martin, 1978), striatal (Kuriyama *et al.*, 1984) or nigral (Arbilla *et al.*, 1979) slices, cortical synaptosomes (Brennan *et al.*, 1981) or synaptosomes from the median eminence and pituitary neurointermediate lobe (Anderson and Mitchell, 1986) was reported to be inhibited by muscimol in a bicuculline- or picrotoxin-sensitive manner. Similar results were observed with δ-aminolaevulinic acid, another GABA$_A$ agonist (Brennan and Cantrill, 1979). On the other hand, Arbilla *et al.* (1979) were unable to find an inhibition by muscimol or exogenous GABA of GABA release from rat occipital cortical slices, and Limberger *et al.* (1986) found muscimol ineffective in the inhibition of the release of labelled GABA from rabbit striatal slices. With baclofen, either no effects were found on GABA release at all, or they were not interpreted as indicating the existence of GABA$_B$ autoreceptors. Potashner (1979) did not find an effect of baclofen on the electrically induced (100 Hz) release of GABA formed from glucose in guinea-pig cortical slices. Johnston *et al.* (1980) observed no effect of this agent on the K^+-evoked release of [^{3}H]GABA from cortical and spinal cord slices. Collins *et al.* (1982) described a reduction by 5 and 25 μmol/l (−)-baclofen of the release of endogenous GABA from slices of the rat olfactory cortex elicited by very low frequency stimulation (0.083 Hz) of the lateral olfactory tract. They rejected the possibility of an effect on autoreceptors mainly because baclofen failed to affect GABA release evoked by 50 mmol/l K^+ in their own experiments and by electrical stimulation at 100 Hz in those of Potashner (1979). They considered it to be a consequence of a reduction of the excitatory (glutamatergic and/or aspartergic) input to GABA-containing interneurons, because baclofen reduced the evoked field potential (P-wave). A small inhibitory effect of baclofen on the electrically induced (5 Hz) release of [^{3}H]GABA from rabbit striatal slices was observed by Limberger *et al.* (1986), but this was not, however, considered to indicate the existence of GABA$_B$ autoreceptors because exogenous GABA failed to have an effect in the presence of the uptake inhibitor, nipecotic acid.

The first mention of $GABA_B$ autoreceptors was by Anderson and Mitchell (1985b). These authors found an inhibitory effect of baclofen on the release of [^{3}H]GABA elicited by 15 mmol/l K^+ from synaptosomes of the median eminence, but not of the pituitary neurointermediate lobe. Following this, Raiteri's group (see Bonanno *et al.*, 1988, 1989a,b; Pittaluga *et al.*, 1987), measuring the K^+-stimulated release of mainly [^{3}H]GABA but also endogenous GABA from rat and human cortical synaptosomes, and our own team (see Baumann *et al.*, 1990; Waldmeier *et al.*, 1988a,b, 1989), measuring electrically and K^+-stimulated release of mainly endogenous GABA but also [^{3}H]GABA from rat brain slices, were able to increase the evidence for the existence of $GABA_B$ autoreceptors and to characterize them. This chapter summarizes our work on slices, attempts to delineate the reasons why many studies have been unable to observe the effects of baclofen on GABA release and describes the effects of $GABA_B$ antagonists.

METHODOLOGY

A brief general description of the methodology is appropriate. Cross-chopped slices (1 mm × 350 μm × 350 μm) were prepared from several areas of the brains of male Tif:RAIf(SPF) rats weighing 160–200 g, obtained from Tierfarm Sisseln (Switzerland). The slices were suspended in physiological buffer containing 2.6 mmol/l Ca^{2+}, if not stated otherwise, and allowed to settle. For the experiments with radiolabelled GABA, the slices were incubated for 5 min in the presence of 100 nmol/l [2,3-^{3}H(N)]GABA (25 Ci/mmol). Appropriate amounts of the slice suspension (10–50 μl, corresponding to about 0.4–2 mg protein) were transferred to stimulation chambers and superfused at 0.4 ml/min with physiological buffer gassed with 5% CO_2 in O_2, in general containing 10 μmol/l of the GABA uptake inhibitor, SK&F 89976 (Yunger *et al.*, 1984), and, if the release of [^{3}H]GABA was measured, 50 μmol of the GABA transaminase inhibitor, aminooxyacetic acid. Fractions of 6 min (i.e. 2.4 ml) were collected, in general beginning 50 min after starting the superfusion. Two to four trains of monophasic pulses (2 ms, 40 mA) were applied at appropriate times, always at the beginning of a fraction. In some experiments, stimulation was also performed by exposure to elevated K^+ (for exact concentrations see legends of figures) for 5 min. K^+ was added to the medium such as to reach the chamber at the beginning of the subsequent fraction. Endogenous GABA was determined by high-performance liquid chromatography with electrochemical detection after derivatization with 2,4,6-trinitrobenzenesulphonic acid, by a modification of the procedure of Yamamoto *et al.* (1985), as described previously (Waldmeier *et al.*, 1988b). Radioactivity in the superfusion medium and in the slices (after solubilization in Irgasolv® and subsequent neutralization) was counted (after addition of Irgascint A300® scintillator). Release was expressed as a percentage of the GABA or [^{3}H]GABA content of the slices at the time the corresponding fraction was collected (fractional release); stimulated release was calculated by subtracting basal from total release during a stimulation period.

DEPENDENCE OF EVOKED GABA RELEASE ON THE FREQUENCY OF STIMULATION

If the release of a transmitter is controlled by a presynaptic autoreceptor, it will be expected to decrease with increasing stimulation frequency, because the transmitter will progressively accumulate in the extracellular space due to insufficient clearance by the superfusion medium, in particular if reuptake is blocked. This statement in general holds true for slices; it does not for synaptosomal preparations, in which superfusion effectively removes the transmitter once it is released. In Figure 4.1, the stimulated fractional release of endogenous GABA per pulse is plotted against stimulation frequency (normalized data from a number of experiments carried out with rat cortical slices over a wide range of frequencies). It should be noted that GABA release was entirely

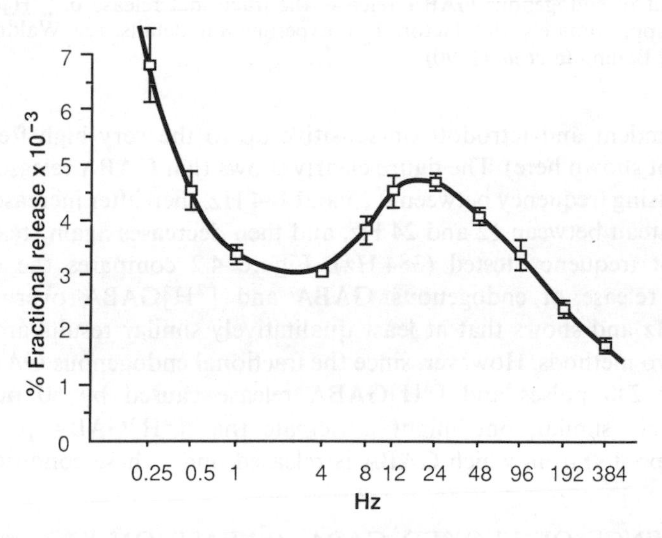

Figure 4.1. Fractional release of endogenous GABA from rat cortical slices per pulse as a function of stimulation frequency. The graph summarizes the results from a number of experiments in which the dependence of the release of endogenous GABA on the frequency of electrical field stimulation was investigated over different ranges of frequency. To ensure comparability, the values obtained from the different experiments were normalized using overlapping points. The number of applied pulses was 240 in some and 360 in most of the experiments. Data are means of the stimulated fractional release per pulse ± SEM. The number of slice preparations investigated at each frequency varied between 4 and 12, except at 4 Hz, which served as a reference in most experiments and where N was 44. Data for frequencies up to 48 Hz are taken from Waldmeier *et al.* (1988a, 1989); those for frequencies above 48 Hz are unpublished

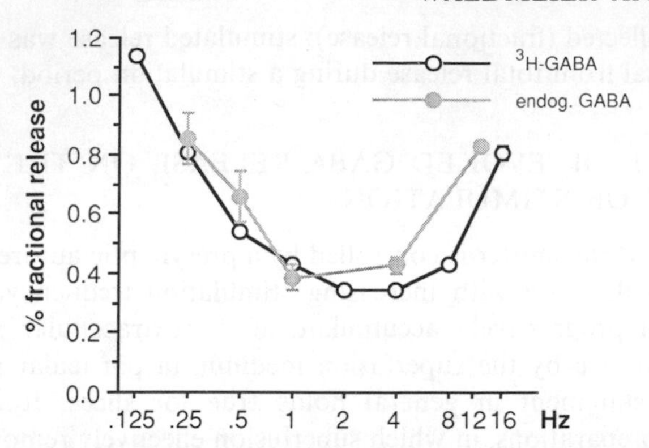

Figure 4.2. Frequency dependence of the electrically stimulated release of [^{3}H]GABA and endogenous GABA from rat cortical slices. The data originate from experiments in which groups of four rat cortical slice preparations, preloaded with [^{3}H]GABA or not, were stimulated at different frequencies by a constant number of pulses (60 if [^{3}H]GABA and 240 if endogenous GABA was measured). Data are means of the stimulated fractional release ± SEM. Note that, since four times more pulses were applied for the measurement of endogenous GABA release, the fractional release of [^{3}H]GABA was higher by approximately this factor. For experimental details, see Waldmeier *et al.* (1988a) and Baumann *et al.* (1990)

Ca^{2+}-dependent and tetrodotoxin-sensitive up to the very high frequency of 384 Hz (not shown here). The figure clearly shows that GABA release decreases with increasing frequency between 0.25 and 1–4 Hz, thereafter increases to reach a short plateau between 12 and 24 Hz, and then decreases again steadily up to the highest frequency tested (384 Hz). Figure 4.2 compares the stimulated fractional release of endogenous GABA and [^{3}H]GABA over the range 0.125–16 Hz and shows that at least qualitatively similar results are obtained with the two methods. However, since the fractional endogenous GABA release elicited by 240 pulses and [^{3}H]GABA release caused by 60 pulses were quantitatively similar, one might anticipate that [^{3}H]GABA preferentially enters the pool(s) from which GABA is released under these conditions.

DEPENDENCE OF EVOKED GABA RELEASE ON K^+ CONCENTRATION

The dependence of the stimulated release of endogenous GABA and [^{3}H]GABA on the concentration of K^+ in the range 10–40 mmol/l was also compared. As Figure 4.3 shows, the resulting curves were quite similar. In particular, the fractional releases elicited by K^+ concentrations above 25 mmol/l were quantitatively the same. Between 10 and 25 mmol/l, however, the fractional release of endogenous GABA was clearly smaller than that of [^{3}H]GABA. This is in some way comparable to what was observed with electrical stimulation (see

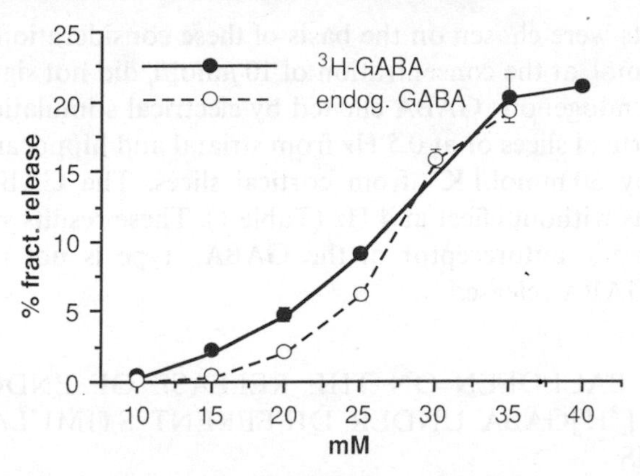

Figure 4.3. Dependence of K^+-stimulated release of [^{3}H]GABA and endogenous GABA from rat cortical slices on K^+ concentration. Groups of four rat cortical slice preparations, preloaded with [^{3}H]GABA or not, were stimulated for 5 min with different K^+ concentrations and the release of [^{3}H]GABA or endogenous GABA was measured. Data are means of the stimulated fractional release ± SEM

above) and again suggests that [^{3}H]GABA preferentially enters the pool(s) from which GABA is released by relatively low K^+ concentrations.

It may be worth mentioning here that the fractional release of endogenous GABA elicited by 30 mmol/l K^+ is clearly higher than what can be obtained by an electrical stimulation of a similar duration of 24 Hz (see Waldmeier *et al.*, 1989). Since the release per pulse at this frequency is almost as high as that at frequencies below 0.5 Hz, this suggests that K^+ stimulation generally releases more GABA than electrical stimulation, which in turn might mean that the former is able to release the amino acid from pools not affected by the latter.

ABSENCE OF EFFECTS OF MUSCIMOL AND BICUCULLINE ON THE RELEASE OF ENDOGENOUS GABA UNDER DIFFERENT STIMULATION CONDITIONS

If the decrease of the release of GABA with increased frequency of stimulation observed between 0.25 and 4 Hz and between 24 and 384 Hz (Figure 4.1) is indeed related to activation of a presynaptic autoreceptor by endogenously released GABA, a potential effect of an exogenous agonist will be most easily detected near the lower ends of these ranges of frequency. This is because under these circumstances the exogenous agonist will not have to compete with the endogenous agonist, and because the extent of the effect it can exert will be greater than at higher frequencies. For a putative antagonist, the reverse will be true, and such compounds are therefore best investigated at frequencies near the upper end of the above ranges. The conditions for the testing of GABA agonists and antagonists were chosen on the basis of these considerations. The $GABA_A$ agonist, muscimol, at the concentration of 10 μmol/l, did not significantly affect the release of endogenous GABA elicited by electrical stimulation at 0.5, 12 or 24 Hz from cortical slices or at 0.5 Hz from striatal and hippocampal slices, nor that evoked by 30 mmol/l K^+ from cortical slices. The $GABA_A$ antagonist, bicuculline, was without effect at 4 Hz (Table 1). These results strongly suggest that a presynaptic autoreceptor of the $GABA_A$ type is not involved in the regulation of GABA release.

EFFECT OF BACLOFEN ON THE RELEASE OF ENDOGENOUS GABA AND [^{3}H]GABA UNDER DIFFERENT STIMULATION CONDITIONS

The effect of (–)-baclofen on the electrically induced release of endogenous GABA was tested at different frequencies in the lower range (0.25–4 Hz) and at a low frequency (24 Hz) in the upper range (12–384 Hz); that on K^+-evoked release was studied at two different K^+ concentrations, 15 and 30 mmol/l. The results are summarized in Table 1. (–)-Baclofen significantly inhibited the electrically induced release of endogenous GABA throughout the low frequency range. The effect of the $GABA_B$ agonist was clearly frequency-dependent, being more marked at 0.25 than at 0.5 Hz, and more marked at the latter frequency than at 4 Hz. At 0.5 Hz, 10 μmol/l (–)-baclofen caused approximately 60% reduction in cortical as well as hippocampal and striatal slices. The (+)-enantiomer was completely ineffective in cortical slices under these conditions. On the other hand, GABA release evoked by stimulation at 24 Hz was not sensitive to (–)-baclofen. Since the inhibition of release upon stimulation of autoreceptors may be mediated by a decrease in the availability of intraneuronal Ca^{2+} at the sites relevant for stimulus–release coupling, and strong stimuli raise axoplasmic Ca^{2+} to high levels which may saturate those sites, uncoupling the release process from control by autoreceptors may occur under such conditions. To check whether this might be the reason for the failure of (–)-baclofen to reduce GABA release at 24 Hz, the experiment was repeated at a low Ca^{2+} concentration (0.3 mmol/l); however, the $GABA_B$ agonist remained inactive under these conditions also, excluding this possibility (for details see Waldmeier *et al.*, 1989).

(–)-Baclofen also inhibited the release of endogenous GABA from cortical slices elicited by a relatively low K^+ concentration (15 mmol/l), but not by 30 mmol/l (Table 1). This result agrees well with the observation made in cortical synaptosomes that (–)-baclofen is more effective at lower K^+ concentrations (Bonanno *et al.*, 1988, 1989a, b). Together with its higher effectiveness at low frequencies, this suggests that GABA release elicited by mild stimuli is more susceptible towards inhibition by $GABA_B$ agonists, and this at least in part explains the failure of previous authors to observe an effect of such agents.

Table 1. Effects of the enantiomers of baclofen, muscimol and bicuculline on the release of endogenous GABA from rat brain slices

	Conc. (μmol/l)	Stimulation	% inhibition of stimulated release: Cortex	Hippocampus	Striatum
		Frequency (Hz)			
(−)-Baclofen	1	0.25	44.4 ± 3.5*	—	—
(−)-Baclofen	1	0.5	25.2 ± 1.4*	—	—
(−)-Baclofen	10	0.5	67.7 ± 7.5*	63.8 ± 2.8*	65.1 ± 5.3*
(−)-Baclofen	10	4	24.1 ± 1.6*	—	—
(−)-Baclofen	10	24	0.0 ± 7.0	—	—
(+)-Baclofen	10	0.5	− 3.9 ± 4.8	—	—
Muscimol	10	0.5	−10.7 ± 4.2	9.1 ± 8.3	12.6 ± 20.5
Muscimol	10	12	−9.6 ± 6.7	—	—
Muscimol	10	24	0.8 ± 4.9	—	—
Bicuculline	10	4	−20.5 ± 8.5	—	—
		K⁺ concn (mmol/l)			
(−)-Baclofen	10	15	59.3 ± 3.1*	—	—
(−)-Baclofen	10	30	−14.0 ± 7.5	—	—
Muscimol	10	30	0.0 ± 11.8	—	—

Data are means ± SEM from groups of three to four slices. For experimental details, see Waldmeier *et al.* (1988a, 1989).
*$P > 0.01$ (Student's *t*-test).

The measurement of the endogenous release of GABA is rather laborious, which makes detailed studies very time-consuming. Fortunately, the experiments with radiolabelled GABA hitherto carried out in parallel (see above) as well as the experience of Raiteri's group with synaptosomes (see Bonanno *et al.*, 1988, 1989a, b) seemed to indicate that at least qualitatively similar results could be expected with both methods. With the faster radiotechnique, it was possible to study the effects of (−)-baclofen on the release of [^{3}H]GABA in more detail.

Figure 4.4 shows the dependence of [^{3}H]GABA release on the stimulation frequency in the presence or absence of 10 μmol/l (−)-baclofen. This high concentration of the GABA$_B$ agonist suppressed the evoked release of [^{3}H]GABA to a low and constant level between 0.125 and 4 Hz. Of course, the percentage of inhibition decreased with increasing frequency in consequence, since in the absence of (−)-baclofen the release of [^{3}H]GABA did so. Interestingly, (−)-baclofen became progressively ineffective when the frequency was raised from 4 to 16 Hz, corroborating the result obtained with endogenous release at 24 Hz.

A concentration–response curve of (−)-baclofen was generated at 0.125 Hz, i.e. at the frequency at which the compound was most effective. It suppressed

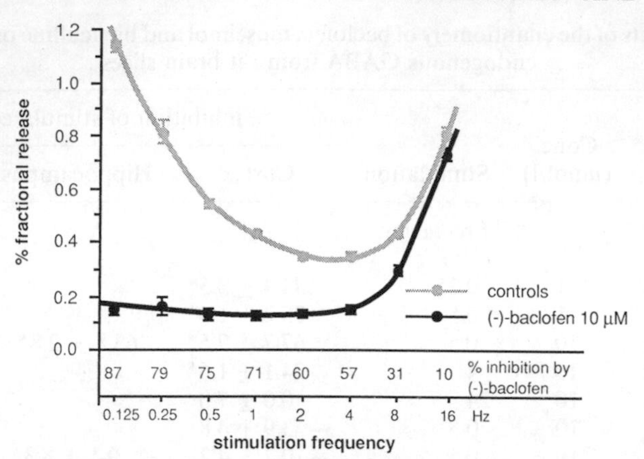

Figure 4.4. Effects of (−)-baclofen on [^{3}H]GABA overflow at different frequencies of electrical stimulation. The data originate from a series of experiments in each of which two groups of four rat cortical slice preparations preloaded with [^{3}H]GABA were stimulated three times by a constant number of 60 pulses. The first stimulation (S_1) served as a reference for the normalization of the results from the different experiments and was always at 2 Hz. The second and third stimulations (S_2 and S_3) were applied at the test frequency. (−)-Baclofen (10 μmol/l) was added to the superfusion medium of one group before S_3. Data are means of the stimulated fractional release ± SEM. In most cases the SEM was smaller than the symbols and was therefore omitted. Data are taken from Baumann *et al.* (1990), where experimental details are also available

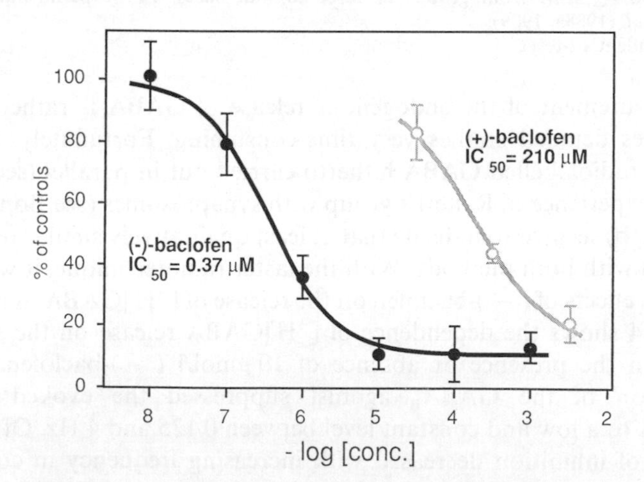

Figure 4.5. Inhibition of the electrically induced release of [^{3}H]GABA from rat cortical slices by (−)- and (+)-baclofen. For each concentration of the drugs, a separate experiment with an own control was carried out (n = 4 for both drug and control), in which the slices were stimulated twice at 0.125 Hz for 8 min (60 pulses). Drug was added to the medium of one group of four slice preparations after the first stimulation (S_1). Data represent means of the S_2/S_1 ratios as a percentage of those of the controls and are taken from Baumann *et al.* (1990). The error bars represent 5% confidence limits

[^{3}H]GABA release to about 10% of the control level at 10 μmol/l and above. An IC_{50} value of 0.37 μmol/l was found (Figure 4.5). It seems that (−)-baclofen suppresses [^{3}H]GABA release in the electrically stimulated slice preparation more effectively and more potently than in synaptosomes stimulated by K^+ concentrations of 9 or 15 mmol/l (cf. Bonanno *et al.*, 1988, 1989a,b; Pittaluga *et al.*, 1987). This may be related to the observation made above that the effect of (−)-baclofen is the more clear-cut the milder the stimulus. It should be noted that, while (+)-baclofen was not inactive, its IC_{50} was about 500-fold higher than that of the (−)-enantiomer. The effect may in fact be due to traces of the former; an impurity of 0.2% would be sufficient.

EFFECT OF $GABA_B$ ANTAGONISTS ON THE RELEASE OF ENDOGENOUS GABA AND [^{3}H]GABA UNDER DIFFERENT STIMULATION CONDITIONS

Of course, the investigation of the regulation of GABA release by putative $GABA_B$ autoreceptors is not complete without information on the effects of antagonists. Such compounds have not been available until 1987, when Kerr *et al.* (1987) reported that the phosphono analogue of baclofen, phaclofen, exhibited such properties, although it had a rather low potency. In the following year, Bonanno *et al.* (1988) showed that the compound was able to antagonize the effects of baclofen on the release of [^{3}H]GABA from cortical synaptosomes. As must be expected with this experimental set-up, they could not demonstrate an enhancing effect of this compound on its own. We have therefore studied phaclofen using our slice preparation in which an antagonist should increase GABA release, provided the appropriate frequency is chosen. In the first set of experiments, the effect of the antagonist at 1 mmol/l on the release of [^{3}H]GABA was studied in relation to the applied frequency, in a very similar way to that done before with the agonist, (−)-baclofen (see Figure 4.4). The results are shown in Figure 4.6. Phaclofen, as expected, did show an enhancement of [^{3}H]GABA release. Surprisingly, however, it did not alter the shape of the curve obtained in its absence, but just displaced it upwards. Correspondingly, and contrary to expectations, the extent of the enhancement did not increase with increasing frequency, but remained relatively stable, at least up to 16 Hz, where the enhancement was almost lost. For the portion of the frequency dependence between 1 and 4 Hz, this can be explained by phaclofen's poor potency, i.e. by its partial and progressive displacement from the putative autoreceptor by endogenously released GABA. However, one would have expected that the antagonist would be less effective at lower frequencies, where there is little endogenous GABA released to activate the autoreceptor. It cannot be excluded that endogenously released GABA already markedly suppressed its own release at this low frequency. The steepness of the frequency dependence in this range does at least not contradict this idea. Moreover, it must be considered

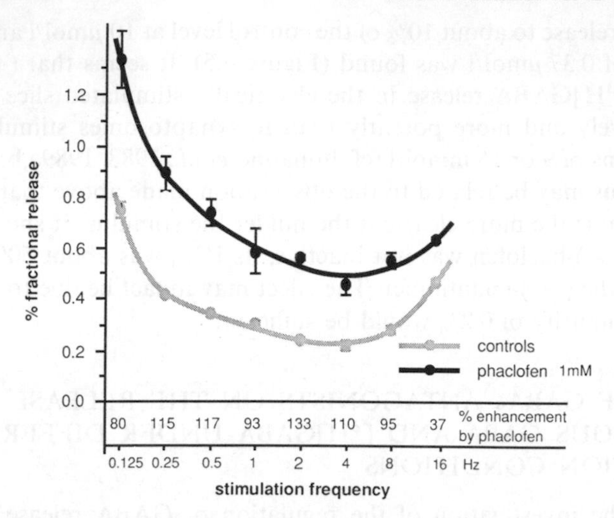

Figure 4.6. Effects of phaclofen on [^{3}H]GABA overflow at different frequencies of electrical stimulation. This series of experiments was carried out exactly as that described in the legend on Figure 4.4, with the exception that the drug used was 1 mmol/l phaclofen. Data are taken from Baumann *et al.* (1990), where experimental details are also available

that all these experiments were carried out in the presence of a GABA uptake inhibitor. This probably also permits 'cross-talk', i.e. activation of presynaptic autoreceptors by transmitter molecules originating from neighbouring terminals. An alternative possibility can also not be dismissed at present. Since our slice preparation is likely to contain intact GABA interneurons with, at least partly, preserved inputs, it is possible that, for example, somatodendritically located $GABA_B$ receptors on such interneurons are involved. If so, to explain our results, one would still have to assume that such receptors are more potently affected by phaclofen than are the terminal ones. The postulation of $GABA_B$ receptors with different sensitivity towards phaclofen is not without precedent: the compound antagonized the postsynaptic hyperpolarization of hippocampal pyramidal cells induced by baclofen much more potently than that by GABA, whereas the presynaptic inhibition was unaffected (Dutar and Nicoll, 1988a,b). The issue still awaits clarification.

Figure 4.7 shows the concentration–response curve of phaclofen with respect to the release of [^{3}H]GABA at a frequency of 2 Hz. The threshold concentration was about 100 μmol/l, and somewhat more than a doubling of the control release was reached at 3 mmol/l; higher concentrations were not tested, and it is therefore not clear whether the maximally possible enhancement is even greater.

Since phaclofen at these high concentrations markedly interferes with our assay for endogenous GABA, corresponding investigations could not be made. Instead, we tested CGP 35348 (3-aminopropane-diethoxymethylphosphinic

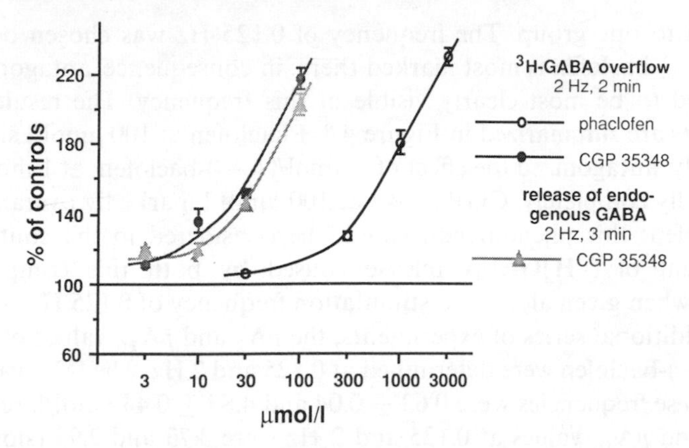

Figure 4.7. Enhancement of the electrically stimulated release of [^{3}H]GABA by phaclofen and CGP 35348, and of that of endogenous GABA by the latter compound. For each concentration of the drugs, a separate experiment with an own control was carried out ($n = 4$ for both drug and control), in which the slices were stimulated twice at 2 Hz ([^{3}H]GABA: 2 min = 240 pulses; endogenous GABA: 3 min = 360 pulses). Drug was added to the medium of one group of four slice preparations after the first stimulation (S_1). Data represent means ± SEM of the S_2/S_1 ratios as a percentage of those of the controls

acid), a phosphinic acid recently discovered in our laboratories, which is markedly more potent as a GABA$_B$ antagonist than is phaclofen and seems to be rather specific (Bittiger *et al.*, Chapter 3). This compound increased the release of both endogenous GABA and [^{3}H]GABA very similarly, with a threshold concentration of about 10 μmol/l (Figure 4.7). The curves were very similar in shape to that for phaclofen; they were displaced to the left, indicating that CGP 35348 is about 10–20 times more potent than phaclofen.

Thus, it seems that it is a common property of GABA$_B$ antagonists to increase the release of GABA in the slice preparation under suitable stimulation conditions. It should be recalled that phaclofen showed an apparently anomalous behaviour at low frequencies which has not as yet been fully explained. According to preliminary data not presented here, CGP 35348 behaves similarly. This phenomenon clearly needs further study.

ANTAGONISM BY GABA$_B$ ANTAGONISTS OF THE INHIBITORY EFFECT OF (−)-BACLOFEN

Measuring [^{3}H]GABA release, phaclofen at 100 μmol/l and 1 mmol/l and CGP 35348 at 100 μmol/l were tested against 1 μmol/l (−)-baclofen at a stimulation frequency of 0.125 Hz. In this series, two groups of slices were stimulated three times, the first time under control conditions and the second and third time in the presence of (−)-baclofen alone. Before the third stimulation, the antagonist was added to one group. The frequency of 0.125 Hz was chosen because the effect of (−)-baclofen is most marked there; in consequence, antagonism could be expected to be most clearly visible at this frequency. The results of these experiments are summarized in Figure 4.8. Phaclofen at 100 μmol/l slightly, but significantly, antagonized the effect of 1 μmol/l (−)-baclofen; at 1 mmol/l it did so practically completely. CGP 35348 at 100 μmol/l markedly overantagonized (−)-baclofen; this phenomenon should be considered in the context of the enhancement of [^{3}H]GABA release caused by both this compound and phaclofen when given alone at a stimulation frequency of 0.125 Hz (see above).

In an additional series of experiments, the pA_2 and pA_{10} values of phaclofen against (−)-baclofen were determined at 0.125 and 2 Hz. The IC_{50} values of the latter at these frequencies were 0.63 ± 0.04 and 4.88 ± 0.44 μmol/l, respectively. The pA_2 and pA_{10} values at 0.125 and 2 Hz were 3.73 and 2.93 (slope = 1.18)

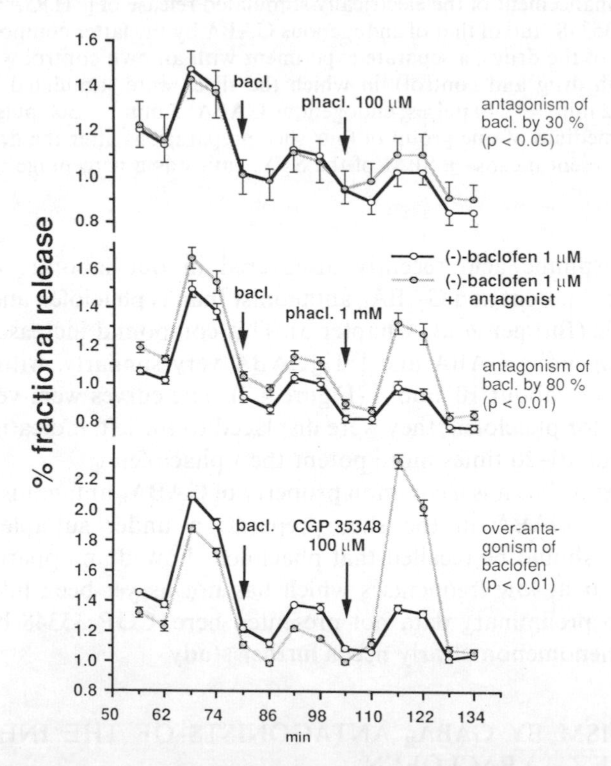

Figure 4.8. Antagonism by phaclofen and CGP 35348 of the inhibitory effect of (−)-baclofen on the release of [^{3}H]GABA. Two groups of four slices preloaded with [^{3}H]GABA were stimulated three times for 8 min at 0.125 Hz in each experiment. During the first stimulation (S_1), no drug was present; during the second and third (S_2 and S_3), 1 μmol/l (−)-baclofen was present in both groups. The antagonists were added to the medium of one group before S_3. Data are means ± SEM of the percentage fractional release. For statistical comparisons, Student's *t*-test was used

Table 2. Antagonism by CGP 35348 of the inhibitory effect of (−)-baclofen on the release of endogenous GABA

	S_1	S_2	S_2/S_1
Controls	1.382 ± 0.032	1.286 ± 0.101	0.928 ± 0.060
(−)-Baclofen	1.240 ± 0.066	0.388 ± 0.073	0.309 ± 0.049*
(−)-Baclofen +CGP 35348	1.201 ± 0.061	0.731 ± 0.039	0.610 ± 0.022**

Three groups of four slice preparations were stimulated twice (S_1 and S_2) at 0.5 Hz for 12 min (2 ms, 25 mA). Drugs were added after the first stimulation ((−)-baclofen: 10 μmol/l; CGP 35348: 100 μmol/l). Data in the first two columns are means ± SEM of the stimulated fractional release; those in the third are means ± SEM of the S_2S_1 ratios.
*$P < 0.01$ vs controls; **$P < 0.01$ vs (−)-baclofen alone (Dunnett's test).

and 3.36 and 2.93 (slope 2.32), respectively (Baumann *et al.*, 1990). These last results suggest a competitive interaction between phaclofen and baclofen at 0.125 Hz, but not at 2 Hz.

CGP 35348 at 100 μmol/l was also able to antagonize by about 50% the effect of 10 μmol/l (−)-baclofen on the release of endogenous GABA elicited by electrical stimulation at 0.5 Hz (Table 2).

CONCLUSIONS

The interaction of baclofen and GABA with presynaptic heteroreceptors of the GABA$_B$ type on peripheral noradrenergic and cholinergic nerve endings is quite well established. Although there is also evidence for the existence of such receptors on central noradrenergic, dopaminergic, serotonergic, cholinergic and glutamatergic nerve endings, the corresponding interactions have not been fully characterized in most cases. Generally, it seems that most of these interactions are relatively weak (maximally achievable reductions of release hardly exceed 30%) and occur only at rather high concentrations of GABA or baclofen. Exceptions to this have been published in the cases of the release of endogenous 5-HT (one paper only!) and of glutamate. With noradrenaline release there are doubts concerning the specificity of baclofen, which may cause its effect by a weak α_2-agonistic action rather than via GABA$_B$ receptors. With the availability of GABA$_B$ antagonists, a better characterization of the role of presynaptic GABA$_B$ heteroreceptors in transmitter release is now possible, and more information is therefore expected in the near future.

The existence of GABA$_B$ receptors regulating GABA release is now well established, on the basis of our investigations into the effects of baclofen and the GABA$_B$ antagonists, phaclofen and CGP 35348, on the electrically and K^+-stimulated release of [^{3}H]GABA and endogenous GABA from brain slices, as well as of the studies by Raiteri's group (see Bonanno *et al.*, 1988, 1989a,b;

Pittaluga *et al.*, 1987) on the K^+-stimulated release of [^{3}H]GABA and endogenous GABA from synaptosomes. It should be pointed out that neither group could find any evidence whatsoever for an involvement of $GABA_A$-type autoreceptors in the control of GABA release, under a variety of different conditions. However, there are still some issues awaiting clarification. One concerns the enhancement of GABA release at low frequencies by $GABA_B$ antagonists. Another is the question of the existence and role of other pools. A third and particularly important one is that concerning the significance of $GABA_B$ autoreceptors in the in-vivo situation. Especially with respect to the last two questions, the availability of a relatively potent antagonist, CGP 35348, may prove beneficial.

REFERENCES

Anderson, R. and Mitchell, R. (1985a) Effects of GABA receptor agonists on [^{3}H]dopamine release from median eminence and pituitary neurointermediate lobe. *Eur. J. Pharmacol.*, **115**, 109–112.

Anderson, R.A. and Mitchell, R. (1985b) Evidence for $GABA_B$ autoreceptors in median eminence. *Eur. J. Pharmacol.*, **118**, 355–358.

Anderson, R. and Mitchell, R. (1986) Uptake and autoreceptor-controlled release of [^{3}H]-GABA by the hypothalamic median eminence and pituitary neurointermediate lobe. *Neuroendocrinology*, **42**, 277–284.

Arbilla, S., Kamal, L. and Langer, S.Z. (1979) Presynaptic GABA autoreceptors on GABAergic nerve endings of the rat substantia nigra. *Eur. J. Pharmacol*, **57**, 211–217.

Baumann, P.A., Wicki, P., Stierlin, C. and Waldmeier, P.C. (1990) Investigations on $GABA_B$ receptor-mediated autoinhibition of GABA release. *Naunyn-Schmiedeberg's Arch. Pharmacol.*, **341**, 88–93.

Bein, H.J. (1972) Pharmacologic differentiation of muscle relaxants. In: W. Birkmayer (ed.) *Spasticity–a topical survey*. Vienna: Hans Huber, pp. 76–82.

Bonanno, G., Fontana, G. and Raiteri, M. (1988) Phaclofen antagonizes GABA at autoreceptors regulating release in rat cerebral cortex. *Eur. J. Pharmacol.*, **154**, 223–224.

Bonanno, G., Cavazzani, P., Andrioli, G.C., Asaro, D., Pellegrini, G. and Raiteri, M. (1989a), Release-regulating autoreceptors of the $GABA_B$-type in human cerebral cortex. *Br. J. Pharmacol.*, **96**, 341–346.

Bonanno, G., Pellegrini, G., Asaro, D., Fontana, G., and Raiteri, M. (1989b) $GABA_B$ autoreceptors in rat cortex synaptosomes: response under different depolarizing and ionic conditions. *Eur. J. Pharmacol.*, **172**, 41–49.

Bowery, N.G. and Hudson, A.L. (1979) γ-Aminobutyric acid reduces the evoked release of [^{3}H]noradrenaline from sympathetic nerve terminals. *Br. J. Pharmacol.*, **66**, 108P.

Bowery, N.G., Hill, D.R., Hudson, A.L., Doble, A., Middlemiss, D.N., Shaw, J. and Turnbull, M. (1980) (−)Baclofen decreases neurotransmitter release in the mammalian CNS by an action at a novel GABA receptor. *Nature*, **283**, 92–94.

Bowery, N.B., Doble, A., Hill, D.R., Hudson, A.L., Shaw, J.S., Turnbull, M.J. and Warrington, R. (1981) Bicuculline-insensitive GABA receptors on peripheral autonomic nerve terminals. *Eur. J. Pharmacol.*, **71**, 53–70.

Brennan, M.J.W. and Cantrill, R.C. (1979) δ-Aminolaevulinic acid is a potent agonist for GABA autoreceptors. *Nature*, **280**, 514–515.

Brennan, M.J.W., Cantrill, R.C., Oldfield, M. and Krogsgaard-Larsen, P. (1981) Inhibition of γ-aminobutyric acid release by γ-aminobutyric acid agonist drugs: pharmacology of the γ-aminobutyric acid autoreceptor. *Mol. Pharmacol.*, **19**, 27–30.

Brown, D.A. and Higgins, A.J. (1979) Presynaptic effects of γ-aminobutyric acid in isolated rat superior cervical ganglia. *Br. J. Pharmacol.*, **66**, 108P–109P.

Burke, S.P. and Nadler, J.V. (1988) Regulation of glutamate and aspartate release from slices of the hippocampal CA_1 area: effects of adenosine and baclofen. *J. Neurochem.*, **51**, 1541–1551.

Collins, G.G.S., Anson, J. and Kelly, E.P. (1982) Baclofen: effects on evoked field potentials and amino acid neurotransmitter release in the rat olfactory cortex slice. *Brain Res.* **238**, 371–383.

Davidoff, R.A. and Sears, E.S. (1974) The effects of Lioresal on synaptic activity in the isolated spinal cord. *Neurology*, **24**, 957–963.

Dutar, P. and Nicoll, R.A. (1988a) A physiological role for $GABA_B$ receptors in the central nervous system. *Nature*, **332**, 156–158.

Dutar, P. and Nicoll, R.A (1988b) Pre- and postsynaptic $GABA_B$ receptors in the hippocampus have different pharmacological properties. *Neuron*, **1**, 585–591.

Fox, S., Krnjevic, K. and Morris, M.E. (1976) Lioresal depresses synaptic transmission in the cuneate nucleus. *Neurosci. Abstr.*, **2**, 1003 (Abstr.).

Fung, S.-C., Swarbrick, M.J. and Fillenz, M. (1985) Effect of baclofen on in vitro noradrenaline release from rat hippocampus and cerebellum: an action at an α_2-adrenoceptor. *Neurochem. Int.*, **7**, 155–163.

Gray, J.A. and Green, A.R. (1987) $GABA_B$-receptor mediated inhibition of potassium-evoked release of endogenous 5-hydroxytryptamine from mouse frontal cortex. *Br. J. Pharmacol.*, **91**, 517–522.

Hill, D.R. and Bowery, N.G. (1981) ^{3}H-baclofen and ^{3}H-GABA bind to bicuculline-insensitive $GABA_B$ sites in rat brain. *Nature*, **290**, 149–152.

Johnston, G.A.R., Hailstone, M.H. and Freeman, C.G. (1980) Baclofen: stereoselective inhibition of excitant amino acid release. *J. Pharm. Pharmacol.*, **32**, 230–231.

Kerr, D.J., Ong, J., Prager, R.H., Gynther, B.D. and Curtis, D.R. (1987) Phaclofen: a peripheral and central baclofen antagonist. *Brain Res.*, **405**, 150–154.

Kuriyama, K., Kanmori, K., Taguchi, J. and Yoneda, Y. (1984) Stress-induced enhancement of suppression of [^{3}H]GABA release from striatal slices by presynaptic autoreceptors. *J. Neurochem.*, **42**, 943–950.

Limberger, N., Spaeth, L. and Starke, K. (1986) A search for receptors modulating the release of γ-[^{3}H]aminobutyric acid in rabbit caudate nucleus slices. *J. Neurochem.*, **46**, 1109–1117.

Mitchell, P.R. and Martin, I.L. (1978) Is GABA release modulated by presynaptic receptors? *Nature*, **274**, 904–905.

Morris, M.E., Krnjevic, K. and Fox, S. (1977) Action of baclofen on cuneate transmission. *Neurosci. Abstr.*, **3**, 517 (Abstr.).

Pierau, F.K., Matheson, G.K. and Wurster, R.D. (1975) Presynaptic action of β-(4-chlorophenyl)-GABA. *Exp. Neurol.*, **48**, 343–351.

Pittaluga, A., Asaro, D., Pellegrini, G. and Raiteri, M. (1987) Studies on [^{3}H]GABA and endogenous GABA release in rat cerebral cortex suggest the presence of autoreceptors of $GABA_B$ type. *Eur. J. Pharmacol.*, **144**, 45–52.

Polc, P. and Haefely, W. (1976) Effects of two benzodiazepines, phenobarbitone, and baclofen on synaptic transmission in the cat arcuate nucleus. *Naunyn-Schmiedeberg's Arch. Pharmacol.*, **294**, 121–131.

Potashner, S.J. (1978) Baclofen: effects on amino acid release. *Can. J. Physiol. Pharmacol.*, **56**, 150–154.

Potashner, S.J. (1979) Baclofen: effects on amino acid release and metabolism in slices of guinea-pig cerebral cortex. *J. Neurochem.*, **32**, 103–109.

Reimann, W., Zumstein, A. and Starke, K. (1982) γ-Aminobutyric acid can both inhibit and facilitate dopamine release in the caudate nucleus of the rabbit. *J. Neurochem.*, **39**, 961–969.

Schlicker, E., Classen, K. and Göthert, M. (1984) $GABA_B$ receptor-mediated inhibition of serotonin release in the rat brain. *Naunyn-Schmiedeberg's Arch. Pharmacol.*, **326**, 99–105.

Spencer, P., Lynch, M.A. and Bliss, T.V.P. (1986) In-vitro release of [^{14}C]glutamate from dentate gyrus is modulated by GABA. *J. Pharm. Pharmacol.*, **38**, 393–395.

Waldmeier, P.C., Wicki, P., Feldtrauer, J.J. and Baumann, P.A. (1988a) Potential involvement of a baclofen-sensitive autoreceptor in the modulation of the release of endogenous GABA from rat brain slices in vitro. *Naunyn-Schmiedeberg's Arch. Pharmacol.*, **337**, 289–295.

Waldmeier, P.C., Wicki, P., Feldtrauer, J.J. and Baumann, P.A. (1988b) The measurement of the release of endogenous GABA from rat brain slices by liquid chromatography with electrochemical detection. *Naunyn-Schmiedeberg's Arch. Pharmacol.*, **337**, 284–288.

Waldmeier, P.C., Wicki, P., Feldtrauer, J.J. and Baumann, P.A. (1989) Ca^{2+}-dependent release of endogenous GABA from rat cortical slices from different pools by different stimulation conditions. *Naunyn-Schmiedeberg's Arch. Pharmacol.*, **339**, 200–207.

Wichmann, T., Illing, R.-B. and Starke, K. (1987) Evidence for a neurotransmitter function of acetylcholine in rabbit superior colliculus. *Neuroscience*, **23**, 991–1000.

Yamamoto, T., Nanjoh, C. and Kuruma, I. (1985) Determination of endogenous GABA released from the cerebral cortex slices of the rat by high-performance liquid chromatography with a series-dual electrochemical detector. *Neurochem. Int.*, **7**, 77–82.

Yunger, L.M., Fowler, P.J., Zarevics, P. and Setler, P.E. (1984) Novel inhibitors of γ-aminobutyric acid (GABA) uptake: anticonvulsant actions in rats and mice. *J. Pharmacol. Exp. Ther.*, **228**, 109–115.

Zhu, X.-Z. and Chuang, D.-M. (1987) Modulation of calcium uptake and D-aspartate release by $GABA_B$ receptors in cultured cerebellar granule cells. *Eur. J. Pharmacol.*, **141**, 401–408.

5 Release-regulating GABA Autoreceptors in Human and Rat Central Nervous System

M. RAITERI, M. T. GIRALT, G. BONANNO, A. PITTALUGA, E. FEDELE and G. FONTANA

INTRODUCTION

Studies in the laboratory animal have shown that autoregulation of transmitter release mediated by receptors sited on the releasing terminals (autoreceptors) appears to be a common feature of the major transmitter systems (Chesselet, 1984; Langer, 1981; Raiteri *et al.*, 1984; Starke, 1981). In fact, the existence of such receptors for noradrenergic, serotonergic and cholinergic neurons is now widely accepted and there is also agreement as to their pharmacological characterization in terms of type or subtype. Thus noradrenaline autoreceptors belong to the α_2-adrenoceptor subtype (for reviews, see Langer, 1981; Starke, 1981), the serotonin (5-hydroxytryptamine, 5-HT) autoreceptors in the rat brain have been classified as 5-HT_{1B} (Bonanno *et al.*, 1986; Engel *et al.*, 1986; Maura *et al.*, 1986) and acetylcholine autoreceptors are muscarinic and appear to belong to the pirenzepine-insensitive M_3 subtype (Marchi and Raiteri, 1985, 1989; Mash and Potter, 1986; Meyer and Otero, 1985).

In contrast with the above systems, autoreceptors for GABA have been less investigated. In the last few years some reports have focused on the existence and, particularly, the pharmacological characterization of GABA autoreceptors in the mammalian central nervous system (CNS). In contrast to previous studies showing that the release of radio-labelled GABA from rat cerebral cortex slices (Mitchell and Martin, 1978) or synaptosomes (Brennan *et al.*, 1981) or from rat striatal slices (Kuriyama *et al.*, 1984) was inhibited by muscimol in a bicuculline-sensitive manner, recent investigations failed to confirm such results. In fact, the $GABA_A$ receptor agonist could not inhibit GABA release from rabbit striatal slices (Limberger *et al.*, 1986a), rat cortical synaptosomes (Pittaluga *et al.*, 1987), slices of rat cerebral cortex, hippocampus or striatum (Waldmeier *et al.*, 1988) or from synaptosomes prepared from human cerebral cortex (Bonanno *et al.*, 1989a). In all these systems the release of GABA was instead inhibited by the $GABA_B$ receptor agonist baclofen (Bowery, 1983) in a stereoselective manner. The findings that (−)-baclofen acted directly on GABAergic nerve terminals (Anderson and Mitchell, 1985; Pittaluga *et al.*, 1987) and, in particular, that its

$GABA_B$ Receptors in Mammalian Function. Edited by N.G. Bowery, H. Bittiger and H.-R. Olpe

inhibitory action was counteracted by phaclofen (Bonanno *et al.*, 1988, 1989b), a $GABA_B$ receptor antagonist (Dutar and Nicoll, 1988a; Kerr *et al.*, 1987), strongly support the idea that GABA autoreceptors in the mammalian CNS belong to the $GABA_B$ type. Among the recent studies, however, that by Floran *et al.* (1988) seems to represent an exception. These authors investigated the effects of GABA agonists and antagonists on the release of [^{3}H]GABA stimulated by high K^+ in the pars compacta and in the pars reticulata respectively of the rat substantia nigra. While in the latter nigral subregion the release of [^{3}H]GABA was decreased by (−)-baclofen, muscimol-sensitive $GABA_A$ autoreceptors were reported to be present in the pars compacta.

In the present chapter we report detailed evidence that, in the rat brain, GABAergic nerve terminals are endowed with autoreceptors belonging to the $GABA_B$ subtype. Results of studies carried out with fresh human cerebral cortex suggest that autoreceptors of the $GABA_B$ subtype also exist in the human brain.

STUDIES ON RELEASE-REGULATING GABA AUTORECEPTORS IN RAT CORTICAL SYNAPTOSOMES

The reasons that prompted us to reinvestigate the problem of GABA autoreceptors in the CNS were essentially: (a) the apparent discrepancies among the data available in the literature; (b) the evidence that the GABA receptor exists in at least two types, $GABA_A$ and $GABA_B$; (c) the recent availability of new potent and selective inhibitors of GABA uptake.

The last point is of great importance in these types of studies. In fact, if one assumes that autoreceptors located on a given family of nerve terminals are involved in a feedback mechanism aimed to regulate the release of the transmitter, these autoreceptors should be activated firstly by the natural neurotransmitter, particularly in view of the existence of receptor subtypes, differentially sensitive to exogenous agonists. To use the natural transmitter as an agonist may not always be easy, however; nerve terminals generally possess a reuptake system for their own transmitter. A carrier-mediated exchange may therefore occur between the external transmitter used as an autoreceptor agonist and the radiolabelled neurotransmitter accumulated in the nerve terminal. This exchange may obscure the effect of autoreceptor activation (inhibition of release) unless a potent blocker of the uptake carrier is used.

Since the exchange process is particularly efficient in the case of GABA (Levi and Raiteri, 1974) and since the GABA uptake inhibitors available until recently, such as nipecotic acid (Martin, 1976), are not suitable uptake blockers because they are themselves substrates for the carrier system, GABA has rarely been used as an agonist in autoreceptor studies. Other drugs, essentially $GABA_A$ receptor agonists, like muscimol, have been the preferred tools.

In our studies we have used the recently proposed GABA uptake inhibitor SK&F 89976A (Larsson *et al.*, 1988; Yunger *et al.*, 1984), which appears also to be very selective for the GABA transporter (Bonanno and Raiteri, 1987). The compound did not modify the spontaneous outflow or depolarization-evoked overflow of [^{3}H]GABA previously taken up in the synaptosomal preparations. Since reuptake of the released [^{3}H]GABA does not occur in the superfusion set-up used (see Raiteri and Levi, 1978 for technical details), the absence of effects on release indicates that SK&F 89976A behaves as a pure GABA uptake inhibitor, devoid of intrinsic releasing properties, and thus it is particularly suitable in autoreceptor studies when using GABA as an exogenous receptor agonist.

Synaptosomes were then exposed to GABA in the presence of SK&F 89976A. Figure 5.1 shows that GABA, added to the superfusion medium, inhibited in a

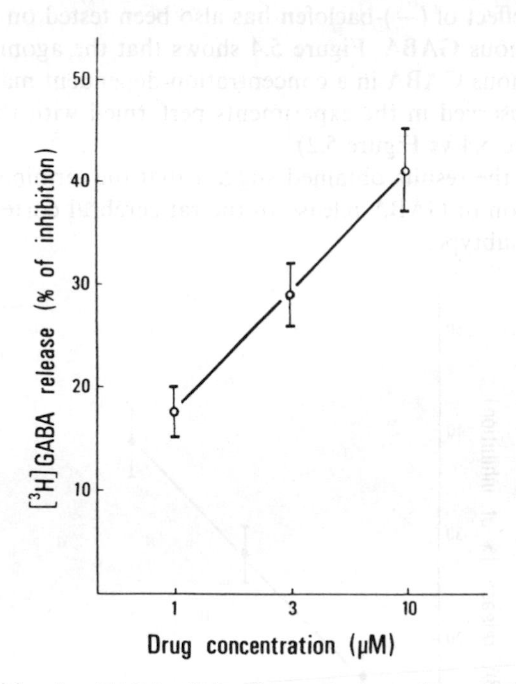

Figure 5.1. Inhibition by GABA of the K^+-evoked overflow of [^{3}H]GABA from rat cortical synaptosomes. Data are means ± SEM of 3–5 experiments in triplicate. Synaptosomes were prepared as previously described (Gray and Whittaker, 1962) and prelabelled with 0.01 μM [^{3}H]GABA. Aliquots of the labelled suspension were then layered on 0.65-μm Millipore filters at the bottom of parellel superfusion chambers and superfused at a rate of 0.6 ml/min for 48 min (Raiteri *et al.*, 1974) using a standard medium having the following composition (mM): NaCl, 125; KCl, 3; $MgSO_4$, 1.2; $CaCl_2$, 1.2; NaH_2PO_4, 1.0; $NaHCO_3$, 22; aminooxyacetic acid, 0.05; glucose, 10 (aeration with 95% O_2/5% CO_2 at 37 °C); pH 7.2–7.4. After 36 min of superfusion, fractions were collected according to the following scheme: two 3-min fractions (basal release) before and after one 6-min fraction (depolarization-evoked release). Synaptosomes were depolarized in superfusion by a 90 s pulse of high KCl at the end of the first fraction collected

concentration-dependent way the release of [^{3}H]GABA evoked by 15 mM KCl. The $GABA_B$ receptor agonist (−)-baclofen also inhibited the K^+-evoked release of [^{3}H]GABA and its concentration–inhibition curve was practically superimposable on that of GABA. The (+)-isomer of baclofen was inactive. The K^+-evoked release of [^{3}H]GABA was not inhibited by the $GABA_A$ receptor agonist muscimol (Figure 5.2).

The inhibitory effect of exogenous GABA was not affected by the $GABA_A$ receptor antagonists bicuculline, picrotoxin or SR 95531 (Figure 5.3) but was concentration-dependently blocked by the novel $GABA_B$ selective antagonist phaclofen (Table 1).

Since newly taken up radiolabelled GABA may sometimes respond to releasing stimuli differently from the endogenous amino acid (see for instance Szerb, 1983), the effect of (−)-baclofen has also been tested on the K^+-evoked release of endogenous GABA. Figure 5.4 shows that the agonist inhibited the release of endogenous GABA in a concentration-dependent manner, similar to what had been observed in the experiments performed with the radiolabelled amino acid (Figure 5.4 vs Figure 5.2).

Taken together the results obtained suggest that the terminal autoreceptors mediating inhibition of GABA release in the rat cerebral cortex belong to the $GABA_B$ receptor subtype.

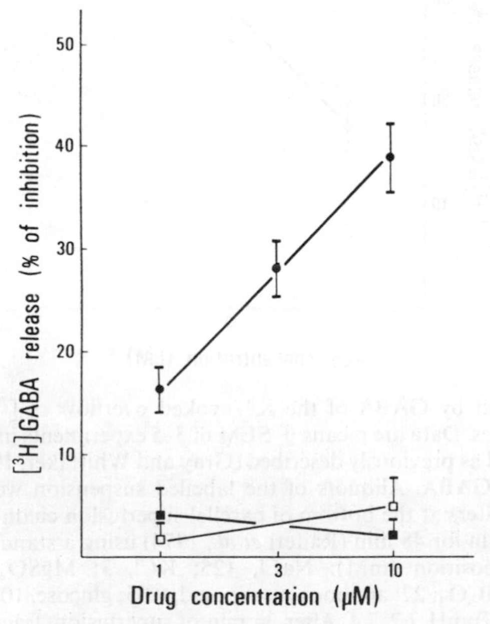

Figure 5.2. Effect of (−)-baclofen (●), (+)-baclofen (□) and muscimol (■) on the K^+-evoked overflow of [^{3}H]GABA from rat cortical synaptosomes. Data are means ± SEM of 4–7 experiments in triplicate

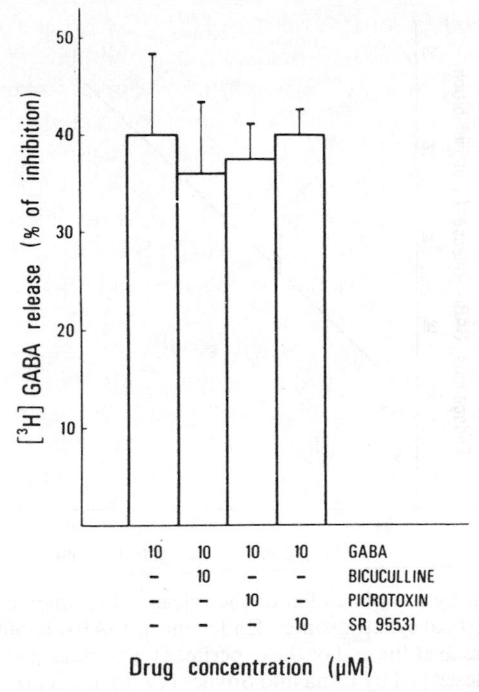

Figure 5.3. Effect of the GABA antagonists bicuculline, picrotoxin and SR 95531 on the GABA-induced inhibition of [^{3}H]GABA release. Data are means ± SEM of 3–5 experiments in triplicate

Table 1. Antagonism by phaclofen of the inhibition of [^{3}H]GABA release caused by GABA in rat cortical synaptosomes

Drugs	% change vs controls
1 μM GABA	−30.2 ± 2.0 (8)
1 μM GABA + 1 μM phaclofen	−31.5 ± 2.7 (3)
1 μM GABA + 10 μM phaclofen	−10.5 ± 4.1 (6)*
1 μM GABA + 100 μM phaclofen	− 7.7 ± 4.0 (6)*
10 μM phaclofen	− 0.8 ± 2.5 (6)
100 μM phaclofen	− 9.3 ± 3.7 (6)

Data shown are means ± SEM of *n* experiments in triplicate, as indicated in parentheses.
*$P < 0.005$, when compared to 1 μM GABA.

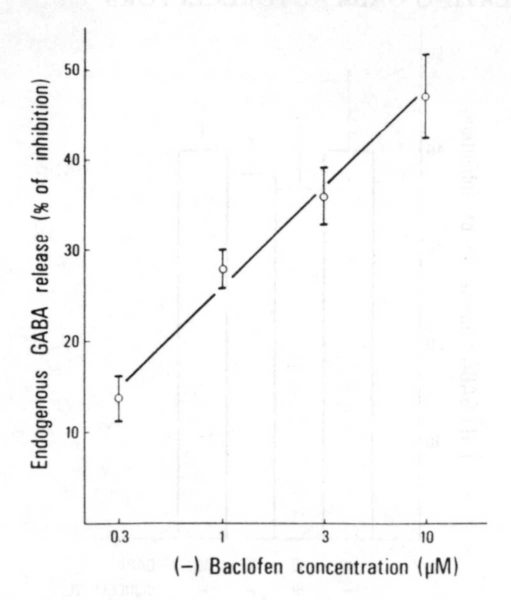

Figure 5.4. Inhibition by (−)-baclofen of the release of endogenous GABA evoked by 15 mM KCl in rat cortical synaptosomes. Endogenous GABA content in the superfusate samples and in the tissue at the end of the experiment was measured with a radioreceptor assay, essentially as described by Enna and Snyder (1976). Data are means ± SEM of 5–7 experiments in triplicate

STUDIES ON RELEASE-REGULATING GABA AUTORECEPTORS IN ELECTRICALLY STIMULATED RAT CORTICAL SLICES

The discrepancy between the findings described in the preceding paragraph and most of the data reported in the literature prompted us to study autoregulation of GABA release using a 'more physiological' approach, i.e with rat cortical slices in which GABA release was elicited using electrical stimulation. Electrically stimulated brain slices have rarely been used to study the release of GABA. In fact, even at relatively high frequencies of stimulation the induced overflow is quite limited (Limberger *et al.*, 1986b). One major reason is likely to be represented by the very efficient recapture of the released amino acid into neurons and glial cells.

In our study we used an experimental approach which differs in several aspects from those previously employed: (a) rat cerebral cortex slices were labelled with [^{3}H]GABA in the presence of the glial uptake inhibitor β-alanine (Schon and Kelly, 1975) to label selectively neuronal terminals; (b) the slices were stimulated electrically in the presence of β-alanine and of SK&F 89976A, to prevent [^{3}H]GABA reuptake into glia and GABAergic nerve terminals, thus enhancing [^{3}H]GABA overflow; (c) the natural transmitter GABA was used as an autoreceptor agonist since GABA homoexchange was minimized by SK&F 89976A; (d) the effects of GABA were challenged not only with the classical $GABA_A$ receptor antagonist bicuculline but also with the novel $GABA_B$ receptor antagonist phaclofen.

Table 2 shows the overflow of tritium evoked by the first period of stimulation (S_1) from rat cortical slices preincubated with the radiolabelled amino acid and stimulated electrically at 5 or 10 Hz in the presence of the above-mentioned GABA uptake inhibitors (see the legend to Figure 5.5 for other details): increasing the frequency of electrical stimulation from 5 to 10 Hz caused acceleration of the release of the amino acid. The presence of SK&F 89976A or β-alanine also increased the tritium overflow; the effects of the two substances were in part additive and led to double the overflow obtained in their absence. Chromatographic analysis of the superfusate samples showed that the tritium released could be almost entirely (>90%) accounted for by authentic [^{3}H]GABA. The release of [^{3}H]GABA evoked electrically using either 5 or 10 Hz, in the presence of 100 μM β-alanine and 30 μM SK&F 89976A, was completely tetrodotoxin-sensitive and largely (>70%) Ca^{2+}-dependent (not shown).

Exogenous GABA added to the superfusion medium decreased, concentration-dependently, the release of [^{3}H]GABA: the effect of GABA was markedly increased when the frequency of the electrical stimulation was lowered from 10 to 5 Hz (Figure 5.5a). The $GABA_B$ receptor antagonist (−)-baclofen behaved similarly to GABA; in contrast, the $GABA_A$ receptor agonist muscimol was completely inactive even at 5 Hz (Figure 5.5b).

Table 2. Effect of GABA uptake inhibitors on [^{3}H]GABA release

Stimulation frequency (Hz)	Drug	S_1 (% of tissue tritium)
5	—	0.338 ± 0.007 (3)
	30 μM SK&F 89976A	0.605 ± 0.002 (4)**
	100 μM β-alanine	0.445 ± 0.017 (3)*
	100 μM β-alanine + 30 μM SK&F 89976A	0.685 ± 0.009 (3)**‡
10	—	0.653 ± 0.041 (3)
	30 μM SK&F 89976A	1.083 ± 0.031 (4)**
	100 μM β-alanine	0.904 ± 0.027 (3)*
	100 μM β-alanine + 30 μM SK&F 89976A	1.247 ± 0.076 (3)*†

When used, SK&F 89976A and/or β-alanine were present from the beginning of superfusion. Data are means ± SEM of n experiments in duplicate, as indicated in parentheses.
*$P < 0.005$, **$P < 0.001$, when compared to the respective control conditions obtained in the absence of SK&F 89976A and β-alanine; †$P < 0.02$, ‡$P < 0.005$, when compared to the corresponding control conditions obtained in the presence of SK&F 89976A.

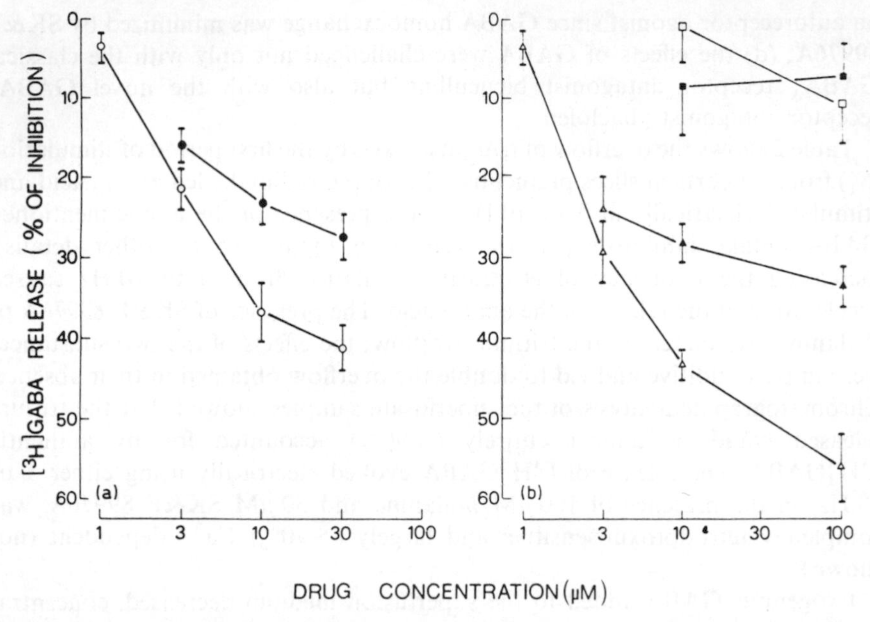

Figure 5.5. Concentration–inhibition curves of (a) GABA (●; ○) and (b) (−)-baclofen (▲; △) and muscimol (■; □) on the electrically evoked overflow of [^{3}H]GABA from rat cortical slices. Solid symbols: stimulation frequency 10 Hz; open symbols: stimulation frequency 5 Hz. Data are means ± SEM of 4–7 experiments in duplicate. Temporoparietal slices (400 μm) were prepared from cortices and prelabelled with 0.04 μM [^{3}H]GABA in a medium having the same composition as that used for synaptosome experiments. Incubation was carried out in the presence of β-alanine (100 μM). Slices were then superfused at a rate of 1 ml/min in the presence of β-alanine (100 μM) and SK&F 89976A (30 μM) and stimulated electrically (5 or 10 Hz; 2 ms, 36 mA, 5 min) after 50 (S_1) and 95 (S_2) min of superfusion. The overflows evoked by S_1 and S_2 were calculated as a percentage of the total tritium, and the S_2/S_1 ratios of the drug-treated slices were compared to the S_2/S_1 ratios of appropriate controls in order to estimate drug effects

Paralleling the data obtained with synaptosomes, the effect of GABA was counteracted by the GABA$_B$ receptor antagonist phaclofen but not by bicuculline or SR 95531 (Figure 5.6). Moreover, phaclofen, when added before the second period of stimulation (S_2), increased on its own the electrically evoked overflow of [^{3}H]GABA. No effect was observed using bicuculline or SR 95531 under the same experimental conditions (Figure 5.7).

To our knowledge this is the first report showing that the natural transmitter GABA can decrease the electrically stimulated release of preaccumulated [^{3}H]GABA. Thus the data obtained using electrically stimulated cortical slices confirm those obtained with synaptosomes and strengthen the idea that GABA autoreceptors are present on GABAergic nerve terminals in the rat cerebral cortex and that they belong to the GABA$_B$ receptor subtype. Finally, the decrease of the potency of GABA and (−)-baclofen observed when the

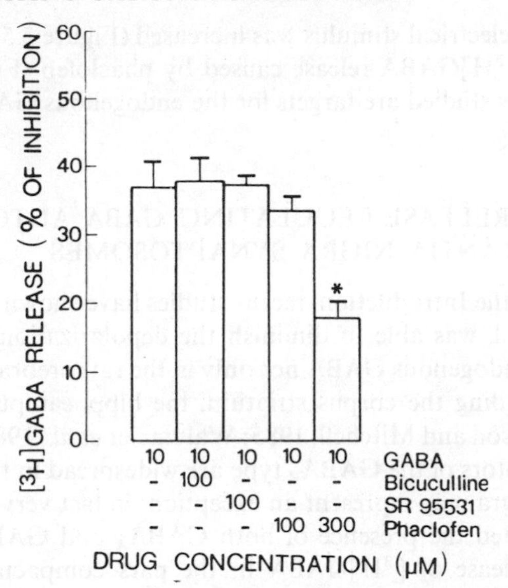

Figure 5.6. Effects of GABA receptor antagonists on the GABA-induced inhibition of the electrically evoked [^{3}H]GABA overflow from rat cortical slices. GABA was added to the superfusion medium 25 min before S_2; bicuculline, SR 95531 or phaclofen was present throughout superfusion. Data are means ± SEM of 3–6 experiments in duplicate. *$P < 0.01$

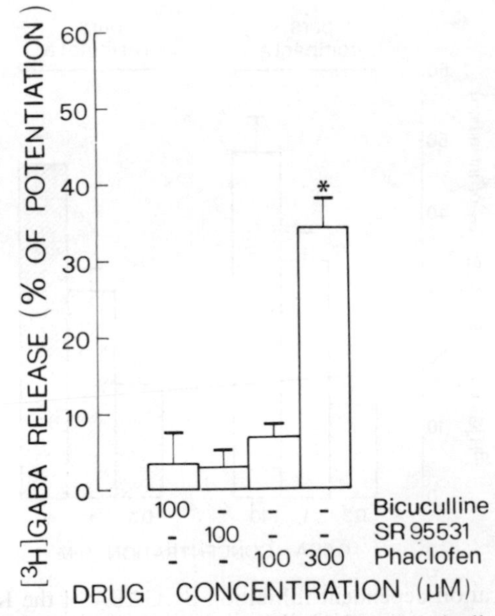

Figure 5.7. Effect of GABA receptor antagonists on the electrically evoked overflow of [^{3}H]GABA from rat cortical slices. Bicuculline, SR 95531 or phaclofen was added to the superfusion medium 25 min before S_2. Data are means ± SEM of 3–5 experiments in duplicate. *$P < 0.01$

frequency of the electrical stimulus was increased (Figure 5.5), together with the potentiation of [^{3}H]GABA release caused by phaclofen (Figure 5.7), suggests that the receptors studied are targets for the endogenous GABA released in the synaptic cleft.

STUDIES ON RELEASE-REGULATING GABA AUTORECEPTORS IN RAT SUBSTANTIA NIGRA SYNAPTOSOMES

As mentioned in the Introduction, recent studies have shown that (−)-baclofen, but not muscimol, was able to diminish the depolarization-evoked release of [^{3}H]GABA or endogenous GABA not only in the rat cerebral cortex but also in other areas including the corpus striatum, the hippocampus and the median eminence (Anderson and Mitchell, 1985; Waldmeier *et al.*, 1988), suggesting that GABA autoreceptors of the $GABA_B$ type are widespread in the CNS. However, the substantia nigra may represent an exception: in fact very recently Floran *et al.* (1988) suggested the presence of both $GABA_A$ and $GABA_B$ autoreceptors inhibiting the release of [^{3}H]GABA in the pars compacta and in the pars reticulata respectively of the rat substantia nigra. Formerly, the autoreceptors regulating [^{3}H]GABA release from slices of rat substantia nigra depolarized with high K^+ had been found to be muscimol-sensitive (Arbilla *et al.*, 1979).

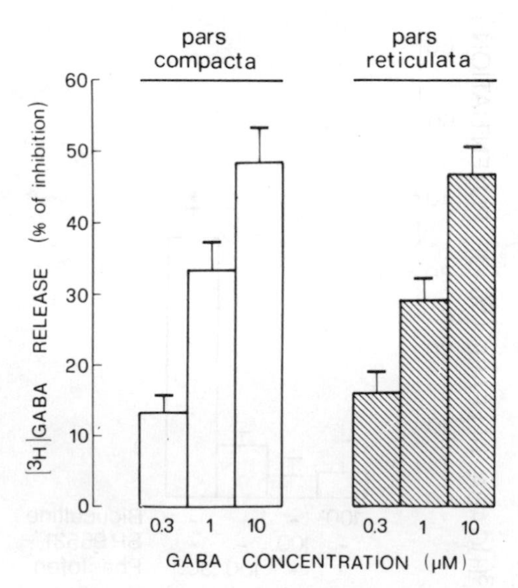

Figure 5.8. Concentration-dependent inhibition by GABA of the K^+-evoked release of [^{3}H]GABA from synaptosomes of the pars compacta and the pars reticulata of the rat substantia nigra. See the legend to Figure 1 for technical details. Data are means ± SEM of 3–10 experiments in triplicate

This area was also investigated by Waldmeier *et al.* (1988) when they studied the effect of (−)-baclofen on the release of endogenous GABA evoked by electrical stimulation in several rat brain areas. Unfortunately, the authors failed to detect any evoked overflow of endogenous GABA in the substantia nigra.

The paucity of the results available in the substantia nigra together with their peculiarity prompted us to address the problem of the pharmacological classification of GABA autoreceptors in this area. We here report the data obtained using superfused synaptosomes prepared from both the pars compacta and the pars reticulata of the rat substantia nigra.

Figure 5.8 shows that GABA concentration-dependently inhibited the K^+-evoked release of [^{3}H]GABA from both pars compacta and pars reticulata synaptosomes.

The data reported in Figure 5.9 and Figure 5.10 concerning experiments with $GABA_A$ or $GABA_B$ selective agonists and antagonists show that the pharmacological profile of the GABA autoreceptor is very similar in the two nigral subregions and suggest that only autoreceptors of the $GABA_B$ subtype are present on GABAergic nerve terminals in the rat substantia nigra. This

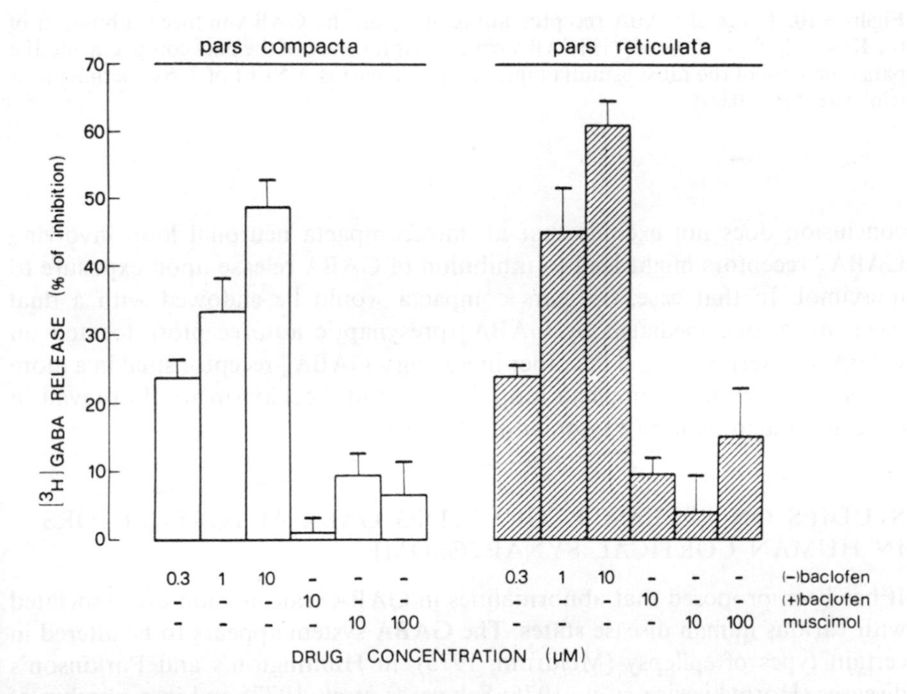

Figure 5.9. Effect of (−)-baclofen, (+)-baclofen and muscimol on the K^+-evoked overflow of [^{3}H]GABA from synaptosomes of the pars compacta and the pars reticulata of the rat substantia nigra. Data are means ± SEM of 3–6 experiments in triplicate

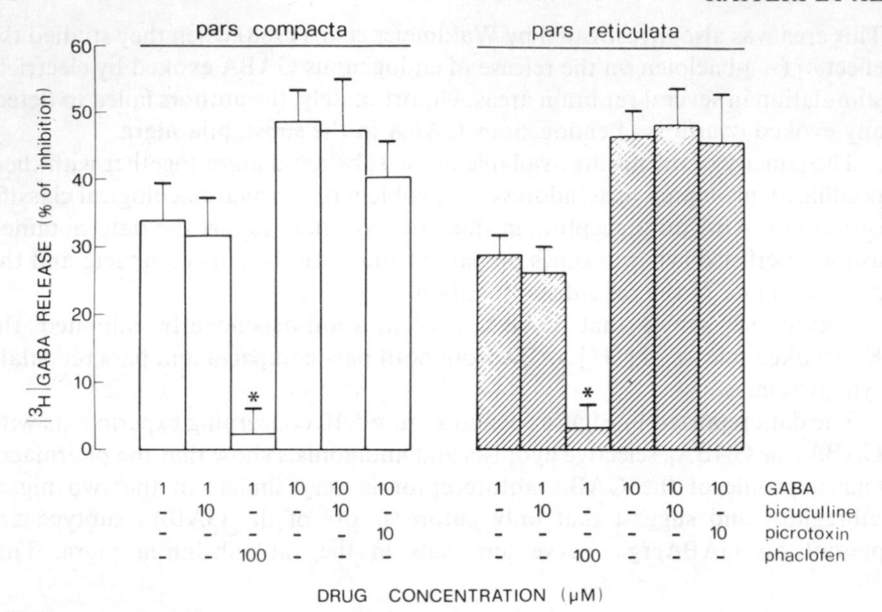

Figure 5.10. Effect of GABA receptor antagonists on the GABA-induced inhibition of the K^+-evoked overflow of [^{3}H]GABA from synaptosomes of the pars compacta and the pars reticulata of the rat substantia nigra. Data are means $\pm$ SEM of 3–6 experiments in triplicate. $*P < 0.001$

conclusion does not exclude that an intracompacta neuronal loop involving $GABA_A$ receptors might lead to inhibition of GABA release upon exposure to muscimol. In that case, the pars compacta would be endowed with a dual mechanism, one mediated by $GABA_B$ presynaptic autoreceptors located on GABAergic terminals and the other involving a $GABA_A$ receptor sited in a more or less complex intracompacta loop. The second mechanism would prevail in the slice system used by Floran *et al.* (1988).

STUDIES ON RELEASE-REGULATING GABA AUTORECEPTORS IN HUMAN CORTICAL SYNAPTOSOMES

It has been proposed that abnormalities in GABA transmission are associated with various human disease states. The GABA system appears to be altered in certain types of epilepsy (Meldrum, 1975), in Huntington's and Parkinson's diseases (Hornykiewicz *et al.*, 1976; Schwarcz *et al.*, 1977) and in a number of neuropsychiatric disorders including depression (Lloyd *et al.*, 1985). In spite of this, studies focused on the modulation of GABA release until now have been performed only using laboratory animals. Due to the potential involvement of GABA in neurological or psychiatric diseases we thought that it was of relevance to determine whether GABA autoreceptors are present in human brain and whether or not their pharmacological properties are similar to those found in the rodent brain.

In our experiments synaptosomes were prepared from fresh human cerebral cortex and superfused as described in the legend to Figure 5.1. As shown in Table 3, when human cortical synaptosomes were exposed during superfusion to 15 mM KC1, an acceleration of [^{3}H]GABA efflux could be observed and the depolarization-evoked release was almost totally Ca^{2+}-dependent. Exogenous GABA, added to the superfusion fluid, in the presence of SK&F 89976A, decreased in a concentration-dependent manner the K^+-evoked overflow of [^{3}H]GABA (Figure 5.11). Figure 5.11 also shows that the (−)-enantiomer of baclofen produced a concentration–inhibition curve almost superimposable on that of GABA. In contrast, muscimol was totally ineffective. The inhibitory effect of GABA was antagonized neither by bicuculline nor by SR 95531 (Table 4).

The effect of (−)-baclofen was also tested on both the basal and the K^+-evoked release of endogenous GABA from human synaptosomes. Table 5 shows that, while the basal release was not affected significantly, the depolarization-evoked overflow was inhibited by about 50% when the agonist was present at 10 μM.

The data obtained demonstrate that $GABA_B$ autoreceptors regulating endogenous and [^{3}H]GABA release are present in human cortical synaptosomes. When the present results in human cerebral cortex are compared with those obtained in the cerebral cortex of the rat, no clear differences appear to exist between the release-regulating GABA autoreceptors present in the two species. Actually, the effects of GABA and (−)-baclofen are strikingly similar in man and rat.

Table 3. Calcium-dependence of the K^+-evoked release of [^{3}H]GABA from human cortical synaptosomes

	Standard medium	Ca^{2+}-free medium
Basal release/min	0.62 ± 0.08 (4)	0.56 ± 0.06 (4)
K^+ (15 mM) -evoked overflow	2.96 ± 0.38 (4)	0.32 ± 0.05 (4)*

The release is expressed as a percentage of the total tritium content of synaptosomes. Data are means ± SEM of *n* experiments in triplicate, as indicated in parentheses. *$P < 0.001$, when compared to the K^+-evoked overflow in standard medium.

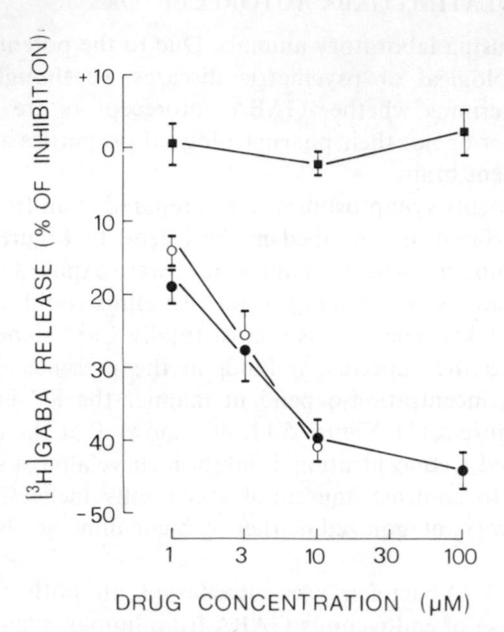

Figure 5.11. Effect of GABA (○), (−)-baclofen (●) and muscimol (■) on the release of [^{3}H]GABA evoked by 15 mM KCl from human cortical synaptosomes. Data are means ± SEM of 3–4 experiments in triplicate. Samples of human cerebral cortex were obtained from patients undergoing neurosurgery. The tissues used were removed by the surgeon to reach deeply located tumours. The samples used represented parts of frontal (2), temporal (3), parietal (2) and occipital (2) lobes and were obtained from 5 female and 4 male patients (aged 45–69 years). The tissues were obtained and processed separately on different days. Immediately after removal the tissue was placed in a physiological salt solution kept at 2–4 °C and a synaptosomal fraction was obtained within 60 min

Table 4. Effects of bicuculline and SR 95531 on the GABA-induced inhibition of the K^+ (15 mM) -evoked release of [^{3}H]GABA from human cortical synaptosomes

	% change vs controls
10 μM GABA	−40.4 ± 2.9 (6)
10 μM GABA + 10 μM bicuculline	−41.2 ± 2.8 (4)
10 μM GABA + 10 μM SR 95531	−37.8 ± 3.2 (4)

Data are means ± SEM of *n* experiments in triplicate, as indicated in parentheses.

Table 5. Effect of (−)-baclofen (10 μM) on the K^+-evoked release of endogenous GABA from human cortical synaptosomes

	Controls	(−)-Baclofen
Basal release/min	28.3 ± 3.2 (5)	25.2 ± 2.7 (5)
K^+ (15 mM) -evoked overflow	149.6 ± 21.7 (5)	76.7 ± 11.5 (5)*

The release is expressed as pmoles of endogenous GABA/mg protein. Data are means ± SEM of *n* experiments in triplicate, as indicated in parentheses.
*$P < 0.05$, when compared to the K^+-evoked overflow under control conditions.

CONCLUSION

The most recent results clearly indicate that the GABA autoreceptors which are present in various regions of the rat brain belong to the $GABA_B$ type. Autoreceptors sensitive to (−)-baclofen could be detected under several experimental conditions in which the release of previously taken up radiolabelled GABA or of endogenous GABA was studied from synaptosomes stimulated with high K^+, or veratrine, as well as from slices stimulated electrically. The fact that autoreceptors display $GABA_B$ pharmacology independently of the experimental set-up chosen makes it quite difficult to explain why, on the basis of studies published about 10 years ago, they should be classified as $GABA_A$. In fact, under the above experimental conditions, GABA autoreceptors appear to be quite insensitive to muscimol and bicuculline.

GABA autoreceptors are $GABA_B$ also in the brain of our species, as indicated by the results obtained studying the release of the amino acid from the nerve terminals of fresh human cerebral cortex. As an important consequence, the laboratory animal appears to be a model useful in further studies on GABA autoreceptors and in the development of new molecules of potential therapeutic value. Interestingly, it has recently been proposed that $GABA_B$ receptors are pharmacologically heterogeneous and, in particular, that pre- and postsynaptic $GABA_B$ receptors are different (Dutar and Nicoll, 1988b). If this is the case, drugs selective for GABA autoreceptors might be useful to regulate GABA transmission.

ACKNOWLEDGEMENTS

This work was supported by grants from the Italian Ministry of University and of Scientific and Technological Research, and from the Italian National Research Council.

REFERENCES

Anderson, R.A. and Mitchell, R. (1985) Evidence for $GABA_B$ autoreceptors in median eminence. *Eur. J. Pharmacol.*, **118**, 355–358.

Arbilla, S., Kamal, L. and Langer, S.Z. (1979) Presynaptic GABA autoreceptors on GABA-ergic nerve endings of the rat substantia nigra. *Eur. J. Pharmacol.*, **57**, 211–217.

Bonanno, G. and Raiteri, M. (1987) Coexistence of carriers for dopamine and GABA uptake on a same nerve terminal in the rat brain. *Br. J. Pharmacol.*, **91**, 237–243.

Bonanno, G., Maura, G. and Raiteri, M. (1986) Pharmacological characterization of release-regulating serotonin autoreceptors in rat cerebellum. *Eur. J. Pharmacol.*, **126**, 317–321.

Bonanno, G., Fontana, G. and Raiteri, M. (1988) Phaclofen antagonizes GABA at autoreceptors regulating release in rat cerebral cortex. *Eur. J. Pharmacol.*, **154**, 223–224.

Bonanno, G., Cavazzani, P., Andrioli, G.C., Asaro, D., Pellegrini, G. and Raiteri, M. (1989a) Release-regulating autoreceptors of the $GABA_B$-type in human cerebral cortex. *Br. J. Pharmacol.*, **96**, 341–346.

Bonanno, G., Pellegrini, G., Asaro, D., Fontana, G. and Raiteri, M. (1989b) $GABA_B$ autoreceptors in rat cortex synaptosomes: response under different depolarizing and ionic conditions. *Eur. J. Pharmacol.* (*Mol. Pharmacol. Section*), **172**, 41–49.

Bowery, N.G. (1983) Classification of GABA receptors. In: S.J. Enna (ed.) *The GABA receptors.* Clifton, NJ: Humana Press, pp. 177–213.

Brennan, M.J.W., Cantrill, R.C., Oldfield, M. and Krogsgaard-Larsen, P. (1981) Inhibition of γ-aminobutyric acid release by γ-aminobutyric acid agonist drugs: pharmacology of the γ-aminobutyric acid autoreceptor. *Mol. Pharmacol.*, **19**, 27–30.

Chesselet, M.-F. (1984) Presynaptic regulation of neurotransmitter release in the brain: facts and hypothesis. *Neuroscience*, **12**, 347–375.

Dutar, P. and Nicoll, R.A. (1988a) A physiological role for $GABA_B$ receptors in the central nervous system. *Nature, Lond.*, **332**, 156–158.

Dutar, P. and Nicoll, R.A. (1988b) Pre- and post-synaptic $GABA_B$ receptors in the hippocampus have different pharmacological properties. *Neuron*, **1**, 585–591.

Engel, G., Göthert, M., Hoyer, D., Schlicker, E. and Hillenbrand, K. (1986) Identity of inhibitory presynaptic 5-hydroxytryptamine (5-HT) autoreceptors in the rat brain cortex with $5\text{-}HT_{1B}$ binding sites. *Naunyn-Schmiedeberg's Arch. Pharmacol.*, **332**, 1–7.

Enna, S.J. and Snyder, S.H. (1976) A simple, sensitive and specific radioreceptor assay for endogenous GABA in brain tissue. *J. Neurochem.*, **26**, 221–224.

Floran, B., Silva, I., Nava, C. and Aceves, J. (1988) Presynaptic modulation of the release of GABA by $GABA_A$ receptors in pars compacta and by $GABA_B$ receptors in pars reticulata of the rat substantia nigra. *Eur. J. Pharmacol.*, **150**, 277–286.

Gray, E.G and Whittaker, V.P. (1962) The isolation of nerve endings from brain: an electron-microscopic study of cell fragments derived by homogenization and centrifugation. *J. Anat., Lond.*, **96**, 79–87.

Hornykiewicz, O., Lloyd, K.G. and Davidson, L. (1976) The GABA system, function of the basal ganglia, and Parkinson's disease. In: E. Roberts, T. Chase and D. Tower (eds) *GABA in nervous system function.* New York: Raven Press, pp. 479–485.

Kerr, D.I.B., Ong, J., Prager, R.H., Gynther, B.D. and Curtis, D.R. (1987) Phaclofen: a peripheral and central baclofen antagonist. *Brain Res.*, **405**, 150–154.

Kuriyama, K., Kanmori, K., Taguchi, J. and Yoneda, Y. (1984) Stress-induced enhancement of suppression of $[^3H]$-GABA release from striatal slices by presynaptic autoreceptor. *J. Neurochem.*, **42**, 943–950.

Langer, S.Z. (1981) Presynaptic regulation of the release of catecholamines. *Pharmacol. Rev.*, **32**, 337–362.

Larsson, O.M., Falch, E., Krogsgaard-Larsen, P. and Schousboe, A. (1988) Kinetic characterization of inhibition of γ-aminobutyric acid uptake into cultured neurons and astrocytes by 4,4-diphenyl-3-butenyl derivatives of nipecotic acid and guvacine. *J. Neurochem.*, **50**, 818–823.

Levi, G. and Raiteri, M. (1974) Exchange of neurotransmitter amino acid at nerve endings can simulate high affinity uptake. *Nature, Lond.*, **250**, 735–737.

Limberger, N., Späth, L. and Starke, K. (1986a) A search for receptors modulating the release of γ-[^{3}H]aminobutyric acid in rabbit caudate nucleus slices. *J. Neurochem.*, **46**, 1109–1117.

Limberger, N., Späth, L. and Starke, K. (1986b) Release of previously incorporated γ-[^{3}H]aminobutyric acid in rabbit caudate nucleus slices. *J. Neurochem.*, **46**, 1102–1108.

Lloyd, K.G., Thuret, F. and Pilc, A. (1985) Upregulation of $GABA_B$ binding sites in rat frontal cortex: a common action of repeated administration of different classes of antidepressants and electroshock. *J. Pharmacol. Exp. Ther.*, **235**, 191–199.

Marchi, M. and Raiteri, M. (1985) On the presence in the cerebral cortex of muscarinic receptor subtypes which differ in neuronal localization, function and pharmacological properties. *J. Pharmacol. Exp. Ther*, **235**, 230–233.

Marchi, M. and Raiteri, M. (1989) Interaction ACh–GLU in rat hippocampus: involvement of two subtypes of M-2 muscarinic receptors. *J. Pharmacol. Exp. Ther.*, **248**, 1255–1260.

Martin, D.L. (1976) Carrier-mediated transport and removal of GABA from synaptic regions. In: E. Roberts, T.N. Chase and D.B. Tower (eds) *GABA in nervous system function*. New York: Raven Press, pp. 347–386.

Mash, D.C. and Potter, L.T. (1986) Autoradiographic localization of M1 and M2 muscarine receptors in the rat brain. *Neuroscience*, **19**, 551–564.

Maura, G., Roccatagliata, E. and Raiteri, M. (1986) Serotonin autoreceptor in rat hippocampus: pharmacological characterization as a subtype of the $5\text{-}HT_1$ receptor. *Naunyn-Schmiedeberg's Arch. Pharmacol.*, **334**, 323–326.

Meldrum, B.S. (1975) Epilepsy and γ-aminobutyric acid-mediated inhibition. *Int. Rev. Neurobiol.*, **17**, 1–36.

Meyer, E.M. and Otero, D.H. (1985) Pharmacological and ionic characterizations of the muscarinic receptors modulating ^{3}H-acetylcholine release from rat cortical synaptosomes. *J. Neurosci.*, **5**, 1202–1207.

Mitchell, P.R. and Martin, I.L. (1978) Is GABA release modulated by presynaptic receptors? *Nature*, **274**, 904–905.

Pittaluga, A., Asaro, D., Pellegrini, G. and Raiteri, M. (1987) Studies on ^{3}H-GABA and endogenous GABA release in rat cerebral cortex suggest the presence of autoreceptors of the $GABA_B$ type. *Eur. J. Pharmacol.*, **144**, 45–52.

Raiteri, M. and Levi, G. (1978) Release mechanisms for catecholamines and serotonin in synaptosomes. In: S. Ehrenpreis and I. Kopin (eds) *Reviews of Neuroscience*, vol. 3. New York: Raven Press, pp. 77–130.

Raiteri, M., Angelini, F. and Levi, G. (1974) A simple apparatus for studying the release of neurotransmitters from synaptosomes. *Eur. J. Pharmacol.*, **25**, 411–414.

Raiteri, M., Marchi, M. and Maura, G. (1984) Release of catecholamines, serotonin and acetylcholine from isolated brain tissue. In: A. Lajtha (ed.) *Handbook of neurochemistry*, vol. 6. New York: Plenum, pp. 431–462.

Schon, F. and Kelly, J.S. (1975) Selective uptake of [^{3}H]β-alanine by glia: association with the glial uptake system for GABA. *Brain Res.*, **86**, 243–257.

Schwarcz, R., Bennett, J.P. and Coyle, J.T. (1977) Inhibitors of GABA metabolism: implications for Huntington's disease. *Ann. Neurol.*, **2**, 299–305.

Starke, K. (1981) Presynaptic receptors. *Annu. Rev. Pharmacol. Toxicol.*, **21**, 7–30.

Szerb, J.C. (1983) The release of ^{3}H-GABA formed from ^{3}H-glutamate in rat hippocampal slices: comparison with endogenous and exogenous labeled GABA. *Neurochem. Res.*, **8**, 341–351.

Waldmeier, P.C., Wicki, P., Feldtrauer, J.-J. and Baumann, P.A. (1988) Potential involvement of a baclofen-sensitive autoreceptor in the modulation of the release of endogenous GABA from rat brain slices in vitro. *Naunyn-Schmiedeberg's Arch. Pharmacol.*, **337**, 289–295.

Yunger, L.M., Fowler, P.J., Zarevics, P. and Setler, P.E. (1984) Novel inhibitors of γ-aminobutyric acid (GABA) uptake: anticonvulsant actions in rats and mice. *J. Pharmacol. Exp. Ther.*, **228**, 109–115.

Part III

PERIPHERAL ASPECTS

6 Review of Peripheral $GABA_B$ Effects

A. GIOTTI, A. BARTOLINI, P. FAILLI, G. GENTILINI,
M. MALCANGIO and L. ZILLETTI

INTRODUCTION

In the preface to a comprehensive article on the classification of GABA receptors, Bowery (1983) suggested that the use of the term 'putative' is no longer necessary when describing the role of GABA as a central inhibitory transmitter. Whether this is true within the mammalian periphery is still an open question. In 1986 Erdö and Bowery published a comprehensive review covering the possible peripheral effects related to activation or deactivation of GABAergic mechanisms, but physiological evidence is still lacking. In the present review we have decided to restrict the topic, limiting ourselves to commenting on papers published after 1985.

One conclusion that can be derived from the history of neurotransmission is that a 'system' which utilizes a main neurotransmitter operates through a family of receptors. This type of organization has been ascribed to the *principium of economy* as applied to the evolution of living organisms. Certainly the '*polyreceptor*' involvement which characterizes neural function creates difficult problems in physiology. Theoretically we need selective antagonists for each of the receptor subtypes to pass from in-vitro 'effects' to the analysis of physiological functions.

In the case of GABA receptors the intuition of Bowery and Hudson (1979) has to be considered as a great intellectual achievement, showing the existence of GABA receptors 'different' from the classical ones, which are blocked by bicuculline or its methohalide salts with high stereoselective resolution.

$GABA_B$ receptors, the discovery of which was 'born' of a negative finding, i.e. the insensitivity of certain GABA effects to bicuculline, have been subjected to detailed pharmacological study by the discoverers and their co-workers, both in the peripheral (Bowery *et al.*, 1981) as well as in the central nervous system (Bowery *et al.*, 1980, 1983; Hill and Bowery, 1981). The new receptors have been revealed to be profoundly different biochemically and pharmacologically from the $GABA_A$ receptors as well as from aminergic and other amino acidergic receptors (Bowery *et al.*, 1989; Nicoll and Dutar, 1989; Wojcik *et al.*, 1989).

$GABA_B$ Receptors in Mammalian Function. Edited by N.G. Bowery, H. Bittiger and H.-R. Olpe

Some agonists and antagonists of $GABA_B$ mechanisms derived from the aforementioned pioneer studies show sufficient selectivity (e.g. (−)-baclofen, a non-$GABA_A$ ligand) to enable them to be used in the pharmacological 'dissection' of the GABA system and the analysis of its receptor composition as well as its role in physiology.

$GABA_B$ EFFECTS

$GABA_B$ Effects on the Circulation

GABAergic agents which cross the blood–brain barrier exert a profound influence on respiratory and circulatory function. Erdö (1985) concludes that 'GABA depresses or facilitates respiratory and sympathetic vasomotor activities through both $GABA_A$ and $GABA_B$ receptors'. In the many experiments reported on the action in vivo of $GABA_A$ agonists and baclofen, control of respiratory gases seems to have been absent or at least very poor. With this limitation the variations of blood pressure and cardiac function become polyphasic and it is practically impossible to discriminate between the role of the primary actions of the drug on the peripheral sympathetic nervous system and any secondary effects due to responses of the centres regulating cardiovascular function in hypoxia. In evaluating this problem, the contributions of anaesthesiologists who have been interested in baclofen for its analgesic effect (Cutting and Jordan, 1975; Wilson and Yaksh, 1978; Yaksh and Reddy, 1981), and in GABA analogues (Cheng and Brunner, 1985) for inducing anaesthesia, are important.

Sill *et al.* (1986) have noted that:

> Baclofen, in dosages that are used clinically in the management of muscle spasm may, during general anaesthesia and surgery, disturb autonomic control of the circulation. We recommend that when anaesthetizing such patients, the possibility of severe bradycardia and hypotension should be anticipated and drugs appropriate for treating this emergency should be readily available.

Since, during clinical anaesthesia, respiration is controlled, the observed effects are pure primary effects of the drug on its cardiovascular and specific central receptors. But even in controlled conditions of respiration it is not easy to differentiate the relative roles of the central and peripheral receptors.

For evaluating the role of the peripheral $GABA_B$ mechanisms in vivo, i.v. GABA (which does not cross the blood–brain barrier) in the presence of bicuculline or picrotoxin is probably preferable to baclofen (which does cross). De Feudis (1983) synthesized the results of in-vivo experiments, concluding that 'in general, hypotension and bradycardia produced by GABAergic agents are reversed by the $GABA_A$ antagonists, bicuculline and picrotoxin'. It is also necessary to recall the powerful antagonistic action of $GABA_A$ antagonists on the respiratory depression provoked by centrally acting depressants like barbiturates in experimental animals and humans (Giotti and Giusti, 1948; Maloney, 1933).

The subject has been examined recently by Giuliani *et al.* (1986), who have studied the influence of GABA effects on anaesthesia (barbitone or urethane) in guinea-pigs and on variations in the neuroendocrine regulation of the cardiovascular system (reserpine pretreatment, hexamethonium, phentolamine plus propranolol, adrenalectomy). They concluded that the 'depressive' cardiovascular changes produced by i.v. GABA may involve activation of peripheral $GABA_B$ receptors since they were substantially bicuculline-insensitive. The doses of GABA employed were the same as those used to activate $GABA_B$ receptors in urethane-anaesthetized guinea-pigs to modulate colonic motility (Giotti *et al.*, 1985; see later for further comments).

An 'excitatory' effect of GABA on the cardiovascular system appears in barbitone-anaesthetized animals after an initial depressive effect. This excitatory effect, but not the depressive effect, is retained in reserpinized animals and is reproduced by $GABA_A$ receptor agonists which cross-desensitize with GABA. It is not mimicked by ($\pm$)-baclofen (which actually has a slight depressant effect), is not prevented by bicuculline or picrotoxin and is not observed in the presence of phentolamine or in adrenalectomized animals. If the excitatory effects of i.v. GABA are mediated through $GABA_B$ mechanisms as seems possible, since they were obtained in the presence of picrotoxin or bicuculline, why are they not mimicked by ($\pm$)-baclofen? The authors concluded that the $GABA_A$ receptors involved in the excitatory effects are 'atypical $GABA_A$ receptors'. It is of note that the authors emphasize the low resting tone of the autonomic nervous system in controlling cardiovascular functions (Maggi and Meli, 1986) and the marked depression of respiratory rate in barbitone-anaesthetized animals (neither parameter is influenced by urethane anaesthesia). It is therefore possible that the 'atypical' nature of the response does not imply the unmasking of a new atypical receptor but rather is the result of alterations produced in cellular mechanisms by hypoxia. The 'two' opposing sets of cardiovascular changes, i.e. depressive (hypotension, bradycardia and negative inotropic effect) and excitatory (hypertension, tachycardia and positive inotropic effect), indicate that further studies are required to evaluate the role of different GABA receptors, despite the recent contributions of Giuliani *et al.* (1986).

When in-vivo experiments present complex answers, in-vitro isolated preparations can frequently offer the possibility of identifying the sites and mechanism(s) of action. One such in-vitro experiment contributed fundamentally to $GABA_B$ history: the use of rat isolated atria (together with mouse and guinea-pig isolated vas deferens) led Bowery *et al.* (1981) to an accurate description of the presence of $GABA_B$ receptors on peripheral autonomic sympathetic nerve terminals and of the structural requirements and stereoselectivity for their activation.

Several reports indicate that GABA relaxes intracranial vessels in different animal species and that the relaxation is antagonized by either bicuculline or picrotoxin (Edvinsson and Krause, 1979; Fujiwara *et al.*, 1975). However, Anwar and Mason (1982) have described a $GABA_B$ receptor-mediated inhibition of electrically evoked sympathetic neurotransmission in the rabbit basilar artery (the effect was not present in isolated mesenteric or carotid arteries). The response to GABA (median effective dose, $ED_{50} \pm SEM = 5.6 \pm 2.1 \times 10^{-7}$ M) was not antagonized by bicuculline and ($\pm$)-baclofen produced a similar inhibition ($ED_{50} = 6.8 \pm 1.4 \times 10^{-7}$ M). GABA in higher concentrations ($2.4 \pm 1.1 \times 10^{-5}$ M) produced a bicuculline-sensitive relaxation if the arterial tone had been increased with 5-hydroxytryptamine (5-HT, serotonin), but was not active on the tone of isolated mesenteric or internal carotid arteries whether or not the tone was increased with 5-HT. Similarly GABA did not change the responses to transmural stimulation in isolated mesenteric or internal carotid arteries.

The suggested $GABA_B$ mechanism may therefore be present together with a less sensitive $GABA_A$ system only in certain arteries although the vasodilating mechanism seems short-lived and may escape detection. An analogous observation of a GABA relaxing effect was obtained in cerebral arteries of the cat (and two observations in small pial arteries taken during neurosurgical tumour operations on human subjects) when the arteries were contracted by prostaglandin $F_{2\alpha}$ ($PGF_{2\alpha}$) or 5-HT. The dilating effect is also present in the dog middle-cerebral artery but absent in one human extracranial vessel (Edvinsson and Krause, 1979). It was not easy to interpret the relaxation followed by contraction produced by GABA in the isolated dog basilar artery since both effects were blocked by bicuculline and picrotoxin. The incongruity of the experiments on arteries, also in vitro, has been discussed previously (Krause, 1986).

The presence of $GABA_B$ prejunctional receptors in the rabbit ear artery has been reported by Manzini *et al.* (1985a). Receptor activation with GABA or baclofen inhibited neurogenic vasoconstrictor responses (selective $GABA_A$ agonists were inactive; Figure 6.1 and Figure 6.2). The inhibition was suppressed by 'selective' $GABA_B$ antagonists like 5-aminovaleric acid (1 mM) or homotaurine (100 μM). However, the action of GABA was also partially reversed by picrotoxin (100 μM) but the authors suggest this to be due to a direct effect of the drug (picrotoxin) producing an increase in the amplitude of stimulation-induced contractions. Negative findings on the relaxing effects of GABA and related drugs on cerebral and mesenteric arteries contracted by $PGF_{2\alpha}$ or K^+ have recently been reported in monkeys, rabbits and rats (Lai *et al.*, 1988).

Transmural field stimulation of the carefully isolated artery at various frequencies, together with measurement of [^{3}H]norepinephrine (noradrenaline) release, may be a better technique for studying peripheral neuronal mechanisms

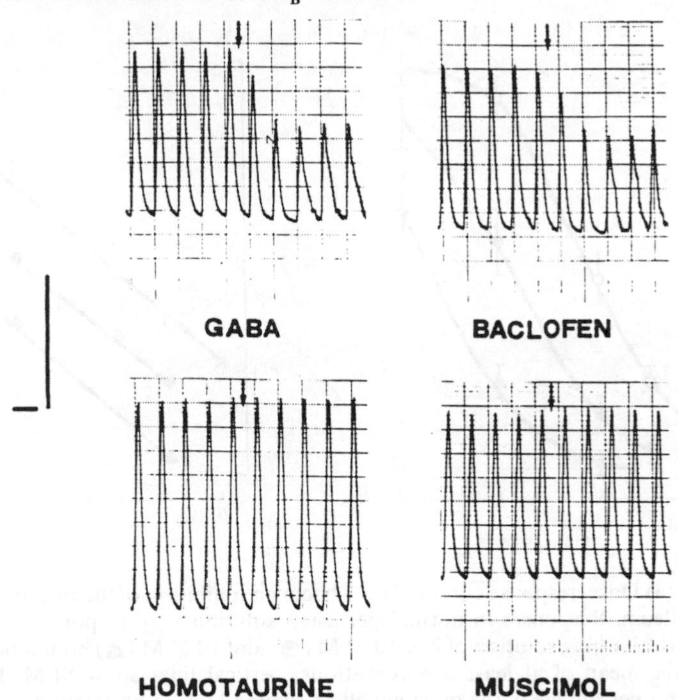

Figure 6.1. Typical tracings showing the effect of GABA (100 μM), baclofen (100 μM), homotaurine (100 μM) and muscimol (100 μM) on field-stimulation (5 Hz) -induced vasoconstrictions of isolated perfused rabbit ear artery. Each drug was intraluminally administered at the arrow. Records were obtained in the same arterial segment at 30 min intervals. Calibration bars: horizontal = 1 min; vertical = 40 mmHg. (For demonstration of the GABA$_B$ antagonistic property of homotaurine, Figure 6.2.) (After Manzini *et al.*, 1985a, with permission.)

at the arterial level. During the preparation of the arteries, consideration has to be given to the presence of many sympathetic axons which ramify over the adventitial surface although axons rarely, if ever, penetrate into the smooth muscle layer. Thus the adventitia must be carefully treated during surgery. As pointed out by Hirst and Edwards (1989), 'almost all mammalian arteries and arterioles are surrounded by sympathetic adrenergic nerve axons. Exceptions are the aorta of some species, some fine pial vessels and the pulmonary arteries of certain species'.

Thus careful experiments with arteries from different vascular beds of the same species and between species should clarify whether the GABA$_B$ mechanisms at the arterial level are a generally diffuse system such as the sympathetic system or if the development of GABAergic control is the natural response to the physiologically different requirements of different vascular beds such as the cerebral circulation and arteries directly implicated in homeothermic regulation

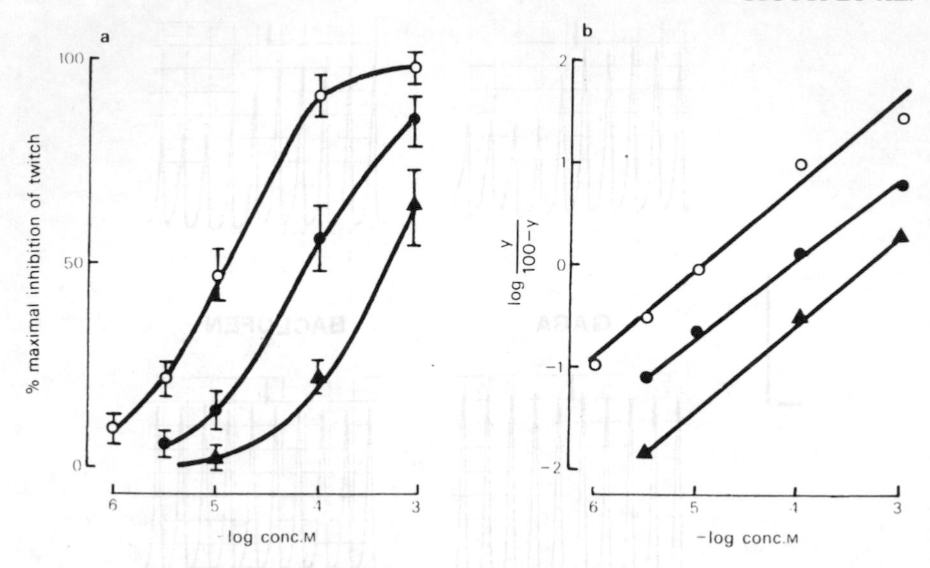

Figure 6.2. (a) Dose–response curves of (−)-baclofen inhibition of the twitch response in guinea-pig ileum. Responses in normal perfusion solution (○); responses 30 min after addition to the bathing solution of 2×10^{-4} M (●) and 10^{-3} M (▲) homotaurine. Each symbol is the mean of at least 5 observations; vertical lines show SEM. Effects are presented as a percentage of the maximal effect elicited by (−)-baclofen; mean maximal inhibition was 29.8. ± 4.3%. (b) Clark plot of homotaurine-induced antagonism on the GABA dose–response curve; symbols are the same as in (a). (After Giotti *et al.*, 1983, with permission.)

(and other local special requirements). Some of the inhibitory $GABA_B$ mechanisms thus far reported have indeed come from such types of arteries (this is even true for the rabbit ear artery, through which the animal regulates thermal dispersion from the practically 'nude' ears).

$GABA_B$ Effects on Respiration

To our knowledge only very few studies have attempted to ascertain whether $GABA_B$ mechanisms control respiration through the peripheral systems. The effects of GABA on neuronally mediated contractions of trachealis smooth muscle (Tamaoki *et al.*, 1987) and on acetylcholine release from the lung (Shirakawa *et al.*, 1987) have been demonstrated.

Coburn (1987) points out that there are large differences between species, particularly in the non-adrenergic, non-cholinergic inhibitory nervous system. In the guinea-pig as well as in preparations taken from humans, field-stimulated relaxations that are not due to release of norepinephrine have been reported, whereas none have been demonstrated in the dog (see Coburn, 1987). Tamaoki *et al.* (1987), using electrical field-stimulated guinea-pig tracheal rings (in the presence of 10^{-6} M indomethacin), have shown that GABA, baclofen and muscimol all inhibit the evoked contractions. None of the drugs had any effect on the baseline tone of the tracheal rings. GABA and baclofen were equiactive and their effects were antagonized by bicuculline, pretreatment with furosemide or substitution of Cl^-. The non-adrenergic, non-cholinergic inhibitory component was not altered by GABA (doses not specified). Tamaoki *et al.* suggest that, in guinea-pig trachea, baclofen appears to act on the bicuculline-sensitive $GABA_A$ receptors and that GABA inhibits the exocytotic release of acetylcholine from the postganglionic neurons. It would appear that this effect is exerted by activating a Cl^--dependent bicuculline-sensitive GABA receptor.

Shirakawa *et al.* (1987) have studied the influence of GABA on acetylcholine release from thin strips of guinea-pig lung tissue preloaded in vitro with [^{3}H]choline. GABA (EC_{50} 1 μM) evoked a release of [^{3}H]acetylcholine which was abolished by tetrodotoxin, by Ca^{2+}-free medium and by bicuculline (10^{-6} M). The effect of GABA was mimicked by muscimol (4×10^{-7} M) but baclofen, even at 10^{-5} M, did not affect spontaneous [^{3}H]acetylcholine release. The concentration–response curve of muscimol was shifted to the right in the presence of bicuculline (10^{-7} M) and furosemide also blocked the effect of muscimol, whereas diazepam and pentobarbital potentiated it.

The conclusion of Shirakawa *et al.* (1987) was that the $GABA_A$ receptor in the guinea-pig lung is coupled to the Cl^- ionophore, as with $GABA_A$ receptors in other tissues. GABA also inhibited the K^+-evoked release of [^{3}H] acetylcholine in the presence of tetrodotoxin and bicuculline and the effect was mimicked by baclofen. The bicuculline-insensitive $GABA_B$ receptor may be located on the cholinergic nerve terminals and participate in the inhibition of acetylcholine release, as in the case of the intestine. The tracheobronchial tree arises from the ventral wall of the foregut and thus shares some of the anatomical and physiological features of the gastrointestinal tract.

The apparent incongruity of the results for GABA on function and those on release requires further experimentation but it should be noted that the release comes from lung tissue, while the function was studied on trachealis smooth muscle.

The possibility of controlling some types of brochospasm in vivo with baclofen has been reported (Luzzi *et al.*, 1986, 1987). The drug is effective in the range 1–15 μmol/kg in anaphylactic bronchospasm (Figure 6.3) but its action extends also to some cases of non-allergic bronchospasm; it is active against the dyspnoea induced by histamine, but not that induced by acetylcholine. The effect is present also in isolated sensitized guinea-pig lungs. The inefficacy of muscimol in vivo and the stereoselectivity of baclofen ((+)-baclofen is ineffective) suggested the possibility of an effect through $GABA_B$ receptors.

Belvisi *et al.* (1988, 1989) studied the effects of GABA and related substances on non-adrenergic, non-cholinergic neurally evoked bronchoconstriction in the

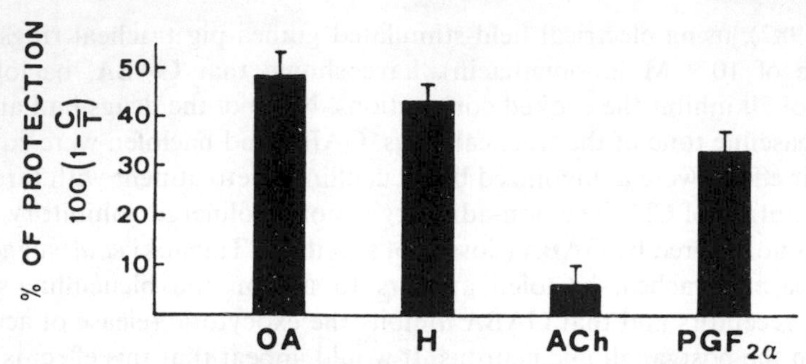

Figure 6.3. Effect of (−)-baclofen (5 μmol/kg) on the bronchospasm induced by an aerosol of histamine (H), acetylcholine (ACh), prostaglandin F_{2a} (PGF_{2a}) and, in sensitized animals, by ovalbumin (OA). The effect of the drug is expressed as percentage of protection (*C* is the preconvulsion time in controls, *T* is the preconvulsion time in treated animals). Each value is the mean of at least 7 observations. (After Luzzi *et al.*, 1987, with permission.)

anaesthetized guinea-pig. This approach, like that of Luzzi *et al.* (1986, 1987), provides a better interpretation of the role of neuropeptides such as substance P and neurokinin A released from intramural sensory nerves in asthma (see Barnes, 1987). The results obtained with GABA and baclofen, THIP (4,5,6,7-tetrahydroisooxazolo[5,4-*c*]pyridin-3-ol) and bicuculline are in line with an inhibitory effect of GABA on the release of transmitter(s) from non-adrenergic, non-cholinergic nerves via an action on $GABA_B$ receptors. The possibility that GABA might play a role in the regulation of physiological and pathological, but not classically mediated, neurogenic responses, which was suggested by these pioneer studies, is provoking great interest in pharmaceutical laboratories. $GABA_B$ peripheral agonists are indeed promising drugs as therapeutic agents in the respiratory field.

$GABA_B$ Effects at the Intestinal Level

Most studies on the intestine concern GABA effects on mechanical function (see Kerr and Ong, 1986) but in the exhaustive review by Cooke (1989), despite the impressive list of ganglionic receptors regulating intestinal secretions (cholinergic, muscarinic, nicotinic, serotonergic, substance P, neurotensin, histamine, norepinephrine, somatostatin, δ-enkephalin), GABA receptors were not mentioned.

Kerr and Ong (1984) reported on the actions of GABA in guinea-pig ileum obtained through the *release* of endogenous GABA by ethylenediamine (EDA). In isolated ileal preparations maintained in Krebs–bicarbonate solution, EDA induced a dose-dependent, transient, cholinergic contractile response (a $GABA_A$ receptor-mediated effect) followed by an after-relaxation (a $GABA_B$ receptor-mediated effect). When tested on repetitive twitch responses to electrical transmural stimulation of the cholinergic neurons, EDA induced a transient superimposed contraction followed by a *depression* of the twitch contractions (Kerr and Ong, 1984).

The $GABA_B$ receptor-mediated after-relaxation and the depression of cholinergic twitch contractions were susceptible to antagonism by 5-aminovaleric acid. Interestingly the pA_2 (according to Schild, 1947) value for bicuculline methochloride antagonism of EDA $GABA_A$-ergic effects was identical to that for GABA. 3-Mercaptopropionic acid (3-MPA), which prevents GABA release from brain cortical slices (Fan *et al.*, 1981), not only prevents the EDA-induced release of [^{3}H]GABA from preloaded myenteric plexus of the guinea-pig ileum but also prevents EDA-induced GABAergic responses in the ileum without affecting those elicited by GABA or its analogues. EDA acts as the EDA monocarbamate formed through the rapid reaction of EDA with bicarbonate-buffered Krebs solution. Preformed EDA monocarbamate is a direct releaser of GABA (Kerr and Ong, 1987). These results are consistent with a role for endogenous GABA in intestinal motility. Interest in these findings is increased by the knowledge that EDA is a specific releasing agent of GABA. In rat striatal slices it releases GABA but not dopamine (Lloyd *et al.*, 1982).

The use of uptake inhibitors of GABA to potentiate GABA-induced (injected or released) contractile responses is another fascinating means by which to follow the perireceptor fate of GABA. The initial observation that the glial cells (of rat sensory ganglia) possess a high-affinity uptake system for [^{3}H]GABA which differs from that present in nerve terminals (in the brain) (Iversen and Kelly, 1975; Schon and Kelly, 1974a,b, 1975) has been fruitfully extended even to the possible therapeutic application of such selective antagonists. Krogsgaard-Larsen *et al.* (1988) stress that 'while substrate inhibitors of neuronal GABA uptake appear to be proconvulsant or convulsant, compounds acting as selective substrate inhibitors of glial uptake system have anticonvulsant effects'. Recent studies (Ong, 1987) using nipecotic acid and other inhibitors confirm that GABA uptake limits GABA-mediated responses also in the intestine. Unfortunately (for $GABA_B$ reviewers!) the work was based on the transient contraction, sensitive to atropine, bicuculline methochloride (10 μM) and tetrodotoxin ($GABA_A$ effect); the delayed, prolonged, bicuculline-insensitive 'after-relaxation' was not included in the investigation.

Ong and Kerr (1987) compared the responses induced by EDA or GABA and its analogues in the duodenum, jejunum and ileum, as well as the distal colon of the guinea-pig. Only the ileum showed significant $GABA_A$ receptor-mediated contractile responses, whereas a $GABA_B$ receptor-mediated relaxation predominated in other segments of the intestine. The release of [^{3}H]GABA induced by EDA was greatest in the distal colon and ileum, with less in the jejunum and duodenum (Ong and Kerr, 1987). The authors point out that, in keeping with their earlier findings in the ileum, the [^{3}H]GABA released from the jejunum, duodenum and distal colon by EDA is very likely of neuronal origin, being Ca^{2+}-dependent and prevented by 3-MPA. The differential effects of $GABA_A$ or

$GABA_B$ agonists and antagonists suggested that 'a $GABA_B$ receptor-mediated depression of transmitter output is a common feature on the excitatory neurons that innervate the smooth muscles throughout the entire intestine as Giotti *et al.* (1985) have proposed in guinea-pig colon' (Ong and Kerr, 1987). Baclofen- and GABA-induced depression of responses to transmural stimulation in the guinea-pig isolated ileum is Ca^{2+}-dependent over the range 0.6–2.4 mM (Ong *et al.*, 1987). The depressive responses are potentiated in the presence of Ruthenium Red (0.1 μM), which prevents transmitter release by binding to the sites closely related to voltage-sensitive Ca^{2+}-channels in nerve terminals (Tapia *et al.*, 1985).

Most studies on GABAergic mechanisms at the intestinal level have used guinea-pig preparations. However, the possibility of using the rat instead has been clearly shown (Krantis and Harding, 1987). From the experiments on longitudinal isolated organ preparations it clearly emerged that GABAergic actions in the rat small intestine are mediated by GABA sites which, in those studies using classical pharmacological 'dissection' methods, appear to be both $GABA_A$ and $GABA_B$. Their distribution and sensitivity differ along the length of the small intestine. $GABA_A$ receptor agonists also induce non-adrenergic, non-cholinergic effects (see later for other reports on this type of action).

The cat is another species in which studies on GABA have contributed fundamentally to the concept of GABA neurotransmission in the intestine. In the early eighties Taniyama *et al.* (1982, 1983b) used the isolated cat colon to demonstrate a frequency (1–10 Hz)-dependent, tetrodotoxin-sensitive, Ca^{2+}-sensitive, maximum-contraction-related [^{3}H]GABA efflux. In recent pharmacological studies a parallelism was shown between GABA-evoked acetylcholine release and GABA-evoked rhythmic contractions of circular muscle; both effects were abolished by bicuculline (10^{-5} M) and furosemide (10^{-6} M) as well as by tetrodotoxin (3×10^{-7} M) or Ca^{2+}-free medium (acetylcholine release was measured; contractility of course was absent). Scopolamine (10^{-6} M) suppressed the mechanical effect but not the acetylcholine release. Hexamethonium was not effective (Taniyama *et al.*, 1987). In all these studies $GABA_B$-ergic mechanisms were ignored.

The presence of substance P neuronal endings within the spinal cord, peripherally in the myenteric plexus and in axons and collaterals of the sensory neurons, together with the demonstration that substance P provokes GABA release from the myenteric plexus of the guinea-pig small intestine (Tanaka and Taniyama, 1985), has suggested two other possible sites (centrally and peripherally) for $GABA_A$ and $GABA_B$ effects. These authors obtained direct evidence that substance P produces a neuronal release of GABA through its receptor located in the guinea-pig small intestine and that the released GABA evokes a neuronal release of acetylcholine which could be blocked by a selective substance P antagonist (D-Pro^2, D-Trp^{7,9}) and by bicuculline (10^{-6} M) (Tanaka and Taniyama, 1985).

A number of events may be interrelated in substance P–cholinergic–GABA processing mechanisms:

(1) activation of sensory axon collaterals with a substance P-ergic terminal;
(2) physiological or pharmacological release of substance P;
(3) activation of substance P receptors on GABAergic neurons;
(4) GABA release;
(5) activation of $GABA_A$ receptors on postsynaptic membranes of cholinergic neurons:
(6) acetylcholine release;
(7) acetylcholine contraction or increase of rhythmical activity;
(8) $GABA_B$ control of substance P secretion.

However, present evidence only permits the possibility of actions mediated through (1) to (7) of the sequence (Bartho *et al.*, 1982; Bartho and Vizi, 1985; Chahl, 1982; Franco *et al.*, 1979; Morita *et al.*, 1980; Tanaka and Taniyama, 1985; Tonini *et al.*, 1986a): no evidence for $GABA_B$-ergic mechanisms has been demonstrated. In (1) and (2) the presence of $GABA_A$ receptors on substance P terminals might explain the involvement of substance P in the excitatory action of $GABA_A$ agonists on cholinergic neurons in the guinea-pig ileum (Tonini *et al.*, 1987). The presence of bicuculline-sensitive $GABA_A$ receptors on unmyelinated and myelinated sensory peripheral nerve fibres explains (according to Maggi *et al.*, 1989) the activation by GABA of peripheral terminals of capsaicin-sensitive nerves through a membrane depolarization brought about by an increase in Cl^- conductance (Morris *et al.*, 1983; see also Bhisitkul *et al.*, 1987).

But $GABA_A$ and $GABA_B$ receptors appear to coexist on the membrane of primary afferents, as demonstrated by intracellular recording from rat dorsal root ganglion neurons (Desarmenien *et al.*, 1984).

In the sequence (1) to (7) a site should also be reserved for neurotensin, a neuropeptide with multiple locations in the ileum, where scattered nerve fibres exist in the smooth muscle. This peptide evokes the neuronal release of GABA, as well as of acetylcholine, and bicuculline partly inhibits the neurotensin-evoked release of [^{3}H]acetylcholine (Nakamoto *et al.*, 1987).

The substance P-ergic–GABAergic–cholinergic neuronal circuit located in the myenteric plexus, as suggested by the findings of Tanaka and Taniyama (1985), still requires close examination for possible $GABA_B$ actions in its first two segments. GABA, at least in the duodenum, may be involved in the activation of intramural non-adrenergic, non-cholinergic neurons (Maggi *et al.*, 1984a,b). Considerable, although indirect, experimental evidence (Manzini *et al.*, 1985b, 1986) suggests that ATP might be the endogenous substance released by GABA. Relaxation was evoked by GABA, with a shift to the right of the concentration–response curve in the presence of bicuculline and picrotoxin, and was not evoked by (±)-baclofen. Furthermore, cross-desensitization developed between GABA and homotaurine but not between GABA and (±)-baclofen. The GABA receptor involved appears to be of the $GABA_A$ subtype. According to Matusak and Bauer (1986), the results of experiments based mainly on cross-desensitization between the responses to transmural nerve stimulation and putative transmitters suggest a possible transmitter role for the substance P-like peptide in excitatory non-adrenergic, non-cholinergic transmission and a role for 5-HT in the activation of non-adrenergic, non-cholinergic neurons. The necessity of delivering supramaximal high-frequency stimulation (20 Hz for at least 10 min) to desensitize against endogenous 'non-adrenergic, non-cholinergic mediators', and the need for histamine to increase the basal tension to obtain responses to neurotransmitters, may introduce bias into this type of experiment.

Matusak and Bauer (1986) report that 'GABA even in a high concentration (300 μM) did not evoke significant changes of the basal tension increased by histamine and of the non-adrenergic non-cholinergic responses elicited by TNS [transmural nerve stimulation]'. The complexity of the neurotransmitter release underlying the experimental situation and the possible role of $GABA_A$-ergic mechanisms in ileal non-adrenergic, non-cholinergic inhibition suggest that studies on $GABA_B$-ergic mechanisms at this level should await basic clarification. To stress the underlying uncertainties, we recall the demonstration of muscarinic M_1 receptors which, when activated, cause the relaxation of the isolated rat duodenum, jejunum and ileum through a release of GABA (Micheletti *et al.*, 1988). The release of GABA has been invoked because the effects are not observed in the presence of bicuculline. Furthermore, α-adrenoceptors also mediate regulation of GABA release from myenteric neurons of the guinea-pig small intestine (Hashimoto *et al.*, 1986). This group of findings suggests the need to study non-adrenergic, non-cholinergic transmission only in the presence of agents selectively blocking the adrenergic and cholinergic terminals or the effector cells without interrupting M_1 cholinergic and α_2-adrenergic steps in the circuit in which GABA release is involved.

The apparent controversy concerning the interaction between 5-HT (a substance which fulfils many of the criteria for a neurotransmitter in Auerbach's plexus; Wood and Mayer, 1978) and intestinal GABAergic mechanisms (Ong and Kerr, 1985; Tonini *et al.*, 1983) seems to have found a solution (Tonini *et al.*, 1986b). 5-HT is not involved in the development of a GABA excitatory response but may possibly operate through the depression of a 5-HT-sensitive synapse in non-cholinergic transmission in the myenteric plexus to which the GABAergic mechanism in cholinergic activation is supposed to be connected. It remains clear (see Ong and Kerr, 1985) that both GABA and 5-HT induce neurally mediated cholinergic contractile responses and that desensitization to 5-HT does not consistently depress contractile responses to GABA. We have no information on whether, in the 5-HT-activated circuit, $GABA_B$ functionally operating receptors are present at sites different from those of the cholinergic pathway.

The pharmacological tools which enable the GABA receptor subtypes involved in simple functional responses to be separated ($GABA_A$ receptors mediate prompt dose-dependent contractions of the ileum; $GABA_B$ receptors mediate depression of the electrical twitch contraction of the ileum) have provided the possibility of studying GABAergic mechanisms in more integrated tests in vitro and in vivo. Krantis and Kerr (1981) and Ong and Kerr (1983) investigated the possible involvement of GABA in intestinal motility in the guinea-pig, using the speed of propulsion of faecal pellets along isolated segments of distal colon. Combined antagonism of $GABA_A$ receptors and desensitization to baclofen slowed pellet propulsion to the same extent as GABA desensitization alone, indicating that both $GABA_A$ and $GABA_B$ receptor sites are involved in the modification of peristalsis by GABA (Krantis and Kerr, 1981; Ong and Kerr, 1983).

A modulatory activity of $GABA_B$ receptors on cholinergic tone in guinea-pig distal colon has been demonstrated in vitro and in vivo (Giotti *et al.*, 1985; see Figure 6.4). The maximal relaxant effects of GABA and (−)-baclofen did not

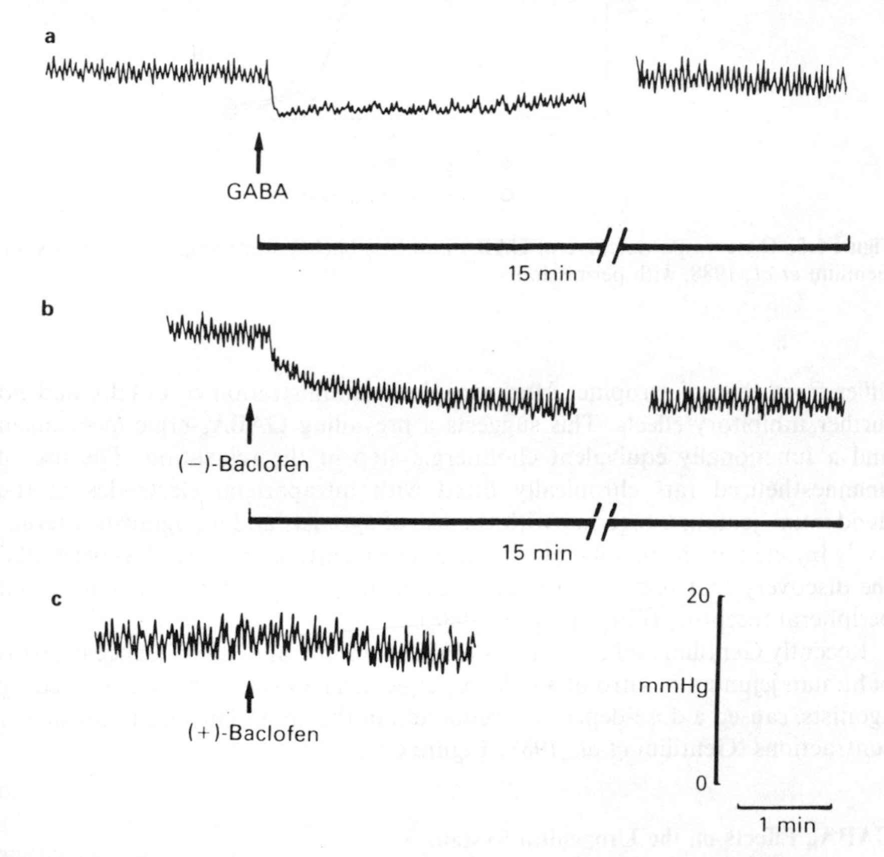

Figure 6.4. Effects of GABA, (−)-baclofen and (+)-baclofen on the resting tone of guinea-pig distal colon i.v. (a) Effect of GABA (10 μmol/kg). (b) Effect of (−)-baclofen (5 μmol/kg). (c) Effect of (+)-baclofen (5 μmol/kg). Tracings (a), (b) and (c) were recorded in three different animals. (After Giotti *et al.*, 1985, with permission.)

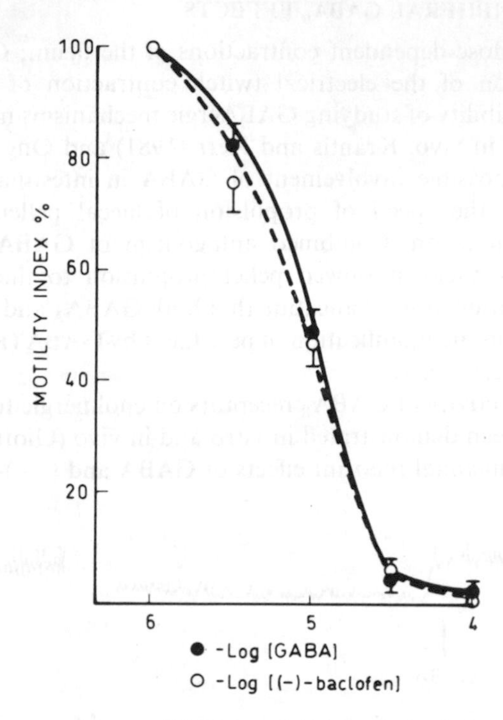

Figure 6.5. Dose–response curve of GABA and (−)-baclofen in human jejunum. (After Gentilini *et al.*, 1988, with permission.)

differ from that of atropine. After atropine, administration of GABA had no further inhibitory effects. This suggests a prevailing $GABA_B$-ergic mechanism and a functionally equivalent cholinergic step in the relaxation. The use of unanaesthetized rats chronically fitted with intraparietal electrodes in the duodenum–jejunum, together with the use of agonists and antagonists alternatively injected by the peripheral or intracerebroventricular routes, has permitted the discovery that baclofen stimulates duodenal motility through central and peripheral receptors (Fargeas *et al.*, 1988).

Recently Gentilini and co-workers (1988) demonstrated the presence in strips of human jejunum in vitro of a $GABA_B$-ergic mechanism which, if activated by agonists, caused a dose-dependent reduction in the amplitude and frequency of contractions (Gentilini *et al.*, 1988; Figure 6.5).

$GABA_B$ Effects on the Urogenital System

In their review article on GABAergic mechanisms and their functional relevance in the urinary bladder Taniyama and Tanaka (1986) conclude that:

GABAergic neurons may be included among the inhibitory systems in the vesical ganglia of the urinary bladder. The $GABA_B$ receptor is linked in the inhibition of bladder motility via a reduction in the release of acetylcholine. The $GABA_B$ receptor may also be involved in the inhibitory effect of GABA via a reduction in the release of acetylcholine and non-cholinergic excitatory neurotransmitters.

Evidence has indeed been presented for the existence in the urinary bladder of guinea-pigs, rabbits and newborn rats (Maggi *et al.*, 1983, 1984b, 1985a,b,c; Santicioli *et al.*, 1984) of a prejunctional $GABA_B$ receptor, activation of which reduced the cholinergic component of postganglionic excitatory neurotransmission to the detrusor muscle. In the guinea-pig isolated bladder, activation of the $GABA_A$ receptor seems also to be involved in regulating bladder motility by modulating both cholinergic and non-cholinergic, non-adrenergic components of the postganglionic excitatory innervation (Kusunoki *et al.*, 1984; Maggi *et al.*, 1985a,b; Taniyama *et al.*, 1983a).

Activation of prejunctional $GABA_B$ receptors inhibits the amplitude of field-stimulated contractions of the mouse urinary bladder (Santicioli *et al.*, 1986). The authors suggest that $GABA_B$ receptor activation operates through a reduction of neurotransmitter release from postganglionic nerve endings. This conclusion results from the following observations: (a) the selective $GABA_B$ receptor agonist, ($\pm$)-baclofen, mimicked the effects of GABA, whereas the $GABA_A$ receptor agonist, homotaurine, did not; (b) the $GABA_B$ receptor antagonists, homotaurine and 5-aminovaleric acid, antagonized the effect of GABA, while picrotoxin did not; (c) GABA had no inhibitory effect on the contractile response induced by electrical stimulation of tetrodotoxin-treated preparations using parameters which produce a direct excitation of smooth muscle cells.

The effects of baclofen have also been studied in vivo. Sillen *et al.* (1985), using cystometric recordings in pentobarbitone-anaesthetized rats, observed an inhibition of the L-dopa-induced hyperactivity (of urinary bladder) after i.c.v. (0.1 μg) and i.v. (10 mg/kg) administration of baclofen. Baclofen appears to depress the hyperactive bladder by a central action which is bicuculline-insensitive and unrelated to substance P or glutamate neurotransmission but possibly related to an interaction with opioid mechanisms (Sillen *et al.*, 1985).

Maggi *et al.* (1987b) investigated the effects of baclofen on micturition reflexes through the transurethral cystometrogram in urethane-anaesthetized rats and concluded that the drug may act on $GABA_B$ receptors to modulate excitatory neurotransmission in pelvic ganglia. Investigation by the same group (Maggi *et al.*, 1987a) on the effects of GABA, picrotoxin and bicuculline methiodide on the voiding cycle of the urinary bladder in urethane-anaesthetized guinea-pigs showed a potentiation of the efficiency of the voiding cycle after i.v. or topical application of the GABA antagonists.

In a double-blind crossover clinical trial with baclofen (5 mg q.d.s. versus placebo) a significant improvement was noted in diurnal and nocturnal frequency of micturition and in the severity of incontinence (Taylor and Bates, 1979).

$GABA_B$ mechanisms in the female genital system have been the object of relatively numerous investigations, apparently inspired by the 'high' concentration of GABA in the rat ovary and oviduct (see Del Rio *et al.*, 1986; Erdö, 1986) and the possible physiological significance of an extraneuronal extraglial compartment of the GABA synthesizing–storing–releasing system, which is sensitive to endocrine regulation (Celotti *et al.*, 1989). In rabbit uterine strips, GABA and baclofen induced frequent contractions with increased amplitude and strongly elevated basal tone. Cross-desensitization between two stimulants occurred and tetrodotoxin did not prevent the action of the GABA agonists (Riesz and Erdö, 1985; see also Erdö *et al.*, 1984).

The dose of bicuculline used in these experiments (10^{-4} M) was too high for studying selective antagonism: indeed it 'evoked strong contractions of spontaneously active specimens'.

The extraneuronal location of uterine $GABA_B$ receptors and their possible location on smooth muscle cells would indeed offer another interesting possibility for studying the cellular mechanisms of $GABA_B$ effects.

CONCLUSION

In recent years peripheral GABAergic mechanisms have been extensively studied in laboratory animals and also in humans.

A relatively large number of studies (see Table 1) have been performed on isolated preparations taken from the respiratory, cardiovascular, gastrointestinal and urogenital systems.

A common experimental procedure has been to observe the effects of drugs related to GABAergic mechanisms in isolated preparations of smooth muscle cells maintaining parts of neural peripheral plexuses, spontaneously active or subjected to chemical, electrical or mechanical (reflex) stimulation.

The combined use of GABA in the presence of low concentrations of bicuculline and/or picrotoxin and of (−)-baclofen has revealed effects which, if antagonized by $GABA_B$ receptor antagonists, have been reasonably defined as $GABA_B$ effects. Usually the $GABA_B$ peripheral effects have been interpreted as the consequence of a depression of the neurotransmitter function of nerve terminals. In very few studies have $GABA_B$ effects been mimicked in the experimental situation in which a release of 'physiological' GABA has been evoked.

The extension of similar studies using preparations in which the sensory transmission can also be examined (like the 'inferior mesenteric ganglion attached to a segment of distal colon in guinea pigs in vitro' as described by Kreulen and Peters (1986)) is desirable.

Table 1. $GABA_B$-ergic peripheral effects

Effects	Reference
(1) Hypotension, bradycardia during general anaesthesia in humans	Sill *et al.* (1986)
(2) Depression of the evoked release of [^{3}H]norepinephrine from rat isolated atria	Bowery *et al.* (1981)
(3) Reduction of the electrically evoked contraction of the basilar artery of rabbit	Anwar and Mason (1982)
(4) Inhibition of vasoconstrictor responses in rabbit ear arteries	Manzini *et al.* (1985a)
(5) Inhibition of acetylcholine release by high K^+ in strips of guinea-pig lungs	Shirakawa *et al.* (1987)
(6) Inhibition of anaphylactic or histamine bronchospasm in guinea-pig in vivo and in vitro	Luzzi *et al.* (1986, 1987)
(7) Inhibition of non-adrenergic, non-cholinergic neuronally evoked bronchoconstriction in the anaesthetized guinea-pig	Belvisi et al. (1988, 1989)
(8) EDA-evoked after-relaxation and depression of cholinergic twitch contraction in guinea-pig intestinal isolated preparations	Kerr and Ong (1984); Ong and Kerr (1987)
(9) Mechanical effects on isolated intestinal preparations of the rat	Krantis and Harding (1987)
(10) Inhibition of the speed of faecal pellet propulsion along isolated segments of guinea-pig distal colon	Krantis and Kerr (1981); Ong and Kerr (1983)
(11) Modulatory activity on cholinergic tone in guinea-pig distal colon in vitro and in vivo	Giotti *et al.* (1985)
(12) Reduction in the amplitude and frequency of contractions in strips of human jejunum	Gentilini *et al.* (1988)
(13) Reduction of the cholinergic component of the postganglionic excitatory transmission to the mouse detrusor muscle	Santicioli *et al.* (1986)
(14) Inhibition of the field-stimulation-induced contraction in mouse urinary bladder	Santicioli *et al.* (1986)
(15) Improvement of diurnal and nocturnal frequency of micturition and of the severity of incontinence in humans	Taylor and Bates (1979)

EDA, ethylenediamine.

Studies with functionally organized preparations could reveal whether a GABAergic system in the periphery is able to modulate cholinergic; noradrenergic; non-adrenergic, non-cholinergic; or other physiological systems as preliminary studies suggest. This will require the development of more potent and selective peripheral $GABA_B$ antagonists.

ACKNOWLEDGEMENTS

The research performed in the Department of Pharmacology, University of Florence and cited in this article was partially funded by grants from the Italian National Research Council and Ministry of Public Instruction.

REFERENCES

Anwar, N. and Mason, D.F.J. (1982) Two actions of γ-aminobutyric acid on the responses of the isolated basilar artery from the rabbit. *Br. J. Pharmacol.*, **75**, 177–181.

Barnes, P.J. (1987) Airway neuropeptides and asthma. *Trends Pharmacol. Sci.*, **8**, 24–27.

Bartho, L. and Vizi, S.E. (1985) Neurochemical evidence for the release of acetylcholine from the guinea-pig ileum myenteric plexus by capsaicin. *Eur. J. Pharmacol.*, **110**, 125–127.

Bartho, L., Holzer, P., Lembeck, F., Szolcsanyi, J. (1982) Evidence that the contractile response of the guinea-pig ileum to capsaicin is due to release of substance P. *J. Physiol.*, **332**, 157–167.

Belvisi, M.G., Ichinose, M. and Barnes, P.J. (1988) Evidence for $GABA_B$ receptors on non-adrenergic non-cholinergic nerves in guinea-pig airways. *Br. J. Pharmacol.*, **95** (Suppl.), 776P.

Belvisi, M.G., Ichinose, M. and Barnes, P.J. (1989) Modulation of non-adrenergic, non-cholinergic neural bronchoconstriction in guinea-pig airways via $GABA_B$ receptors. *Br. J. Pharmacol.*, **97**, 1225–1231.

Bhisitkul, R.B., Villa, J.E and Kocsis, J.D. (1987) Axonal GABA receptors are selectively present on normal and regenerated sensory fibres in rat peripheral nerve. *Exp. Brain Res*, **66**, 659–663.

Bowery, N.G. (1983) Classification of GABA receptors. In: S.J. Enna (ed.) *The GABA receptors.* Clifton, NJ: Human Press, pp. 177–213.

Bowery, N.G. and Hudson, A.L. (1979) Gamma-aminobutyric acid reduces the evoked release of ^{3}H-noradrenaline from sympathetic nerve terminals. *Br. J. Pharmacol.*, **66**, 108P.

Bowery, N.G., Hill, D.R. and Hudson, A.L. (1980) (−)Baclofen decreases neurotransmitter release in the mammalian CNS by an action at a novel GABA receptor. *Nature*, **283**, 92–94.

Bowery, N.G., Doble, A., Hill, D.R., Hudson, A.L., Shaw, J.S., Turnbull, M.J. and Warrington, R. (1981) Bicuculline-insensitive GABA receptors on peripheral autonomic nerve terminals. *Eur. J. Pharmacol.*, **71**, 53–70.

Bowery, N.G., Hill, D.R. and Hudson, A.L. (1983) Characteristics of $GABA_B$ receptor binding sites on rat whole brain synaptic membranes. *Br. J. Pharmacol.*, **78**, 191–206.

Bowery, N.G., Hill, D.R. and Moratalla, R. (1989) Neurochemistry and autoradiography of $GABA_B$ receptors in mammalian brain: second messenger system(s). In E.A. Barnard and E. Costa (eds) *Allosteric modulation of amino acid receptors: therapeutic implications.* New York: Raven Press, pp. 159–172.

Celotti, F., Moore, G.D., Rovescalli, A.C., Solimena, M., Negri-Cesi, P. and Racagni, G. (1989) The GABAergic system in rat oviduct: presence, endocrine regulation and possible localization. *Pharmacol. Res. Commun.*, **21**, 103–104.

Chahl, L.A. (1982) Evidence that the contractile response of the guinea-pig ileum to capsaicin is due to substance P release. *Naunyn-Schmiedeberg's Arch. Pharmacol.*, **319**, 212–215.

Cheng, S.C. and Brunner, E.A. (1985) Inducing anaesthesia with a GABA analog, THIP. *Anesthesiology*, **63**, 147–151.

Coburn, R.F. (1987) Peripheral airway ganglia. *Annu. Rev. Physiol.*, **49**, 573–582.

Cooke, H.J. (1989) Role of the 'little brain' in the gut in water and electrolyte homeostasis. *Faseb. J.*, **3**, 127–318.

Cutting, D.A. and Jordan, C.C. (1975) Alternative approaches to analgesia: baclofen as a model compound. *Br. J. Pharmacol.*, **54**, 171–179.

De Feudis, F.V. (1983) Involvement of gamma-aminobutyric acid in cardiovascular regulation. *Trends Pharmacol. Sci.*, **4**, 356–358.

Del Rio, R.M., Orensanz, L.M. and Fernandez, I. (1986) The GABA system in the ovary. In: S.L. Erdö and N.G. Bowery (eds) *GABAergic mechanisms in the mammalian periphery*. New York: Raven Press, pp. 241–247.

Desarmenien, M., Feltz, P., Occhipinti, G., Santangelo, F. and Schlichter, R. (1984) Co-existence of $GABA_A$ and $GABA_B$ receptors on Aδ and C primary afferents. *Br. J. Pharmacol.*, **81**, 327–333.

Edvinsson, L. and Krause, D.N. (1979) Pharmacological characterization of GABA receptors mediating vasodilation of cerebral arteries in vitro. *Brain Res.*, **173**, 89–97.

Erdö, S.L. (1985) Peripheral GABAergic mechanisms. *Trends Pharmacol Sci.*, **6**, 205–208.

Erdö, S.L. (1986) GABAergic mechanisms and their possible role in the oviduct and the uterus. In: S.L. Erdö and N.G. Bowery (eds) *GABAergic mechanisms in the mammalian periphery*. New York: Raven Press, pp. 205–222.

Erdö, S.L. and Bowery, N.G. (eds) (1986) *GABAergic mechanisms in the mammalian periphery*. New York: Raven Press.

Erdö, S.L., Reisz, M., Karpati, E. and Szporny, L. (1984) $GABA_B$ receptor-mediated stimulation of the contractility of isolated rabbit oviduct. *Eur. J. Pharmacol.*, **99**, 333–336.

Fan, S.G., Wusteman, M. and Iversen L.L. (1981) 3-Mercaptopropionic acid inhibits GABA release from rat brain slices in vitro. *Brain Res.*, **229**, 371–377.

Fargeas, M.J., Fioramonti, J. and Bueno, L. (1988) Central and peripheral action of $GABA_A$ and $GABA_B$ agonists on small intestine motility in rats. *Eur. J. Pharmacol.*, **150**, 163–169.

Franco, R., Costa, M. and Furness, J.B. (1979) Evidence that axons containing substance P in the guinea-pig ileum are of intrinsic origin. *Naunyn-Schmiedeberg's Arch. Pharmacol.*, **307**, 57–63.

Fujiwara, M., Muramatsu, I. and Shibata, S. (1975) γ-Aminobutyric acid receptor on vascular smooth muscle of dog cerebral arteries. *Br. J. Pharmacol.*, **55**, 561–562.

Gentilini, G., Luzzi, S., Franchi-Micheli, S., Pantalone, D., Cortesini, C. and Zilletti, L. (1988) Effect of gamma-aminobutyric acid on human jejunum 'in vitro'. *Pharmacol. Res. Commun.*, **20**, 423–424.

Giotti, A. and Giusti, G. (1948) L'associazone cardiazolo-picrotossina nella pratica tossicologica e in medicina interna. *Riv. Clin. Med.*, **48**, 353–364.

Giotti, A., Luzzi, S., Shagnesi, S. and Zilletti, L. (1983) Homotaurine: a $GABA_B$ antagonist in guinea-pig ileum. *Br. J. Pharmacol.*, **79**, 855–862.

Giotti, A., Luzzi, S., Maggi, C.A., Spagnesi, S. and Zilletti, L. (1985) Modulatory activity of $GABA_B$ receptors on cholinergic tone in guinea-pig distal colon. *Br. J. Pharmacol.*, **84**, 883–895.

Giuliani, S., Maggi, C.A. and Meli, A. (1986) Differences in cardiovascular responses to peripherally administered GABA as influenced by basal conditions and type of anaesthesia. *Br. J. Pharmacol.*, **88**, 659–670.

Hashimoto, S., Tanaka, C. and Taniyama, K. (1986) Presynaptic muscarinic and α-adrenoceptor-mediated regulation of GABA release from myenteric neurones of the guinea-pig small intestine. *Br. J. Pharmacol.*, **89**, 787–792.

Hill, D.R. and Bowery, N.G. (1981) ^{3}H-Baclofen and ^{3}H-GABA bind to bicuculline-insensitive $GABA_B$ sites in rat brain. *Nature*, **290**, 149–152.

Hirst, G.D.S. and Edwards, F.R. (1989) Sympathetic neuroeffector transmission in arteries and arterioles. *Physiol. Rev.*, **69**, 546–595.

Iversen, L.L. and Kelly, J.S. (1975) Uptake and metabolism of γ-aminobutyric acid by neurones and glial cells. *Biochem. Pharmacol.*, **24**, 933–938.

Kerr, D.I.B. and Ong, L. (1984) Evidence that ethylenediamine acts in the isolated ileum of the guinea-pig by releasing endogenous GABA. *Br. J. Pharmacol.*, **83**, 169–177.

Kerr, D.I.B. and Ong, J. (1986) GABAergic mechanisms in the gut: their role in the regulation of gut motility. In: S.L. Erdö and N.G. Bowery (eds) *GABAergic mechanisms in the mammalian periphery*. New York: Raven Press, pp. 153–174.

Kerr, D.I.B. and Ong, J. (1987) Bicarbonate-dependence of responses to ethylenediamine in the guinea-pig isolated ileum: involvement of ethylenediamine-monocarbamate. *Br. J. Pharmacol.*, **90**, 763–769.

Krantis, A. and Harding, R.K. (1987) GABA-related actions in isolated in vitro preparations of the rat small intestine. *Eur. J. Pharmacol.*, **141**, 291–298.

Krantis, A. and Kerr, D.I.B. (1981) The effects of GABA antagonism on propulsive activity of the guinea-pig large intestine. *Eur. J. Pharmacol.*, **76**, 111–114.

Krause, D.N. (1986) 'Involvement of local GABA mechanisms in vascular regulation'. In: S.L. Erdö and N.G. Bowery (eds) *GABAergic mechanisms in the mammalian periphery*. New York: Raven Press, pp. 193–203.

Kreulen, D.L. and Peters, S. (1986) Non-cholinergic transmission in a sympathetic ganglion of the guinea-pig elicited by colon distension. *J. Physiol.*, **374**, 315–334.

Krogsgaard-Larsen, P., Hjeds, H., Falch, E., Jorgensen, F.S. and Nielsen, L. (1988) Recent advances in GABA agonists, antagonists and uptake inhibitors: structure-activity relationships and therapeutic potential. *Adv. Drug. Res.*, **17**, 382–456.

Kusunoki, M., Taniyama, K. and Tanaka, C. (1984) Neuronal GABA release and GABA inhibition of ACh release in guinea-pig urinary bladder. *Am J. Physiol.*, **246**, R502–R509.

Lai, F.M., Tanikella, T. and Cervoni, P. (1988) Effect of γ-aminobutyric acid (GABA) on vasodilation in resistance-sized arteries isolated from the monkey, rabbit and rat. *J. Cardiovasc. Pharmacol.*, **12**, 372–376.

Lloyd, H.G.E., Perkins, M.N. and Stone, T.W. (1982) Ethylenediamine as a specific releasing agent of gamma-aminobutyric acid in rat striatal slices. *J. Neurochem.*, **38**, 1168–1169.

Luzzi, S., Franchi-Micheli, S., Ciuffi, M. and Zilletti, L. (1986) Effects of GABA agonists on Herxheimer microshock in guinea-pigs. *Agents Actions*, **18**, 245–247.

Luzzi, S., Franchi-Micheli, S., Folco, G., Ciuffi, M. and Zilletti, L. (1987) Effect of baclofen on different models of bronchial hyperreactivity in the guinea-pig. *Agents Actions*, **20**, 307–309.

Maggi, C.A. and Meli, A. (1986) Suitability of urethane anaesthesia for physiopharmacological investigations in various systems: 2. Cardiovascular system. *Experientia*, **12**, 292–297.

Maggi, C.A., Santicioli, P., Grimaldi, G. and Meli, A. (1983) The effect of peripherally administered GABA on spontaneous contractions of the rat urinary bladder in vivo. *Gen. Pharmacol.*, **14**, 455–460.

Maggi, C.A., Manzini, S. and Meli, A. (1984a) Evidence that $GABA_A$ receptors mediate relaxation of rat duodenum by activating intramural, non-adrenergic, non-cholinergic neurones. *J. Auton. Pharmacol.*, **4**, 77–85.

Maggi, C.A., Santicioli, P. and Meli, A. (1984b) $GABA_B$ receptor mediated inhibition of field stimulation induced contractions of detrusor strips from newborn rats. *J. Auton. Pharmacol.*, **4**, 45–51.

Maggi, C.A., Santicioli, P. and Meli, A. (1985a) $GABA_A$ and $GABA_B$ receptors in detrusor strips from guinea-pig bladder dome. *J. Auton. Pharmacol.*, **5**, 55–64.

Maggi, C.A., Santicioli, P. and Meli, A. (1985b) Dual effect of GABA on the contractile activity of the guinea-pig isolated urinary bladder. *J. Auton. Pharmacol.*, **5**, 131–141.

Maggi, C.A., Santicioli, P. and Meli, A. (1985c) Pharmacological evidence for the existence of two components in the twitch response to field stimulation of detrusor strips from the rat urinary bladder. *J. Auton. Pharmacol.*, **5**, 221–230.

Maggi, C.A., Meli, A. and Santicioli, P. (1987a) Neuroeffector mechanisms in the voiding cycle of the guinea-pig urinary bladder. *J. Auton. Pharmacol.*, **7**, 295–308.

Maggi, C.A., Santicioli, P., Giuliani, S., Furio, M., Conte, B., Meli, P., Gragnani, L. and Meli, A. (1987b) The effects of baclofen on spinal and supraspinal micturition reflexes in rats. *Naunyn-Schmiedeberg's Arch. Pharmacol.*, **336**, 197–203.

Maggi, C.A., Giuliani, S., Manzini, S. and Meli, A. (1989) $GABA_A$ receptor-mediated positive inotropism in guinea-pig isolated left atria: evidence for the involvement of capsaicin-sensitive nerves. *Br. J. Pharmacol.*, **97**, 103–110.

Maloney, A.H. (1933) A comparative study of the antidotal action of picrotoxin, strychnine and cocaine in acute intoxication by the barbiturates. *J. Pharmacol. Exp. Ther.*, **49**, 133–140.

Manzini, S., Maggi, C.A. and Meli, A. (1985a) Inhibitory effect of GABA on sympathetic neurotransmission in rabbit ear artery. *Arch. Int. Pharmacodyn.*, **273**, 100–109.

Manzini, S., Maggi, C.A. and Meli, A. (1985b) Further evidence for involvement of adenosine-5-triphosphate in non-adrenergic non-cholinergic relaxation of the isolated rat duodenum. *Eur. J. Pharmacol.*, **113**, 399–408.

Manzini, S., Maggi, C.A. and Meli, A. (1986) Pharmacological evidence that at least two different non-adrenergic non-cholinergic inhibitory systems are present in the rat small intestine. *Eur. J. Pharmacol.*, **123**, 229–236.

Matusak, O. and Bauer, V. (1986) Effect of desensitization induced by adenosine 5-triphosphate, substance P, bradykinin, serotonin, gamma-aminobutyric acid and endogenous non-cholinergic, non-adrenergic transmitter in the guinea-pig ileum. *Eur. J. Pharmacol.*, **126**, 199–209.

Micheletti, R., Schiavone, A. and Giachetti, A. (1988) Muscarinic M1 receptors stimulate a non-adrenergic, non-cholinergic inhibitory pathway in the isolated rat duodenum. *J. Pharmacol. Exp. Ther.*, **244**, 680–684.

Morita, K., North, R.A. and Katayama, Y. (1980) Evidence that substance P is a neurotransmitter in the myenteric plexus. *Nature*, **287**, 151–152.

Morris, M.E., Di Costanzo, G.A., Fox, S. and Werman, R. (1983) Depolarizing action of GABA on myelinated fibres of peripheral nerves. *Brain Res.*, **278**, 117–126.

Nakamoto, M., Tanaka, C. and Taniyama, K. (1987) Release of gamma-aminobutyric acid and acetylcholine by neurotensin in guinea-pig ileum. *Br. J. Pharmacol.*, **90**, 545–551.

Nicoll, R.A. and Dutar, P. (1989) Physiological roles of $GABA_A$ and $GABA_B$ receptors in synaptic transmission in the hippocampus. In: E.A. Barnard and E. Costa (eds) *Allosteric modulation of amino acid receptors: therapeutic implications*. New York: Raven Press, pp. 195–205.

Ong, J. (1987) Uptake inhibitors potentiate γ-aminobutyric acid-induced contractile responses in the isolated ileum of the guinea-pig. *Br. J. Pharmacol.*, **91**, 9–15.

Ong, J. and Kerr, D.I.B. (1983) $GABA_A$ and $GABA_B$ receptor mediated modification of intestinal motility. *Eur. J. Pharmacol.*, **86**, 9–17.

Ong, J. and Kerr, D.I.B. (1985) Evidence that 5-hydroxytryptamine does not mediate GABA-induced contractile responses in the guinea-pig proximal ileum. *Eur. J. Pharmacol.*, **106**, 665–668.

Ong, J. and Kerr, D.I.B. (1987) Comparison of GABA-induced responses in various segments of the guinea-pig intestine. *Eur. J. Pharmacol.*, **134**, 349–353.

Ong, J., Kerr, D.I.B. and Johnston, G.A.R. (1987) Calcium dependence of baclofen- and

GABA-induced depression of responses to transmural stimulation in the guinea-pig isolated ileum. *Eur. J. Pharmacol.*, **134**, 369–372.

Riesz, M. and Erdö, S.L. (1985) $GABA_B$ receptors in the rabbit uterus may mediate contractile responses. *Eur. J. Pharmacol.*, **119**, 199–204.

Santicioli, P., Maggi, C.A. and Meli, A. (1984) The $GABA_B$ receptor mediated inhibition of field stimulation induced contractions of rabbit bladder muscle in vitro. *J. Pharm. Pharmacol.*, **36**, 378–381.

Santicioli, P., Maggi, C.A. and Meli, A. (1986) The postganglionic excitatory innervation of the mouse urinary bladder and its modulation by prejunctional $GABA_B$ receptors. *J. Auton. Pharmacol.*, **6**, 53–66.

Schild, H.O. (1947) pA, a new scale for the measurement of drug antagonism. *Br. J. Pharmacol.*, **2**, 189–206.

Schon, F. and Kelly, J.S. (1974a) Autoradiographic localization of [^{3}H]GABA and [^{3}H]glutamate over satellite glial cells. *Brain Res.*, **66**, 275–288.

Schon, F. and Kelly, J.S. (1974b) The characterization of [^{3}H]GABA uptake into the satellite glial cells of the sensory ganglia. *Brain Res.*, **66**, 289–300.

Schon, F. and Kelly, J.S. (1975) Selective uptake of [^{3}H]beta-alanine by glia: association with the glial uptake system for GABA. *Brain Res.*, **86**, 243–257.

Shirakawa, J., Taniyama, K. and Tanaka, C. (1987) Gamma-aminobutyric acid-induced modulation of acetylcholine release from the guinea-pig lung. *J. Pharmacol. Exp. Ther.*, **243**, 364–369.

Sill, J.C., Schumacher, K., Southorn, P.A., Reuter, J. and Yaksh, T.L. (1986) Bradycardia and hypotension associated with baclofen used during general anaesthesia. *Anesthesiology*, **64**, 255–258.

Sillen, V., Persson, B. and Rubenson, A. (1985) Central effects of baclofen on the L-dopa induced hyperactive urinary bladder of the rat. *Naunyn-Schmiedeberg's Arch. Pharmacol.*, **330**, 175–178.

Tamaoki, J., Graf, P.D. and Nadel, J. A. (1987) Effect of γ-aminobutyric acid on neurally mediated contraction of guinea-pig trachealis smooth muscle. *J. Pharmacol. Exp. Ther.*, **243**, 86–90.

Tanaka, C. and Taniyama, K. (1985) Substance P provoked γ-aminobutyric acid release from the myenteric plexus of the guinea-pig small intestine. *J. Physiol.*, **362**, 319–329.

Taniyama, K. and Tanaka, C. (1986) GABAergic mechanisms and their functional relevance in the urinary bladder. In: S.L. Erdö and N.G. Bowery (eds) *GABAergic mechanisms in the mammalian periphery*. New York: Raven Press, pp. 175–183.

Taniyama, K., Kusunoki, M., Saito, N. and Tanaka, C. (1982) Release of γ-aminobutyric acid from cat colon. *Science.*, **217**, 1038–1040.

Taniyama, K., Kusunoki, M. and Tanaka, C. (1983a) γ-aminobutyric acid inhibits motility of the isolated guinea-pig urinary bladder. *Eur. J. Pharmacol.*, **89**, 163–166.

Taniyama, K., Miki, Y., Kusunoki, M., Saito, N. and Tanaka, C. (1983b) Release of endogenous and labelled GABA from isolated guinea-pig ileum. *Am. J. Physiol.*, **245**, G717–G721.

Taniyama, K., Saito, N., Miki, Y. and Tanaka, C. (1987) Enteric γ-aminobutyric acid-containing neurons and the relevance to motility of the cat colon. *Gastroenterology*, **93**, 519–525.

Tapia, R., Arias, C. and Morales, E. (1985) Binding of lanthanum ions and ruthenium red to synaptosomes and its effects on neurotransmitter release. *J. Neurochem.*, **45**, 1464–1470.

Taylor, M.C. and Bates, C.P. (1979) A double-blind crossover trial of baclofen: a new treatment for the unstable bladder syndrome. *Br. J. Urol.*, **51**, 504–505.

Tonini, M., Onori, L., Lecchini, S., Frigo, G.M., Perucca, E., Saltarelli, P. and Crema, A.

(1983) 5-HT mediated GABA excitatory responses in the guinea-pig proximal ileum. *Naunyn-Schmiedeberg's Arch. Pharmacol.*, **324**, 180–184.

Tonini, M., Onori, L., Rizzi, C.A., De Ponti, F., Perucca, E., Budassi, P. and Crema, A. (1986a) Inhibitory action of capsaicin on cholinergic responses induced by $GABA_A$ agonists in the guinea-pig ileum. *Eur. J. Pharmacol.*, **128**, 273–276.

Tonini, M., Saltarelli, P., Onori, L., De Ponti, F., Garlaschelli, L., Rizzi, C.A. and Crema, A. (1986b) A re-appraisal of the mode of action of 5-HT in inhibiting GABA-induced cholinergic contractions in the guinea-pig ileum. *Eur. J. Pharmacol.*, **127**, 267–270.

Tonini, M., Onori, L., Rizzi, C.A., Perucca, E., Manzo, L. and Crema, A. (1987) Involvement of substance P in the excitatory action of $GABA_A$ agonists on cholinergic neurons in the guinea-pig ileum. *Naunyn-Schmiedeberg's Arch. Pharmacol.*, **335**, 629–635.

Wilson, P.R. and Yaksh, T.L. (1978) Baclofen is antinociceptive in the spinal intrathecal space of animals. *Eur. J. Pharmacol.*, **51**, 323–330.

Wojcik, W.J., Paez, X. and Ulivi, M. (1989) A transduction mechanism for $GABA_B$ receptors. In: E.A. Barnard and E. Costa (eds) *Allosteric modulation of amino acid receptor: therapeutic implications.* New York: Raven Press, pp. 173–193.

Wood, J.D. and Mayer, C.J. (1978) Slow synaptic excitation mediated by serotonin in Auerbach's plexus. *Nature*, **276**, 836–837.

Yaksh, T.L. and Reddy, S.V.R. (1981) Studies in the primate on the analgetic effects associated with intrathecal actions of opiates, alpha-adrenergic agonists and baclofen. *Anesthesiology*, **54**, 451–467.

Part IV

BIOCHEMICAL ASPECTS

7 Evidence for the Existence of $GABA_B$ Receptor Subpopulations in Mammalian Central Nervous System

E. W. KARBON, S. H. ZORN,
R. J. NEWLAND and S. J. ENNA

INTRODUCTION

A common feature of many neurotransmitter substances is their ability to interact with multiple receptors. For example, norepinephrine (noradrenaline) activates both α- and β-adrenoceptors, each of which may be subdivided further on the basis of functional properties and pharmacological selectivity (Lefkowitz and Caron, 1988). Similarly, acetylcholine acts upon nicotinic and as many as five different muscarinic receptors (Bonner *et al.*, 1987). The existence of receptor subtypes allows for a limited number of neurotransmitter and neuromodulatory substances to regulate an extensive array of cellular processes.

Amino-acid neurotransmitters such as GABA and glutamic acid also interact with multiple receptors. For GABA, there appear to be at least two receptor subtypes, $GABA_A$ and $GABA_B$ (Enna and Karbon, 1986; Hill and Bowery, 1981). The $GABA_A$ sites mediate changes in chloride ion conductance, are activated by muscimol and THIP (4,5,6,7-tetrahydroisoxazolo [5,4-*c*] pyridin-3-ol), and inhibited by bicuculline and picrotoxin. In contrast, $GABA_B$ receptors are bicuculline- and THIP-insensitive, and are stereoselectively activated by β-*p*-chlorophenyl-GABA (baclofen). Indeed, as described in the present report, there are now data suggesting a multiplicity of pharmacologically distinct $GABA_B$ receptors.

PROPERTIES OF $GABA_B$ RECEPTORS

The $GABA_B$ receptor was initially proposed when it was found that GABA inhibited neurotransmitter release from peripheral tissues such as atria, and attenuated electrically induced smooth muscle contractions in a bicuculline-insensitive manner (Bowery *et al.*, 1980, 1981; Kaplita *et al.*, 1982). Moreover, these effects are mimicked by baclofen, but not by THIP. While the physiological relevance of peripheral $GABA_B$ receptors is unclear since most organs do not

$GABA_B$ Receptors in Mammalian Function. Edited by N.G. Bowery, H. Bittiger and H.-R. Olpe

Table 1. Central and peripheral effects of baclofen

Centrally-mediated actions
- antinociceptive
- antispastic
- sedative
Effects on organ systems
- reduces intestinal motility in vitro
- inhibits airway smooth muscle contractility in vitro
- blocks vagally mediated bronchoconstriction in vivo
- reduces uterine and bladder contractions in vitro

possess GABAergic neurons, these studies pointed to the possible existence of a population of GABA receptors that differed pharmacologically from the classical GABA binding site. Shortly after their discovery in peripheral tissues, $GABA_B$ receptors were identified in the central nervous system (CNS) when baclofen was shown to inhibit potassium-stimulated neurotransmitter release from brain slices (Bowery *et al.*, 1980; Schlicker *et al.*, 1984). Clinically, baclofen is employed as an antispastic agent in the treatment of multiple sclerosis (Table 1). Baclofen has also been reported to display analgesic properties (see Zorn and Enna, 1985), although it sedates at doses that significantly elevate the pain threshold.

Both in vivo and in vitro, baclofen displays a number of effects on various organ systems (Table 1). The ability of baclofen to inhibit electrically induced intestinal smooth muscle contractions is particularly interesting inasmuch as GABA is highly concentrated in enteric neurons that synapse upon acetylcholine-containing cells, regulating acetylcholine release (Tanaka, 1985). Both baclofen and GABA inhibit electrically evoked contractions of isolated guinea-pig trachea, and attenuate vagally mediated bronchoconstriction in anesthetized guinea-pigs (Belvisi *et al.*, 1989; Tamaoki *et al.*, 1987). These findings suggest that selective $GABA_B$ receptor agents might by useful for treating certain pulmonary or gastrointestinal disorders, as well as for modifying CNS activity. The variety of effects produced by baclofen raises the question as to whether they are mediated by a single population of receptors, or by pharmacologically distinct subpopulations of $GABA_B$ sites.

MULTIPLICITY OF $GABA_B$ RECEPTORS

Following the identification of $GABA_B$ receptors, $GABA_B$ binding sites were characterized in brain membranes using radioligand binding assays (Hill and Bowery, 1981; Bowery *et al.*, 1983; Karbon *et al.*, 1983). These studies revealed that [^{3}H]GABA and [^{3}H]($\pm$)-baclofen recognized both low- and high-affinity $GABA_B$ sites. Whereas the ratio of low- to high-affinity $GABA_B$ binding does not differ substantially among rat brain regions, destruction of the dorsal noradrenergic bundle selectively reduces the number of lower affinity GABA$_B$ binding sites (Karbon *et al.*, 1983), suggesting kinetically distinct populations of GABA$_B$ receptors.

Additional evidence favoring the existence of multiple GABA$_B$ receptor subpopulations was obtained from studies aimed at determining the effector mechanism(s) associated with these sites (Table 2). In rat brain membranes, baclofen inhibits basal and forskolin-stimulated adenylate cyclase activity, decreasing cyclic AMP formation (Wojcik and Neff, 1984). Thus, GABA$_B$ receptors appear similar to α_2-adrenergic, muscarinic cholinergic, and adenosine A_1, which reduce adenylyl cyclase activity by coupling with G_i, the inhibitory guanine nucleotide binding protein (Drummond, 1984). A distinguishing characteristic of G-protein-coupled receptors is that agonist binding affinity is reduced in the presence of GTP, which attaches to a regulatory site on the α-subunit of the G-protein (Gilman, 1987). As would be predicted from the cyclase data, GABA$_B$ receptor binding is attenuated by GTP, which reduces the affinity of GABA$_B$ recognition sites for the radioligand (Hill *et al.*, 1984).

Additional evidence favoring an association of GABA$_B$ receptors with G_i was provided by the finding that islet-activating protein (IAP, pertussis toxin), which prevents receptor–G_i interactions, prevents baclofen from inhibiting adenylyl cyclase (Xu and Wojcik, 1986). Likewise, treatment of brain membranes with activated IAP inhibits GABA$_B$ receptor binding, an effect that is reversed by the addition of purified G_i (Asano *et al.*, 1985).

While these results suggest that GABA$_B$ receptors are negatively coupled to adenylyl cyclase, studies performed with brain slices suggest that the GABA can increase brain cyclic AMP levels (Table 2). Thus, in many rat brain regions, including cerebral cortex, hippocampus, and striatum, baclofen enhances neurotransmitter-stimulated cyclic AMP accumulation while having no effect on second messenger formation itself (Karbon and Enna, 1985; Karbon *et al.*, 1984). The response to baclofen is restricted to the (−)-isomer, is mimicked by

Table 2. Evidence for multiple GABA$_B$ receptor systems

Binding
– [^{3}H]GABA and [^{3}H]baclofen label both low- and high-affinity GABA$_B$ binding sites
Biochemical
– inhibits adenylyl cyclase activity in brain membranes
– reduces forskolin-stimulated cyclic AMP accumulation in brain slices
– augments neurotransmitter-stimulated cyclic AMP accumulation in brain slices
– attenuates neurotransmitter-stimulated inositol phosphate formation in brain slices
Electrophysiological
– hyperpolarization resulting from increased K^+ conductance
– reduces voltage-sensitive Ca^{2+} conductance

GABA, and is bicuculline-insensitive (Figure 7.1). This augmenting response is observed using a variety of agents to stimulate cyclic AMP production, including isoproterenol, norepinephrine, adenosine, 2-chloroadenosine, and vasoactive intestinal peptide (Karbon and Enna, 1985; Suzdak and Gianutsos, 1986; Watling and Bristow, 1986). While the precise mechanism responsible for the augmentation is unknown, the presence of extracellular calcium ion appears necessary (Enna and Karbon, 1987; Karbon and Enna, 1985).

In contrast to its effect on neurotransmitter-stimulated cyclic AMP accumulation, baclofen inhibits forskolin-stimulated cyclic nucleotide accumulation in cerebral cortical slices (Karbon and Enna, 1985) (Figure 7.2). Therefore in the same tissue preparation, baclofen may either enhance or inhibit cyclic AMP accumulation, depending upon the agent used to stimulate production of the second messenger. Interestingly, like $GABA_B$ sites, activation of α_2-adrenergic receptors causes inhibition of adenylyl cyclase in brain membranes but augments cyclic AMP production in brain slices (Duman *et al.*, 1986). Therefore, it is possible that functionally distinct $GABA_B$ receptors are present in the brain, just as the existence of subpopulations of α_2-adrenergic receptors has been proposed (Bylund and Ray-Preger, 1989).

Recently, GABA and baclofen have been reported to inhibit histamine (H_1) and serotonin (5-hydroxytryptamine; 5-HT_2) receptor-mediated inositol phos-

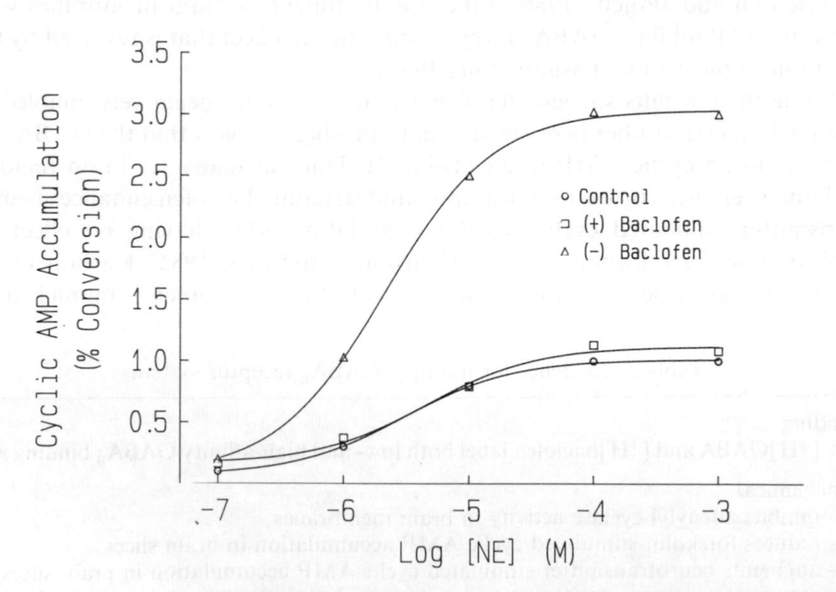

Figure 7.1. Stereoselective effect of baclofen on norepinephrine-stimulated cyclic AMP accumulation in rat cerebral cortical slices. Tissue slices were preincubated in the absence or presence of either (+)-baclofen (100 μM) or (−)-baclofen (100 μM) for 10 min prior to the addition of norepinephrine (NE). Each point represents the mean of three experiments

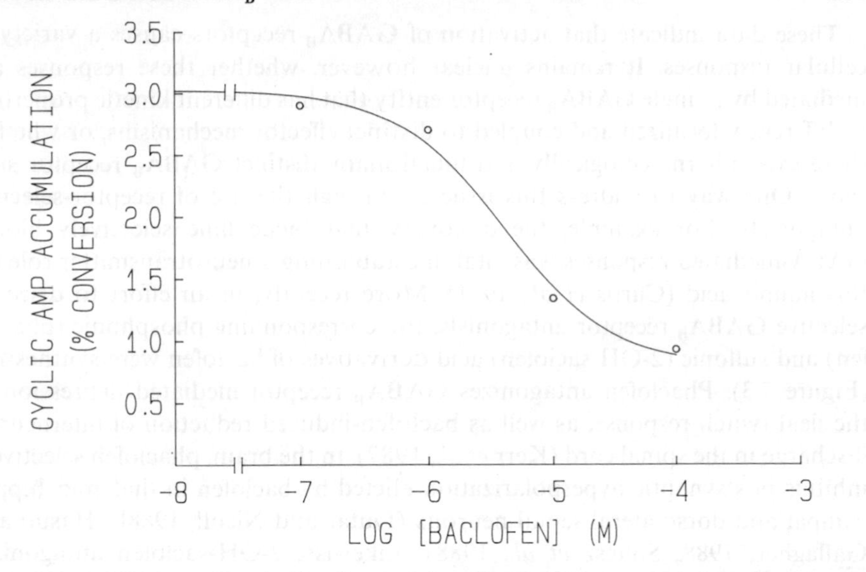

Figure 7.2. Inhibition of forskolin-stimulated cyclic AMP accumulation by baclofen in rat cerebral cortical slices. Tissue slices were preincubated in the presence of varying concentrations of baclofen for 10 min prior to the addition of forskolin (5 μM). Basal per cent conversion was equal to 0.15. Each point represents the mean of three experiments

phate accumulation in slices of rat and mouse cerebral cortex, respectively (Crawford and Young, 1988; Godfrey *et al.*, 1988) (Table 2). Whether this represents a direct effect, or is mediated indirectly as a consequence of changes in cyclic AMP production, is unknown. None the less, these findings provide further evidence supporting a neuromodulatory role for GABA acting through $GABA_B$ receptors, and should be considered when evaluating the possible existence of multiple $GABA_B$ receptor subtypes.

Electrophysiological studies of $GABA_B$ receptors support the existence of multiple $GABA_B$ receptor subtypes (Table 2). In cultured embryonic sensory neurons, GABA and baclofen elicit a bicuculline-insensitive reduction in the duration of the calcium-dependent action potential by decreasing calcium current (Dunlap, 1981). A similar mechanism might also account for the ability of GABA and baclofen to reduce neurotransmitter release from primary afferent terminals.

When applied to rat hippocampal pyramidal cells, baclofen elicits a postsynaptic hyperpolarizing response resulting from an increase in potassium conductance (Gähwiler and Brown, 1985; Newberry and Nicoll, 1984). $GABA_B$ receptors located presynaptically in the hippocampus also inhibit synaptic transmission (Dutar and Nicoll, 1988a). Therefore, it appears that $GABA_B$ receptors are located both pre- and postsynaptically, and influence both calcium and potassium ion conductances.

These data indicate that activation of $GABA_B$ receptors causes a variety of cellular responses. It remains unclear however, whether these responses are mediated by a single $GABA_B$ receptor entity that has different kinetic properties, is differently localized and coupled to distinct effector mechanisms, or whether there exist pharmacologically and functionally distinct $GABA_B$ receptor subtypes. One way to address this issue is through the use of receptor-selective antagonists. For example, the discovery that bicuculline selectively blocks GABA-mediated responses was vital in establishing a neurotransmitter role for this amino acid (Curtis *et al.*, 1971). More recently, in an effort to discover selective $GABA_B$ receptor antagonists, the corresponding phosphonic (phaclofen) and sulfonic (2-OH-saclofen) acid derivatives of baclofen were synthesized (Figure 7.3). Phaclofen antagonizes $GABA_B$ receptor-mediated depression of the ileal twitch response, as well as baclofen-induced reduction of interneuron discharge in the spinal cord (Kerr *et al.*, 1987). In the brain, phaclofen selectively inhibits postsynaptic hyperpolarization elicted by baclofen in thalamic, hippocampal and dorsolateral septal neurons (Dutar and Nicoll, 1988b; Hasuo and Gallagher, 1988; Soltesz *et al.*, 1988). Likewise, 2-OH-saclofen antagonizes GABA- and baclofen-induced depression of electrically stimulated smooth muscle contractions (Kerr *et al.*, 1988). Although the utility of these compounds

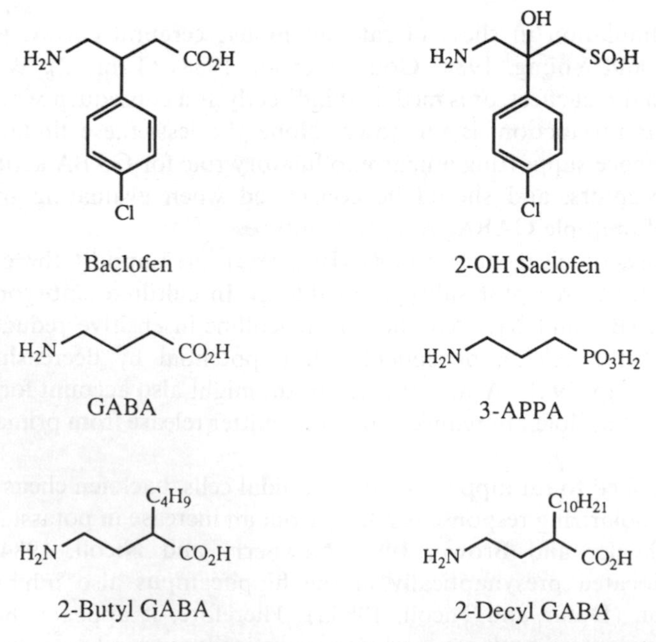

Figure 7.3. Chemical structure of GABA derivatives. 3-APPA, 3-aminopropylphosphonic acid

is limited by their lack of potency ($pA_2 = 4$–5), these studies have contributed to establishing a physiological role for $GABA_B$ receptors.

Using 2-OH-saclofen, efforts were made to determine whether the receptors mediating the inhibitory effect of baclofen on adenylyl cyclase differed from those responsible for augmenting second messenger accumulation in brain tissue. The results of these experiments revealed that 2-OH-saclofen reduces the potency of baclofen to enhance isoproterenol-stimulated cyclic AMP accumulation (Figure 7.4), and blocks the adenylyl cyclase inhibitory response to baclofen (Figure 7.5). This finding suggests that 2-OH-saclofen is incapable of differentiating between these two receptor responses.

Evidence for the existence of pharmacologically distinct $GABA_B$ receptor subtypes was provided by the finding that, like baclofen, 3-aminopropylphosphonic acid (3-APPA) reduced forskolin-stimulated cyclic AMP accumulation, but unlike baclofen, did not enhance catecholamine-stimulated cyclic AMP production (Table 3). The inhibition of the forskolin response by 3-APPA was not additive with baclofen, consistent with the notion that the two amino-acid receptor agonists act at the same site. While these findings suggested that 3-APPA is a selective $GABA_B$ receptor agonist, additional studies revealed that it antagonizes the effect of baclofen on catecholamine-stimulated cyclic AMP

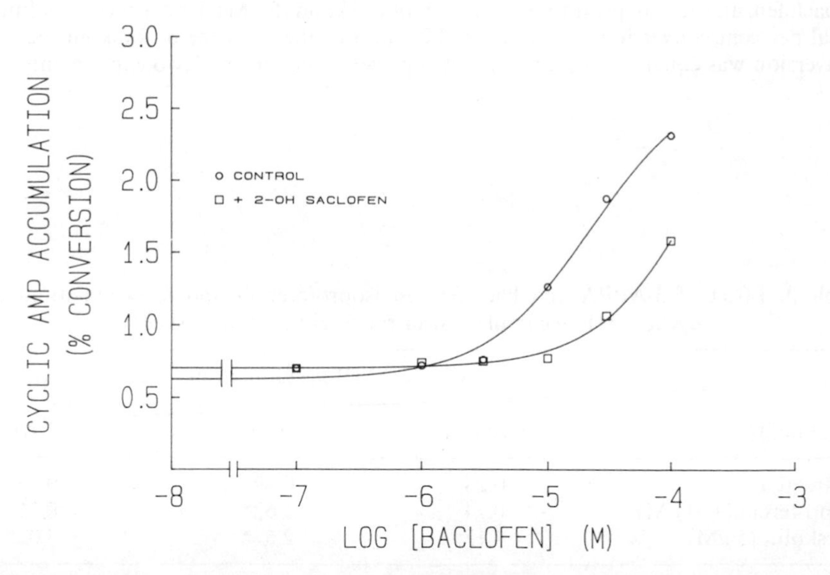

Figure 7.4. Effect of 2-OH-saclofen on baclofen-induced augmentation of isoproterenol-stimulated cyclic AMP accumulation in rat cerebral cortical slices. Tissue slices were pre-exposed to 2-OH-saclofen (100 μM) for 5 min prior to the addition of baclofen, and 15 min prior to the addition of isoproterenol (10 μM). Under these conditions, basal per cent conversion was equal to 0.20, while in the presence of isoproterenol alone per cent conversion was equal to 0.64. Each point represents the mean of two experiments

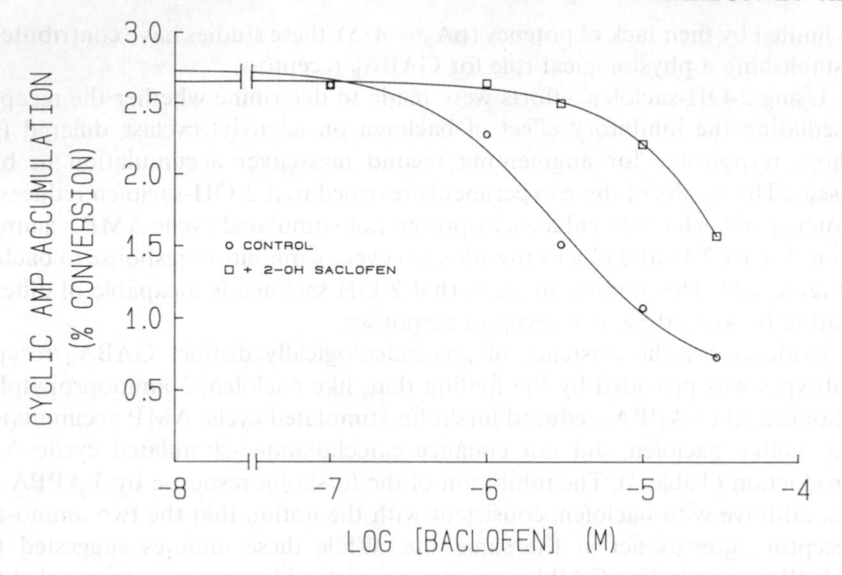

Figure 7.5. Effect of 2-OH-saclofen on baclofen-induced inhibition of forskolin-stimulated cyclic AMP accumulation in rat cerebral cortical slices. Tissue slices were preincubated in the presence of 2-OH-saclofen (100 μM) for 5 min prior to the addition of baclofen, and 15 min prior to the addition of forskolin (5 μM). Under these conditions, basal per cent conversion was equal to 0.22, and in the presence of forskolin per cent conversion was equal to 2.74. Each point represents the mean of two experiments

Table 3. Effect of 3-APPA and baclofen on isoproterenol- and forskolin-stimulated cyclic AMP accumulation in rat cerebral cortical slices

	Cyclic AMP accumulation (% conversion)		
Treatment	Control	+ Baclofen	+ 3-APHA
Untreated	0.28	0.49	0.25
Isoproterenol (10 μM)	0.74	1.62*	0.71
Forskolin (5 μM)	4.17	2.67*	3.02*

Rat cerebral cortical slices were exposed to either 3-APPA (3-aminopropylphosphonic acid; 100 μM) or baclofen (50 μM) for 10 min prior to the addition of either isoproterenol or forskolin. Values represent the mean of at least four determinations.
*Significantly different from the corresponding control ($P < 0.05$, Student's two-tailed t-test). (Adapted from Scherer *et al.*, 1988.)

accumulation (Figure 7.6). Interestingly, 3-APPA has been reported to be an antagonist in guinea-pig airway (Rizzo and Kreutner, 1989). These findings suggest that, in addition to being functionally distinct, the GABA$_B$ receptors associated with forskolin and catecholamine effects on cyclic AMP production are pharmacologically discrete. In addition, it appears that GABA$_B$ receptors located in the CNS may differ from those located in the periphery.

Various GABA derivatives have been tested for their ability to interact with cyclic AMP-generating systems in rat brain slices (Figure 7.7). For example, both 2-butyl- and 2-decyl-GABA inhibit the baclofen augmenting response but, unlike 2-OH-saclofen, 2-decyl-GABA has no effect on forskolin-stimulated cyclic AMP accumulation (Figure 7.8).

While these results appear to support the existence of multiple GABA$_B$ receptor subtypes, it is troubling that the concentrations of these compounds required to activate or inhibit GABA$_B$ sites are quite high. Thus, in the cyclic AMP studies, in general it is necessary to examine concentrations of test compound above 100 μM, enhancing the possibility of observing a non-specific effect. For this reason, alternative approaches have been used to discriminate

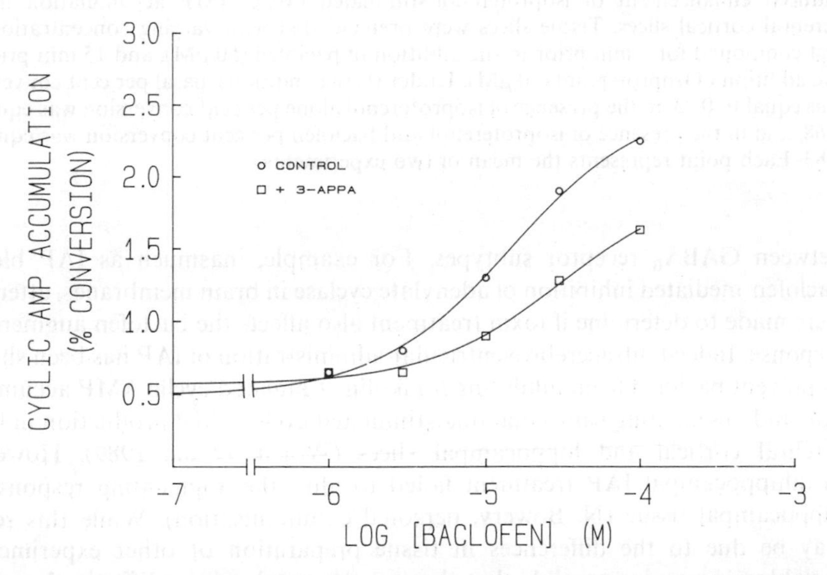

Figure 7.6. Effect of 3-aminopropylphosphonic acid (3-APPA) on baclofen-induced enhancement of isoproterenol-stimulated cyclic AMP accumulation in rat cerebral cortical slices. Tissue slices were pre-exposed to 3-APPA (300 μM) 5 min prior to the addition of baclofen, and 15 min prior to the addition of isoproterenol (10 μM). Under these conditions, basal per cent conversion was equal to 0.18, and in the presence of isoproterenol per cent conversion was equal to 0.50. Each point represents the mean of three experiments

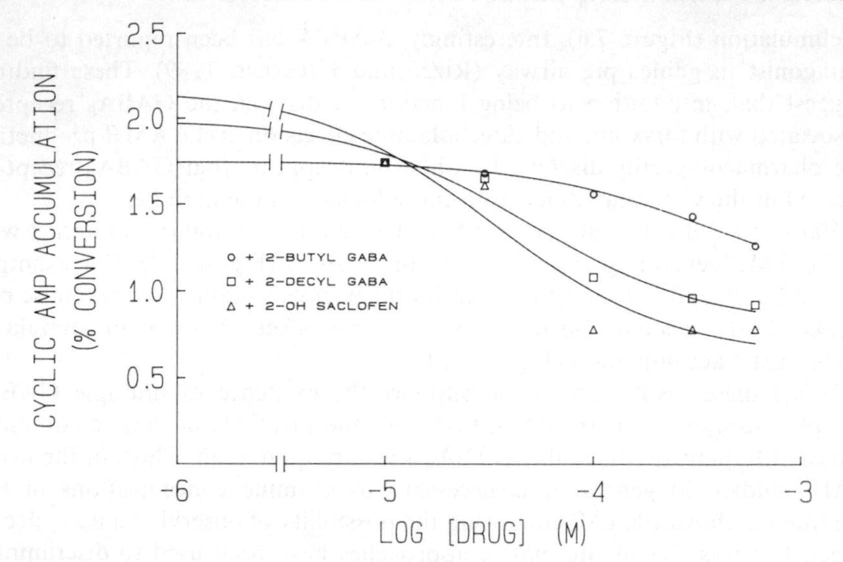

Figure 7.7. Effect of 2-butyl-GABA, 2-decyl-GABA and 2-OH-saclofen on baclofen-induced enhancement of isoproterenol-stimulated cyclic AMP accumulation in rat cerebral cortical slices. Tissue slices were preincubated with varying concentrations of test compound for 5 min prior to the addition of baclofen (50 μM), and 15 min prior to the addition of isoproterenol (10 μM). Under these conditions, basal per cent conversion was equal to 0.25, in the presence of isoproterenol alone per cent conversion was equal to 0.68, and in the presence of isoproterenol and baclofen per cent conversion was equal to 1.83. Each point represents the mean of two experiments

between $GABA_B$ receptor subtypes. For example, inasmuch as IAP blocks baclofen-mediated inhibition of adenylate cyclase in brain membranes, attempts were made to determine if toxin treatment also affects the baclofen augmenting response. Indeed intracerebroventricular administration of IAP has been shown to prevent baclofen from inhibiting forskolin-stimulated cyclic AMP accumulation and augmenting catecholamine-stimulated cyclic AMP production in both cerebral cortical and hippocampal slices (Wojcik *et al.*, 1989). However, intrahippocampal IAP treatment failed to alter the augmenting response in hippocampal tissue (N. Bowery, personal communication). While this result may be due to the differences in tissue preparation or other experimental variables, it remains possible that the receptor mechanisms differ in these two brain regions. In any event, the results fail to prove whether IAP-sensitive G-proteins mediate both the augmenting and the inhibitory responses to baclofen.

Several approaches have been taken to determine whether the pre- and postsynaptic events elicited by baclofen in CA1 hippocampal pyramidal cells are mediated by the same receptor. Phaclofen blocks postsynaptic events, including baclofen-induced hyperpolarizing response and the slow inhibitory postsynaptic

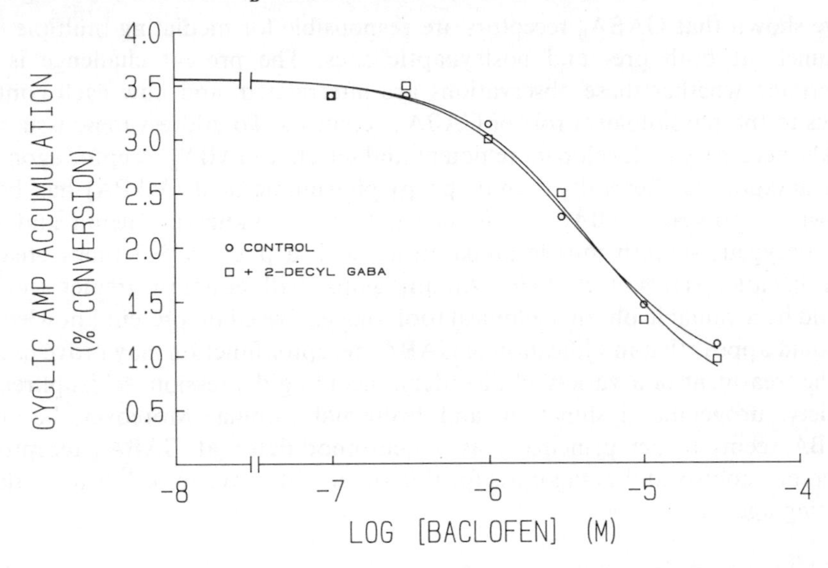

Figure 7.8. Lack of effect of 2-decyl-GABA on baclofen-induced inhibition of forskolin-stimulated cyclic AMP accumulation in rat cerebral cortical slices. Tissue slices were preincubated in the presence of 2-decyl-GABA (100 μM) for 5 min prior to the addition of baclofen, and 15 min prior to the addition of forskolin (5 μM). Under these conditions, basal per cent conversion was equal to 0.20, and in the presence of forskolin per cent conversion was equal to 3.58. Each point represents the mean of two experiments

potential seen following CA1 afferent stimulation (Dutar and Nicoll, 1988b). In contrast, baclofen-induced presynaptic suppression of the excitatory postsynaptic potential elicited by afferent stimulation is phaclofen-insensitive (Dutar and Nicoll, 1988b). Likewise, IAP treatment reduces the postsynaptic, but not the presynaptic, response to baclofen (Dutar and Nicoll, 1988b). These data also support the existence of distinct GABA$_B$ receptors in terms of their pharmacological selectivity and effector coupling mechanisms.

SUMMARY AND CONCLUSIONS

Much has been learned about the pharmacological, biochemical and physiological properties of GABA$_B$ receptors. These studies have demonstrated that GABA plays a neuromodulatory role through its interaction with GABA$_B$ receptors, and suggest that the GABA$_B$ receptor system is complex. The available evidence seems to favor the existence of pharmacologically and functionally distinct GABA$_B$ receptor subpopulations. Thus, receptor binding experiments revealed that [^{3}H]GABA binds to both low- and high-affinity GABA$_B$ sites, biochemical analyses indicate the involvement of GABA$_B$ receptors in a variety of second messenger pathways, and electrophysiological studies have shown that $GABA_B$ receptors are responsible for mediating multiple ion channels at both pre- and postsynaptic sites. The present challenge is to determine whether these observations are interrelated, and how each contributes to the physiological role of $GABA_B$ receptors. To address these issues, it will be necessary to develop more potent and selective $GABA_B$ receptor agonists and antagonists. Recently, 3-aminopropylphosphinic acid (3-APA) has been shown to possess $GABA_B$ agonist-like activity in guinea-pig ileum and rat anococcygeus smooth muscle preparations with a potency 5–7 times greater than baclofen (Hills *et al.*, 1989). An antagonist with equal or greater affinity would be a valuable pharmacological tool. Indeed, based on present knowledge, it would appear that modification of $GABA_B$ receptor function may prove useful in the treatment of a variety of disorders, including depression, schizophrenia, anxiety, urogenital dysfunction, and bronchial asthma. Moreover, because GABA seems to act principally as a neuromodulator at $GABA_B$ receptors, receptor agonists and antagonists for this site might prove to be less toxic than existing agents.

REFERENCES

Asano, T., Ui, M. and Ogasawara, N. (1985) Prevention of the agonist binding to gamma-aminobutyric acid B receptors by guanine nucleotides and islet activating protein, pertussis toxin, in bovine cerebral cortex. *J. Biol. Chem.*, **260**, 12653–12658.

Belvisi, M.G., Ichinose, M. and Barnes, P.J. (1989) Modulation of non-adrenergic, non-cholinergic neural bronchoconstriction in guinea-pig airways via $GABA_B$ receptors. *Br. J. Pharmacol.*, **97**, 1225–1231.

Bonner, T.I., Buckley, N.J., Young, A.C. and Brann, M.R. (1987) Identification of a family of muscarinic acetylcholine receptor genes. *Science*, **237**, 527–532.

Bowery, N.G., Hill, D.R., Hudson, A.L., Doble, A., Middlemiss, D.N., Shaw, J. and Turnbull, M. (1980) (−)Baclofen decreases neutrotransmitter release in the mammalian CNS by an action at a novel GABA receptor. *Nature*, **283**, 92–94.

Bowery, N.G., Hill, D.R. and Hudson, A.L. (1983) Characteristics of $GABA_B$ receptor binding sites on rat whole brain synaptic membranes. *Br. J. Pharmacol.*, **78**, 191–206.

Bylund, D.B. and Ray-Preger, C. (1989) Alpha-2A and alpha-2B adrenergic receptor subtypes: attenuation of cyclic AMP production in cell lines containing only one receptor subtype. *J. Pharmacol. Exp. Ther.*, **251**, 640–644.

Crawford, M.L.A and Young, J.M (1988) $GABA_B$ receptor mediated inhibition of histamine-H_1-induced inositol phosphate formation in slices of rat cerebral cortex. *J. Neurochem.*, **51** 1441–1447.

Curtis, D.R., Duggan, A.W., Felix, D. and Johnston, G.A.R. (1971) Bicuculline, an antagonist of GABA and synaptic inhibition in the spinal cord of the cat. *Brain Res.*, **6**, 1–8.

Drummond, G.I. (ed.) (1984) *Cyclic nucleotides in the nervous system.* New York: Raven Press.

Duman, R.S., Karbon, E.W., Harrington, C. and Enna S.J. (1986) An examination of the involvement of phospholipases A_2 and C in the alpha-adrenergic and gamma-aminobutyric acid receptor modulation of cyclic AMP accumulation in rat brain slices. *J. Neurochem.*, **47**, 800–810.

Dunlap, K. (1981) Two types of gamma-aminobutyric acid receptor on embryonic sensory neurones. *Br. J. Pharmacol.*, **74**, 579-585.

Dutar, P. and Nicoll, R.A. (1988a) Pre- and postsynaptic $GABA_B$ receptors have different pharmacological properties. *Neuron*, **1**, 585-591.

Dutar, P. and Nicoll, R.A. (1988b) A physiological role for $GABA_B$ receptors in the central nervous system. *Nature*, **332**, 156-158.

Enna, S.J. and Karbon, E.W. (1986) GABA receptors: an overview. In J.C. Venter and R.W. Olsen (eds) *Benzodiazepine/GABA receptors and chloride channels: structural and functional properties*. New York: Alan Liss, pp. 41-56.

Enna, S.J. and Karbon, E.W. (1987) Receptor regulation: evidence for a relationship between phospholipid metabolism and neurotransmitter receptor-mediated cAMP formation in brain. *Trends Pharmacol. Sci.*, **8**, 21-25.

Gähwiler, B.H. and Brown, D.A. (1985) $GABA_B$-receptor-activated K^+ current in voltage-clamped CA_3 pyramidal cells in hippocampal cultures. *Proc. Natl Acad. Sci. USA*, **82**, 1558-1562.

Gilman, A.G. (1987) G proteins: transducers of receptor-generated signals. *Annu. Rev. Biochem.*, **56**, 615-649.

Godfrey, P.P., Grahame-Smith, D.G. and Gray, J.A. (1988) $GABA_B$ receptor activation inhibits 5-hydroxytryptamine-stimulated inositol phospholipid turnover in mouse cerebral cortex. *Eur. J. Pharmacol.*, **152**, 185-188.

Hasuo, H. and Gallagher, J.P. (1988) Comparison of antagonism by phaclofen of baclofen induced hyperpolarizations and synaptically mediated late hyperpolarizing potentials recorded intracellularly from rat dorsolateral septal neurons. *Neurosci. Lett.*, **86**, 77-81.

Hill, D.R. and Bowery, N.G. (1981) ^{3}H-Baclofen and ^{3}H-GABA bind to bicuculline-insensitive $GABA_B$ sites in rat brain membranes. *Nature*, **290**, 149-152.

Hill, D.R., Bowery, N.G and Hudson, A.L. (1984) Inhibition of $GABA_B$ receptor binding by guanyl nucleotide. *J. Neurochem.*, **42**, 652-657.

Hills, J.M., Dingsdale, R.A., Parsons, M.E., Dolle, R.E. and Howson, W. (1989) 3-Aminopropylphosphinic acid: a potent, selective $GABA_B$ receptor agonist in the guinea pig ileum and rat anococcygeus muscle. *Br. J. Pharmacol.*, **97**, 1292-1296.

Kaplita, P.V., Waters, D.H. and Triggle, D.J. (1982) Gamma-aminobutyric acid action in guinea pig ileal myenteric plexus. *Eur. J. Pharmacol.*, **79**, 43-51.

Karbon, E.W. and Enna, S.J. (1985) Characterization of the relationship between gamma-aminobutyric acid B agonists and transmitter-coupled cyclic mucleotide generating systems in rat brain. *Mol. Pharmacol.*, **27**, 53-59.

Karbon, E.W., Duman, R.S. and Enna, S.J. (1983) Biochemical identification of multiple $GABA_B$ binding sites: association with noradrenergic terminals in rat forebrain. *Brain Res.*, **274**, 393-396.

Karbon, E.W., Duman, R.S. and Enna, S.J. (1984) $GABA_B$ receptors and norepinephrine-stimulated cAMP production in rat brain cortex. *Brain Res.*, **306**, 327-332.

Kerr, D.I.B., Ong, J., Prager, R.H., Gynther, B.D. and Curtis, D.R. (1987) Phaclofen: a peripheral and central baclofen antagonist. *Brain Res.*, **405**, 150-154.

Kerr, D.I.B., Ong, J., Johnston, G.A.R., Abbenante, J. and Prager, R.H. (1988) 2-Hydroxysaclofen: an improved antagonist at central and peripheral $GABA_B$ receptors. *Neurosci. Lett.*, **92**, 92-96.

Lefkowitz, R.J. and Caron, M. (1988) Adrenergic receptors. *Adv. Sec. Mess. Phosphopro. Res.*, **21**, 1-10.

Newberry, N.R. and Nicoll, R.A. (1984) Baclofen directly hyperpolarizes hippocampal cells. *Nature*, **308**, 450-452.

Rizzo, C. and Kreutner, W. (1989) Prejunctional GABA-B inhibition of electrically-stimulated contraction in isolated guinea pig trachea. *FASEB J.*, **3**, 1241.

Scherer, R.W., Ferkany, J.W. and Enna, S.J. (1988) Evidence for pharmacologically distinct subsets of $GABA_B$ receptors. *Brain Res. Bull.*, **21**, 439–443.
Schlicker, E., Classen, K. and Gothert, M. (1984) $GABA_B$ receptor-mediated inhibition of serotonin release in the rat brain. *Naunyn-Schmiedeberg's Arch. Pharmacol.*, **326**, 99–105.
Soltesz, I., Haby, M., Leresche, N. and Crunelli, V. (1988) The $GABA_B$ antagonist phaclofen inhibits the late K^+-dependent IPSP in cat and rat thalamic and hippocampal neurones. *Brain Res.*, **448** 351–354.
Suzdak, P.D. and Gianutsos, G. (1986) Effect of chronic imipramine or baclofen on GABA-B binding and cyclic AMP production in cerebral cortex. *Eur. J. Pharmacol.*, **131**, 129–133.
Tamaoki, J., Graf, P.D. and Nadel, J.A. (1987) Effect of gamma-aminobutyric acid on neurally mediated contraction of guinea pig trachealis smooth muscle. *J. Pharmacol. Exp. Ther.*, **243**, 86–90.
Tanaka, C. (1985) Gamma-aminobutyric acid peripheral tissues. *Life Sci.*, **37**, 2221–2235.
Watling, K.J. and Bristow, D.R. (1986) $GABA_B$ receptor-mediated enhancement of vasoactive intestinal peptide-stimulated cyclic AMP production in slices of rat cerebral cortex. *J. Neurochem.*, **46**, 1755–1762.
Wojcik, W.J. and Neff, N.H. (1984) Gamma-aminobutyric acid B receptors are negatively coupled to adenylate cyclase in brain, and in the cerebellum these receptors are associated with granule cells. *Mol. Pharmacol.*, **25**, 24–28.
Wojcik, W.J., Ulivi, M., Paez, X. and Costa, E. (1989) Islet activating protein inhibits the β-adrenergic receptor facilitation elicited by gamma-aminobutyric acid B receptors. *J. Neurochem.*, **53**, 753–758.
Xu, J. and Wojcik, W.J. (1986) Gamma-aminobutyric acid B receptor-mediated inhibition of adenylate cyclase in cultured cerebellar granule cells: blockade by islet activating protein. *J. Pharmacol. Exp. Ther.*, **239**, 568–573.
Zorn, S.H. and Enna, S.J. (1985) GABA uptake inhibitors produce a greater antinociceptive response in the mouse tail immersion assay than other types of GABAergic drugs. *Life Sci.*, **37**, 1901–1912.

8 GABA$_B$ Receptors and Inhibition of Cyclic AMP Formation

W. J. WOJCIK, M. BERTOLINO, R. A. TRAVAGLI, S. VICINI and M. ULIVI

GABA$_B$ RECEPTORS INHIBIT ADENYLYL CYCLASE

Agonist-activated transmitter receptors transduce information into neurons by various metabolotropic mechanisms, one of which is a reduction in adenylate cyclase activity. The GABA$_B$ receptor uses such a receptor–effector system in primary culture of cerebellar granule cells and in other neuronal models (Asano *et al.*, 1985; Wojcik and Neff, 1984). Because many similarities have been observed among the different receptor-linked effectors that inhibit adenylate cyclase, a set of criteria has been proposed to define new receptors functioning through negative modulation of adenylate cyclase (for reviews, see Casey and Gilman, 1988; Jakobs *et al.*, 1981; Neer and Clapham, 1988). These criteria include the following provisions: (a) receptor-mediated inhibition of adenylate cyclase requires GTP; (b) the half-maximal inhibitory concentration (IC_{50}) of GTP should be approximately 1 μM; (c) pertussis toxin should uncouple/prevent receptor-mediated inhibition of adenylate cyclase; and (d) the affinity of an agonist for the receptor recognition site (as seen by labeled ligand binding methodology) should show GTP dependency. The following discussion documents that GABA$_B$ receptors fulfill these requirements and therefore belong to the class of receptors using inhibition of adenylate cyclase as an effector system.

A receptive protein located in the neuronal membrane can interact with the membrane-bound adenylate cyclase only indirectly; it must first couple to one of the many guanine nucleotide binding proteins (G-proteins) (Casey and Gilman, 1988; Katada *et al.*, 1986; Mumby *et al.*, 1988; Neer and Clapman, 1988). According to this view, the inhibitory G-protein (G_i) appears to be the most common coupler for receptors which inhibit adenylate cyclase, while the stimulatory G-protein (G_s) appears to couple receptors which stimulate adenylyl cyclase. Both G-proteins are composed of three subunits, α, β, and γ. The α-subunit of both G_i and G_s contains a GTPase. In the case of the G_i-protein, the agonist-bound receptor binds this G-protein when GTP is bound to the α_i-subunit and results in the separation of the trimer state (α_i-, β- and γ-subunits) into α_i- and β/γ-subunits. Probably, the GTP-bound α_i-subunit directly inhibits adenylyl cyclase. The β/γ-subunits may also contribute to the adenylyl cyclase

GABA$_B$ Receptors in Mammalian Function. Edited by N.G. Bowery, H. Bittiger and H.-R. Olpe

inhibition indirectly, by binding to free α_s-subunits to form the inactive trimer state of G_s (Katada *et al.*, 1986). Inactivation of the α_i-subunit which inhibits adenylyl cyclase occurs when the GTP is hydrolyzed to GDP and inorganic phosphate. Activation of $GABA_B$ receptors increases GTPase activity (Bowery *et al.*, 1989; Wojcik *et al.*, 1985) and the agonist activation of $GABA_B$ receptors inhibits adenylyl cyclase only in the presence of GTP. The IC_{50} of GTP that is necessary for baclofen, a $GABA_B$ receptor agonist, to inhibit adenylyl cyclase is 0.7 μM (Xu and Wojcik, 1986). In general, this IC_{50} is consistent with that described for other receptors which inhibit adenylyl cyclase and differs from the GTP concentrations (100 nM range) that are needed for a receptor agonist-activation of adenylyl cyclase.

Pertussis toxin can uncouple from G-proteins those receptors which inhibit adenylyl cyclase (Asano *et al.*, 1985; Katada and Ui, 1982). The toxin inhibits the GTPase present on the α-subunit by catalyzing the ADP-ribosylation of this subunit from NAD. In fact, as a result of using [^{32}P]NAD, scientists have tagged these subunits and isolated the α_i and another novel protein called α_O (the other G-protein). Because G_i and G_O are ADP-ribosylated after pertussis toxin treatment, both G-proteins were thought to be involved in coupling receptors in an inhibitory manner to adenylyl cyclase. The inhibition of adenylate cyclase elicited by $GABA_B$ receptor activation was in fact attenuated after a prior treatment with pertussis toxin (Asano *et al.*, 1985; Xu and Wojcik, 1986). Our studies included pretreatment of primary cultures of cerebellar granule cells with pertussis toxin. Back-[^{32}P]ADP-ribosylation studies that were performed after the toxin pretreatment indicated that the toxin ADP-ribosylated approximately 50% of the α-subunits having a molecular mass of 40–41 KDa. The average masses of α_O and α_i are 39 and 41 KDa, respectively. Both α_O- and α_i-subunits are thought to be involved in signal transduction for receptors that inhibit adenylyl cyclase. These molecular masses differ from that of the α_s, which is approximately 45 KDa (Casey and Gilman, 1988; Neer and Clapham, 1988).

Radiolabeled GABA and/or baclofen binding to the $GABA_B$ recognition site also show different affinity states in the presence of GTP (Hill *et al.*, 1984). A high-affinity site is observed in the absence of GTP and a low-affinity binding site is observed in its presence. These GTP-dependent affinity states are characteristic of receptors that couple to G-proteins. However, this evidence does not rigorously support the view that the $GABA_B$ receptor couples to G-proteins important specifically in the inhibition of adenylyl cyclase. Other $GABA_B$ receptor-mediated events that are mediated by G-protein couplers are discussed later.

Consistent with the aforementioned ability of $GABA_B$ receptors to interact with adenylyl cyclase, its receptive protein might be part of a group of receptors with seven transmembrane spanning regions (Bonner *et al.*, 1987; Bunzow *et al.*, 1988; Kobilka *et al.*, 1987). The isolation and purification of the $GABA_B$ protein has not been accomplished due, in part, to the low density of receptive proteins in most tissues and to a limited pharmacological repertoire of specific, high-affinity agonists and/or antagonists.

Thus, the data presented on the $GABA_B$ receptor clearly fulfill the criteria for receptors which inhibit adenylyl cyclase activity. However, one important issue remains. If the adenylyl cyclase activity is reduced after application of $GABA_B$ agonist, then the intracellular content of cyclic AMP should also be diminished. In order for these receptors to have any action, one must assume that adenylyl cyclase is actively converting ATP to cyclic AMP. However, in many situations, adenylyl cyclase may not be activated. Thus, in many experimental paradigms, forskolin, a direct activator of adenylyl cyclase, is used to activate adenylyl cyclase and therefore insure detection of those receptors that inhibit the adenylyl cyclase. Activation of $GABA_B$ receptors will reduce the cyclic AMP accumulation that was induced by forskolin (Karbon and Enna, 1984; Xu and Wojcik, 1986). Another means of increasing cyclic AMP formation would be to activate receptors that stimulate the adenylyl cyclase. A classic example of this approach is found in studies of the pituitary in which the dopamine D_2 receptor reduced the cyclic AMP formation that had been increased by vasoactive intestinal peptide (VIP) (Onali *et al.*, 1981). However, studies in slices of brain showed that the activation of $GABA_B$ receptors further increases the accumulation of cyclic AMP that is elicited by receptors which stimulate adenylyl cyclase such as VIP, β-adrenergic and adenosine A_2 receptors (Karbon and Enna, 1984). This $GABA_B$ receptor-mediated activation of adenylyl cyclase is not observed in a membrane preparation. The $GABA_B$-mediated potentiation of receptor-stimulated cyclic AMP content in brain slices may not be occurring through a direct coupling of the $GABA_B$ receptor to G_s-proteins. Rather, these studies provide evidence that the $GABA_B$ receptor activates phospholipase A_2 to release arachidonic acid. Through a series of events that were induced by arachidonic acid, an indirect facilitation of receptors which stimulate adenylyl cyclase results (Duman *et al.*, 1986).

The initial question still remains. Will activation of the $GABA_B$ receptor reduce intracellular content of cyclic AMP, even in the presence of agonists that activate receptors which stimulate adenylyl cyclase? Because of the complexity of the tissue slice preparation, this question cannot be answered unless a homogeneous population of cells, grown in tissue culture, is found to contain both the $GABA_B$ receptor which inhibits adenylyl cyclase and another receptor which stimulates adenylyl cyclase. This cell culture would mimic tissue slices but would possibly have an added advantage of monitoring (and controlling) various cellular events. One such event could be to determine whether these cells also contain a $GABA_B$ receptor-mediated activation of phospholipase A_2.

MULTIPLE EFFECTORS FOR $GABA_B$ RECEPTORS

In addition to the $GABA_B$ receptor-mediated inhibition of adenylyl cyclase, there are other effectors for $GABA_B$ receptor function. To date, four additional effectors have been described: (1) facilitation of receptor-mediated activation of adenylyl cyclase, which was briefly discussed in the preceding text (Hill, 1985; Karbon *et al.*, 1984; Suzdak and Gianutsos, 1986; Watling and Bristow, 1986; Wojcik *et al.*, 1989b); (2) activation of K^+ channels (Andrade *et al.*, 1986; Dutar and Nicoll, 1988; Gähwiler and Brown, 1985; Ogata *et al.*, 1987; Pinnock, 1984); (3) inhibition of Ca^{2+} channels (Dunlap, 1981; Holtz *et al.*, 1986; Konnerth and Heinemann, 1983; Scott and Dolphin, 1986; Zhu and Chuang, 1987); and, most recently, (4) inhibition of histamine- and serotonin-stimulated inositol triphosphate (IP_3) synthesis (Crawford and Young, 1988; Godfrey *et al.*, 1988). All of these responses can be attenuated by a prior exposure to pertussis toxin, except for the inhibition of receptor-mediated stimulation of IP_3 synthesis, whose coupling to G-proteins has not been investigated thoroughly. More detailed information will be found on these various $GABA_B$ receptor–effectors in other chapters. Thus, one could infer the existence of either one $GABA_B$ receptive protein which will couple to $G_{i/o}$-proteins to affect one of the nearby effector systems or multiple $GABA_B$ receptor subtypes.

DOES $GABA_B$-MEDIATED INHIBITION OF ADENYLYL CYCLASE AFFECT CALCIUM INFLUX AND TRANSMITTER RELEASE?

These various effectors, linked either directly or indirectly to $GABA_B$ recognition sites, are the means whereby endogenous GABA or exogenous baclofen can transduce information into a neuron via the $GABA_B$ receptor. The significance of $GABA_B$ receptor activation on the whole neuron has also been studied. The most common neuronal response is a decrease in depolarization-induced release of transmitter (Bowery *et al.*, 1980; Fox *et al.*, 1978; Gray and Green, 1987; Potashner, 1979; Swartzwelder *et al.*, 1986). Many factors could account for this response, such as $GABA_B$-mediated activation of K^+ channels to hyperpolarize the cell and hinder cell depolarization. Another possibility is that a $GABA_B$ receptor-mediated decrease in Ca^{2+} conductance could hinder the Ca^{2+}-dependent steps leading to a release of transmitter and/or reduce the action potential duration. Our laboratory and many others are attempting to identify whether one or several $GABA_B$ receptor–effector event(s) mediate the inhibition of transmitter release. It is of interest that other receptors which inhibit adenylyl cyclase, similar to the $GABA_B$ receptor, also inhibit transmitter release and, in many cases, reduce Ca^{2+} influx as well (e.g. adenosine receptors; Dolphin *et al.*, 1986; Dolphin and Prestwich, 1985). These similarities appear to be more than coincidental and experiments in our laboratory were designed to determine whether the $GABA_B$ receptor-mediated inhibition of adenylyl cyclase mediated the decrease in depolarization-induced Ca^{2+} influx and transmitter release; we also asked whether or not the decrease in Ca^{2+} influx and decrease in transmitter release were related events.

In order to test this hypothesis, we studied as a model primary cultures of cerebellar granule cells. These cell cultures are 90–95% homogeneous for this neuronal cell type (Nicoletti *et al.*, 1986). Furthermore, the granule cells contain GABA$_B$ receptors which inhibit adenylyl cyclase. Activation of GABA$_B$ receptors also decreased the accumulation of intracellular cyclic AMP when cyclic AMP phosphodiesterase was blocked (Figure 8.1.). GABA$_B$ receptor-mediated effects on neuronal Ca^{2+} homeostasis were studied using depolarization-induced influx of $Ca^{2+}/^{45}Ca^{2+}$ and by measuring Ca^{2+} currents in the whole cell by patch-clamp technology. It has previously been reported that these cerebellar granule cell cultures contain dihydropyridine-sensitive voltage-dependent Ca^{2+} channels (DHP-VDCCs), also known as L-type channels (Carboni and Wojcik, 1988). Figure 8.2 shows that depolarization-induced influx of Ca^{2+} can be blocked by DHP antagonists such as nifedipine and nitrendipine, but stimulated by BAY K 8644, a DHP agonist. Patch-clamp analysis has also demonstrated the presence of a DHP-VDCC or L-type Ca^{2+} current (Figure 8.3). Electrophysiological studies have identified this current; it is blocked by cobalt, has a reversal potential of +60 mV and does not

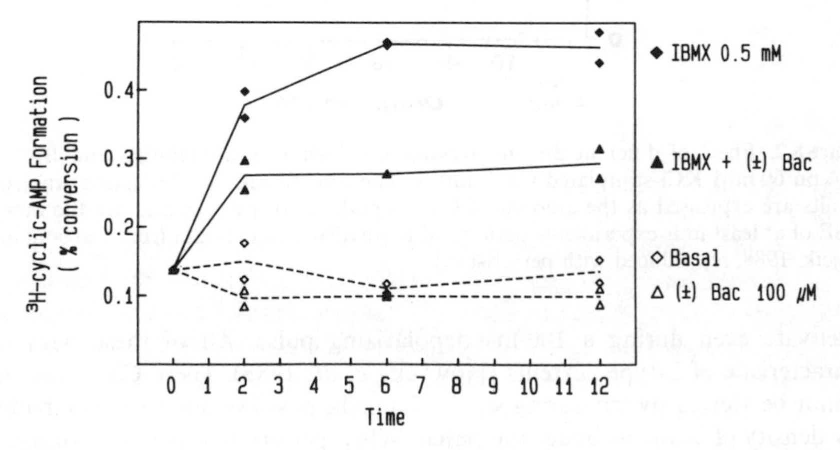

Figure 8.1. Time-course (in minutes) of (±)-baclofen (100 μM)-mediated inhibition of cyclic AMP formation in cerebellar granule cells. The basal adenylyl cyclase can be measured by blocking the cyclic AMP degrading enzyme, cyclic AMP phosphodiesterase, with 3-isobutyl-1-methylxanthine (IBMX). Thus, in the presence of IBMX, the cyclic AMP accumulates in the cell. The results show individual duplicate values for one representative experiment

Method

[^{3}H]Adenine (1 μCi/ml) was added to the cell culture for at least 1 h before the start of the experiment. Afterwards, the medium was changed to a buffered balanced salt solution at 37 °C. Baclofen and/or IBMX were added after 20 min of preincubation. After various times (in minutes), incubations were stopped with the addition of an SDS-containing solution. The [^{3}H]cyclic AMP was separated from other labeled adenine catabolites by ion chromatography according to the method of Salomon (1978).

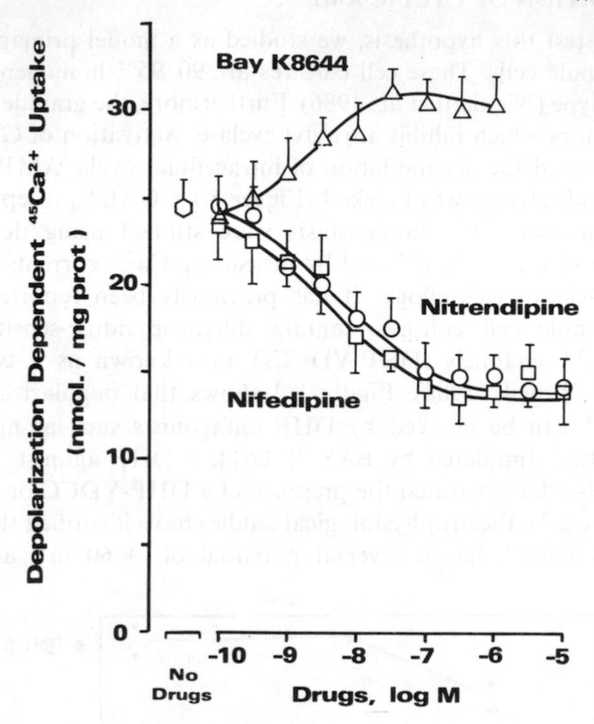

Figure 8.2. Effect of different dihydropyridine Ca^{2+}-channel antagonists and BAY K 8644 on 60 mM KCl-stimulated Ca^{2+} influx. The time of cell depolarization was 20 s. Results are expressed as the amount of Ca^{2+} uptake into the cells and are the means $\pm$ SE of at least four experiments performed in quadruplicate. (Data from Carboni and Wojcik, 1988, reproduced with permission)

inactivate even during a 100-ms depolarizing pulse. All of these data are characteristic of L-type currents (Nowycky *et al.*, 1985). These Ca^{2+} currents cannot be viewed by measuring single channels, possibly due to the extremely low density of channels under the patch. When performing these recordings, it appears necessary to reduce run-down of the Ca^{2+} currents by including ATP and cyclic AMP in the patch pipette (Reuter, 1983), since recent evidence shows that these compounds are needed for protein kinase A (cyclic AMP-dependent kinase) to phosphorylate the DHP-VDCCs. When phosphorylated, DHP-VDCCs conduct Ca^{2+} during membrane depolarization (Armstrong and Eckert, 1987).

Because cerebellar granule cells are glutamatergic, the effects of $GABA_B$ receptors on the release of this neurotransmitter by depolarization were studied. Endogenous glutamate was separated from other substances by cation-exchange high-performance liquid chromatography (HPLC) and detected fluorometrically by postcolumn derivatization reaction of the primary amino group of

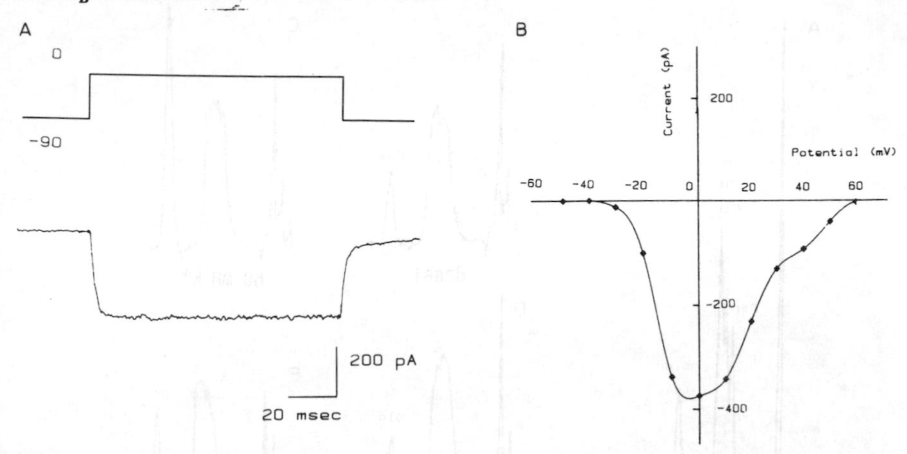

Figure 8.3. Whole-cell Ca^{2+} currents were obtained from cerebellar granule cells in primary culture using the patch-clamp technique (Hamill *et al.*, 1981). (A) Representative Ca^{2+} current trace showing a voltage step from a holding potential of -90 mV to 0 mV for a duration of 100 ms. (B) Plot of the current–voltage relationship obtained for the peak Ca^{2+} current in these cells. For these studies, the holding potential was -90 mV and the step potential started at -50 mV. The reversal potential appeared to be $+60$ mV

Method

Patch pipettes were filled with (in mM): CsCl, 100; TEA, 20; glucose, 10; HEPES, 10; BAPTA, 10; ATP-Mg, 5; cyclic AMP, 0.25; where pH was adjusted to 7.3 with CsOH. Currents were measured by using barium as the main carrier (20 mM BaCl was in the outside balanced salt solution). The traces shown were obtained after subtraction of the leakage currents.

glutamate with ortho-phthaldialdehyde (Schmid *et al.*, 1980). Figure 8.4A shows a typical HPLC chromatogram of both amino-acid standards and Figure 8.4B and C show endogenous compounds that were released into the incubation buffer during a 1-min exposure to cerebellar granule cell cultures. Figure 8.4D and E demonstrate the Ca^{2+}-dependent release of glutamate upon depolarization with 60 mM KCl.

Thus, measurements of adenylyl cyclase activity (or cyclic AMP accumulation), Ca^{2+} influx and Ca^{2+} currents, and glutamate release have been performed in order to address four basic questions. (1) Do ($\pm$)-baclofen and GABA mediate the same responses? If so, the half-maximal effective (EC_{50}) concentrations for each agonist should be the same for all responses that are subsequent to the initial receptor coupling. Moreover, this similarity and a consistent, relative potency for GABA and baclofen would support the view that these responses are mediated via one subtype of GABA$_B$ receptor. (2) Are all GABA$_B$ responses attenuated after pertussis toxin treatment of the cell cultures? (3) Would an exogenous cyclic AMP analog reverse these GABA$_B$-mediated responses? If the GABA$_B$ receptor-mediated inhibition of adenylyl cyclase is responsible for the decrease in Ca^{2+} influx/Ca^{2+} current and glutamate release,

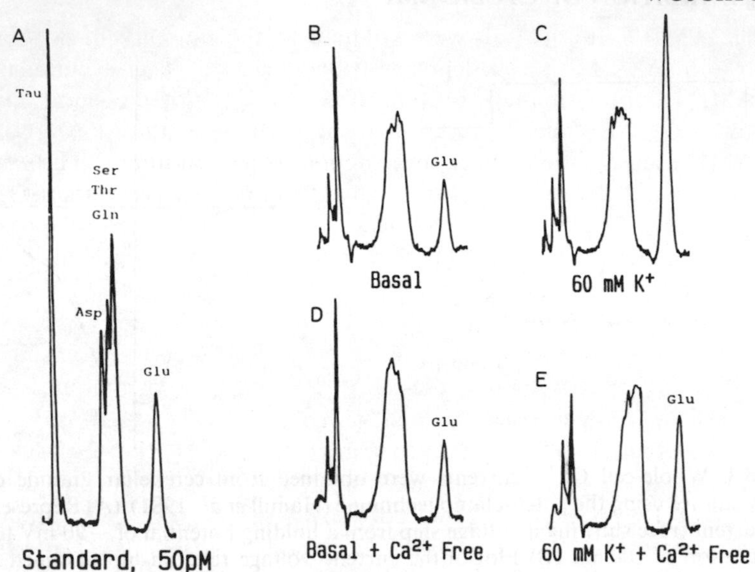

Figure 8.4. High-performance liquid chromatography (HPLC) separation of glutamate (Glu) from other compounds with primary amino groups. (A) Chromatogram showing the separation of glutamate from other known amino-acid standards. Tau, taurine; Asp, aspartic acid; Ser, serine; Thr, threonine; Gln, glutamine. (B), (C) Chromatograms of endogenous amino acids that are released from cerebellar granule cell cultures into a buffered balanced salt solution. Cell depolarization for 1 min induced a three- to four-fold stimulation of glutamate release from these cells. (D), (E) When the buffered salt solution does not contain Ca^{2+}, cell depolarization does not evoke the release of glutamate. The retention time for glutamate was 30 min. (The HPLC method for separation and detection of glutamate was previously described by Schmid *et al.*, 1980.)

then the analog should reverse the effects of $GABA_B$ agonists. (4) Do $GABA_B$ agonists and DHP antagonists produce either additive or non-additive effects on Ca^{2+} influx and glutamate release? If the responses are non-additive, then the $GABA_B$ receptor and the DHP-VDCC might be linked together through a cascade of events.

(1) Are GABA- and Baclofen-mediated Effects Similar? (See Table 1)

($\pm$)-Baclofen and GABA have already been shown to inhibit by about 40% basal adenylyl cyclase and to reduce the formation of cyclic AMP intracellularly in cerebellar granule cells (Figure 8.1; Wojcik and Neff, 1984; Xu and Wojcik, 1986). The EC_{50} for both compounds is similar, approximately 10–20 μM. Because the experimental design for studying inhibition of Ca^{2+} influx and inhibition of transmitter release involves depolarization of the cells with a balanced salt solution containing 60 mM KCl, the effects of cell depolarization on cyclic AMP accumulation were studied. In the presence of 3-isobutyl-1-methylxanthine (IBMX), cell depolarization increased basal accumulation of cyclic AMP by approximately fourfold over non-depolarizing conditions (data not shown). (±)-Baclofen induced a strong inhibition (almost 80%) of basal cyclic AMP accumulation under these depolarizing conditions. The EC_{50} for (±)-baclofen did not change substantially from that observed in non-depolarizing conditions.

In cerebellar granule cell cultures, (±)-baclofen, and to a lesser extent GABA, inhibits depolarization-induced $^{45}Ca^{2+}$ influx by 30–40%. Surprisingly, the EC_{50} for (±)-baclofen was 4 nM, while GABA was approximately three orders of magnitude less potent. Furthermore, 200 nM (±)-baclofen can reduce the Ca^{2+} current from whole-cell recordings of patched granule cells by 60–70%. These channels are activated with one voltage step from a holding membrane potential of − 90 mV to 0 mV. This effect of (±)-baclofen at 200 nM is maximal since (±)-baclofen at 30 μM does not further block the Ca^{2+} current. It appears that the effects of baclofen on both $^{45}Ca^{2+}$ influx and Ca^{2+} conductance are due to a similar mechanism. However, the relatively higher potency of baclofen over GABA in blocking $^{45}Ca^{2+}$ influx differs from the typical profile that these compounds display at the $GABA_B$ receptors. These data do not exclude the possibility that baclofen acts at a $GABA_B$ receptor, but

Table 1. Summary of baclofen/$GABA_B$-mediated responses in cerebellar granule cells

Baclofen/$GABA_B$-mediated responses	Maximal response (%)	EC_{50}
Inhibition of cyclic AMP synthesis		
GABA	40–50	15 μM
(±)-baclofen	40–50	14 μM
Inhibition of $^{45}Ca^{2+}$ influx		
GABA	30–40	10 μM
(±)-baclofen	30–40	4 nM
Inhibition of Ca^{2+} current		
(±)-baclofen	60–70	(max. 200 nM)
Inhibition of glutamate release		
(±)-baclofen	40–50	7 μM

Summary table of concentration–response curves for both GABA and (±)-baclofen for various receptor-mediated responses observed in intact cerebellar granule cell cultures. Studies on $^{45}Ca^{2+}$ influx and glutamate release followed similar experimental designs: 1-min exposure to drug followed by 1-min cell depolarization (60 mM KCl) in the presence of drug (and $^{45}Ca^{2+}$). In the $^{45}Ca^{2+}$ influx studies, the Petri dish was washed four times with two volumes of balanced salt solution before cell lysis with 5% trichloroacetic acid to release intracellular $^{45}Ca^{2+}$. The $^{45}Ca^{2+}$ was quantitated by liquid scintillation spectroscopy. In the glutamate release studies, the endogenous glutamate was measured by high-performance liquid chromatography and fluorometric detection in both resting and depolarized cells. Results are the average of no less than three experiments conducted in triplicate. Ca^{2+} currents were measured by whole-cell patch-clamp technique. The current was activated by a step voltage from the holding membrane potential of − 90 mV to 0 mV. The legend to Figure 8.3 contains additional information on the method.

suggest that baclofen may act on its own receptor. Such a receptor for baclofen has already been proposed by Sawynok and Dickson (1985).

Depolarization of cerebellar granule cells releases endogenous glutamate. In the presence of (±)-baclofen, the amount of glutamate released is reduced by 40–50%, with an EC_{50} for (±)-baclofen of 7 μM. This concentration is quite similar to the concentration of (±)-baclofen needed to inhibit adenylyl cyclase, but was three orders of magnitude higher than the concentration of baclofen required to reduce Ca^{2+} conductance. Because the application of exogenous GABA confounds the HPLC separation and detection of released glutamate, the EC_{50} for GABA could not be determined. Other means of obtaining this information are being sought. Thus, the reduced Ca^{2+} influx that results from low concentrations of baclofen does not alter transmitter release, even though the glutamate release is Ca^{2+}-dependent. One possible explanation for these findings is that baclofen's action on Ca^{2+} currents might occur on the cell body, while the baclofen-mediated inhibition of glutamate release, and possibly inhibition of adenylyl cyclase, may occur in the nerve terminal. Possibly, the amount of $Ca^{2+}/^{45}Ca^{2+}$ that enters the nerve terminal is not substantially great in comparison to that which enters the cell body. Thus, any change in the Ca^{2+} influx in the terminal region may not be detectable after activation of $GABA_B$ receptors (possibly of the low-affinity type). However, there is no evidence to support such different cell localization of these $GABA_B$/baclofen receptors.

(2) Are $GABA_B$ Responses Pertussis Toxin Sensitive? (See Figure 8.5)

Pretreatment of the cerebellar granule cell culture with pertussis toxin (1 μg/ml) for 14–16 h results in approximately 50% labeling of G-proteins having a molecular mass of 40–41 kDa (Xu and Wojcik, 1986). As previously mentioned,

---

Figure 8.5. Baclofen concentration–response data from control and pertussis toxin (PT) -treated (1 μg/ml, 14–16 h) cerebellar granule cell cultures. (A) Baclofen inhibits adenylyl cyclase activity in membranes from these cell cultures. Results are expressed as a percentage of basal adenylyl cyclase activity and shown as the mean ± SE for three experiments. (Data recalculated from Xu and Wojcik, 1986, with permission.) (B) Baclofen reduces Ca^{2+} current in whole-cell recordings by patch-clamp methodology. B1–3 were obtained from control cells and B4–6 were obtained from pertussis toxin-treated cells. B1 and B4, normal Ca^{2+} currents seen during a 100 ms depolarizing pulse from −90 mV to 0 mV. B2 and B5, Ca^{2+} currents seen in the presence of 200 nM (±)-baclofen B3 and B6 Ca^{2+} currents seen in the presence of 30 μM (±)-baclofen. Data shown are one representative experiment of three. Baclofen displayed this action only when placed in the Petri dish for about 5 min before patching a cell. (C) Baclofen concentration–response curve for depolarization-induced release of glutamate from control and pertussis toxin-treated cerebellar granule cell cultures. Baclofen is shown to reduce the average amount of glutamate that was released with a 1 min cell depolarization with 60 mM KCl-containing solution. Results are the average of at least three experiments performed in duplicate and are shown as mean ±SE

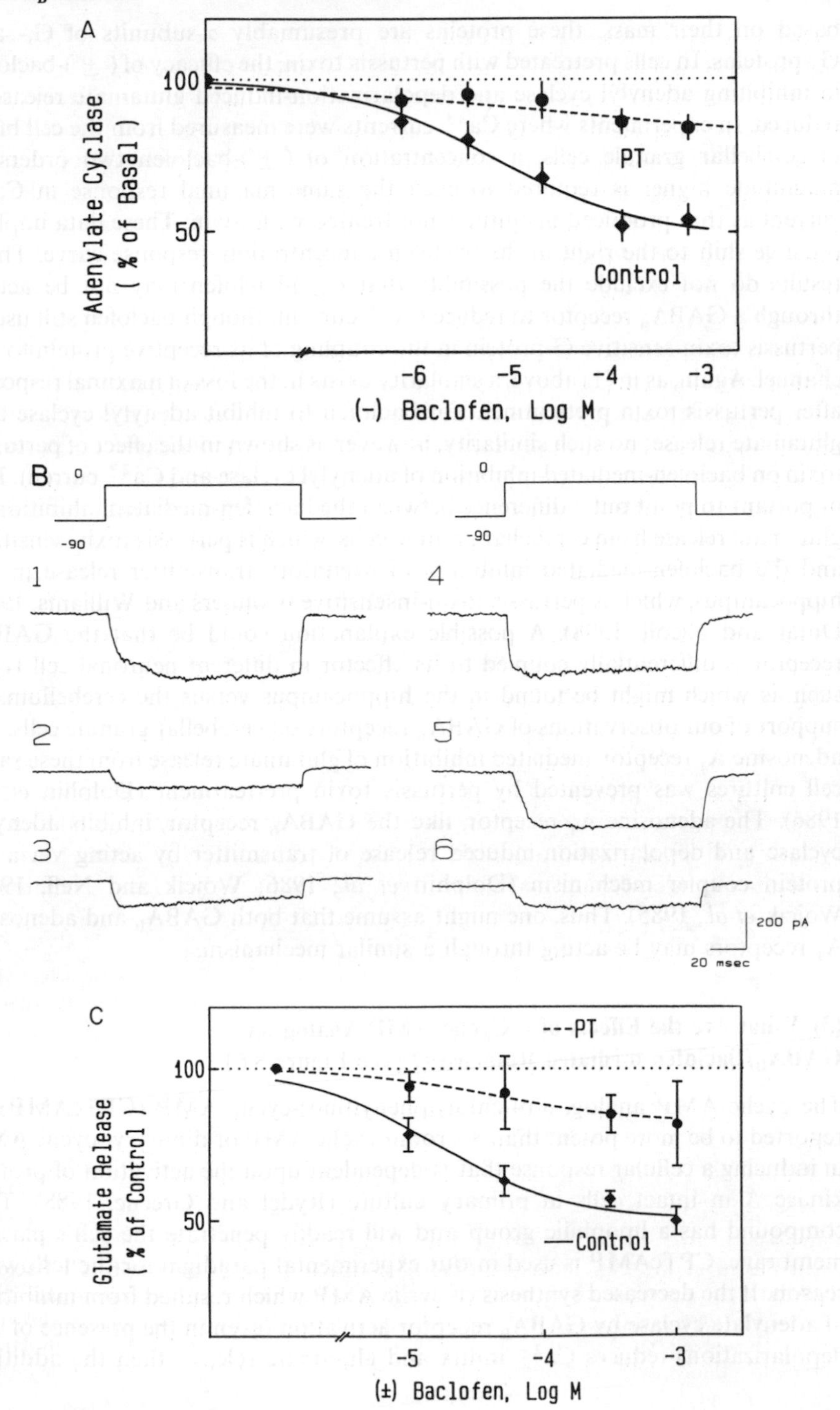
A
100
50
Adenylate Cyclase
(% of Basal)
PT
Control
-6
-5
-4
-3
(-) Baclofen, Log M
B
0
-90
1
2
3
4
5
6
200 pA
20 msec
C
---PT
100
50
Glutamate Release
(% of Control)
—Control
-5
-4
-3
(±) Baclofen, Log M

based on their mass, these proteins are presumably α-subunits of G_i- and G_o-proteins. In cells pretreated with pertussis toxin, the efficacy of (±)-baclofen in inhibiting adenylyl cyclase and depolarization-induced glutamate release is reduced. In experiments where Ca^{2+} currents were measured from the cell body of cerebellar granule cells, a concentration of (±)-baclofen two orders of magnitude higher is required to elicit the same maximal response in Ca^{2+} current as that produced in cultures not treated with toxin. These data imply a possible shift to the right in the baclofen concentration–response curve. These results do not exclude the possibility that (±)-baclofen may not be acting through a $GABA_B$ receptor to reduce Ca^{2+} current, though baclofen still uses a pertussis toxin-sensitive G-protein in the coupling of its receptive protein to the channel. Again, as in (1) above, a similarity exists in the loss of maximal response after pertussis toxin pretreatment for baclofen to inhibit adenylyl cyclase and glutamate release; no such similarity, however, is shown in the effect of pertussis toxin on baclofen-mediated inhibition of adenylyl cyclase and Ca^{2+} current. It is important to point out a difference between the baclofen-mediated inhibition of glutamate release from cerebellar granule cells, which is pertussis toxin-sensitive, and the baclofen-mediated inhibition of excitatory transmitter release in the hippocampus, which is pertussis toxin-insensitive (Colmers and Williams, 1988; Dutar and Nicoll, 1988). A possible explanation could be that the $GABA_B$ receptor is differentially coupled to its effector in different neuronal cell types such as which might be found in the hippocampus versus the cerebellum. In support of our observations of $GABA_B$ receptors on cerebellar granule cells, the adenosine A_1 receptor-mediated inhibition of glutamate release from these same cell cultures was prevented by pertussis toxin pretreatment (Dolphin *et al.*, 1986). The adenosine A_1 receptor, like the $GABA_B$ receptor, inhibits adenylyl cyclase and depolarization-induced release of transmitter by acting via a G-protein coupler mechanism (Dolphin *et al.*, 1986; Wojcik and Neff, 1982; Wojcik *et al.*, 1985). Thus, one might assume that both $GABA_B$ and adenosine A_1 receptors may be acting through a similar mechanism.

(3) What Are the Effects of a Cyclic AMP Analog on $GABA_B$/Baclofen-mediated Responses? (See Figure 8.6)

The cyclic AMP analog, 8-(4-chlorophenylthio)-cyclic AMP (CPTcAMP), is reported to be more potent than 8-bromo-cyclic AMP or dibutyryl-cyclic AMP in inducing a cellular response that is dependent upon the activation of protein kinase A in intact cells in primary culture (Rydel and Greene, 1988). This compound has a lipophilic group and will readily penetrate the cell's plasma membrane. CPTcAMP is used in our experimental paradigm for the following reason. If the decreased synthesis of cyclic AMP which resulted from inhibition of adenylate cyclase by $GABA_B$ receptor activation (even in the presence of cell depolarization) reduces Ca^{2+} influx and glutamate release, then the addition

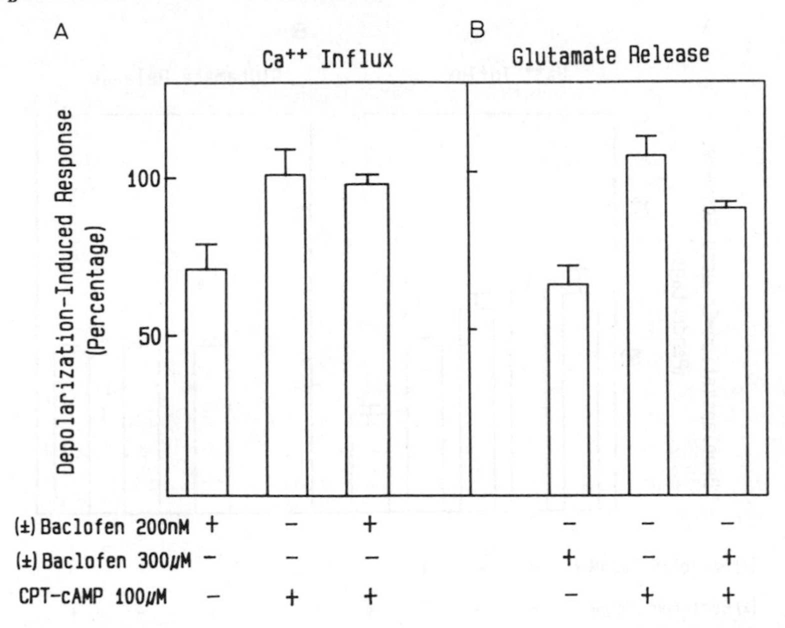

Figure 8.6. In cerebellar granule cells 8-(4-chlorophenylthio)-cyclic AMP (CPTcAMP), the cyclic AMP analog, reverses the baclofen-mediated inhibition of depolarization-induced $^{45}Ca^{2+}$ influx (A) and glutamate release (B). Cell depolarization was evoked by 60 mM KCl-containing buffered salt solution for 1 min. Results are the average ±SE for at least three experiments performed in triplicate. Experimental design is similar to those previously described

of an exogenous cyclic AMP analog should reverse the effects of a GABA$_B$ agonist. In cerebellar granule cell cultures, CPTcAMP is shown to reverse the (±)-baclofen-mediated inhibition of $^{45}Ca^{2+}$ influx and glutamate release. CPTcAMP at 100 μM was effective only if the cell culture was incubated for 10 min before its depolarization, while it was ineffective if the preincubation lasted only 1 min before depolarization (data not shown). CPTcAMP did not produce any action by itself under normal or depolarizing conditions. Moreover, CPTcAMP failed to reverse the effect of nifedipine on Ca^{2+} influx and glutamate release observed after cell depolarization. It is proposed that the 10-min delay required for CPTcAMP to reverse baclofen's inhibition of Ca^{2+} current and glutamate release is due to CPTcAMP's activation of protein kinase A. A more detailed discussion will be provided later.

(4) Effects of Nifedipine on Responses Elicited by Baclofen (see Figure 8.7)

Nifedipine belongs to the dihydropyridine class of compounds that are reported to block the L-type VDCC. This Ca^{2+} channel is present in primrary cultures of

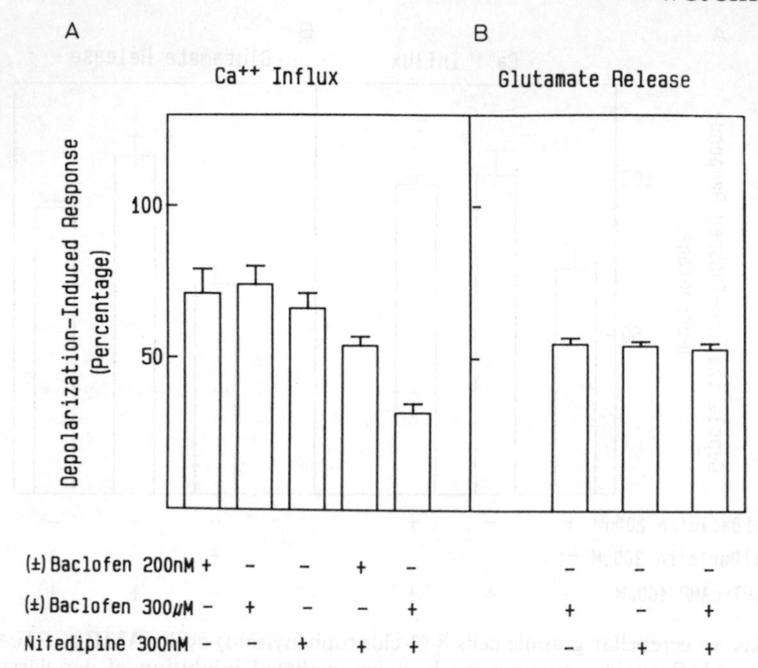

Figure 8.7. In cerebellar granule cells, baclofen and nifedipine inhibit, in an additive manner, depolarization-induced $^{45}Ca^{2+}$ influx and, in a non-additive manner, depolarization-induced glutamate release. Both drugs were added 1 min before the cell depolarization with 60 mM KCl-containing buffered salt solution. Results are the average $\pm$SE for three experiments performed in triplicate

cerebellar granule cells, since nifedipine is shown to block depolarization-induced $^{45}Ca^{2+}$ influx and Ca^{2+} currents in these cells (Carboni and Wojcik, 1988; Kingsbury and Balazs, 1987). Because nifedipine does not completely block Ca^{2+} currents, there might be more than one subtype of L-type VDCC present in cerebellar granule cells. In the presence of ($\pm$)-baclofen, maximal concentrations of nifedipine produce nearly additive effects on $^{45}Ca^{2+}$ influx. These data suggest that baclofen and nifedipine might be acting on different populations of VDCCs. It is also possible that the observed additivity results from an uneven cellular location of the baclofen-mediated inhibition of Ca^{2+} influx and the DHP-VDCC.

These DHP-VDCCs appear to be important for the depolarization-induced release of glutamate. Nifedipine or ($\pm$)-baclofen alone reduces the depolarization-induced release of glutamate by 50%. Incubation with nifedipine and ($\pm$)-baclofen together, however, results in non-additive effects. Thus, in the case of transmitter release, the actions of baclofen and nifedipine might be directly or indirectly linked to the same mechanism.

DISCUSSION

These results show that there are basically two categories of baclofen-mediated responses: one mediated by low-affinity receptors for baclofen and the other mediated by high-affinity receptors for baclofen. Activation of the low-affinity receptor by micromolar concentrations of baclofen results in an inhibition of both adenylyl cyclase activity and depolarization-induced glutamate release, while nanomolar concentrations of baclofen activate a high-affinity receptor that inhibits VDCCs. The VDCC might be an L-type VDCC, but further characterization is needed. It is still unclear whether these receptors for baclofen should be termed $GABA_B$, since the potency of GABA does not parallel that of baclofen in reducing cyclic AMP formation, Ca^{2+} influx and glutamate release. A reason that nanomolar concentrations of GABA have been found inactive could be that such low concentrations of GABA are more susceptible to cellular uptake and metabolism than are high micromolar concentrations, where uptake and metabolism may be saturated. Additional studies must be performed to address whether a $GABA_B$ receptor that is more sensitive to baclofen than to GABA really exists.

We have shown that micromolar concentrations of ($\pm$)-baclofen result in an inhibition of cyclic AMP formation and depolarization-induced glutamate release. Whether these two events are related – that is, whether the inhibition of cyclic AMP is the first step in a cascade of events resulting in reduced transmitter release – is still debatable. We have provided two lines of evidence that could support a relationship between these two events. The first is that CPTcAMP, the cyclic AMP analog, reverses baclofen's inhibition of glutamate release. The second is that baclofen and nifedipine are non-additive in inhibiting depolarization-induced glutamate release. An interpretation of these results could be that the mechanism whereby baclofen inhibits glutamate release includes a DHP-VDCC (Wojcik *et al.*, 1989a). Baclofen can elicit neuronal responses via the inhibition of adenylyl cyclase, but only if the adenylyl cyclase activity is elevated. Consequently, if in cerebellar granule cells the adenylyl cyclase activity is low, then Ca^{2+} entry during depolarization may occur exclusively through the N-type of VDCC (Hirning, *et al*, 1988). Since transmitter release is dependent on Ca^{2+} influx, Ca^{2+} entry through N-type channels may be operative in mediating glutamate release in granule cells. However, when the adenylyl cyclase activity is high, protein kinase A becomes activated and thereby phosphorylates the DHP-VDCC (Curtis and Caterall, 1985). In the phosphorylated state, the DHP-VDCC is reported to be functional; that is, it can conduct Ca^{2+} as a result of membrane depolarization (Armstrong and Ekert, 1987;

Yatani *et al.*, 1988). Cell depolarization will allow additional Ca^{2+} to enter the nerve terminal through both N- and L-type (DHP-) VDCCs, and we predict that more transmitter than normal will be released. When the adenylyl cyclase activity is high, then activation of the $GABA_B$ receptors that inhibit adenylyl cyclase should reduce the formation of cyclic AMP. As a result of $GABA_B$ receptor activation, protein kinase A activity is reduced and the DHP-VDCCs can no longer be maintained in the active phosphorylated state. Thus, in the presence of $GABA_B$ receptor activation, cell depolarization results in the opening of the N-type of VDCCs and in a normal release rate of glutamate.

This model, however, shows some inconsistencies with our observation that baclofen-mediated inhibition of Ca^{2+} influx is additive with the inhibition by nifedipine. However, if our model were correct, both compounds should reduce the Ca^{2+} entry through the same mechanism and therefore the two effects should not be additive. Moreover, nanomolar concentrations of baclofen inhibited Ca^{2+} influx, but micromolar concentrations of baclofen, which affected glutamate release, did not inhibit further Ca^{2+} influx. Thus, the mechanism whereby $GABA_B$ receptors inhibit transmitter release and the physiological role of $GABA_B$ receptors are still unclear No information is currently available for those other receptors which also inhibit adenylyl cyclase and reduce transmitter release.

Some aspects of practical importance of the high-affinity baclofen-mediated inhibition of Ca^{2+} influx/Ca^{2+} current may be apparent. ($\pm$)-Baclofen is currently used therapeutically as an antispastic agent in individuals with multiple sclerosis or spinal cord injury (Young and Delwaide, 1981). It is believed that baclofen acts by blocking the mono- and polysynaptic reflex loops at the level of the afferent terminals in the spinal cord (Fox *et al.*, 1978). A conflict exists between the high nanomolar concentration of baclofen in cerebrospinal fluid of the treated individuals and the micromolar concentration of baclofen needed to activate most described receptor–effector systems such as the inhibition of adenylyl cyclase. Interestingly, in the cerebellar granule cells, ($\pm$)-baclofen inhibits Ca^{2+} influx/Ca^{2+} current in nanomolar concentrations. It is tempting to suggest that this receptor–effector system is involved in the therapeutic mechanism of action as an antispastic agent. However, various reports describe nonomolar concentrations of baclofen inhibiting serotonin release from mouse cerebral cortex and reducing serotonin-stimulated IP_3 synthesis (Godfrey *et al.*, 1988; Grey and Green, 1987). The physiological mechanisms that involve a micromolar, low-affinity receptor-mediated response in inhibiting glutamate release in the cerebellar granule cell culture and the high-affinity receptor-mediated response in inhibiting transmitter release in mouse cerebral cortex are unclear. Moreover, while we report a nanomolar, high-affinity receptor for baclofen which acts to modulate Ca^{2+} channels, Dolphin and Scott (1986) report a micromolar, low-affinity receptor for baclofen in dorsal root ganglion cell cultures. Such discrepancies in the affinity of baclofen for eliciting various responses make for a stimulating scientific search as to the mechanisms of baclofen/GABA$_B$ receptor-mediated responses and their correlation to physiological events.

REFERENCES

Andrade, R., Malenka, R.C. and Nicoll, R.A. (1986) A G protein couples serotonin and GABA$_B$ receptors to the same channels in hippocampus. *Science*, **234**, 1261–1265.

Armstrong, D. and Eckert, R. (1987) Voltage-activated calcium channels that must be phosphorylated to respond to membrane depolarization. *Proc. Natl Acad. Sci. USA*, **84**, 2518–2522.

Asano, T., Ui, M. and Ogasawara, N. (1985) Prevention of the agonist binding to γ-aminobutyric acid B receptors by guanine nucleotides and islet-activating protein, pertussis toxin, in bovine cerebral cortex. *J. Biol. Chem.*, **260**, 12653–12658.

Bonner, T.I., Buckley, N.J., Young, A.C. and Brann, M.R. (1987) Identification of a family of muscarinic acetylcholine receptor genes. *Science*, **237**, 527–532.

Bowery, N.G., Hill, D.R. and Hudson, A.L. (1980) (–)Baclofen decreases neurotransmitter release in the mammalian CNS by an action at a novel GABA receptor. *Nature*, **283**, 92–94.

Bowery, N.G., Hill, D.R. and Moratalla, R. (1989) Neurochemistry and autoradiography of GABA$_B$ receptors in mammalian brain: second-messenger system(s). In: E.A. Barnard and E. Costa (eds) *Allosteric modulation of amino acid receptors: therapeutic implications.* New York: Raven Press, pp. 159–172.

Bunzow, J.R., Van Tol, H.H.M., Grandy, D.K., Albert, P., Salon, J., McDonald, C., Machida, C.A., Neve, K.A. and Civelli, O. (1988) Cloning and expression of a rat D$_2$ dopamine receptor cDNA. *Nature*, **336**, 783–787.

Carboni, E. and Wojcik, W.J. (1988) Dihydropyridine binding sites regulate calcium influx through specific voltage-sensitive calcium channels in cerebellar granule cells. *J. Neurochem.*, **50**, 1279–1286.

Casey, P.J. and Gilman, A.G. (1988) G protein involvement in receptor–effector coupling. *J. Biol. Chem.*, **263**, 2577–2580.

Colmers, W. and Williams, J.T. (1988) Pertussis toxin pretreatment discriminates between pre- and postsynaptic actions of baclofen in rat dorsal raphe nucleus in vitro. *Neurosci. Lett.*, **93**, 300–306.

Crawford, M.L.A. and Young, J.M. (1988) GABA$_B$ receptor-mediated inhibition of histamine H$_1$-receptor-induced inositol phosphate formation in slices of rat cerebral cortex. *J. Neurochem.*, **51**, 1441–1447.

Curtis, B.M. and Catterall, W.A. (1985) Phosphorylation of the calcium antagonist receptor of the voltage-sensitive calcium channel by cAMP-dependent protein kinase. *Proc. Natl Acad. Sci. USA*, **82**, 2528–2532.

Dolphin, A.C. and Prestwich, S.A. (1985) Pertussis toxin reverses adenosine inhibition of glutamate release. *Nature*, **316**, 148–150.

Dolphin, A.C. and Scott, R.H. (1986) Inhibition of calcium currents in cultured rat dorsal root ganglion neurones by (–)-baclofen. *Br. J. Pharmacol.*, **88**, 213–220.

Dolphin, A.C., Forda, S.R. and Scott, R.H. (1986) Calcium-dependent currents in cultured rat dorsal root ganglion neurons are inhibited by an adenosine analogue. *J. Physiol*, **373**, 47–61.

Duman, R.S., Karbon, E.W., Harrington, C. and Enna, S.J. (1986) An examination of the involvement of phospholipases A_2 and C in the α-adrenergic and γ-aminobutyric acid receptor modulation of cyclic AMP accumulation in rat brain slices. *J. Neurochem.*, **47**, 800–810.

Dunlap, K. (1981) Two types of γ-aminobutyric acid receptor on embryonic sensory neurones. *Br. J. Pharmacol.*, **74**, 579–585.

Dutar, P. and Nicoll, R.A. (1988) Pre- and postsynaptic $GABA_B$ receptors in the hippocampus have different pharmacological properties. *Neuron*, **1**, 585–591.

Fox, S., Krnjevic, K., Morris, M.E., Puil, E. and Werman, R. (1978) Action of baclofen on mammalian synaptic transmission. *Neuroscience*, **4**, 495–515.

Gähwiler, B.H. and Brown, D.A. (1985) $GABA_B$-receptor-activated K^+ current in voltage-clamped CA_3 pyramidal cells in hippocampal cultures. *Proc. Natl Acad. Sci. USA*, **82**, 1558–1562.

Godfrey, P.P., Grahame-Smith, D.G. and Gray, J.A. (1988) $GABA_B$ receptor activation inhibits 5-hydroxytryptamine-stimulated inositol phospholipid turnover in mouse cerebral cortex. *Eur. J. Pharmacol.*, **152**, 185–188.

Gray, J.A. and Green, A.R. (1987) $GABA_B$-receptor mediated inhibition of potassium-evoked release of endogenous 5-hydroxytryptamine from mouse frontal cortex. *Br. J. Pharmacol.*, **91**, 517–522.

Hamill, O.P., Marty, A., Neher, E., Sakmann, B. and Sigworth, F.J. (1981) Improved patch clamp techniques for high resolution current recording from cells and cell free membrane patches. *Pflügers Arch.*, **391**, 85–100.

Hill, D.R. (1985) $GABA_B$ receptor modulation of adenylate cyclase activity in rat brain slices. *Br. J. Pharmacol.*, **84**, 249–257.

Hill, D.R., Bowery, N.G. and Hudson, A.L. (1984) Inhibition of $GABA_B$ receptor binding by guanyl nucleotides. *J. Neurochem.*, **42**, 652–657.

Hirning, L.D., Fox, A.P., McCleskey, E.W., Olivera, B.M., Thayer, S.A., Miller, R.J. and Tsien, R.W. (1988) Dominant role of N-type Ca^{2+} channels in evoked release of norepinephrine from sympathetic neurons. *Science*, **239**, 57–61.

Holtz, G.G., Rane, S.G. and Dunlap, K. (1986) GTP-binding proteins mediate transmitter inhibition of voltage-dependent calcium channels. *Nature*, **319**, 670–672.

Jakobs, K.H., Aktories, K. and Schultz, G. (1981). Inhibition of adenylate cyclase by hormones and neurotransmitters. *Adv. Cycl. Nucleot. Res.*, **14**, 173–187.

Karbon, E.W. and Enna, S.J. (1984) Characterization of the relationship between γ-aminobutyric acid B agonists and transmitter-coupled cyclic nucleotide-generating systems in rat brain. *Mol. Pharmacol.*, **27**, 53–59.

Karbon, E.W., Duman, R.S. and Enna, S.J. (1984) $GABA_B$ receptors and norepinephrine-stimulated cAMP production in rat brain cortex. *Brain Res.*, **306**, 327–332.

Katada, T. and Ui, M. (1982) Direct modification of the membrane adenylate cyclase system by islet-activating protein due to an ADP-ribosilation of a membrane protein. *Proc. Natl Acad. Sci. USA*, **79**, 3129–3133.

Katada, T., Oinuma, M. and Ui, M. (1986) Mechanisms for inhibition of the catalytic activity of adenylate cyclase by the guanine nucleotide-binding proteins serving as the substrate of islet-activating protein, pertussis toxin. *J. Biol. Chem.*, **261**, 5215–5221.

Kingsbury, A. and Balazs, R. (1987) Effect of calcium agonists and antagonists on cerebellar granule cells. *Eur. J. Pharmacol.*, **140**, 275–283.

Kobilka, B.K., Matsui, H., Kobolka, T.S., Yang-Feng, T.L., Francke, U., Caron, M.G., Lefkowitz, R.J. and Regan, J.W. (1987) Cloning, sequencing and expression of the gene coding for the human platelet $alpha_2$-adrenergic receptor. *Science*, **238**, 650–656.

Konnerth, A. and Heinemann, U. (1983) Effects of GABA on presumed presynaptic Ca^{2+} entry in hippocampal slices. *Brain Res.*, **270**, 185–189.

Mumby, S., Pang, I.-H., Gilman, A.G. and Sternweis, P.C. (1988) Chromatographic resolution and immunologic identification of the alpha$_{40}$ and alpha$_{41}$ subunits of guanine nucleotide-binding regulatory proteins from bovine brain. *J. Biol. Chem.*, **263**, 2020–2026.

Neer, E.J. and Clapham, D.E. (1988) Roles of G protein subunits in transmembrane signalling. *Nature*, **333**, 129–134.

Nicoletti, F., Wroblewski, J., Novelli, A., Alho, H., Guidotti, A. and Costa, E. (1986) The activation of inositol phospholipid metabolism as a signal-transducing system for excitatory amino acids in primary cultures of cerebellar granule cells. *J. Neurosci.*, **6**, 1905–1911.

Nowycky, M.C., Fox, A.P. and Tsein, R.W. (1985) Three types of neuronal calcium channel with different calcium agonist sensitivity. *Nature*, **316**, 440–443.

Ogata, N., Inoue, M. and Matsuo, T. (1987) Contrasting properties of K^+ conductances induced by baclofen and γ-aminobutyric acid in slices of the guinea pig hippocampus. *Synapse*, **1**, 62–69.

Onali, P., Schwartz, J.P. and Costa, E. (1981) Dopaminergic modulation of adenylate cyclase stimulation by vasoactive intestinal peptide in anterior pituitary. *Proc. Natl Acad. Sci. USA*, **78**, 6531–6534.

Pinnock, R.D. (1984) Hyperpolarizing action of baclofen on neurons in the rat substantia nigra slice. *Brain Res.*, **322**, 337–340.

Potashner, S.J. (1979) Baclofen: effects on amino acid release and metabolism in slices of guinea pig cerebral cortex. *J. Neurochem.*, **32**, 103–109.

Reuter, H. (1983) Calcium channel modulation by neurotransmitters, enzymes and drugs. *Nature*, **301,** 569–574.

Rydel, R.E. and Greene, L.A. (1988) cAMP analogs promote survival and neurite outgrowth in cultures of rat sympathetic and sensory neurons independently of nerve growth factor. *Proc. Natl Acad. Sci. USA*, **85,** 1257–1261.

Salomon, Y. (1978) Adenylate cyclase assay. In: G. Brooker, P. Greengard and G.A. Robison (eds) *Advances in cyclic nucleotide research*, pp. 35–55.

Sawynok, J. and Dickson, C. (1985) D-Baclofen is an antagonist at baclofen receptors mediating antinociception in the spinal cord. *Pharmacology*, **31**, 248–259.

Schmid, R., Hong, J.S., Meek, J. and Costa, E. (1980) The effects of kainic acid on the hippocampal content of putative transmitter amino acids. *Brain Res.*, **200,** 355–362.

Scott, R.H. and Dolphin, A. C. (1986) Regulation of calcium currents by a GTP analogue: potentiation of (−) baclofen-mediated inhibition. *Neurosci. Lett.*, **69**, 59–64.

Suzdak, P.D. and Gianutsos, G. (1986) Effects of chronic imipramine or baclofen on GABA-B binding and cyclic AMP production in cerebral cortex. *Eur. J. Pharmacol.*, **131**, 129–133.

Swartzwelder, H.S., Bragden, A.C., Sutch, C.P., Adult, B. and Wilson, W.A. (1986) Baclofen suppresses hippocampal epileptiform activity at low concentrations without suppressing synaptic transmission. *J. Pharmacol. Exp. Ther.*, **237**, 881–887.

Watling, K.J. and Bristow, D.R. (1986) GABA$_B$ receptor-mediated enhancement of vasoactive intestinal peptide-stimulated cyclic AMP production in slices of rat cerebral cortex. *J. Neurochem.*, **46**, 1755–1762.

Wojcik, W.J. and Neff, N.H. (1982) Adenosine A$_1$ receptors are associated with cerebellar granule cells. *J. Neurochem.*, **41**, 759–763.

Wojcik, W.J. and Neff, N.H. (1984) γ-aminobutyric acid B receptors are negatively coupled to adenylate cyclase in brain and in the cerebellum; these receptors may be associated with granule cells. *Mol. Pharmacol.*, **25**, 24–28.

Wojcik, W.J., Cavalla, D. and Neff, N.H. (1985) Co-localized adenosine A$_1$ and gamma-aminobutyric acid B ($GABA_B$) receptors of cerebellum may share a common adenylate cyclase catalytic unit. *J. Pharmacol. Exp. Ther.*, **232**, 62–66.

Wojcik, W.J., Paez, X. and Ulivi, M. (1989a) A transduction mechanism for $GABA_B$ receptors. In: E.A. Barnard and E. Costa (eds) *Allosteric modulation of amino acid receptors: therapeutic implications.* New York: Raven Press, pp. 173–193.

Wojcik, W.J., Ulivi, M., Paez, X. and Costa, E. (1989b) Islet-activating protein inhibits the beta-adrenergic receptor facilitation elicited by gamma-aminobutyric $acid_B$ receptors. *J. Neurochem.*, **53**, 753–758.

Xu, J. and Wojcik, W.J. (1986) Gamma aminobutyric acid B receptor-mediated inhibition of adenylate cyclase in cultured cerebellar granule cells: blockade by islet activating protein. *J. Pharmacol. Exp. Ther.*, **239**, 568–573.

Yatani, A., Imoto, Y., Codina, J., Hamilton, S.L., Brown, A.M. and Birnbaumer, L. (1988) The stimulatory G protein of adenylyl cyclase, G_s, also stimulates dihydropyridine-sensitive Ca^{2+} channels. *J. Biol. Chem.*, **263**, 9887–9895.

Young, R.R. and Delwaide, P.J. (1981) Spasticity. *N. Engl. J. Med.*, **304**, 28–33 and 96–99.

Zhu, X.-Z. and Chuang, D.-M. (1987) Modulation of calcium uptake and D-aspartate release by $GABA_B$ receptors in cultured cerebellar granule cells. *Eur. J. Pharmacol.*, **141**, 401–408.

9 Modulation by GABA of Agonist-induced Inositol Phospholipid Metabolism in Guinea-pig and Rat Brain

M. L. A. CRAWFORD and J. M. YOUNG

AGONIST-INDUCED INOSITOL PHOSPHOLIPID HYDROLYSIS

Agonist-induced inositol phospholipid breakdown is now firmly established as one of the major transmembrane signalling mechanisms in mammalian cells (for reviews, see Berridge, 1988; Berridge and Irvine, 1989; Fisher and Agranoff, 1987; see also Berridge and Michell, 1988). The prime event in the pathway is the combination of agonist with receptor, leading to the activation of a G-protein-coupled phosphatidylinositol-4,5-bisphosphate (PIP_2) -selective phospholipase C (phosphoinositidase C, PIC) with the consequent hydrolysis of the lipid and the formation of diacylglycerol (DAG) and inositol 1,4,5-trisphosphate (1,4,5-IP_3).The power and subtlety of the so-called PI pathway derives from the fact that both DAG and 1,4,5-IP_3 can act as second messengers in the cell, respectively causing Ca^{2+} release from the endoplasmic reticulum and activating a calcium- and phospholipid-dependent protein kinase, protein kinase C (PKC) (Figure 9.1).

PKC represents a family of closely related kinases (for review see Nishizuka, 1988), at least one of which is activated with almost equal potency by arachidonic acid, itself a possible product of the PI cycle in cells possessing DAG lipase activity. PKC can phosphorylate a variety of cell proteins, leading for example to activation or inactivation of ion channels or to inactivation of receptors, both those coupled to adenylyl cyclase and those coupled to PIC. PKC-induced inactivation of PIC is clearly a simple mechanism of autoregulation of the cycle, but the bifurcating pathways and the amplifying effect of enzyme or channel activation offer a variety of mechanisms for the control of cellular function. It would be surprising indeed if the activity of such a system were not regulated by neuromodulatory agents.

The metabolism of 1,4,5-IP_3 also presents a surprisingly complex picture (for reviews, see Irvine *et al.*, 1988; Majerus *et al.*, 1988; Shears, 1989), which is shown in simplified form in Figure 9.1. Stepwise hydrolysis yields 1,4-IP_2, 4-IP_1

$GABA_B$ Receptors in Mammalian Function. Edited by N.G. Bowery, H. Bittiger and H.-R. Olpe

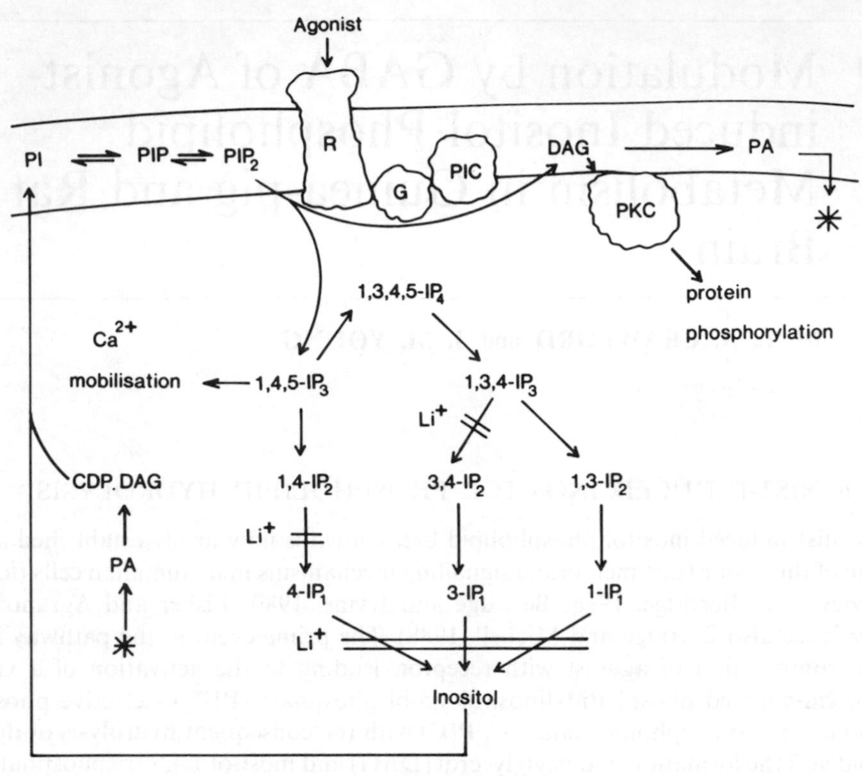

Figure 9.1. Pathways of inositol phospholipid metabolism. The first step in the pathway is the hydrolysis of PIP_2 by activated PIC (shown schematically as an agonist-R-G-PIC association). The scheme for 1,4,5-IP_3 breakdown shows what are believed to be the major pathways in mammalian cells. The route of PA recycling is indicated by the asterisks. DAG, diacylglycerol; G, GTP-binding protein (G_P); IP_n, inositol mono-, bis-, tris- or tetrakisphosphate ($n = 1$, 2, 3, or 4, respectively); PA, phosphatidic acid; PIC, phosphoinositidase C; PI, phosphatidylinositol; PIP, phosphatidylinositol-4-phosphate; PIP_2, phosphatidylinositol-4,5-bisphosphate; PKC, protein kinase C; R, receptor. Direct hydrolysis of PI by PIC would yield 1-IP_1 and hydrolysis of PIP would give 1,4-IP_2. The major action of 10 mM Li^+ is to block the hydrolysis of the monophosphates

and inositol, but 1,4,5-IP_3 can also be phosphorylated to yield 1,3,4,5-IP_4, which is then hydrolysed to 1,3,4-IP_3 by the same 5-phosphatase responsible for 1,4,5-IP_3 hydrolysis. 1,3,4-IP_3 then undergoes further stepwise hydrolysis to inositol, the particular bis- and monophosphate isomers formed depending on the cells involved. Of the various metabolic products, only 1,3,4,5-IP_4 has been proposed to have a physiological role in mediating Ca^{2+} entry in certain tissues (Irvine *et al.*, 1988). Some 1:2-cyclic-4,5-IP_3 may be formed on PIC hydrolysis of PIP_2 but it now seems unlikely that it has any significant role in cell function. Similarly, higher phosphates have been detected in certain cells and a neuroeffector role has been proposed for 1,3,4,5,6-IP_5 and IP_6 (Vallejo *et al.*, 1987), but the levels of these phosphates change only slowly in stimulated cells and it seems that their synthesis may take place by a pathway independent of the PI cycle.

Measurement of the [^{3}H] inositol phosphates formed in cells/tissues in which the phosphoinositides have been prelabelled by incubation with [^{3}H] inositol has become much the most widely used assay of PIC activation, particularly since the realization that the inositol cycle is blocked by lithium salts in the millimolar concentration range (Berridge *et al.*, 1982, and references therein). The major site of action of lithium is at the level of the monophosphatase, leading to the accumulation of IP_1, but the 1-phosphatase enzyme which hydrolyses 1,4-IP_2 and 1,3,4-IP_3 is also inhibited to some extent. Two experimental approaches are commonly employed. In studies aimed at establishing biochemical pathways or relating PIP_2 hydrolysis to some functional response, measurements have normally been made at very short times, seconds, in the absence of Li^+ and the inositol phosphate isomers formed separated by high-performance liquid chromatography. However, in most pharmacological studies measurements have been made at much longer times, 15–60 min, in the presence of 5–10 mM Li^+ and either total phosphates measured or the mono-, bis- and trisphosphates (isomers unresolved) separated on anion-exchange columns. Since the hydrolysis of [^{3}H]IP_1 is blocked, the amount of [^{3}H]inositol phosphates formed over a given time is a measure of the amount of PIP_2 broken down. After an extended period of stimulation [^{3}H]IP_1 is normally much the major product, and the parameters characterizing concentration–response curves for agonist-induced [^{3}H]IP_1 provide very good estimates of those for total [^{3}H]IP formation. However, it must always be borne in mind that it is radioactivity that is normally measured and not mass. Cells in culture can be labelled to equilibrium with [^{3}H]inositol (24–48 h in culture are usually allowed), but this is not a feasible proposition for brain slices. It should also be noted that PIC activity measured over very short time periods, seconds, may not be the same as average activity measured over much longer periods, when modulation of PIC activity or activation of secondary pathways of PI, PIP or PIP_2 breakdown could significantly modify [^{3}H]inositol phosphate formation. However, in investigations in which the modulation of PI responses is the prime interest, the use of extended periods of incubation with agonist has some advantages, at least as a primary screen.

GABA-INDUCED INOSITOL PHOSPHATE FORMATION

GABA (2 mM), acting alone, induced the formation of [^{3}H]inositol phosphates in cross-chopped slices of rat cerebral cortex prelabelled with [^{3}H]-inositol (Table 1). The increase in [^{3}H]IP_1 was small and was only apparent after approximately 30 min incubation (Figure 9.2c). Concentrations of GABA below 0.5 mM had no significant effect on [^{3}H]IP_1 formation. The percentage

Table 1. Effect of GABA on [^{3}H]inositol phosphate formation in guinea-pig and rat brain

	GABA-induced [^{3}H]IP$_n$ accumulation (% of basal)		
	[^{3}H]IP$_1$	[^{3}H]IP$_2$	[^{3}H]IP$_3$
Rat cerebral cortex	137 ± 4 (87)	191 ± 8 (47)	180 ± 5 (47)
Guinea-pig cerebellum	71 ± 3 (19)	83 ± 3 (7)	59 ± 8 (7)
Guinea-pig cerebral cortex	102 ± 3 (11)	120 ± 7 (6)	107 ± 1 (6)

Cross-chopped slices (350 × 350 μm) of rat (Wistar strain, males, 200–350 g) cerebral cortex or guinea-pig (Dunkin–Hartley strain, males, 350–500 g) cerebellum or cerebral cortex were incubated with [^{3}H]inositol (1 μCi per incubation) in Krebs–Henseleit medium, containing 10 mM LiCl, at 37 °C for 30 min before addition of 2 mM GABA (or, in later experiments, other agonists) and further incubation for 60 min. The reaction was terminated and [^{3}H]inositol phosphates were extracted and separated as described by Crawford and Young (1988, 1990). Values are means ± SE with the number of determinations in parentheses. IP$_n$, inositol mono-, bis- or trisphosphate (n = 1, 2 or 3, respectively).

increase in the amounts of the higher phosphates, [^{3}H]IP$_2$ and [^{3}H]IP$_3$, over basal levels induced by 2 mM GABA was greater than for [^{3}H]IP$_1$, but the higher phosphates form only a small portion of the total phosphates (*c.* 10%), the greater bulk being [^{3}H]IP$_1$. The stimulatory action of GABA on [^{3}H]IP formation was abolished when Ca^{2+} was omitted from the medium (see Table 3 below for data from one series of experiments). This suggests that the stimulatory action is indirect and could involve the release of a directly acting agonist.

The stimulatory effect of GABA was a consistent feature of experiments with rat cerebral cortical slices, but was not observed in the two other brain tissues that we have examined (Table 1). In guinea-pig cerebral cortex 2 mM GABA had no effect on [^{3}H]IP$_1$ accumulation, whereas in guinea-pig cerebellum the usual effect was inhibition. These differences between tissues may well reflect differences in the nature of basal accumulation, which is usually highest in guinea-pig cerebellum (Carswell and Young, 1986). The inhibitory effect of GABA in this tissue might well reflect inhibition of the response to an endogenous agonist, as discussed later. In all experiments on the effect of agonist-induced responses in rat cerebral cortex and guinea-pig cerebellum the assumption has been made that the effect on basal [^{3}H]IP formation is independent of any effect on the response of the agonist.

GABA INHIBITION OF HISTAMINE-INDUCED INOSITOL PHOSPHATE FORMATION

There is a particular interest in effects of GABA on histamine-induced responses in brain. The presence in rat brain of a system of histamine-containing nerve fibres is now well established (for review, see Pollard and Schwartz, 1987). The perikarya are located in the posterior hypothalamus, particularly in the caudal magnocellular mammillary nuclei, while the fibres form a diffuse ascending system innervating higher areas. However, almost all the magnocellular nuclei in the posterior hypothalamus that contain histidine decarboxylase, the histamine synthesizing enzyme, also stain for glutamic acid decarboxylase (Takeda *et al.*, 1984). The histaminergic fibres, like other monoamine fibres, appear to make few close synaptic contacts and the presumption must be that the action of histamine is that of a neuromodulator. Whether GABA is co-released with histamine is unknown, but the possibility of a functional interaction between the two is clear.

Histamine, acting at H_1 receptors, induced a 4.6 ± 0.2-fold increase in $[^3H]IP_1$ accumulation in rat cerebral cortical slices, which was inhibited in a non-competitive fashion by 2 mM GABA, i.e. the maximum response to histamine was reduced without any significant change in the median effective concentration (EC_{50}). The time-course of the inhibition of $[^3H]IP_1$ formation by 2 mM GABA, together with the time-course for histamine acting alone, is shown in Figure 9.2. The lag of approximately 10 min before the maximum rate of $[^3H]IP_1$ formation occurs is typical for histamine. This makes it difficult to measure inhibition by GABA at early times, but the effect is apparent at times after 10 min.

EFFECT OF BACLOFEN AND OTHER GABA MIMETICS ON HISTAMINE-INDUCED [^{3}H]INOSITOL PHOSPHATE FORMATION

The inhibitory effect of GABA on histamine-induced $[^3H]$inositol phosphate formation was mimicked by (−)-baclofen (Figure 9.3). The effect of baclofen was exerted at low concentrations (EC_{50} $0.69 \pm 0.04\ \mu M$) and was also non-competitive, as indicated by the coincidence of the inhibition curves for the responses to 0.2 and 1 mM histamine. The effect was strongly stereospecific, with (+)-baclofen having an inhibitory action only at concentrations 400-fold greater than those required with the (−)-isomer (Figure 9.3). (−)-Baclofen also inhibited histamine-stimulated formation of $[^3H]IP_2$ and $[^3H]IP_3$ to a similar extent to that observed with $[^3H]IP_1$. Isoguvacine, a $GABA_A$-selective agonist, had no significant effect on the histamine response at concentrations up to 1 mM and neither 100 μM isoguvacine nor 1 μM (−)-baclofen altered basal $[^3H]IP_1$ accumulation. Muscimol did inhibit the histamine response (Figure 9.4), but only at concentrations above those expected for a $GABA_A$ response. Taking the stereospecific action of baclofen to define the $GABA_B$ receptor, then the inhibitory action of GABA on histamine-induced $[^3H]IP_1$ formation can be described as $GABA_B$ receptor mediated.

The only other system in which a $GABA_B$-receptor-mediated inhibition of

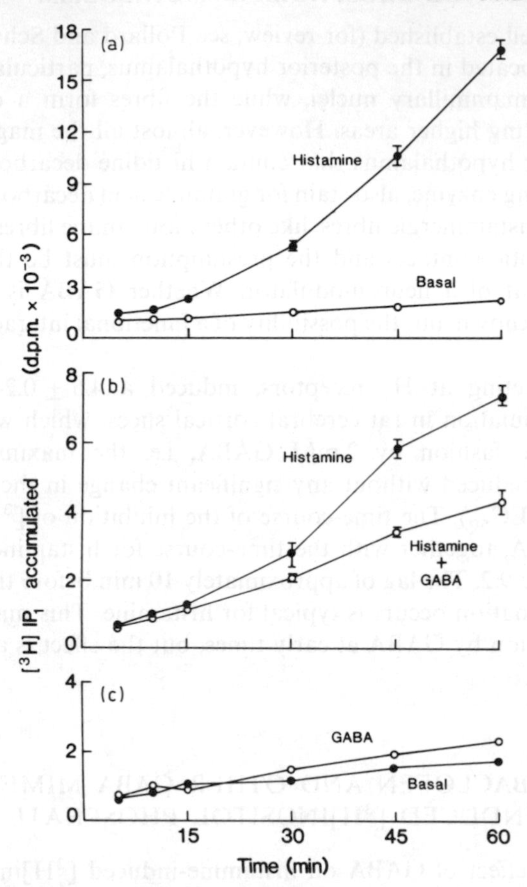

Figure 9.2. Time-course of histamine-induced [^{3}H]inositol monophosphate [^{3}H]IP$_1$ formation in rat cerebral cortical slices in the presence and absence of GABA. Accumulation of [^{3}H]IP$_1$: (a) in the presence of 1 mM histamine (●) or without drug addition (basal) (○); (b) in the presence of 1 mM histamine in the absence (●) or presence (○) of 2 mM GABA; (c) without drug addition (basal) (●) or in the presence of 2 mM GABA (○). GABA was added at the same time as histamine. Curves in (a), (b) and (c) are from independent experiments on different slice preparations. Each point is the mean ± SE of 3 replicate determinations. Each set of experiments was repeated with essentially similar results. (From Crawford and Young, 1988, by permission of the International Society for Neurochemistry Ltd.)

agonist-stimulated phosphoinositide metabolism has been described is 5-hydroxytryptamine (serotonin) receptor (5-HT$_2$) -induced [^{3}H]IP$_1$ formation in slices of rat cerebral cortex (Godfrey *et al.*, 1988). Qualitatively the effects of GABA and (−)-baclofen are similar to those described above against histamine in rat cerebral cortex, but there are differences in the concentrations required. The EC$_{50}$ for (−)-baclofen inhibition of total [^{3}H]IP formation in

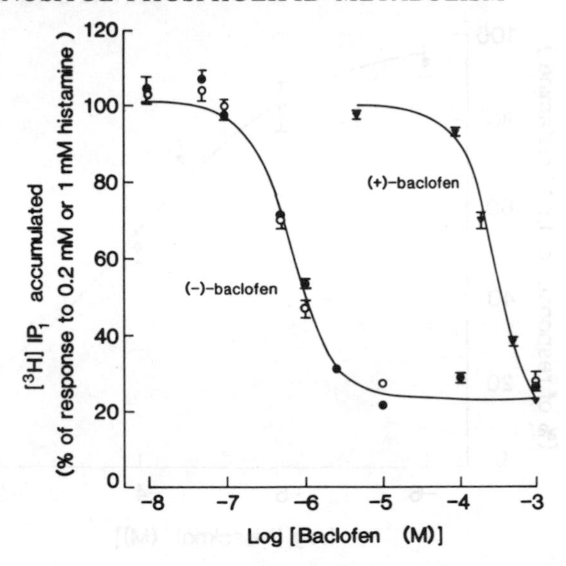

Figure 9.3. Stereospecificity of the inhibition by baclofen of histamine-induced formation of $[^3H]$inositol monophosphate ($[^3H]IP_1$). Incubation with 0.2 or 1 mM histamine and a set concentration of (–)- or (+)-baclofen was for 60 min. The basal level (no histamine present, but with the appropriate concentration of (–)- or (+)-baclofen) was substracted before expression of the results as a percentage of the response to 0.2 mM or 1 mM histamine in the absence of baclofen. (–)-Baclofen (●, ○); points are the means ± approximate SE from 2–6 determinations with 0.2 mM (○) or 1 mM (●) histamine. (+)-Baclofen (▼); points are the weighted means from 2 or 3 independent determinations with 1 mM histamine. (From Crawford and Young, 1988, by permission of the International Society for Neurochemistry Ltd.)

mouse cerebral cortex was 100 nM, compared with 690 nM against histamine in the rat. However, the difference was much bigger with GABA, *c.* 10 μM for 50% inhibition against 5-HT and 300 μM against histamine, both measured in the absence of inhibitors of uptake or metabolism. We have not examined the effect of uptake inhibitors against histamine. Nipecotic acid, which markedly potentiated GABA action in mouse cerebral cortex, has other actions in rat cerebral cortex (see below) and the use of the competitive non-transported inhibitors such as SK&F 89976A, 100330A or 100561A is ruled out against histamine by the fact that they are also potent H_1-receptor antagonists dissociation constants: 0.25 ± 0.01 μM (2 measurements), 0.23 ± 0.01 μM (3 measurements) and 0.24 ± 0.03 μM (1 measurement), respectively, determined from curves of inhibition of $[^3H]$ mepyramine binding to H_1-receptors in guinea-pig cerebral cortical homogenates, as described by Treherne and Young, 1988.

It may be noted that, both in the mouse against 5-HT and in the rat against histamine, GABA and (–)-baclofen were unable to cause a complete inhibition of the response (maximum inhibition by (–)-baclofen in mouse cerebral cortex

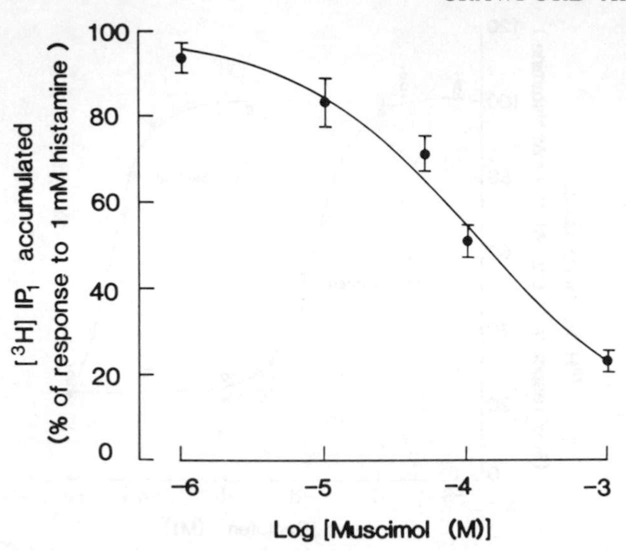

Figure 9.4. Inhibition of histamine-induced [^{3}H]inositol monophosphate ([^{3}H]IP$_1$) accumulation by muscimol. Values are expressed as the percentage of the response to 1 mM histamine alone and are the means $\pm$ approximate SE from 3 determinations

42 $\pm$ 5%; Godfrey *et al.*, 1988). This seems to be a general observation for inhibition of inositol phosphate formation, for example adenosine versus histamine in mouse cerebral cortex (Kendall and Hill, 1988) and rat striatum (Petcoff and Cooper, 1987). It should also be noted, in view of the apparent complexity of the actions of GABA in rat cortex described below, that in this series of experiments the best-fit maximum inhibition of the histamine response was similar for GABA and for ($-$)-baclofen (69 $\pm$ 2 and 76 $\pm$ 1%, respectively).

EFFECT OF GABA ON CARBACHOL- AND NORADRENALINE-STIMULATED [^{3}H]IP FORMATION

It has been reported (Brown *et al.*, 1984) that 1 mM GABA has no effect on [^{3}H]IP formation induced by carbachol in rat cerebral cortex. We have confirmed and extended this observation. Concentration–response curves for carbachol-induced [^{3}H]IP$_1$ formation in the presence and absence of 2 mM GABA are shown in Figure 9.5. Clearly 2 mM GABA has no effect on either the EC$_{50}$ (32 $\pm$ 2 μM) or the maximum attainable response. This is a valuable control, in that it indicates that 2 mM GABA does not have a non-selective toxic action on the tissue. It may also be noted that the carbachol concentration–response curve approximates well to a hyperbola (Hill coefficient 0.97 $\pm$ 0.05). This was not the case for noradrenaline-induced [^{3}H]IP$_1$ accumulation (Hill coefficient 2.05 $\pm$ 0.18), even though the magnitude of the response to the two

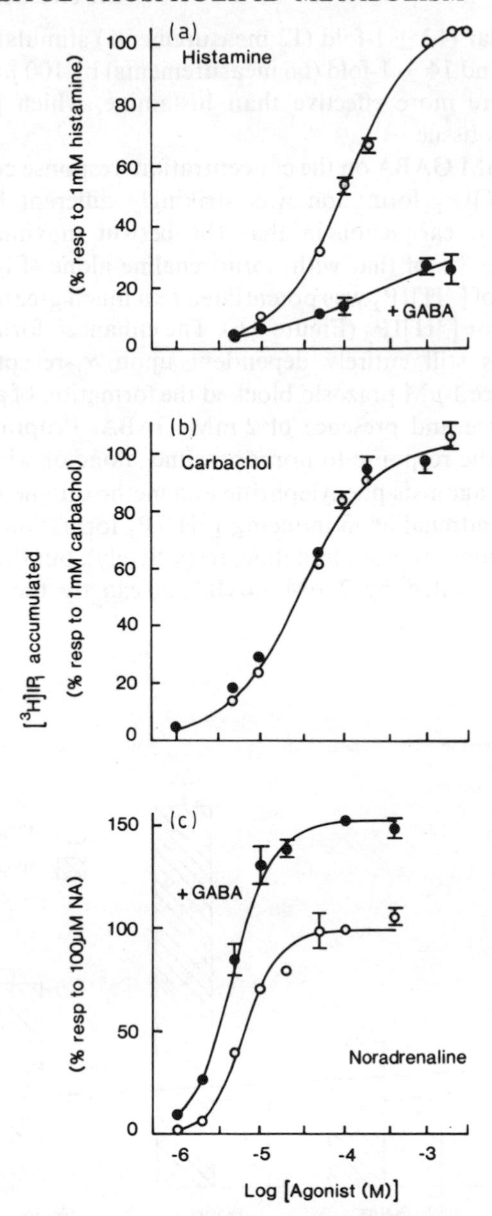

Figure 9.5. Effect of 2 mM GABA on concentration–response curves for [^{3}H]inositol monophosphate ([^{3}H]IP$_1$) accumulation stimulated by (a) histamine, (b) carbachol and (c) noradrenaline (NA) in rat cerebral cortical slices. Points are the weighted means ± approximate SE from 2–8 experiments and are expressed as a percentage of the response to: (a) 1 mM histamine alone; (b) 1 mM carbachol alone; and (c) 100 μM noradrenaline alone. GABA and agonist were added together. (○), Agonist alone; (●), agonist plus 2 mM GABA

agonists was similar (12 ± 1-fold (12 measurements) stimulation over basal by 1 mM carbachol and 14 ± 1-fold (66 measurements) by 100 μM noradrenaline). Both agonists were more effective than histamine, which gave a 4 ± 1-fold stimulation in this tissue.

The effect of 2 mM GABA on the concentration–response curve for noradrenaline-induced [^{3}H]IP$_1$ formation was strikingly different from that against either histamine or carbachol, in that the best-fit maximum response was increased to $153 \pm 3\%$ of that with noradrenaline alone (Figure 9.5). Further, the accumulation of [^{3}H]IP$_2$ was potentiated to a much greater extent than that of either [^{3}H]IP$_1$ or [^{3}H]IP$_3$ (Figure 9.6). The enhanced formation of [^{3}H]IP$_2$ and [^{3}H]IP$_1$ was still entirely dependent upon α_1-receptor activation by noradrenaline, since 1 μM prazosin blocked the formation of all the phosphates, both in the absence and presence of 2 mM GABA. Propranolol (1 μM) was without effect on the response to noradrenaline, alone or with GABA.

The α_1-selective agonists phenylephrine and methoxamine were partial agonists relative to noradrenaline in inducing [^{3}H]IP$_1$ formation (42% and 49% of the maximum response to noradrenaline, respectively), but the response to both agonists was potentiated by 2 mM GABA in exactly the same way as for

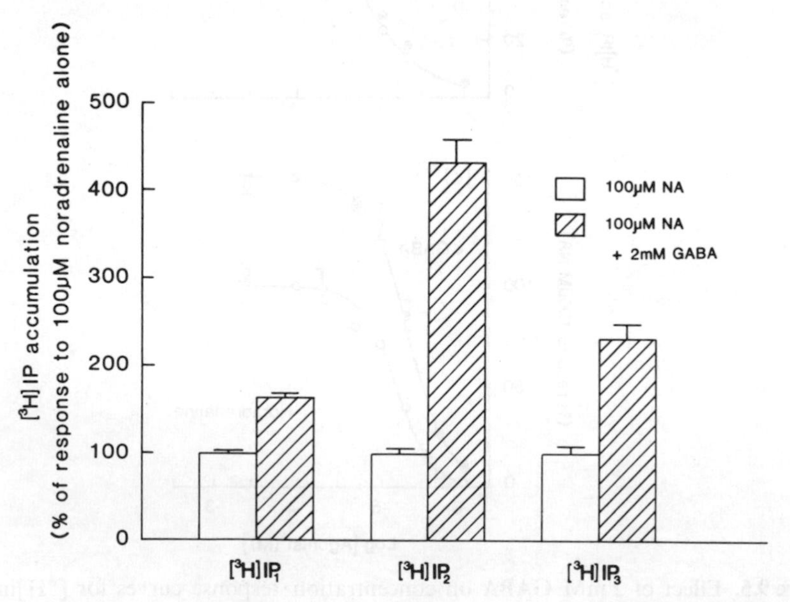

Figure 9.6. GABA potentiation of noradrenaline (NA) -induced formation of the [^{3}H]inositol phosphates, [^{3}H]IP$_1$, [^{3}H]IP$_2$ and [^{3}H]IP$_3$. The amount of each phosphate accumulated in the presence of 100 μM noradrenaline plus 2 mM GABA is expressed as a percentage of the response to 100 μM noradrenaline alone. Values are weighted means $\pm$ approximate SE from 39 ([^{3}H]IP$_1$) or 27 determinations. Open columns, noradrenaline alone; hatched columns, noradrenaline plus GABA

noradrenaline, with a much greater effect on [^{3}H]IP$_2$ than on [^{3}H]IP$_1$ (Crawford and Young, 1990). Methoxamine is neither a substrate for uptake nor for monoamine oxidase and this, together with the reported insensitivity of the noradrenaline response to blockers of these systems, indicates that the action of GABA is on some component of the PI response.

The EC$_{50}$ for noradrenaline-induced [^{3}H]IP$_2$ formation in the presence of 2 mM GABA, 9.1 $\pm$ 0.5 μM, was similar to that without GABA, 8.7 $\pm$ 0.1 μM, and of the same order as that for noradrenaline-induced [^{3}H]IP$_1$ formation, 6.3 $\pm$ 0.4 μM. However, there were differences in concentration–response curves for GABA potentiation of [^{3}H]IP$_1$ and [^{3}H]IP$_2$ accumulation (Figure 9.7). The curve for [^{3}H]IP$_1$ appeared to be near maximal at 2 mM GABA, confirmed in two experiments with 5 mM GABA, and had an EC$_{50}$ of 0.51 $\pm$ 0.07 mM (cf. 0.30 $\pm$ 0.03 mM for the half-maximal inhibitory concentration, IC$_{50}$, for inhibition of the response to histamine). In contrast, the curve for [^{3}H]IP$_2$ showed a similar increase to that for [^{3}H]IP$_1$ up to approximately 0.5 mM GABA, but thereafter increased sharply. Two comparative measurements at 2 mM and 5 mM GABA failed to show any further increase at the higher GABA concentration, which would put the EC$_{50}$ for [^{3}H]IP$_2$ at *c.* 1.5 mM. This difference between the two curves raises the possibility that at higher concentrations GABA is acting at more than a single site to modify the response to noradrenaline. Alternatively, the response to noradrenaline may be more complex than schemes such as that shown in Figure 9.1 would suggest.

The suggestion has been made (Maier and Rutledge, 1987) that, in slices of rat cerebral cortex, noradrenaline may stimulate the hydrolysis of PI as well as PIP$_2$. We have no indication that this is the case with the experimental protocol that we have used. The time-course of noradrenaline-induced [^{3}H]IP$_1$ formation in the presence or absence of GABA shows the same lag as observed with histamine, whereas GABA enhancement of [^{3}H]IP$_2$ formation is apparent at early times (Crawford and Young, 1990). In any case [^{3}H]IP$_2$ could only be derived from PIP or PIP$_2$, not from PI. We have looked for evidence of GABA enabling α_1-receptor-mediated PIP hydrolysis (bearing in mind the prazosin sensitivity) by comparing the ratios [^{3}H]IP$_2$:[^{3}H]IP$_1$ and [^{3}H]IP$_3$:[^{3}H]IP$_1$ in the presence and absence of GABA at various times after addition of 100 μM noradrenaline (Figure 9.8). At all times after 5 min (at which time the stimulated accumulation is small and the errors are consequently large) the ratio [^{3}H]IP$_2$: [^{3}H]IP$_1$ is greater in the presence of 2 mM GABA (Figure 9.8a). If this 'excess' [^{3}H]IP$_2$ in the presence of GABA were related to PIP breakdown, then since this cannot lead to [^{3}H]IP$_3$ formation the ratio [^{3}H]IP$_3$:[^{3}H]IP$_1$ should fall. This assumes, of course, that any [^{3}H]IP$_2$ formed from PIP would, at least in part, be hydrolysed to [^{3}H]IP$_1$. However, there is no indication that the ratio [^{3}H]IP$_3$:[^{3}H]IP$_1$ changes in the presence of GABA (Figure 9.8b) and hence there is no evidence for noradrenaline-induced hydrolysis of any lipid other than PIP$_2$.

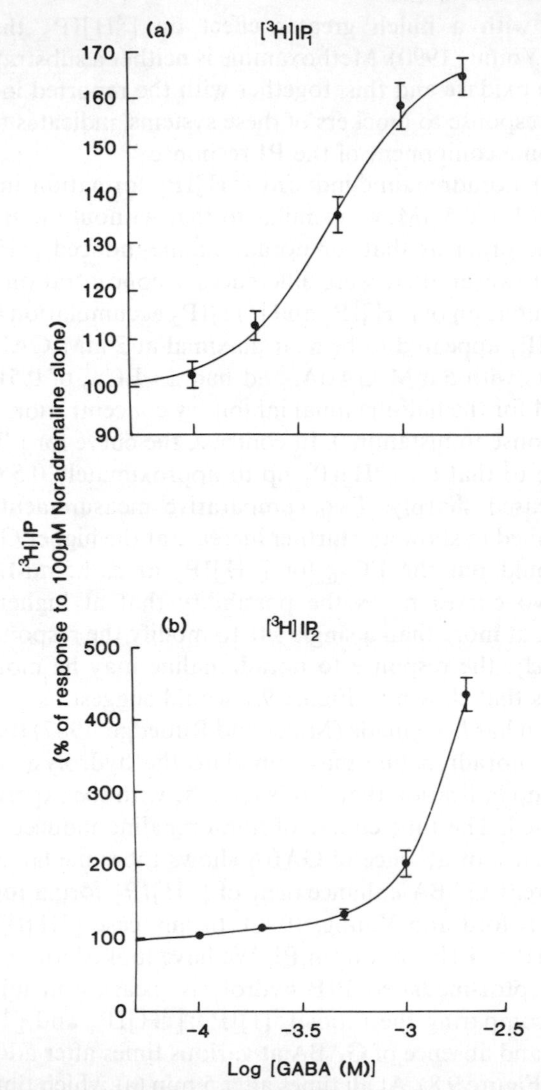

Figure 9.7. Concentration–effect curves for GABA potentiation of the response to noradrenaline. [^{3}H]Inositol mono- and bisphosphate ([^{3}H]IP_1 and [^{3}H]IP_2) formation induced by 100 μM noradrenaline in the presence of GABA is expressed as a percentage of the response to noradrenaline alone. (a) [^{3}H]IP_1. Each point is the mean ± approximate SE from 4–6 determinations (39 determinations at 2 mM GABA). The line drawn was obtained by fitting the points to a Hill equation using non-linear regression. Best-fit parameters: Hill coefficient 1.80 ± 0.26, EC_{50} 0.51 ± 0.07 mM and maximum response 170 ± 5%. (b) [^{3}H]IP_2. Points are the means ± approximate SE from 3 experiments (27 determinations for 2 mM GABA). Note the different scales of the ordinate in (a) and (b). (From Crawford and Young, 1990, by permission of the International Society for Neurochemistry Ltd.)

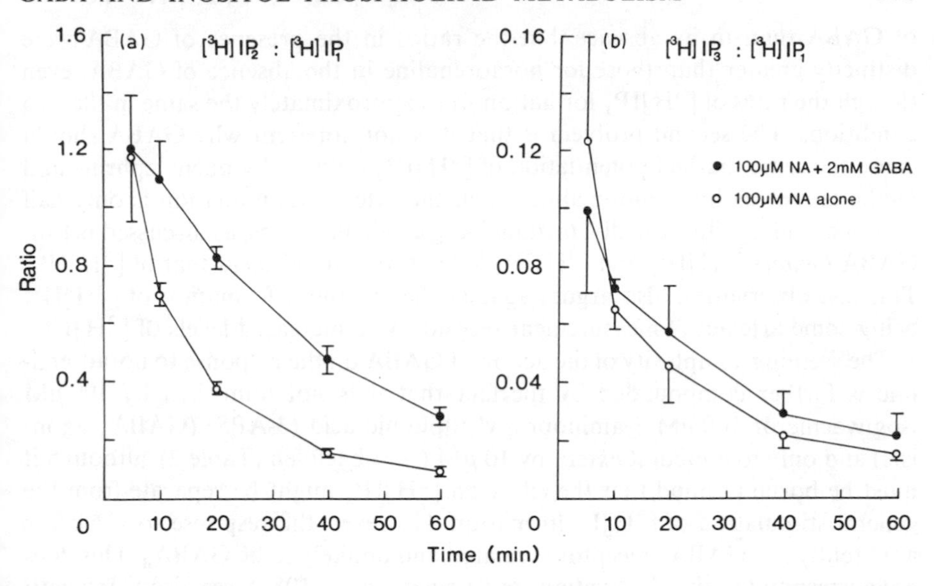

Figure 9.8. Effect of GABA on the [^{3}H]inositol phosphate ratios (a) [^{3}H]IP$_2$:[^{3}H]IP$_1$ and (b) [^{3}H]IP$_3$:[^{3}H]IP$_1$ in response to noradrenaline (NA). Points are means ± SE from 5 experiments with 100 μM noradrenaline alone and 4 experiments with 100 μM noradrenaline plus 2 mM GABA. Basal levels of [^{3}H]IP$_n$ ($n = 1$, 2 or 3) accumulation in the presence or absence of GABA have been subtracted before calculation of ratios. (○) Noradrenaline; (●) noradrenaline plus GABA. (From Crawford and Young, 1990, by permission of the International Society for Neurochemistry Ltd.)

Another possibility for explaining the enhanced accumulation of [^{3}H]IP$_2$ is that GABA is acting as an inhibitor of [^{3}H]IP$_2$ metabolism. However, if this were simple blockade at the level of the phosphatase (which would imply an intracellular action), then this should also be apparent in the absence of Li^+. This is not so. In three experiments in the absence of LiCl, noradrenaline-induced accumulation of all the phosphates was greatly reduced, but the effect of 2 mM GABA was to enhance the accumulation of [^{3}H]IP$_1$ (421 ± 20% of the response to noradrenaline alone) much more than that of [^{3}H]IP$_2$ (187 ± 7%) (Crawford and Young, 1990). If there is an action of GABA at the level of a phosphatase, then it has to be in concert with Li^+.

The simplest explanation for the action of GABA on [^{3}H]IP$_2$ levels would be that it is a consequence of increased flux through the phosphatase pathways in the presence of an uncompetitive inhibitor, Li^+, which will increase in effect as the substrate concentration rises (Inhorn and Majerus, 1987). This simple flux explanation is attractive, but there are three pieces of evidence, albeit incomplete or indirect, which cast some doubt on whether it is correct. The first problem is that in preliminary experiments with 100 μM phenylephrine and 100 μM methoxamine, the ratios [^{3}H]IP$_2$:[^{3}H]IP$_1$ not only were bigger in the presence of GABA than in its absence, but the ratios in the presence of GABA were distinctly greater than those for noradrenaline in the absence of GABA, even though the rates of $[^3H]IP_1$ formation were approximately the same in the two conditions. The second problem is that it is not apparent why GABA should cause the same marked potentiation of $[^3H]IP_2$ induced by phenylephrine and methoxamine as by noradrenaline, when the rate of accumulation is only half that of noradrenaline. Finally, in guinea-pig cerebellar slices, as discussed below, GABA *inhibits* $[^3H]IP_1$ and $[^3H]IP_3$ formation but *enhances* that of $[^3H]IP_2$. This last observation also argues against the enhanced formation of $[^3H]IP_2$ being some artefact of measurement secondary to increased levels of $[^3H]IP_1$.

The seeming complexity of the action of GABA on the response to noradrenaline is further compounded by the fact that it is not mimicked by 100 μM isoguvacine or 100 μM 3-aminopropylsulphonic acid (3-APS) ($GABA_A$ agonists) and only to a modest extent by 10 μM (−)-baclofen (Table 2), although it must be borne in mind that the effect on $[^3H]IP_2$ might be separate from the general stimulation of $[^3H]IP$ formation. However, the response to GABA is apparently not $GABA_A$ receptor mediated and unlikely to be $GABA_B$. This does not correspond to the observations of Ruggiero *et al.* (1987) (see also Corradetti *et al.*, 1987), who have previously reported that GABA enhances noradrenaline-induced $[^3H]IP$ formation in vibratome-cut slices of rat hippocampus. The action of GABA was mimicked by 3-APS with an EC_{50} of *c.* 10 μM and was blocked by 10 μM bicuculline. Muscimol (1 μM) produced the same potentiation as 100 μM 3-APS. The evidence for a $GABA_A$ action in rat hippocampus is

Table 2. Effect of GABA analogues on noradrenaline-stimulated $[^3H]$inositol phosphate formation in rat cerebral cortex

	$[^3H]IP_n$ accumulation in the presence of GABA analogue (% of the response to noradrenaline alone)		
	$[^3H]IP_1$	$[^3H]IP_2$	$[^3H]IP_3$
GABA (2 mM)	164 ± 4 (39)	434 ± 24 (27)	233 ± 16 (27)
3-APS (100 μM)	101 ± 5 (4)	104 ± 11 (3)	76 ± 3 (3)
Isoguvacine (100 μM)	97 ± 5 (4)	96 ± 5 (3)	101 ± 6 (3)
(−)-Baclofen (10 μM)	127 ± 5 (5)	148 ± 10 (3)	132 ± 6 (3)
Nipecotic acid (2 mM)	128 ± 5 (4)	234 ± 2 (3)	205 ± 16 (3)

Values are means ± approximate SE with the number of determinations in parentheses. Incubations were with 100 μM noradrenaline in the presence or absence of GABA analogue. Basal values (with or without GABA analogue, as appropriate) were subtracted before calculation of the ratios. IP_n, inositol mono-, bis- or trisphosphate (n = 1, 2 or 3, respectively); 3-APS, 3-aminopropylsulphonic acid.

therefore strong. However, 100 μM bicuculline failed to block the actions of GABA in our system and preliminary measurements with 10 and 100 μM muscimol failed to show any significant increase in [^{3}H]IP$_1$ formation. However, it is important to note that the experimental protocol used in the experiments with hippocampus differs in certain respects from that which we have used. In particular, inositol lipids were 'pulse-labelled' with [^{3}H]inositol, which was not present during the incubation with noradrenaline, which was for 10 min. However, the stimulated incorporation of radioactivity into PI, particularly by muscimol, observed by Ruggiero *et al.* (1987) indicates that lipid labelling was still taking place during agonist stimulation. The incorporation of [^{3}H]inositol into total inositol phospholipids in rat cerebral cortex is enhanced numerically, but not significantly, by 100 μM noradrenaline (121 $\pm$ 8% over basal; 4 measurements, 60 min incubation) and possibly decreased, but not significantly, by 2 mM GABA (13 $\pm$ 5% inhibition; 5 measurements) so that in the presence of noradrenaline and GABA the labelling was unchanged compared to basal (105 $\pm$ 3%; 3 measurements). However, we have not separated the individual phospholipids and since PI is much the most abundant inositol lipid, it is possible that significant changes in the labelling of PIP and PIP$_2$ may have occurred. However, it has again to be borne in mind that the response to carbachol is not altered by GABA in rat cerebral cortex (or by 3-APS in hippocampus) and the response to histamine is inhibited. If the effect of GABA on the response to noradrenaline were due to a change in lipid labelling, then this change would have to be confined to the lipids accessible to noradrenaline. In any circumstances if there were a GABA$_A$ component in rat cortex, isoguvacine and 3-APS might have been expected to have given at least some response. The conclusion must be that, although GABA potentiates noradrenaline-induced [^{3}H]IP formation in both cerebral cortical and hippocampal preparations from the rat, the mechanism apparently differs in the two tissues.

The lack of any simple receptor pharmacology of the response to GABA in rat cerebral cortex and the high concentrations required are discouraging. Nipecotic acid, an inhibitor of GABA uptake, itself mimicked, in part, the action of GABA and in particular stimulated the accumulation of [^{3}H]IP$_2$ to a markedly greater extent than [^{3}H]IP$_1$ (Table 2), although still less than 2 mM GABA. Nipecotic acid also had a very small, but significant, stimulatory effect on basal [^{3}H]IP$_1$ accumulation (14 $\pm$ 3%; 4 measurements). The action of GABA and nipecotic acid is not additive, since 2 mM nipecotic acid failed to cause any significant increase in [^{3}H]IP$_1$ or [^{3}H]IP$_2$ accumulation induced by 100 μM noradrenaline plus 2 mM GABA.

The potentiating effect of nipecotic acid, which is a substrate for the GABA uptake system (Johnston *et al.*, 1976), raises the question of whether the action of GABA and nipecotic acid might be intracellular. We have attempted to test this with SK&F 100330A. However, although SK&F 100330A is a weaker inhibitor of α_1 receptors than of H_1 receptors (see above), at the concentrations required (100 μM) to ensure marked inhibition of the uptake of 2 mM GABA (assuming the values of the K_m for GABA and the K_i for SK&F 100330A given by Larsson *et al.*, 1988) there was significant inhibition of the incorporation of [^{3}H]inositol into inositol phospholipids and inhibition of both basal and noradrenaline-stimulated [^{3}H]IP$_1$ formation. However, there was no indication of a greater inhibition of [^{3}H]IP$_2$ accumulation induced by 100 μM noradrenaline plus 2 mM GABA than by noradrenaline alone. This gives no indication of an additional site of SK&F 100330A action in the presence of GABA and, conversely, gives no indication that GABA has any intracellular effects.

The pharmacology of the action of GABA on the response to noradrenaline in rat cerebral cortex thus remains unclear. There have been reports of responses to GABA in brain which are neither GABA$_A$ nor GABA$_B$ mediated and the existence of a GABA$_C$ receptor has been proposed (Johnston, 1986). However, our current evidence is insufficient to allow the proposition that a novel GABA receptor type is involved. Indeed, in view of the high concentration of GABA required for the effects on noradrenaline-induced [^{3}H]IP formation it remains to be established that the effect of GABA is on a receptor that is primarily a receptor for GABA.

IS CALCIUM INVOLVED IN GABA MODULATION OF RESPONSES TO HISTAMINE AND NORADRENALINE?

The differing effects of GABA on the response to carbachol, histamine and noradrenaline in the same tissue could well mean that muscarinic, H_1 and α_1 receptors are located on different cell types. However, they might equally well reflect differences in the mechanism of agonist-induced [^{3}H]IP$_1$ formation. The possibility of noradrenaline-induced PI or PIP breakdown has been discussed above, but there might also be a secondary phase of enhanced PIP$_2$ breakdown, which may be a target for modulators of agonist-stimulated [^{3}H]IP formation (for review, see Linden and Delahunty, 1989). Ca^{2+} in particular has been implicated in a secondary phase of the response to thyrotropin releasing hormone in rat pituitary lactotroph cells and blockade of Ca^{2+} entry appears to be involved in the inhibition of the response by dopamine (Vallar *et al.*, 1988). Raised levels of intracellular Ca^{2+} can lead to increased levels of all [^{3}H]inositol phosphates, but particularly [^{3}H]IP$_2$. Conversely, Mn^{2+}, which has been used as a blocker of voltage-dependent Ca^{2+} channels, but which may also enter cells, inhibits noradrenaline-induced phosphoinositide breakdown in rat cerebral cortex, but the inhibition of [^{3}H]IP$_1$ and [^{3}H]IP$_3$ formation is greater than that of [^{3}H]IP$_2$ (Maier and Rutledge, 1987). If GABA promoted Ca^{2+} entry, then this might be the explanation for the enhanced levels of [^{3}H]IP$_2$ with noradrenaline in rat cerebral cortex. However, blockade of L-type voltage-dependent Ca^{2+} channels, which mimicks the inhibitory effect of dopamine on

[^{3}H]IP formation in rat pituitary lactotrophs (Vallar *et al.*, 1988), had no effect on the responses to histamine or noradrenaline in rat cerebral cortex. Thus 0.1 μM PY 108068 had no effect on histamine-induced [^{3}H]IP_1 formation and 1 μM nifedipine was similarly without effect. Interestingly, a higher concentration of nifedipine, 10 μM, caused some stimulation of the responses to 1 mM histamine (150 $\pm$ 14% of the response to histamine alone; 5 measurements), but did not reverse the inhibition of the histamine response by 2 mM GABA (2 experiments). The high concentration of nifedipine (10 μM) also stimulated the response to 100 μM noradrenaline (134 $\pm$ 11% of the response to noradrenaline alone; 3 measurements), but failed to prevent a further potentiation by 2 mM GABA (120 $\pm$ 4%; 3 measurements). Thus, there is no evidence that L-type Ca^{2+} channels are involved in the response to histamine or noradrenaline in rat cerebral cortex or in the modulatory effects of GABA.

The lack of effect of nifedipine on agonist responses does not rule out Ca^{2+} involvement. We have made a further test of the involvement of Ca^{2+} by examining the effect of omitting Ca^{2+} from the Krebs–Henseleit solution. In the protocol we have used, all slices were initially incubated in Krebs–Henseleit solution without Ca^{2+}. Ca^{2+} (2.5 mM) was then added to half the slices and labelling with [^{3}H]inositol and exposure to agonist continued for both sets as in the standard protocol. Omission of Ca^{2+} from the medium led to a considerable reduction in the response to noradrenaline. Thus [^{3}H]IP_1 formation was 8.6 $\pm$ 1.6-fold over basal (3 measurements) with Ca^{2+} added back, but only 2.4 $\pm$ 0.1-fold (4 measurements) without added Ca^{2+}. [^{3}H]IP_2 formation was also less in 'Ca^{2+}-free' medium (1.9 $\pm$ 0.1-fold over basal (4 measurements), compared with 4.6 $\pm$ 0.2-fold). Basal accumulation was somewhat less in the slices without added Ca^{2+} (87 $\pm$ 6%; 5 measurements), but as noted above 2 mM GABA alone no longer caused a statistically significant simulation of any of the phosphates (Table 3). However, the response to noradrenaline was still potentiated by 2 mM GABA in the absence of added Ca^{2+} (Table 3), although it is notable that in this series of experiments the increase in the level of [^{3}H]IP_2 in the slices with Ca^{2+} added back was not as great as normally observed in slices in Krebs–Henseleit solution throughout (cf. Figure 9.6).

Thus Ca^{2+} dependence apparently distinguishes the effect of GABA acting alone and its effect on noradrenaline-induced formation. However, it must be borne in mind that slices incubated for any period in the absence of Ca^{2+} do not usually function as well as those incubated in normal Krebs–Henseleit throughout. We have not measured the Ca^{2+} concentration of the medium containing slices but with no added Ca^{2+}. Measurements with slices from guinea-pig cerebral cortex indicate that it can be as high as 26 μM (Carswell *et al.*, 1985). We have carried out a limited number of measurements of the effect of omitting Ca^{2+} on the response to histamine and its inhibition by GABA. However, in slices without added Ca^{2+} the response to histamine is small – at most *c.* 2.5-fold over basal – so that where the stimulation by histamine is much less than this the

Table 3. Ca^{2+}-dependence of GABA potentiation of basal and noradrenaline-induced $[^3H]IP_n$ formation

	$[^3H]IP_n$ accumulated (Response with GABA × 100/response without GABA)	
	No added Ca^{2+}	Ca^{2+} added back
Basal		
$[^3H]IP_1$	96 ± 3 (4)	152 ± 9 (3)
$[^3H]IP_2$	109 ± 6 (4)	172 ± 3 (4)
$[^3H]IP_3$	116 ± 6 (4)	190 ± 28 (4)
100 μM noradrenaline induced		
$[^3H]IP_1$	219 ± 3 (4)	119 ± 14 (3)
$[^3H]IP_2$	291 ± 48 (4)	184 ± 12 (4)
$[^3H]IP_3$	187 ± 38 (4)	160 ± 19 (4)

Rat cerebral cortical slices were incubated for 60 min in Kerbs-Henseleit medium from which Ca^{2+} had been omitted. At the end of this period 2.5 mM $CaCl_2$ was added back to one half of the slice suspension. LiCl (10 mM) and 1 μCi $[^3H]$inositol were then added to each fraction and incubations continued as described in the legend to Table 1, in the presence or absence of 100 μM noradrenaline and in the presence or absence of 2 mM GABA. Values are the means ± approximate SE from 4 paired measurements. IP_n, inositol mono-, bis- or trisphosphate (n = 1, 2 or 3, respectively).

errors preclude observing statistically significant changes and the Ca^{2+} dependence or otherwise of the response remains unclear.

GABA INHIBITION OF HISTAMINE-INDUCED $[^3H]$INOSITOL PHOSPHATE FORMATION IN GUINEA-PIG CEREBELLUM

GABA has different effects on the responses to different agonists in the same tissue. However, even for a given agonist the nature of the effect of GABA can differ between tissues, as illustrated by the apparently different mechanisms of the potentiation of noradrenaline-induced $[^3H]$IP formation in rat cerebral cortex and hippocampus discussed above. This is further underlined by differences in the action of GABA on the response to histamine in rat cerebral cortex and guinea-pig cerebellum.

Guinea-pig cerebellum has the highest density of histamine H_1 receptors, as indicated by $[^3H]$mepyramine binding, in any region of guinea-pig or rat brain that we have examined (Hill and Young, 1980). The receptors are localized predominantly in the molecular layer (as are $GABA_B$ receptors; see Bowery *et al.*, Chapter 1) and appear to be largely functional, as illustrated by the good correlation between receptor density and the magnitude of histamine-stimulated $[^3H]IP_1$ formation in regions of guinea-pig brain (Carswell and Young, 1986;

Daum *et al.*, 1983). However, guinea-pig cerebellum contains little histamine and little histidine decarboxylase and it remains unknown whether the H_1 receptors have any significant role to play in cerebellar function. Rat cerebellum contains few H_1 receptors (Hill and Young, 1980).

GABA inhibited histamine-induced [^{3}H]IP$_1$ accumulation in guinea-pig cerebellar slices (Crawford *et al.*, 1987) with an estimated IC_{50} (0.8 ± 0.2 mM) only a little higher than that in rat cerebral cortex. However, the mechanism of action of GABA is clearly different in the two tissues, since (−)-baclofen was without effect in cerebellar slices (Table 4). Isoguvacine and muscimol were also without appreciable effect, so the action of GABA is apparently not mediated by either $GABA_A$ or $GABA_B$ receptors. However, the inhibitory action of GABA was mimicked by 1 mM nipecotic acid (Table 4), reminiscent of the activity of nipecotic acid on noradrenaline-induced [^{3}H]IP$_1$ accumulation in rat cerebral cortex (Table 2). The particularly interesting feature of the comparison is that in guinea-pig cerebellum both GABA and nipecotic acid increase the accumulation of [^{3}H]IP$_2$, even though the amount of [^{3}H]IP$_1$ is decreased. Whether this enhancement is by the same mechanism as that by which [^{3}H]IP$_2$ generated by noradrenaline in rat cerebral cortex is increased and whether the more modest effect in guinea-pig cerebellum is the result of simultaneous inhibitory and enhancing mechanisms are unknown.

We have not made an extensive investigation of the effect of GABA on other agonists in guinea-pig cerebellum, but [^{3}H]IP$_1$ accumulation induced by endothelin-1 is also inhibited. This does not provide sufficient evidence to suggest that the action of GABA is 'non-specific' and would decrease the

Table 4. Effect of (−)-baclofen and other GABA analogues on histamine-stimulated [^{3}H]inositol phosphate accumulation in guinea-pig cerebellum

	[^{3}H]IP$_n$ accumulated in the presence of GABA analogue (% of response to histamine alone)		
	[^{3}H]IP$_1$	[^{3}H]IP$_2$	[^{3}H]IP$_3$
GABA (2 mM)	50 ± 2 (22)	127 ± 4 (11)	77 ± 8 (11)
Nipecotic acid (1 mM)	62 ± 4 (5)	121 ± 8 (3)	87 ± 7 (3)
Isoguvacine (100 μM)	104 ± 4 (5)	110 ± 1 (3)	108 ± 3 (3)
(−)-Baclofen (100 μM)	103 ± 5 (4)	105 ± 5 (3)	106 ± 7 (3)
Muscimol (100 μM)	98 ± 3 (4)	102 ± 1 (3)	100 ± 2 (3)

Values are the means ± approximate SE from the number of experiments given in parentheses. Incubations were with 200 μM histamine in the presence or absence of GABA analogue. Basal values (with or without GABA analogue, as appropriate) were subtracted before calculation of the ratios. IP$_n$, inositol mono-, bis- or trisphosphate (n = 1, 2 or 3, respectively).

response to all agonists, but it is interesting to note that both GABA (Table 1) and nipecotic acid inhibit basal accumulation of [^{3}H]IP$_1$ in this tissue. Basal levels of [^{3}H]IP$_1$ in guinea-pig cerebellum are high (Carswell and Young, 1986) and might well reflect the release of some endogenous PI-mobilizing agonist. This could well be the component inhibited by GABA and nipecotic acid.

CONCLUSIONS

Of the several and varied actions of GABA on phosphoinositide metabolism that we have observed in this study or which have been reported in the literature, only the inhibition of histamine-induced [^{3}H]IP accumulation in rat cerebral cortex or that of 5-HT in mouse cerebral cortex appear to be clear-cut GABA$_B$-receptor-mediated effects. On the evidence in the literature, GABA potentiation of noradrenaline-induced [^{3}H]IP$_1$ accumulation in rat hippocampul slices is a GABA$_A$ effect, but the effect or effects on the noradrenaline response in rat cerebral cortex cannot be described as either GABA$_A$ or GABA$_B$. High concentrations of GABA, 0.1–1 mM, are required for the latter action, although this might reflect depletion by uptake processes or metabolism. However, the response to carbachol in the same tissue preparation is seemingly untouched by 2 mM GABA and the effect on the response to histamine is inhibition and not potentiation. Similarly, it would be premature to dismiss the effect of GABA in guinea-pig cerebellar slices as 'non-specific'. A non-selective action on another, as yet unidentified, receptor must be a possibility. However, even where the effect of GABA is mediated via GABA$_B$ receptors, we have little indication of what the mechanism of the inhibitory action might be. Evidence has been presented recently for the existence of G-proteins in rat cerebral cortex which when activated inhibit PIC action (Litosch, 1989), but whether GABA$_B$ receptors might couple with such G-proteins in certain cells is unknown. Potentiation of agonist-induced cyclic AMP formation (histamine H$_2$ or β receptors) needs to be eliminated, but it may be noted that H$_2$-receptor antagonists do not modify the response to histamine (Claro *et al.*, 1986) and the response to noradrenaline is unaffected by the β-receptor antagonist propranolol. If either GABA$_B$-receptor-mediated opening of K$^+$ channels or closing of Ca^{2+} channels is involved, then it would imply that the mechanism of histamine- and 5-HT-induced [^{3}H]IP formation is not as simple as the scheme in Figure 9.1 would imply. A proper understanding of the mechanism of either component of the agonist/GABA interaction is clearly bound up with an understanding of the other.

ACKNOWLEDGEMENTS

We are grateful to the Medical Research Council for financial support. Our thanks are also due to CIBA-Geigy and Smith, Kline and French for the gifts of compounds.

REFERENCES

Berridge, M.J. (1988) Inositol lipids and calcium signalling. *Proc. R. Soc. Lond. B*, **234**, 359–378.

Berridge, M.J. and Irvine, R.F. (1989) Inositol phosphates and cell signalling. *Nature, Lond.*, **341**, 197–205.

Berridge, M.J. and Michell, R.H. (eds) (1988) *Inositol lipids and transmembrane signalling*. London: The Royal Society. (Reprinted from *Phil. Trans. R. Soc. Lond. B*, **320**, 237–436.)

Berridge, M.J., Downes, C.P. and Hanley, M.R. (1982) Lithium amplifies agonist-dependent phosphatidylinositol hydrolysis in brain and salivary glands. *Biochem. J.*, **206**, 587–595.

Brown, E.A., Kendall, D.A. and Nahorski, S.R. (1984) Inositol phospholipid hydrolysis in rat cerebral cortical slices: I. Receptor classification. *J. Neurochem.*, **42**, 1379–1387.

Carswell, H. and Young, J.M. (1986) Regional variation in the characteristics of histamine H_1-agonist mediated breakdown of inositol phospholipids in guinea-pig brain. *Br. J. Pharmacol.*, **89**, 809–817.

Carswell, H., Daum, P.R. and Young, J.M. (1985) Histamine H_1-agonist stimulated breakdown of inositol phospholipids. In: C.R. Ganellin and J.-C. Schwartz (eds) *Advances in the biosciences*, vol. 51: *Frontiers in histamine research*. Oxford: Pergamon Press, pp. 27–38.

Claro, E., Arbones, L., Garcia, A. and Picatoste, F. (1986) Phosphoinositide hydrolysis mediated by histamine H_1-receptors in rat brain cortex. *Eur. J. Pharmacol.*, **123**, 187–196.

Corradetti, R., Ruggiero, M., Chiarugi, V.P. and Pepeu, G. (1987) GABA-receptor stimulation enhances norepinephrine-induced polyphosphoinositide metabolism in rat hippocampal slices. *Brain Res.*, **411**, 196–199.

Crawford, M.L.A. and Young, J.M. (1988) $GABA_B$ receptor mediated inhibition of histamine H_1-receptor-induced inositol phosphate formation in slices of rat cerebral cortex. *J. Neurochem.*, **51**, 1441–1447.

Crawford, M.L.A. and Young, J.M. (1990) Potentiation by GABA of α_1-agonist-induced accumulation of inositol phosphates in slices of rat cerebral cortex. *J. Neurochem.*, **54**, 2100–2109.

Crawford, M.L.A., Carswell, H. and Young, J.M. (1987) GABA inhibits histamine-induced inositol phospholipid breakdown in guinea-pig cerebellum. *Br. J. Pharmacol.*, **91**, 304P.

Daum, P.R., Downes, C.P. and Young, J.M. (1983) Histamine-induced inositol phospholipid breakdown mirrors H_1-receptor density in brain. *Eur. J. Pharmacol.*, **87**, 497–498.

Fisher, S.K. and Agranoff, B.W. (1987) Receptor activation and inositol lipid hydrolysis in neural tissues. *J. Neurochem.*, **48**, 999–1017.

Godfrey, P.P., Grahame-Smith, D.G. and Gray, J.A. (1988) $GABA_B$ receptor activation inhibits 5-hydroxytryptamine-stimulated inositol phospholipid turnover in mouse cerebral cortex. *Eur. J. Pharmacol.*, **152**, 185–188.

Hill, S.J. and Young, J.M. (1980) Histamine H_1-receptors in the brain of the guinea-pig and the rat: differences in ligand binding properties and regional distribution. *Br. J. Pharmacol.*, **68**, 687–696.

Inhorn, R.C. and Majerus, P.W. (1987) Inositol polyphosphate 1-phosphatase from calf brain: purification and inhibition by Li^+, Ca^{2+} and Mn^{2+}. *J. Biol. Chem.*, **262**, 15946–15952.

Irvine, R.F., Moor, R.M., Pollock, W.K., Smith, P.M. and Wreggett, K.A. (1988) Inositol phosphates: proliferation, metabolism and function. In: M.J. Berridge and R.H. Michell (eds) *Inositol lipids and transmembrane signalling*. London: The Royal Society, pp. 45–61. (Reprinted from *Phil. Trans. R. Soc. B*, **320**, 281–298.)

Johnston, G.A.R. (1986) Multiplicity of GABA receptors. In: R.W. Olsen and J.C. Venter (eds) *Benzodiazepine/GABA receptors and chloride channels: structural and functional properties*. New York: Alan R. Liss, pp. 57–71.

Johnston, G.A.R., Stephanson, A.L. and Twitchin, B. (1976) Uptake and release of nipecotic acid by brain slices. *J. Neurochem.*, **26**, 83–87.

Kendall, D.A. and Hill, S.J. (1988) Adenosine inhibition of histamine-stimulated inositol phospholipid hydrolysis in mouse cerebral cortex. *J. Neurochem.*, **50**, 497–502.

Larsson, O.M., Falch, E., Krogsgaard-Larsen, P. and Schousboe, A. (1988) Kinetic characterization of inhibition of γ-aminobutyric acid uptake into cultured neurones and astrocytes by 4, 4-diphenyl-3-butenyl derivatives of nipecotic acid and guvacine. *J. Neurochem.*, **50**, 818–823.

Linden, J. and Delahunty, T.M. (1989) Receptors that inhibit phosphoinositide breakdown. *Trends Pharmacol. Sci.*, **10**, 114–120.

Litosch, I. (1989) Guanine nucleotides mediate stimulatory and inhibitory effects on cerebral-cortical membrane phospholipase C activity. *Biochem. J.*, **261**, 245–251.

Maier, K.U. and Rutledge, C.O. (1987) Comparison of norepinephrine- and veratrine-induced phosphoinositide hydrolysis in rat brain. *J. Pharmacol. Exp. Ther.*, **240**, 729–736.

Majerus, P.W., Connolly, T.M., Bansai, V.S., Inhorn, R.C., Ross, T.S. and Lips, D.L. (1988) Inositol phosphates: synthesis and degradation. *J. Biol. Chem.*, **263**, 3051–3054.

Nishizuka, Y. (1988) The molecular heterogeneity of protein kinase C and its implications for cellular regulation. *Nature, Lond.*, **334**, 661–665.

Petcoff, D.W. and Cooper, D.M.F. (1987) Adenosine receptor agonists inhibit inositol phosphate accumulation in rat striatal slices. *Eur. J. Pharmacol.*, **137**, 269–271.

Pollard, H. and Schwartz, J.-C. (1987) Histamine neuronal pathways and their functions. *Trends Neurosci.*, **10**, 86–89.

Ruggiero, M., Corradetti, R., Chiarugi, V. and Pepeu, G. (1987) Phospholipase C activation induced by noradrenaline in rat hippocampal slices is potentiated by GABA-receptor stimulation. *EMBO J.*, **6**, 1595–1598.

Shears, S.B. (1989) Metabolism of the inositol phosphates produced upon receptor activation. *Biochem. J.*, **260**, 313–324.

Takeda, N., Inagaki, S., Shiosaka, S., Taguchi, Y., Oertel, W., Tohyama, M., Watanabe, T. and Wada, H. (1984) Immunohistochemical evidence for the coexistence of histidine-decarboxylase-like and glutamate decarboxylase-like immunoreactivities in nerve cells of the posterior hypothalamus of rats. *Proc. Natl Acad. Sci. USA*, **81**, 7647–7650.

Treherne, J.M. and Young, J.M. (1988) [^{3}H]-(+)-*N*-methyl-4-methyldiphenhydramine, a quaternary radioligand for the histamine H_1-receptor. *Br. J. Pharmacol.*, **94**, 797–810.

Vallar, L., Vicentini, L.M. and Meldolesi, J. (1988) Inhibition of inositol phosphate production is a late, Ca^{2+}-dependent effect of D_2 dopaminergic receptor activation in rat lactotroph cells. *J. Biol. Chem.*, **263**, 10127–10134.

Vallejo, M., Jackson, T.R., Lightman, S.L. and Hanley, M.R. (1987) Occurrence and extracellular actions of inositol pentakis- and hexakisphosphate in mammalian brain. *Nature, Lond.*, **330**, 656–658.

10 Solubilization and Partial Purification of Cerebral $GABA_B$ Receptors

K. KURIYAMA and Y. OHMORI

INTRODUCTION

Receptors for GABA, the major inhibitory neurotransmitter in mammalian brain, have been divided into $GABA_A$ and $GABA_B$ subtypes on the basis of their pharmacological properties (Hill and Bowery, 1981). The $GABA_A$ receptor has been found to be coupled structurally as well as functionally with the benzodiazepine receptor and chloride channels (Kuriyama and Taguchi, 1987; Taguchi and Kuriyama, 1984) and reconstitution of the purified $GABA_A$ receptor complex with phospholipid vesicles indicates functional couplings (Hirouchi *et al.*, 1987). On the other hand, the $GABA_B$ receptor has been found to be coupled with calcium ion channels (Dunlap, 1981) and the adenylyl cyclase system via GTP-binding proteins (Wojcik and Neff, 1983). It is not yet clear, however, whether these reactions occur sequentially or independently and which of these components is directly coupled to the $GABA_B$ receptor. In order to elucidate these unknown characteristics of the cerebral $GABA_B$ receptor, solubilization and purification of the receptor is important.

In the present study, we have, therefore, attempted to solubilize the $GABA_B$ receptor and to examine any alteration in its functional coupling with intracellular transducing systems following solubilization. We have also attempted to develop a new affinity gel for the purification of the $GABA_B$ receptor and have tried to apply the affinity column for purification with a gel filtration chromatographic procedure.

MATERIALS AND METHODS

Materials

[2,3-^{3}H]GABA (70.0 Ci/mmol) and [^{3}H]ATP ([2,8-^{3}H]adenosine 5-triphosphate, 49 Ci/mmol) were purchased from New England Nuclear Inc. and Amersham International, respectively. Epoxy-activated Sepharose 6B and Sephacryl S-300 were obtained from Pharmacia Fine Chemicals.

$GABA_B$ Receptors in Mammalian Function. Edited by N.G. Bowery, H. Bittiger and H.-R. Olpe

Preparation of Baclofen–Epoxy-activated Sepharose 6B

Epoxy-activated Sepharose 6B (1.0 g) was suspended in 10 ml distilled water and allowed to stand for 10 min. The moist gel was transferred to a glass filter and washed extensively with 100 ml distilled water and finally washed with 100 ml 0.1 M $NaHCO_3$. Baclofen (1.0 g) was dissolved in 5 ml 0.1 M $NaHCO_3$, and the pH of the solution adjusted to 13. The epoxy-activated Sepharose 6B gel, prepared as described above, was transferred to the baclofen solution and shaken overnight to react the aliphatic epoxy group of the Sepharose with the amino group of the ligand.

Preparation of Solubilized Fraction by Various Detergents

Frozen crude synaptic membrane prepared by the method of Zukin *et al.* (1974) was thawed and washed three times by suspending and centrifuging at 48 000 × ***g*** for 20 min with ice-cold 50 nM Tris–HCl buffer (pH 7.4). The final pellet obtained was suspended in 10 volumes of 50 mM Tris–HCl buffer containing the detergent under test, 0.02% asolectin and 40 μg/ml bacitracin. The suspension was incubated at 2 °C for 90 min and centrifuged for 120 min at 105 000 × ***g***. The resulting supernatant was then dialysed extensively against 50 mM Tris–HCl buffer containing 1/10 volume of the detergent used for solubilization and 0.02% asolectin. The detergents used are listed in Table 1.

Procedures for Affinity Column Chromatography

Baclofen–epoxy-activated Sepharose 6B (15 ml) was packed into a 1 cm × 20 cm glass column and equilibrated with 50 mM Tris–HCl buffer containing 0.02% sodium deoxycholate, 1 M KCl and 0.02% asolectin (buffer A). The solubilized fraction (90 ml), prepared by treatment with 0.2% deoxycholate in the presence of 1 M KCl, was applied to the affinity column, and washed with 50 ml buffer A at a flow rate of 16.6 ml/h. The receptor protein retained in the column was then eluted by a linear gradient of NaSCN (50 ml; 1–2 M) in buffer A, and the eluted fractions were immediately dialysed overnight against buffer A at 4 °C.

Procedures for Gel Filtration Column Chromatography

Sephacryl S-300 (70 ml) was packed into a 1.4 cm × 50 cm glass column and equilibrated with buffer A. The fraction eluted from the baclofen–epoxy-activated Sepharose 6B (1 ml) and concentrated by Amicon 8MC was applied to a Sephacryl S-300 column at a flow rate of 16.6 ml/h. Following the equilibrium of the column, it was eluted with buffer A.

All column chromatographic procedures were carried out at 4 °C.

Assay of $GABA_B$ Receptor Binding

Each receptor preparation was incubated with 5 nM [^{3}H]GABA at 2 °C for 30 min. Following the incubation, 25 μl bovine γ-globulin solution (20 mg/ml) and 0.3 ml 50% (w/v) polyethyleneglycol (PEG) were added to a final concentration of 0.5 mg/ml and 15%, respectively. After allowing to stand for 5 min, the mixture was filtered rapidly under vacuum through a Whatman GF/B glass-fibre filter. The filter was then washed three times at 2 °C with 3 ml 50 mM Tris–HCl buffer containing 8% (w/v) PEG. The specific binding of [^{3}H]GABA to $GABA_B$ receptors was defined as the total binding minus the binding obtained in the presence of 1 mM baclofen. The radioactivity trapped on each filter was then measured by a liquid scintillation spectrometer at a counting efficiency of 40–50%.

Assay of Adenylyl Cyclase Activity

Adenylyl cyclase activity was determined by the enzymatic conversion of [^{3}H]ATP to cyclic [^{3}H]AMP according to the method of Lynch *et al.* (1977). The crude synaptic membrane fraction was incubated at 37 °C for 10 min with 140 μl 28.6 mM Tris–HCl buffer (pH 7.4) containing [^{3}H]ATP (70 μM), magnesium phosphate (2.9 mM), sodium creatine phosphate (7.8 mM), creatine phosphokinase (16 units) and isobutylmethylxanthine (0.9 mM). The reaction was terminated by immersing in a boiling water-bath, and the mixture was then centrifuged at 1000 × ***g*** for 5 min. The supernatant thus obtained was lyophilized and the residue dissolved in 50% ethanol. This solution was spotted onto cellulose plate and the cyclic [^{3}H]AMP formed was separated by thin-layer chromatography using a mixture of methanol, *n*-butanol and 1 M ammonium acetate (5:5:4). The spot corresponding to authentic cyclic AMP was scraped off and its radioactivity determined by liquid scintillation spectrometry.

RESULTS

Solubilization of the $GABA_B$ Receptor

The solubilization of the $GABA_B$ receptors from the crude synaptic membrane fraction of bovine cerebral cortex was performed using various detergents as shown in Table 1. The $GABA_B$ receptor was easily solubilized by 0.2% DOC-Na (sodium deoxycholate) in the presence of 1 M KCl. Treatment with Triton X-100, Nonidet P-40, DOC-Na in the absence of 1 M KCl and 3-[(3-cholamido-propyl)dimethylammonio]-1-propanesulphonate, Tween 80 (CHAPS) also solubilized the receptor. In contrast, Briji 35, Digitonin, Lubrol-PX and lysolecithin failed to solubilize the receptor.

Table 1. Solubilization of $GABA_A$ and $GABA_B$ receptors with various detergents

Detergent	Protein (mg)	Specific [^{3}H]GABA binding*				
		Total binding (fmol)		Specific binding (fmol/mg protein)		
		A site	B site	A site	B site	A/B
P_2 membrane	19.4	105.5	29.4	5.4	1.5	3.5
DOC-Na (0.2%) + KCl (1 M)	7.1	99.4	56.5	13.8	7.8	1.7
Triton X-100 (1%)	10.2	63.7	31.4	6.2	3.0	2.0
Nonidet P-40 (1%)	14.5	43.9	24.1	3.0	1.6	1.8
DOC-Na (0.2%)	6.0	36.3	15.9	5.9	2.6	2.2
CHAPS (1.2%)	4.7	29.2	13.9	6.1	2.9	2.0
Tween 80 (1%)	2.3	19.4	7.9	8.1	3.3	2.4
Briji 35 (1%)	3.3	0.7	0.4	0.2	0.1	1.5
Digitonin (1%)	4.4	33.0	0	7.5	0	—
Lubrol-PX (0.5%)	7.1	11.4	0	1.5	0	—
Lysolecithin (0.25%)	2.6	6.7	0	2.5	0	—

* determined in the supernatant fraction.

P_2 synaptic membrane suspension was treated with various detergents, and [^{3}H]GABA binding t $GABA_A$ and $GABA_B$ receptors in the solubilized fraction obtained by the method described i Materials and Methods was examined. The value for each detergent represents the results from a sing experiment performed in triplicate. DOC-Na, Sodium deoxycholate; CHAPS, 3-[(3-cholamidopropyl dimethylammonio]-l-propanesulphonate.

Alteration in the Coupling Mechanism of the $GABA_B$ Receptor with Various Components Following Solubilization

In the solubilized fraction obtained with DOC-Na in the presence of 1 M KCl, calcium ions did not induce a significant increase of $GABA_B$ receptor binding. Calcium ion channel blockers, such as nicardipine and diltiazem, also failed to enhance $GABA_B$ receptor binding in the presence of calcium ions (Figure 10.1). Furthermore, GTP and its analogues had no significant effect on $GABA_B$ receptor binding as shown in Table 2. Forskolin has been known to stimulate basal adenylyl cyclase activity (Nishikawa and Kuriyama, 1989). Although forskolin stimulated basal adenylate cyclase activity in the solubilized fraction, GABA and baclofen were ineffective on the forskolin-stimulated adenylyl cyclase activity (Figure 10.2). These results strongly suggest that the functional coupling of $GABA_B$ receptors with various cellular components such as calcium ion channels, GTP-binding protein (Ohmori *et al.*, 1988) and adenylyl cyclase (Nishikawa and Kuriyama, 1989) may be eliminated following solubilization.

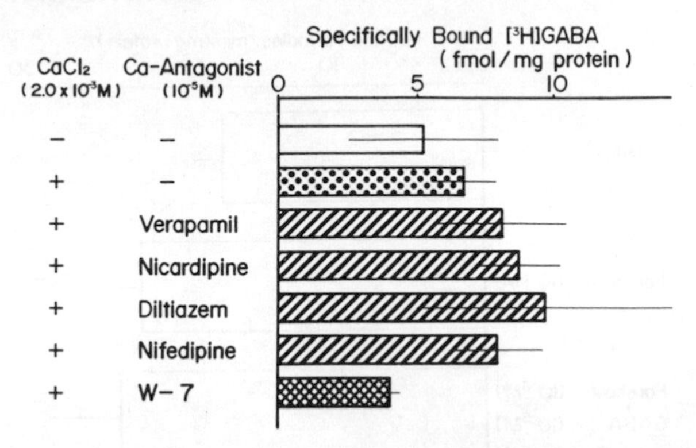

Figure 10.1. Effect of calcium ion channel blockers on [^{3}H]GABA binding to $GABA_B$ receptors in a solubilized fraction from crude synaptic membranes. [^{3}H]GABA binding was assayed in 50 mM Tris-HCl buffer (pH 7.4) containing 2 mM $CaCl_2$ and 1.5 mM $MgSO_4$. Each value represents the mean $\pm$ SEM of three separate experiments

Affinity Gel Chromatography

Following the elution of the solubilized fraction from baclofen-epoxy-activated Sepharose 6B (Figure 10.3), approximately 60% of the protein applied to the column was recovered in the run-through fraction. No $GABA_B$ receptor binding was detected in the fraction. The application of NaSCN (1-2 M) resulted in the elution of 34.78% of total $GABA_B$ receptor binding. The specific activity of the binding was found to be 19.43 fmol/mg protein, which corresponded to a purification of 21.8-fold (Table 3 and Figure 10.4).

Table 2. Effect of GTP analogues on baclofen-sensitive [^{3}H]GABA binding to a DOC-Na-solubilized fraction from synaptic membranes of bovine cerebral cortex

Drug	[^{3}H]GABA binding (fmol/mg protein)
Control	7.99 $\pm$ 0.53
GTP (10^{-3} M)	11.04 $\pm$ 3.35
GTPγS (10^{-4} M)	6.18 $\pm$ 1.63
GppNHp (10^{-3} M)	8.80 $\pm$ 3.29
GDP (10^{-3} M)	9.23 $\pm$ 3.60

DOC-Na, 0.2% sodium deoxycholate (+ 1 M KCl).
GTPγS, guanosine-5'-*O*-3-thiotriphosphate.
GppNHp, 5'-guanylyl imidodiphosphate.

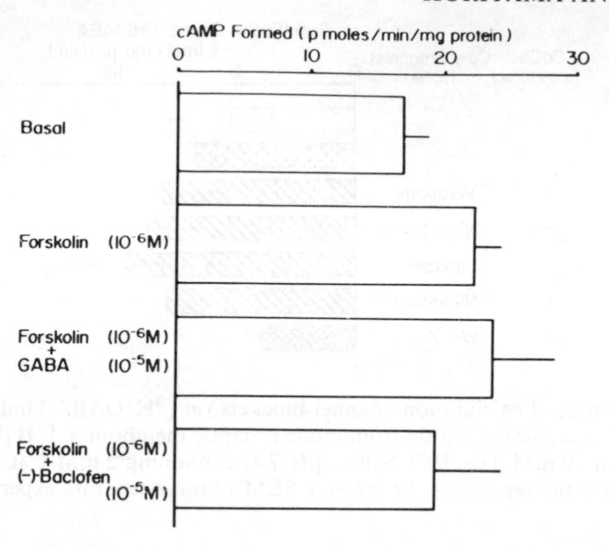

Figure 10.2. Effect of GABA and (−)-baclofen on forskolin-stimulated adenylyl cyclase activity in solubilized fraction from crude synaptic membranes. Each value represents the mean ± SEM of five separate experiments

Gel Filtration Column Chromatography

The elution pattern of protein from the Sephacryl S-300 column showed a broad distribution, while that of $GABA_B$ receptor binding exhibited two peaks (A and B as shown in Figure 10.5). The molecular weight of peak A was calculated to be approximately 240 000, while that of the other peak, B, was approximately 80 000. The purification and specific activity of peak A were found to be 141-fold and 125.60 fmol/mg protein, respectively. On the other hand, the purification and specific activity of peak B were found to be 70-fold and 62.27 fmol/mg protein, respectively.

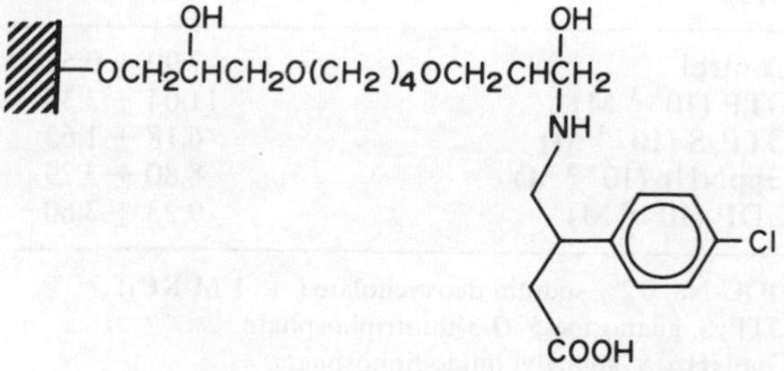

Figure 10.3. Chemical structure of baclofen–epoxy-activated Sepharose 6B

Table 3. Summary of chromatographic separation of DOC-Na-solubilized $GABA_B$ receptor from bovine cerebral cortex

	Volume (ml)	Protein content (mg)	[^{3}H]GABA binding (5 nM)			
			Specific binding (fmol/mg prot.)	Purification (fold)	Total binding (fmol)	Recovery (%)
I Solubilized fraction	85	163.2	0.89	1.0	145.2	—
II Baclofen–Sepharose 6B (1–2 M NaSCN)	13	2.6	19.43	21.8	50.5	(II/I) 34.78
Sephacryl S-300						
III Peak A	4	0.114	125.60	141.1	14.3	(III/II) 28.32
IV Peak B	6	0.173	62.27	70.0	10.8	(IV/II) 21.39

DOC-Na, 0.2% sodium deoxycholate (+ 1 M KCl).

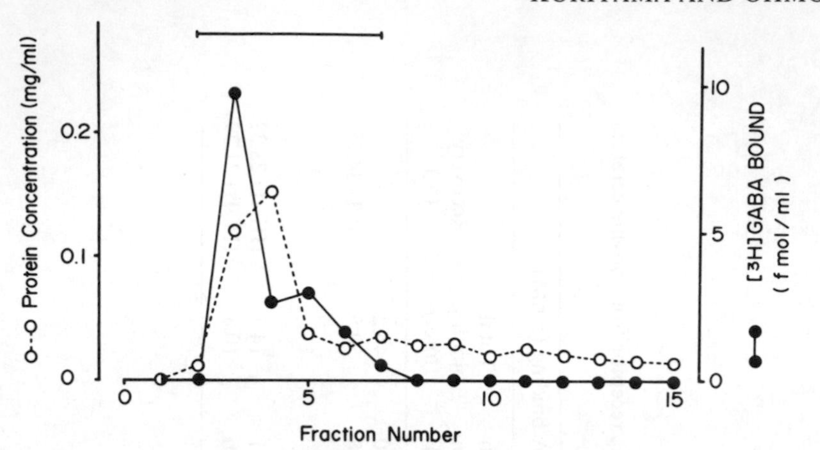

Figure 10.4. Affinity column chromatography using baclofen–epoxy-activated Sepharose 6B of a solubilized fraction from crude synaptic membranes. Each elution volume was 3 ml. The specific binding of [^{3}H]GABA to the solubilized receptor fraction applied to the column was 0.89 fmol/mg protein. The fractions from No. 3 to No. 7 (indicated by the horizontal bar in the upper part of the figure) were collected, concentrated to approximately 1 ml by Amicon 8MC, and then applied to Sephacryl S-300 gel filtration chromatography as shown in Figure 10.5

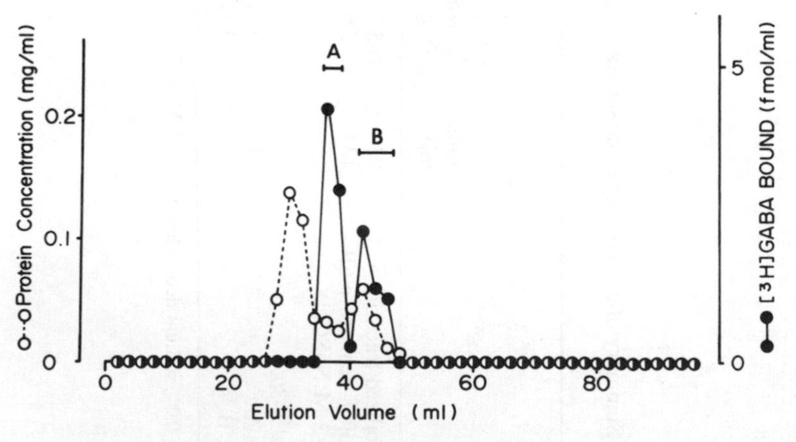

Figure 10.5. Gel filtration using Sephacryl S-300 of the fraction obtained by baclofen-affinity column chromatography. One ml of the concentrated sample obtained from the baclofen-affinity column (see Figure 10.4) was applied to the Sephacryl S-300 column (1.4 × 50.0 cm). Chromatography was performed at a flow rate of 16.6 ml/h, and 2 ml of each aliquot was collected for the determination of protein and [^{3}H]GABA binding in the presence of 1 mM baclofen. (A) The peak A of [^{3}H]GABA binding; (B) the peak B of [^{3}H]GABA binding

DISCUSSION

We have recently reported that $GABA_B$ receptor binding in crude synaptic membranes is significantly stimulated by the addition of calcium ions as well as magnesium ions, and the calcium ion-induced stimulation of $GABA_B$ receptor binding is further enhanced by calcium ion channel blockers such as nicardipine and diltiazem (Ohmori *et al.*, 1988). These results suggest that the $GABA_B$ receptor may be functionally coupled with the calcium ion channel, and $GABA_B$ receptor binding may be negatively modulated by the channel. It has also been reported that the $GABA_B$ receptor is coupled to GTP-binding protein (Hill *et al.*, 1984). In our previous study, we also found that $GABA_B$ receptor binding in the crude synaptic membrane fraction is significantly inhibited by GTP and GTP analogues, and this inhibition is eliminated by pretreatment of the membranes with islet-activating protein (IAP). In addition, it has been found that GABA and baclofen significantly inhibit not only basal adenylyl cyclase activity but also forskolin-stimulated activity in the crude synaptic membrane fraction (Nishikawa and Kuriyama, 1989). Furthermore, the inhibition in both cases has been found to be abolished by phaclofen, which is known to be a $GABA_B$ receptor antagonist (Kerr *et al.*, 1987), as well as by the pretreatment with IAP. These pharmacological and biochemical data suggest that IAP-sensitive GTP-binding protein may be involved in the coupling of the $GABA_B$ receptor with the adenylyl cyclase system.

We have attempted to solubilize the $GABA_B$ receptor from a cerebral synaptic membrane fraction as a preliminary step to purifying the receptor. Treatment of the crude synaptic membrane fraction with DOC-Na in the presence of 1 M KCl resulted in the highest solubilization of $GABA_B$ receptors achieved with any of the detergents tested. It was found, however, that the enhancement of $GABA_B$ receptor binding by calcium ions and calcium ion channel blockers which occurred in cerebral synaptic membrane preparations was abolished following solubilization. Similarly, solubilization produced a loss of the inhibitory effect of GTP and its analogues on $GABA_B$ receptor binding as well as of the inhibition of adenylyl cyclase activity induced by GABA and baclofen. These results suggest that the functional coupling of the $GABA_B$ receptor with these components may also be eliminated, at least in part, during solubilization of the $GABA_B$ receptor.

Since activation of the $GABA_B$ receptor resulted in the inhibition of adenylyl cyclase activity (Nishikawa and Kuriyama, 1989), possible involvement of $GABA_B$ receptors in the regulation of the function of inhibitory GTP-binding protein (G_i-protein) is suggested. However, direct evidence that the $GABA_B$ receptor interacts with G_i-protein has not been reported. To clarify this issue, the development of adequate purification procedures for the $GABA_B$ receptor must be considered. In this study, we have, therefore, attempted to synthesize a new affinity gel which comprises baclofen as an immobilized $GABA_B$ receptor ligand, as shown in Figure 10.3. Approximately 22-fold purification was achieved by the use of the affinity column. Since baclofen has two groups capable of coupling with the Sepharose gel, we have prepared two types of baclofen-affinity gel. One is the gel of CH-Sepharose 4B coupled via the carboxyl group of baclofen, whilst the other is the gel of epoxy-activated Sepharose 6B coupled via the amino group of baclofen. The purification and specific activity of the $GABA_B$ receptor obtained from these two affinity columns were almost the same, but were not high enough, possibly due to contamination by other non-specific protein. Therefore, to remove proteins other than the $GABA_B$ receptor, the fraction eluted from the baclofen-affinity column by NaSCN was further fractionated using gel filtration chromatography of Sephacryl S-300. The elution profile obtained following gel filtration indicated the presence of two peaks of $GABA_B$ receptor binding. Although the exact nature of these two peaks is not clear at present, the peak A having a molecular weight (240 000) higher than that of the second peak, B (80 000), may reflect the aggregation or attachment of other cellular components associated with the $GABA_B$ receptor such as GTP-binding protein or calcium ion channels, since the coupling of these components disappears, at least in part, during solubilization (Ohmori *et al.*, 1988). In any event, it would seem that approximately 140-fold purification of the $GABA_B$ receptor was achieved in peak A by the combined use of a baclofen-affinity gel and Sephacryl S-300 gel chromatography.

There are several reports describing the alteration of receptor coupling with intracellular signal transduction systems following solubilization, especially in the case of metabotropic receptors. For example, it has been reported that the functional coupling of D_2 dopamine receptors with GTP-binding protein is easily eliminated during solubilization (Ohara *et al.*, 1988). On the other hand, opioid (Wong *et al.*, 1989) and A_1 adenosine (Nakata, 1989) receptors have been reported to maintain the coupling mechanisms with GTP-binding protein following solubilization. This evidence, therefore, suggests that the cerebral $GABA_B$ receptor may be of the D_2 dopamine receptor type of metabotropic receptor in terms of coupling with GTP-binding protein.

In an earlier study, we also attempted to purify the cerebral $GABA_B$ receptor using a GABA-affinity column after removing the $GABA_A$-benzodiazepine receptor complex by benzodiazepine-affinity column chromatography (Kuriyama and Taguchi, 1987; Taguchi and Kuriyama, 1984), because GABA is an agonist for both $GABA_B$ and $GABA_A$ receptors. Similarly, an affinity gel of Sepharose 4B coupled with 3-aminopropylphosphonic acid (3-APPA), which is considered to be a weak agonist for $GABA_B$ receptors (Kerr *et al.*, 1987), was also prepared and used for purification of the $GABA_B$ receptor. These procedures, however, were not satisfactory for purification. Although a satisfactory level of purification of cerebral $GABA_B$ receptors has not yet been achieved, the combined use of a baclofen-affinity column and Sephacryl S-300 gel column chromatography may provide an initial step for further development for $GABA_B$ receptor purification procedures, and may be useful for partial purification of the $GABA_B$ receptor within the brain.

ACKNOWLEDGEMENTS

This work was supported in part by a Grant-in-Aid for Scientific Research on General Areas, from the Ministry of Education, Science and Culture, Japan.

REFERENCES

Dunlap, K. (1981) Two types of γ-aminobutyric acid receptor on embryonic sensory neurons. *Br. J. Pharmacol.*, **74**, 579–585.

Hill, D.R. and Bowery, N.G. (1981) ^{3}H-Baclofen and ^{3}H-GABA bind to bicuculline-insensitive $GABA_B$ sites in rat brain. *Nature*, **290**, 149–152.

Hill, D.R., Bowery, N.G. and Hudson, A.L. (1984) Inhibition of $GABA_B$ receptor binding by guanyl nucleotides. *J. Neurochem.*, **42**, 652–657.

Hirouchi, M., Taguchi, J., Ueha, T. and Kuriyama, K. (1987) GABA-stimulated $^{36}Cl^-$ influx into reconstituted vesicles with purified $GABA_A$/benzodiazepine receptor complex. *Biochem. Biophys. Res. Commun.*, **146**, 1471–1477.

Kerr, D.I.B., Ong, J., Prager, R.H., Gynther, B.D. and Curtis, D.R. (1987) Phaclofen: a central baclofen antagonist. *Brain Res.*, **405**, 150–154.

Kuriyama, K. and Taguchi, J. (1987) Purification of γ-aminobutyric acid receptor, benzodiazepine receptor and Cl^- channel from bovine cerebral cortex by benzodiazepine affinity gel column chromatography. *Neurochem. Int.*, **10**, 253–263.

Lynch, T.J., Tallant, E.A. and Cheung, W.Y. (1977) Rat brain adenylate cyclase: further studies on its stimulation by Ca^{2+}. *Arch. Biochem. Biophys.*, **182**, 124–133.

Nakata, H. (1989) Affinity chromatography of A_1 adenosine receptors of rat brain membranes. *Mol. Pharmacol.*, **35**, 780–786.

Nishikawa, M. and Kuriyama, K. (1989) Functional coupling of cerebral γ-aminobutyric acid (GABA$_B$ receptor with adenylate cyclase system: effect of phaclofen. *Neurochem. Int.*, **14**, 85–90.

Ohara, K., Haga, K., Berstein, G., Haga, T., Ichiyama, A. and Ohara, K. (1988) The interaction between D-2 dopamine receptor and GTP-binding protein. *Mol. Pharmacol.*, **33**, 290–296.

Ohmori, Y., Nishikawa, M., Taguchi, J. and Kuriyama, K. (1988) Functional coupling of $GABA_B$ receptor with Ca^{2+} channel and GTP-binding protein and its alteration following solubilization of $GABA_B$ receptor. *Neurochem. Res.*, **13**, 1087–1093.

Taguchi, J. and Kuriyama, K. (1984) Purification of γ-aminobutyric acid (GABA) receptor from rat brain by affinity column chromatography using a new benzodiazepine, 1012-S, as an immobilized ligand. *Brain Res.*, **323**, 219–226.

Wojcik, W.J. and Neff, N.H. (1983) γ-Aminobutyric acid B receptors are negatively coupled to adenylate cyclase in brain, and in the cerebellum these receptors may be associated with granule cells. *Mol. Pharmacol.*, **25**, 24–28.

Wong, Y.H., Demolinon-Mason, C.D. and Barnard, E.A. (1989) Opioid receptors in magnesium-digitonin-solubilized rat brain membrane are tightly coupled to a pertussis toxin-sensitive guanine nucleotide-binding protein. *J. Neurochem.*, **52**, 999–1009.

Zukin, S.R., Young, A.B. and Snyder, S.H. (1974) Gamma-aminobutyric acid binding to receptor sites in the rat central nervous system. *Proc. Natl Acad. Sci. USA*, **71**, 4802–4807.

Part V

ELECTROPHYSIOLOGICAL ASPECTS

11 Pre- and Postsynaptic $GABA_B$ Responses in the Hippocampus

R. A. NICOLL and P. DUTAR

INTRODUCTION

Until the early 1980s, it was believed that all of the actions of GABA were mediated by bicuculline- and picrotoxin-sensitive GABA receptors which increase chloride conductance. These receptors are referred to as $GABA_A$ receptors. However, in neurochemical studies it was found that GABA can presynaptically inhibit the release of other neurotransmitters and this effect was entirely resistant to bicuculline (Hill and Bowery, 1981). This action was mimicked by the GABA analogue, baclofen, which had no action on $GABA_A$ receptors. The receptors which mediate this presynaptic inhibition have been termed $GABA_B$ receptors. While baclofen is a very selective $GABA_B$ agonist, it is not certain that all of its actions can be attributed to $GABA_B$ receptor activation. Baclofen has been shown to have very potent pre- and postsynaptic actions in a variety of areas of the central nervous system, including the hippocampus. In this chapter we discuss the actions of baclofen in the hippocampus and their possible physiological significance.

POSTSYNAPTIC $GABA_B$ RECEPTORS

Postsynaptic Action of Baclofen

Application of baclofen causes a large hyperpolarization in virtually all pyramidal cells (see B in Figure 11.1B) (Gähwiler and Brown, 1985; Inoue *et al.*, 1985a; Misgeld *et al.*, 1984; Newberry and Nicoll, 1984b, 1985). In 5.4 mM external K^+ the reversal potential for the response was approximately -85 mV and the reversal potential shifted with changes in extracellular K^+ in a close to Nernstian manner. Thus the receptor activated by baclofen is coupled to a K^+ channel. This K^+ conductance activated by baclofen also shows inward rectification. Serotonin (5-hydroxytryptamine, 5-HT), acting on a 5-HT_{1a} receptor, activates a K^+ conductance with very similar properties. Maximal activation of the K^+ conductance by baclofen prevented any further activation by serotonin, strongly suggesting that the two classes of receptor converge onto the same K^+ channels (Andrade *et al.*, 1986). In addition, pertussis toxin, which

$GABA_B$ Receptors in Mammalian Function. Edited by N.G. Bowery, H. Bittiger and H.-R. Olpe

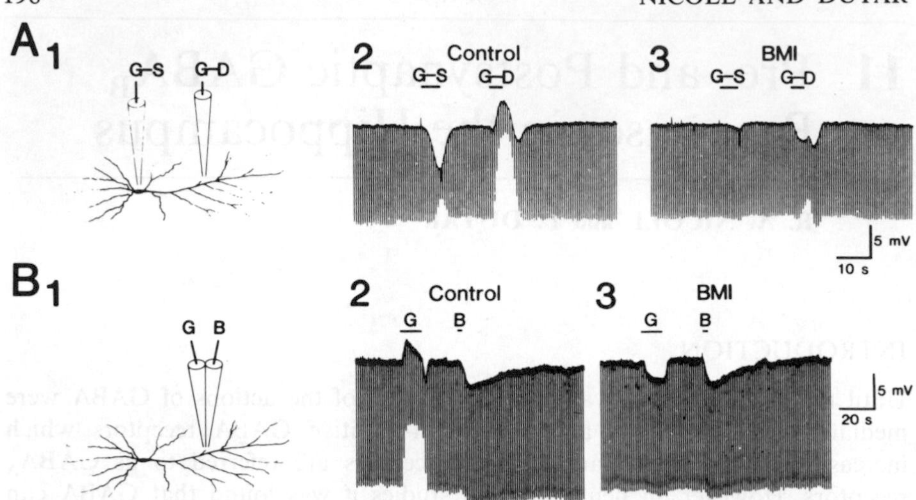

Figure 11.1. Actions of GABA on hippocampal pyramidal cells. (A_1) Placement of iontophoretic electrodes. (A_2) and (A_3) Somatic (G-S) and dendritic (G-D) GABA applications were repeated after superfusion with bicuculline methiodide (BMI) (100 μM) for 20 min. Downward deflections represent the response to constant-current hyperpolarizing pulses applied throughout the recording. (B_1) Placement of double-barrelled iontophoretic electrode. Responses to GABA (G) (60 nA) and baclofen (B) (70 nA) before (B_2) and 25 min after (B_3) superfusing BMI (100 μM). Membrane potentials −55 and −62 mV. (From Newberry and Nicoll, 1985.)

ADP-ribosylates and inactivates some G-proteins, blocks the responses to baclofen (Andrade *et al.*, 1986; Thalmann, 1988b; Colmers and Pittman, 1989). Finally, the hydrolysis-resistant GTP analogue, GTPγS, activates the same conductance and the hydrolysis-resistant GDP analogue, GDPβS, prevents agonist activation of the K^+ conductance (Andrade *et al.*, 1986; Thalmann, 1988b). A direct coupling of the G-protein to the channel may occur since we were able to exclude a number of likely second messenger systems. Phaclofen, which has been found to antagonize baclofen responses in the spinal cord and gut (Kerr *et al.*, 1987), also antagonizes the hyperpolarizing responses in the hippocampus (Figure 11.2A) (Dutar and Nicoll, 1988b; Soltesz *et al.*, 1988). In addition, baclofen has been found to hyperpolarize interneurons in the hippocampus and this is quite likely mediated by a similar conductance mechanism (Madison and Nicoll, 1988; Misgeld *et al.*, 1989).

Activation of protein kinase C by phorbol esters blocks both the serotonin and baclofen response (Andrade *et al.*, 1986). Muscarinic receptor activation has also been reported to depress baclofen responses, possibly by activating phosphatidylinositol turnover (Müller and Misgeld, 1989; Worley *et al.*, 1987). In addition, the K^+ channel blocker, 4-aminopyridine (4-AP), has been reported to depress baclofen (Inoue *et al.*, 1985a).

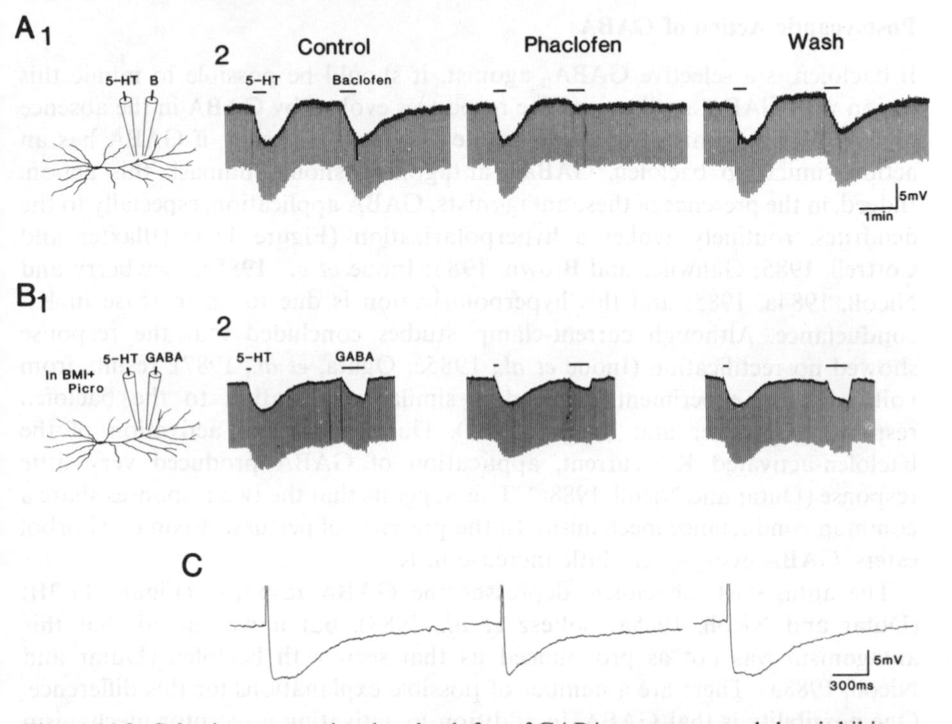

Figure 11.2. Selective antagonism by phaclofen of GABA$_B$ responses and slow inhibitory postsynaptic potential (IPSP) in CA1 pyramidal cells. (A_1) The serotonin (5-hydroxytryptamine, 5-HT) and baclofen (B) electrodes were positioned at the surface of the stratum radiatum, 200 μm away from the pyramidal cell layer. (A_2) 5-HT (220 nA) and baclofen (120 nA) iontophoresed from these electrodes induced hyperpolarizing responses as shown by intracellular recording from the pyramidal cell. In the presence of phaclofen (0.5 mM), superfused for 6 min, the baclofen is strongly antagonized. In striking contrast, the 5-HT response remains unchanged. The baclofen response recovers 6 min after washing phaclofen from the bath. (B_1) GABA and 5-HT are iontophoresed to the dendrites of the pyramidal cell in the presence of the GABA$_A$ antagonists, bicuculline methiodide (BMI) (40 μM) and picrotoxin (Picro) (20 μM). (B_2) intracellular recordings from the pyramidal cell show the hyperpolarizing responses induced by GABA (120 nA) and 5-HT (100 nA). In the presence of phaclofen (0.5 mM) superfused for 15 min, the bicuculline-resistant GABA response is depressed. In contrast, the 5-HT response is unchanged. The GABA response recovers 20 min after washing phaclofen from the bath. Records in (A) and (B) are from two different cells. Membrane potentials − 62 mV and − 68 mV, respectively. The drugs were iontophoresed for the periods indicated by the bars. Calibration in (A) applies to (B). (C) Synaptically evoked potential induced by orthodromic stimulation in the stratum radiatum was recorded from a CA1 pyramidal cell. As shown in the control trace, the electrical stimulation evokes, following the excitatory synaptic potential (truncated in this record), a biphasic hyperpolarizing response contrasting of a fast and a slow IPSP. In the presence of phaclofen (0.2 mM), superfused for 5 min, the slow IPSP is abolished (middle trace). In contrast, the fast IPSP remains unaffected. The slow IPSP recovers 15 min after washing the phaclofen from the bath. Resting potential, − 56 mV. (From Dutar and Nicoll, 1988a.)

Postsynaptic Action of GABA

If baclofen is a selective $GABA_B$ agonist, it should be possible to mimic this action with GABA application. The responses evoked by GABA in the absence of $GABA_A$ antagonists are usually quite complex. However, if GABA has an action similar to baclofen, $GABA_A$ antagonists should unmask this action. Indeed, in the presence of these antagonists, GABA application, especially to the dendrites, routinely evokes a hyperpolarization (Figure 11.1) (Blaxter and Cottrell, 1985; Gähwiler and Brown, 1985; Inoue *et al.*, 1985c; Newberry and Nicoll, 1984a, 1985) and this hyperpolarization is due to an increase in K^+ conductance. Although current-clamp studies concluded that the response showed no rectification (Inoue *et al.*, 1985c; Ogata, *et al.*, 1987), results from voltage-clamp experiments showed a similar rectification to the baclofen response (Gähwiler and Brown, 1985). During maximal activation of the baclofen-activated K^+ current, application of GABA produced very little response (Dutar and Nicoll, 1988a). This suggests that the two responses share a common conductance mechanism. In the presence of pertussis toxin or phorbol esters, GABA evoked very little increase in K^+.

The antagonist, phaclofen, depresses the GABA response (Figure 11.2B) (Dutar and Nicoll, 1988a; Soltesz *et al.*, 1988), but it was noted that this antagonism was not as pronounced as that seen with baclofen (Dutar and Nicoll, 1988a). There are a number of possible explanations for this difference. One possibility is that GABA, in addition to activating a receptor mechanism shared with baclofen, also activates, to a varying degree, some other response. This explanation might also help explain the reported difference in sensitivity of the two responses to carbachol (Müller and Misgeld, 1989) and 4-AP(Ogata *et al.*, 1987).

POSTSYNAPTIC INHIBITORY POTENTIALS

Recordings from hippocampal pyramidal cells (Alger, 1984; Kehl and McLennan, 1985; Newberry and Nicoll, 1984a; Thalmann, 1988a,b) and from dentate granule cells (Thalmann and Ayala, 1982) under normal conditions demonstrate two types of inhibitory postsynaptic potentials (IPSPs) (Figure 11.2C). A fast IPSP that peaks at approximately 50 ms and lasts for 200–300 ms results from stimulation of the alveus, which antidromically activates the axons of CA_1 pyramidal cells. This IPSP is blocked by $GABA_A$ antagonists, has a reversal potential of approximately -70 mV, and is altered by changes in the chloride gradient across the membrane. Thus it is proposed that this IPSP results from GABA acting on somatic $GABA_A$ receptors. An IPSP similar in every respect to that evoked by antidromic stimulation can be evoked by stimulation of afferent fibres in the stratum radiatum at low intensities. However, stronger stimulation evokes a slower and later component which peaks at about 200 ms and lasts about 1 s. A number of terms have been used to characterize this potential, including the slow or late IPSP or the late hyperpolarizing potential. This slow IPSP is resistant to $GABA_A$ antagonists and is due to an increase in K^+ conductance. It is also selectively blocked by pertussis toxin pretreatment (Dutar and Nicoll, 1988b; Thalmann, 1988b) and shares the K^+ conductance activated by GTPγS (Thalmann, 1988b). These features are identical to the $GABA_B$ receptor-mediated action of GABA and raise the possibility that GABA may be the neurotransmitter responsible for the slow IPSP. Two pharmacological manipulations support this conclusion. First, the GABA uptake blocker, *cis*-4-hydroxynipecotic acid can prolong the slow IPSP in the hippocampus (Dingledine and Korn, 1985; Nicoll and Dutar, 1989) and, second, phaclofen selectively antagonizes the slow IPSP (Figure 11.2C) (Dutar and Nicoll, 1988a; Soltesz *et al.*, 1988). On the other hand, it has recently been reported that low concentrations of carbachol reduce the baclofen response, but not the slow IPSP or the bicuculline-resistant action of GABA (Müller and Misgeld, 1989). Further studies with more potent antagonists should resolve this issue.

PRESYNAPTIC $GABA_B$ RECEPTORS

Application of baclofen to hippocampal slices causes a large reduction in the size of the excitatory postsynaptic potential (EPSP) recorded extracellularly (Ault and Nadler, 1982; Lanthorn and Cotman, 1981) and intracellularly (Blaxter and Carlen, 1985; Dutar and Nicoll, 1988b; Inoue *et al.*, 1985b). Baclofen also causes presynaptic inhibition of IPSPs evoked in cultured hippocampal neurons (Harrison *et al.*, 1988). Very little evidence is available concerning whether GABA in the absence of $GABA_A$ activity has an identical presynaptic effect. One might argue that the reduction in the EPSP is due to the postsynaptic conductance increase caused by baclofen. However, the inhibitory action is much greater than the postsynaptic conductance change, and manipulations, such as pertussis toxin treatment, which entirely block the postsynaptic action can have little effect on the EPSP depression. Whether the pharmacology of this presynaptic action of baclofen is different from the postsynaptic action is not entirely clear. Pertussis toxin, injected into the lateral ventricle 3 days before the acute experiment, blocks the postsynaptic response but leaves the presynaptic response intact (Figure 11.3B) (Colmers and Pittman, 1989; Dutar and Nicoll, 1988b). Similar differential effects have been reported for baclofen in the dorsal raphé nucleus (Colmers and Williams, 1988) and the lateral septal nucleus (Gallagher *et al.*, this meeting). In addition, intraventricular injection of pertussis toxin has the same differential effect on the action of adenosine in the hippocampus (Fredholm *et al.*, 1989). However, there is reason to believe that this apparant resistance of the presynaptic inhibitory action to pertussis toxin may in some way be due to limited access of the toxin to the G-protein. Thus,

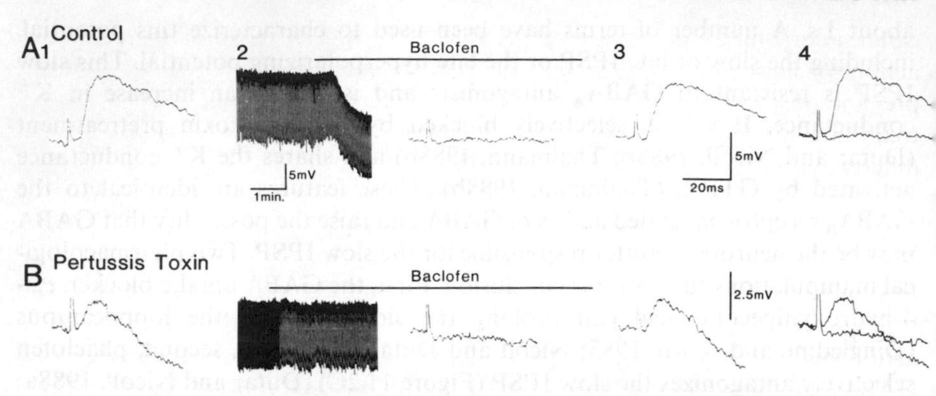

Figure 11.3. The presynaptic inhibitory action of baclofen is not blocked by pertussis toxin. (A) The pre- and postsynaptic inhibitory action of baclofen recorded in a cell from a control slice. (A_1) The excitatory postsynaptic potential (EPSP) was elicited by electrical stimulation of afferents in the stratum radiatum. (A_2) Baclofen (30 μM) applied by superfusion in the bath induced a hyperpolarization and a blockade of the EPSP. (A_3) The size of the EPSP recovered after washing from the bath. (A_4) The traces in (A_1) and (A_2) were superimposed. Resting membrane potential is −62 mV. All of the EPSPs were elicited at a negative potential (− 75 mV) to avoid action potential generation by the cell. (B) The same experiment using a cell from a rat treated 3 days prior to experiment with pertussis toxin (1.5 μg). The postsynaptic action of baclofen (40 μM) applied by superfusion in the bath was blocked. In contrast, baclofen still induced the presynaptic depression of the EPSP. Resting membrane potentials is − 65 mV. All of the EPSPs were elicited from a holding potential of −74 mV. Time calibration in (A) also applies to (B). In (B_2) the chart speed was changed just after the beginning of the perfusion of baclofen. During the baclofen application, the time calibration is the same as in (A_2). (From Dutar and Nicoll, 1988b.)

while the pertussis toxin resistance of the presynaptic inhibitory action of baclofen on inhibitory neurons in cell culture (Harrison *et al.*, 1988) suggests that the access is not an issue, the inhibition by baclofen of glutamate release, measured biochemically, in cell culture, can be blocked by petussis toxin (Wojcik *et al.*, Chapter 8; Dolphin *et al.*, Chapter 15). Furthermore, it has recently been reported that the presynaptic inhibitory action of adenosine and baclofen can be blocked by pertussis toxin pretreatment in vivo, if the pertussis toxin is injected directly into the hippocampus (Stratton *et al.*, 1989). A difference in the pre- and postsynaptic action has also been reported for the action of phaclofen, which has little effect on the presynaptic inhibition of EPSPs (Dutar and Nicoll, 1988b) and IPSPs recorded in culture (Harrison, 1988). Given the low potency of phaclofen on the postsynaptic action, it will be valuable to examine this differential effect with more potent antagonists, to determine if this difference is quantitative or qualitative. Indeed, evidence presented by Harrison *et al.* (Chapter 12) indicates that the presynaptic action can be antagonized by saclofen, although it is less sensitive than the postsynaptic action. In summary, while there is clearly a difference in sensitivity between the pre- and postsynaptic action of baclofen to intraventricular injection of pertussis toxin and to $GABA_B$ antagonists, it remains to be determined whether these differences reflect a difference in receptor subtype or not.

CONCLUSION

In summary, baclofen, which is considered to be a selective $GABA_B$ agonist has both post- and presynaptic inhibitory actions. The postsynaptic inhibitory action involves a selective increase in K^+ conductance. GABA, in the presence of $GABA_A$ antagonists, can also increase a K^+ conductance which has many features in common with the baclofen response. However, whether they are identical is not entirely clear, since differences in the sensitivity of these two responses to phaclofen, 4-AP and carbachol have been reported. The selective blockade of the slow IPSP by both phaclofen and saclofen suggests a role for $GABA_B$ receptors in this form of inhibition. However, detailed comparison of the pharmacological properties of GABA and baclofen responses will be of great importance, not only in clarifying the definition of $GABA_B$ receptors in the hippocampus, but also elucidating their physiological role in synaptic transmission.

The presynaptic inhibitory action of baclofen in the hippocampus as well as at a variety of other sites is unaffected by intraventricular injections of pertussis toxin, whereas the postsynaptic action is readily blocked. However, a recent report on a study using intrahippocampal injections suggests that local high concentrations can block this presynaptic action. In addition, while phaclofen was ineffective against the presynaptic actions, saclofen, a more potent antagonist, can block this action. It will be important to determine whether these quantitative differences between the pre- and postsynaptic actions of baclofen result from different receptor subtypes.

REFERENCES

Alger, B.E. (1984) Characteristics of a slow hyperpolarizing synaptic potential in rat hippocampal pyramidal cells in vitro. *J. Neurophysiol.*, **52**, 892–910.

Andrade, R., Malenka, R.C. and Nicoll, R.A. (1986) A G protein couples serotonin and $GABA_B$ receptors to the same channels in hippocampus. *Science*, **234**, 1261–1265.

Ault, B. and Nadler, J.V. (1982) Baclofen selectively inhibits transmission at synapses made by axons of CA_3 pyramidal cells in hippocampal slice. *J. Pharmacol. Exp. Ther.*, **223**, 291–297.

Blaxter, T.J. and Carlen, P.L. (1985) Pre- and postsynaptic effects of baclofen in the rat hippocampal slice. *Brain Res.*, **341**, 195–199.

Blaxter, T.J. and Cottrell, G.A. (1985) Actions of GABA and ethylenediamine on CA_1 pyramidal neurones of the rat hippocampus. *Q. J. Exp. Physiol.*, **70**, 75–93.

Colmers, W.F. and Pittman, Q.J. (1989) Presynaptic inhibition by neuropeptide Y and baclofen in hippocampus: insensitivity to pertussis toxin treatment. *Brain Res.*, **498**, 99–104.

Colmers, W.F. and Pittman, Q.J. (1988) Pertussis toxin pretreatment discriminates between pre- and postsynaptic actions of baclofen in rat dorsal raphé nucleus in vitro. *Neurosci. Lett.*, **93**, 300–306.

Dingledine, R. and Korn, S.J. (1985) γ-Aminobutyric acid uptake and the termination of inhibitory synaptic potentials in the rat hippocampal slice. *J. Physiol.*, **366**, 387–410.

Dutar, P. and Nicoll, R.A. (1988a) A physiological role for $GABA_B$ receptors in the CNS. *Nature*, **332**, 156–158.

Dutar, P. and Nicoll, R.A. (1988b) Pre- and postsynaptic $GABA_B$ receptors in the hippocampus have different pharmacological properties. *Neuron*, **1**, 585–591.

Fredholm, B.B., Proctor, W., Van der Ploeg, I. and Dunwiddie, T.V. (1989) In vivo pertussis toxin treatment alternates some, but not all, adenosine A effects in slices of the rat hippocampus. *Eur. J. Pharmacol.*, **172**, 249–262.

Gähwiler, B.H. and Brown, D.A. (1985) $GABA_B$-receptor-activated K^+ current in voltage clamped CA_3 pyramidal cells in hippocampus. *Proc. Natl Acad. Sci. USA*, **82**, 1558–1562.

Harrison, N.L. (1988) Baclofen decreases synaptic inhibition in cultured hippocampal neurons by a presynaptic mechanism that is insensitive to pertussis toxin. *Neurosci. Abstr.*, **14**, 1092.

Harrison, N.L., Lange, G.D. and Barker, J.L. (1988) (−)-Baclofen activates presynaptic $GABA_B$ receptors on GABAergic inhibitory neurons from embryonic rat hippocampus. *Neurosci. Lett.*, **85**, 105–109.

Hill, D.R. and Bowery, N.G. (1981) ^{3}H-Baclofen and ^{3}H-GABA bind to bicuculline-insensitive $GABA_B$ sites in rat brain. *Nature*, **290**, 149–152.

Inoue, M., Matsuo, T. and Ogata, N. (1985a) Baclofen activates voltage-dependent and 4-aminopyridine sensitive K^+ conductance in guinea-pig hippocampal pyramidal cells maintained in vitro. *Br. J. Pharmacol.*, **84**, 833–841.

Inoue, M., Matsuo, T. and Ogata, N. (1985b) Characterization of pre- and postsynaptic actions of (−)-baclofen in the guinea-pig hippocampus in vitro. *Br. J. Pharmacol.*, **84**, 843–851.

Inoue, M., Matsuo, T. and Ogata, N. (1985c) Possible involvement of K^+ conductance in the action of γ-aminobutyric acid in the guinea-pig hippocampus. *Br. J. Pharmacol.*, **86**, 515–524.

Kehl, S.J. and McLennan, H. (1985) An electrophysiological characterization of inhibitions and postsynaptic potentials in rat hippocampal CA_3 neurones in vitro. *Exp. Brain. Res.*, **60**, 229–308.

Kerr, D.I.B., Ong, J., Prager, R.H., Gynther, B.D. and Curtis, D.R. (1987) Phaclofen: a peripheral and central baclofen antagonist. *Brain Res.*, **405**, 150–154.

Lanthorn, T.H. and Cotman, C.W. (1981) Baclofen selectively inhibits excitatory synaptic transmission in the hippocampus. *Brain Res.*, **225**, 171–178.

Madison, D.V. and Nicoll, R.A. (1988) Enkephalin hyperpolarizes interneurones in the rat hippocampus. *J. Physiol.*, **398**, 123–130.

Misgeld, U., Klee, M.R. and Zeise, M.L. (1984) Differences in baclofen-sensitivity between CA_3 neurons and granule cells of the guinea pig hippocampus in vitro *Neurosci. Lett.*, **47**, 307–311.

Misgeld, U., Müller, W. and Brunner, H. (1989) Effects of (−)-baclofen on inhibitory neurons in the guinea pig hippocampal slice. *Pflügers Arch.*, **414**, 139–144.

Müller, W. and Misgeld, U. (1989) Carbachol reduces $I_{K,Baclofen}$, but not $I_{K,GABA}$ in guinea pig hippocampal slices. *Neurosci. Lett.*, **102**, 229–234.

Newberry, N.R. and Nicoll, R.A. (1984a) A bicuculline-resistant inhibitory postsynaptic potential in rat hippocampal pyramidal cells in vitro. *J. Physiol.*, **348**, 239–254.

Newberry, N.R. and Nicoll, R.A. (1984b) Direct hyperpolarizing action of baclofen on hippocampal pyramidal cells. *Nature*, **308**, 450–452.

Newberry, N.R. and Nicoll, R.A. (1985) Comparison of the action of baclofen with γ-aminobutyric acid on rat hippocampal pyramidal cells in vitro. *J. Physiol.*, **360**, 162–185.

Nicoll, R.A. and Dutar, P. (1989) Physiological roles of $GABA_A$ and $GABA_B$ receptors in synaptic transmission in the hippocampus. In: E.A. Barnard and E. Costa (eds) *Allosteric modulation of amino acid receptors: therapeutic implications.* New York: Raven Press, pp. 195–204.

Ogata, N., Inoue, M. and Matsuo, T. (1987) Contrasting properties of K^+ conductances induced by baclofen and γ-aminobutyric acid in slices of the guinea pig hippocampus. *Synapse*, **1**, 62–69.

Soltesz, I., Haby, M., Leresche, N. and Crunelli, V. (1988) The $GABA_B$ antagonist phaclofen inhibits the late K^+-dependent IPSP in cat and rat thalamic and hippocampal neurones. *Brain Res.*, **448**, 351–354.

Stratton, K.R., Cole, A.J., Pritchett, J., Eccles, C.U., Worley, P.F. and Baraban, J.M. (1989) Intrahippocampal injection of pertussis toxin blocks adenosine suppression of synaptic responses. *Brain Res.*, **494**, 359–364.

Thalmann, R.H. (1988a) Blockade of a late inhibitory postsynaptic potential in hippocampal CA_3 neurons in vitro reveals a late depolarizing potential that is augmented by pentobarbital. *Neurosci. Lett.*, **95**, 155–160.

Thalmann, R.H. (1988b) Evidence that guanosine triphosphate (GTP)-binding proteins control a synaptic response in brain: effect of pertussis toxin and GTPγS on the late inhibitory postsynaptic potential of hippocampal CA_3 neurons. *J. Neurosci.*, **8**, 4589–4602.

Thalmann, R.H. and Ayala, G. (1982) A late increase in potassium conductance follows synaptic stimulation of granule neurons of the dentate gyrus. *Neurosci. Lett.*, **29**, 243–248.

Worley, P.F., Baraban, J.M., McCarren, M., Snyder, S.H. and Alger, B.E. (1987) Cholinergic phosphatidylinositol modulation of inhibitory, G-protein-linked, neurotransmitter action: electrophysiological studies in rat hippocampus. *Proc. Natl Acad. Sci. USA*, **84**, 3467–3471.

Newberry, N.R. and Nicoll, R.A. (1984a) A bicuculline-resistant inhibitory post-synaptic potential in rat hippocampal pyramidal cells in vitro. *J. Physiol.*, **348**, 239–254.
Newberry, N.R. and Nicoll, R.A. (1984b) Direct hyperpolarizing action of baclofen on hippocampal pyramidal cells. *Nature*, **308**, 450–452.
Newberry, N.R. and Nicoll, R.A. (1985) Comparison of the action of baclofen with γ-aminobutyric acid on rat hippocampal pyramidal cells in vitro. *J. Physiol.*, **360**, 161–185.
Nicoll, R.A. and Dutar, P. (1989) Physiological roles of GABA$_A$ and GABA$_B$ receptors in synaptic transmission in the hippocampus. In: T.A. Barnard and E. Costa (eds) *Allosteric modulation of amino acid receptors: therapeutic implications*. New York: Raven Press, pp. 195–204.
Ogata, N., Inoue, M. and Matsuo, T. (1987) Contrasting properties of K$^+$ conductances induced by baclofen and γ-aminobutyric acid in slices of the guinea pig hippocampus. *Synapse*, **1**, 62–69.
Soltesz, I., Haby, M., Leresche, N. and Crunelli, V. (1988) The GABA$_B$ antagonist phaclofen inhibits the late K$^+$-dependent IPSP in cat and rat thalamic and hippocampal neurones. *Brain Res.*, **448**, 351–354.
Stratton, K.R., Cole, A.J., Pritchett, J., Eccles, C.U., Worley, P.F. and Baraban, J.M. (1989) Intrahippocampal injection of pertussis toxin blocks adenosine suppression of synaptic responses. *Brain Res.*, **494**, 359–364.
Thalmann, R.H. (1988a) Blockade of a late inhibitory postsynaptic potential in hippocampal CA3 neurons in vitro reveals a late depolarizing potential that is augmented by pentobarbital. *Neurosci. Lett.*, **95**, 155–160.
Thalmann, R.H. (1988b) Evidence that guanosine triphosphate (GTP)-binding proteins control a synaptic response in brain: effect of pertussis toxin and GTPγS on the late inhibitory postsynaptic potential of hippocampal CA3 neurons. *J. Neurosci.*, **8**, 4589–4602.
Thalmann, R.H. and Ayala, G.F. (1982) A late increase in potassium conductance follows synaptic stimulation of granule neurons of the dentate gyrus. *Neurosci. Lett.*, **29**, 243–248.
Worley, P.F., Baraban, J.M., McCarren, M., Snyder, S.H. and Alger, B.E. (1987) Cholinergic phosphatidylinositol modulation of inhibitory, G protein-linked, neurotransmitter actions: electrophysiological studies in rat hippocampus. *Proc. Natl Acad. Sci. USA*, **84**, 3467–3471.

12 Presynaptic $GABA_B$ Receptors on Rat Hippocampal Neurons

N. L. HARRISON, N. A. LAMBERT and D. M. LOVINGER

INTRODUCTION

The primary inhibitory neurotransmitter in the brain is GABA. Pharmacological evidence suggests that the mammalian brain contains two classes of GABA receptor (Bowery *et al.*, 1980, 1987). The 'classical' GABA receptor is activated by GABA and muscimol and is blocked by bicuculline. This has been designated the $GABA_A$ receptor; this receptor is a multimeric protein that forms an integral Cl^- ion-selective channel (Bormann *et al.*, 1987; Olsen, 1982; Schofield *et al.*, 1987). The $GABA_A$ receptor is involved in generating 'fast' inhibitory postsynaptic potentials (IPSPs) in the mammalian brain. A second class of receptor, designated the $GABA_B$ receptor, is activated by GABA and baclofen (Bowery *et al.*, 1980; Hill and Bowery, 1981). The $GABA_B$ receptor is known to interact with GTP-binding proteins ('G-proteins'; Hill *et al.*, 1984; Holz *et al.*, 1986; Karbon and Enna, 1984; Wojcik and Neff, 1984). Activation of postsynaptic $GABA_B$ receptors on central neurons mediates an increase in K^+ conductance (Andrade *et al.*, 1986; Colmers and Williams, 1988; Conners *et al.*, 1988; Gähwiler and Brown, 1985; Gallagher *et al.*, 1984; Howe *et al.*, 1987; Lacey *et al.*, 1988; Newberry and Nicoll, 1984; Stevens *et al.*, 1985) that underlies a second, slower IPSP (Dutar and Nicoll, 1988a; Hasuo and Gallagher, 1988; Karlsson and Olpe, 1989; McCormick, 1989; Soltesz *et al.*, 1988). This receptor is blocked by the antagonist phaclofen (Kerr *et al.*, 1987). In addition to these postsynaptic $GABA_B$ receptors, there exist in the central nervous system presynaptic $GABA_B$ receptors (Ault and Nadler, 1982; Colmers and Williams, 1988; Davies, 1981; Dutar and Nicoll, 1988b; Harrison, 1990; Stirling *et al.*, 1989) whose pharmacology has yet to be described in detail.

We have studied the effects of $GABA_B$ receptor agonists and antagonists at the synapses formed between cultured hippocampal neurons from the embryonic rat by making simultaneous recordings from both pre- and postsynaptic neurons. In addition, we have investigated presynaptic $GABA_B$ receptors in the rat hippocampal slice preparation, using both extracellular and intracellular recording techniques.

$GABA_B$ Receptors in Mammalian Function. Edited by N.G. Bowery, H. Bittiger and H.-R. Olpe

METHODS

Cell Culture Electrophysiology

Cultured hippocampal neurons from embryonic (E19–20) rats were dissociated using papain and mechanical trituration, and plated onto confluent cortical astrocytes on poly-D-lysine coated 35-mm culture dishes by methods similar to those previously described for the culture of postnatal cortical neurons (Huettner and Baughmann, 1986). Experiments were carried out at 22 °C after 3–4 weeks in culture. Simultaneous recordings were made from pre- and postsynaptic neurons using the whole-cell patch-clamp technique (Hamill *et al.*, 1981). Patch pipettes contained (mM): potassium gluconate, 145; $MgCl_2$, 2; K_2ATP, 5; HEPES/KOH (pH 7.2), 5; $CaCl_2$, 0.1; EGTA, 1.1; Pipette-to-bath resistance was 4–5 MΩ. The extracellular medium contained (mM): NaCl, 125; KCl, 3; $CaCl_2$, 4; $MgCl_2$, 8; D-glucose, 6; HEPES/NaOH (pH 7.4), 10. The high divalent cation levels attenuated spontaneous synaptic activity and facilitated study of monosynaptic pairs (Barker and Harrison, 1988; Segal and Barker, 1984). IPSPs or excitatory PSPs (EPSPs) were elicited in postsynaptic elements by electrically evoking action potentials in the presynaptic neuron (stimuli 2–5 ms, 1 nA. 0.1 Hz), and were identified as monosynaptic by their ability to follow 10–20 Hz presynaptic trains on a one-for-one basis. To record EPSPs, the postsynaptic cell was held at -60 mV in current clamp. To record inhibitory signals, the postsynaptic cell was held at -40 mV in voltage clamp, at which potential outwardly-directed, exponentially decaying synaptic currents were evoked (Barker and Harrison, 1988). Evoked PSPs and postsynaptic currents (PSCs) were digitized at 5–10 kHz, low pass filtered at 1–5 kHz and stored on a PDP 11/23 computer. Baclofen and combinations of baclofen and 2-hydroxy-saclofen (2-OH-S) were applied to the postsynaptic cell body by pressure (< 2 p.s.i.) from identical blunt-tipped micropipettes. 3-Aminopropylphosphinic acid (3-APA) and 2-OH-S were synthesized and kindly donated by Drs R.H. Prager, G. Hofer, D.I.B. Kerr and J. Ong, of Adelaide, Australia. Other chemicals were from Sigma.

Hippocampal Slice Experiments: Field Potentials

Hippocampal slices for extracellular recordings were prepared from adult (200–300 g) Sprague–Dawley rats using a vibrating blade (Vibraslice; Campden Instruments, UK) by methods similar to those previously described (Alger and Nicoll, 1982). Experiments were carried out at 31–32 °C on slices fully submerged in medium. Extracellular recordings of synaptic field potentials were made in the CA1 dendritic field in the stratum radiatum; recordings were also made from the stratum pyramidale, using glass microelectrodes filled with isotonic NaCl. The extracellular medium contained (mM): NaCl, 124; KCl, 4.5; $CaCl_2$, 2; $MgCl_2$, 1.5; $NaHCO_3$, 26; Na_2HPO_4, 1.2; D-glucose, 10, adjusted to pH 7.4 by continuous bubbling with 95% O_2/5% CO_2.

Population EPSPs (pEPSPs) were evoked by electrical stimuli to the Schaffer collateral pathway (0.02 ms, 15–30 V, 0.05 Hz). Stimulating electrodes were twisted bipolar electrodes (tip ~60 μm), and were placed in the stratum radiatum near the CA2/CA1 border. Evoked potentials were digitized at 5 kHz using a Nicolet storage oscilloscope, low-pass filtered at 3 kHz and stored using a PDP 11/23 computer for off-line averaging and analysis. Baclofen, 8-OH-DPAT and 2-OH-S were applied to the slice by bath superfusion (flow rate 2.5–3 ml/min). To test the activity of baclofen, various concentrations of the drug were applied in a 10 ml volume (contact time 3–4 min); this was sufficient to obtain an equilibrium response under our perfusion conditions. To test the antagonist activity of 2-OH-S, the drug was applied for 2 min prior to the application of a combination of 2-OH-S and baclofen, after which the perfusion was returned to 2-OH-S alone for 2 min. Baclofen and 8-hydroxy-2-(di-*n*-propylamino)tetralin hydrogen bromide (8-OH-DPAT) were obtained from Research Biochemicals Inc. (RBI; Natick, MA, USA). Statistical tests were done using a two-tailed *t*-test.

Hippocampal Slice Experiments: Intracellular Recordings

Hippocampal slices for intracellular recordings were prepared from adult (200–300 g) Sprague–Dawley rats using a tissue chopper by methods similar to those previously described (Teyler, 1980). Experiments were carried out at 31–32 °C in a modified interface chamber. Intracellular recordings were made in the CA1 cell body layer, using glass microelectrodes filled with 4 M potassium acetate or 2 M potassium methylsulphate. The extracellular medium contained (mM): NaCl, 124; KCl, 4.5; $CaCl_2$, 2; $MgCl_2$, 1.5; $NaHCO_3$, 26; Na_2HPO_4, 1.2; D-glucose, 10, adjusted to pH 7.4 by continuous bubbling with 95% O_2/5% CO_2.

PSPs were evoked by electrical stimuli to the Schaffer collateral pathway (0.05 ms, 2–20 V, 0.033 Hz). Stimulating electrodes were concentric bipolar electrodes, and were placed in the stratum radiatum. Evoked monosynaptic EPSPs were digitized at 10 kHz, low-pass filtered at 5 kHz and stored for off-line analysis. EPSP slope was measured at half-maximal amplitude on the rising phase of the EPSP. Baclofen and 2-OH-S were applied to the slice by bath superfusion (flow rate 1 ml/min). To test the antagonist activity of 2-OH-S, the drug was applied for 10 min prior to the application of a combination of 2-OH-S and baclofen, after which the perfusion was returned to 2-OH-S alone for 5 min. Baclofen and 8-OH-DPAT were purchased from RBI. Statistical tests were done using a two-tailed *t*-test.

RESULTS

GABA-mediated Monosynaptic Inhibition in Cultured Rat Hippocampal Neurons

The rat hippocampus in situ contains many GABAergic inhibitory interneurons that mediate postsynaptic inhibition of the hippocampal pyramidal cells (Alger and Nicoll, 1982). When neurons from the embryonic rat hippocampus are dissociated and placed in tissue culture, many of these inhibitory interneurons survive and form synaptic contacts with other neurons (Segal and Barker, 1984). By making dual whole-cell patch-clamp recordings from neighbouring neurons, we have found that monosynaptic inhibitory connections are common. A single action potential in a presynaptic cell is followed within 1–2 ms by an IPSP in the postsynaptic cell (Figure 12.1). The IPSP is most easily seen if the membrane is held depolarized from the resting potential; in our experiments, the IPSP is usually recorded at – 50 mV, and reverses polarity at membrane potentials more negative than – 60 mV (Barker and Harrison, 1988).

Presynaptic $GABA_B$ Receptors are Present on the Terminals of GABAergic Inhibitory Neurons

Application of the $GABA_B$ agonist (–)-baclofen to the area of the postsynaptic cell body reduces IPSP amplitude (Figure 12.1). Unlike bicuculline and other $GABA_A$ receptor antagonists, (–)-baclofen does not reduce pharmacological GABA responses (Harrison *et al.*, 1988). Application of baclofen to the cell body

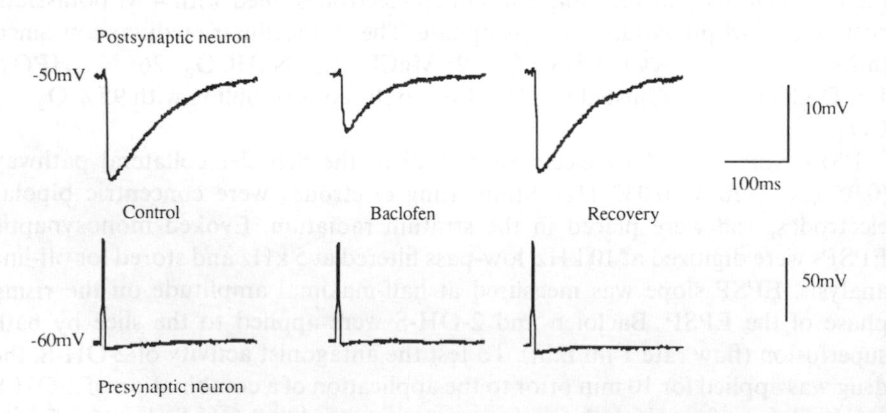

Figure 12.1. Baclofen decreases the amplitude of inhibitory postsynaptic potentials (IPSPs) recorded at synapses between cultured rat hippocampal neurons. In this experiment, baclofen was applied from a micropipette containing 10 μM baclofen. Note that the action potential in the presynaptic neuron is unaltered

did not produce membrane hyperpolarization or increase membrane conductance in these cultured cells and did not alter the threshold current required for spike generation (Harrison *et al.*, 1988).

When the postsynaptic neuron is voltage-clamped at −40 mV, the synaptic signal triggered by a presynaptic stimulus appears as an outward-going (corresponding to *inward* flux of Cl^- ions) synaptic current (IPSC) that reverses in polarity between − 60 and − 80 mV (Barker and Harrison, 1988). The IPSC grows rapidly (2–4 ms), and then decays with exponential kinetics (τ = 10–40 ms).

Voltage-clamp experiments on cultured neurons showed that the IPSC is also depressed by baclofen (Figures 12.2 and 12.3). The synaptic depressant action of baclofen closely resembles that of low-Ca^{2+} medium, and its locus of action is believed to be presynaptic (Harrison, 1990). Other investigators who showed that baclofen depressed GABA-mediated IPSPs in brain slices (Howe *et al.*, 1987; Peet and McLennan, 1986) have suggested that this effect was indirect, i.e. that baclofen suppressed the IPSP by virtue of reduced excitation, or by hyperpolarization of GABAergic interneurons (Misgeld *et al.*, 1989). While at least some of the depressant action of baclofen on IPSPs may be due to hyperpolarization of the interneurons, it is clear from the present results, and from studies of GABA release (Bonanno *et al.*, 1989; Pittaluga *et al.*, 1987; Waldmeier *et al.*, 1988), that the GABAergic neurons also express presynaptic

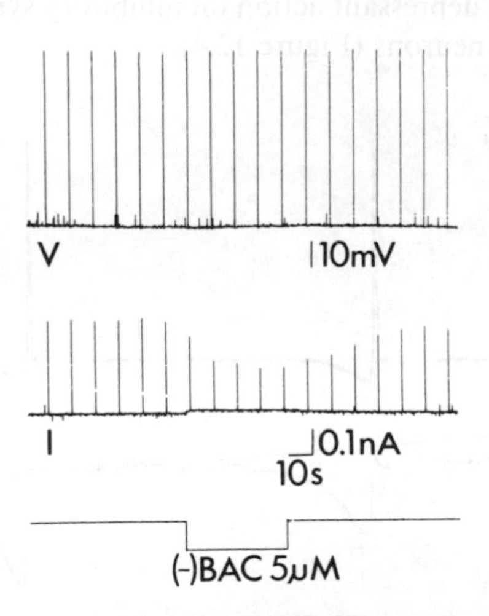

Figure 12.2. The time-course of the synaptic depressant effect of baclofen. The upper trace (pen recording) shows the action potentials elicited at regular intervals in the presynaptic neuron. The centre record shows that the inhibitory postsynaptic current (IPSC; recorded from the postsynaptic neuron under voltage clamp at −40 mV) is depressed by (−)-baclofen over a period of many seconds. Recovery is also relatively slow. The lower trace shows the period of drug application. BAC, baclofen; I, current; V, voltage

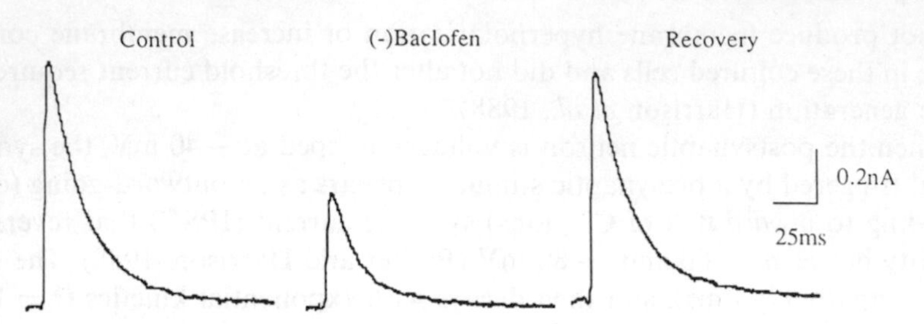

Figure 12.3. The effect of baclofen on individual inhibitory postsynaptic currents (IPSCs) recorded from the postsynaptic neuron under voltage clamp at −40 mV

$GABA_B$ receptors, and that these receptors may play an important role in mediating the sensitivity of hippocampal inhibition to low doses of baclofen.

3-Aminopropylphosphinic Acid is a Potent Agonist at Presynaptic $GABA_B$ Receptors on Cultured Rat Hippocampal Neurons

3-APA closely resembles GABA in structure and is a potent displacer of baclofen binding to $GABA_B$ receptors, being 100-fold more potent than (−)-baclofen. This makes 3-APA the most potent ligand for $GABA_B$ receptors yet found, but little is known of its activity in functional pharmacological assays. 3-APA has a profound depressant action on inhibitory synaptic transmission in cultured hippocampal neurons (Figure 12.4).

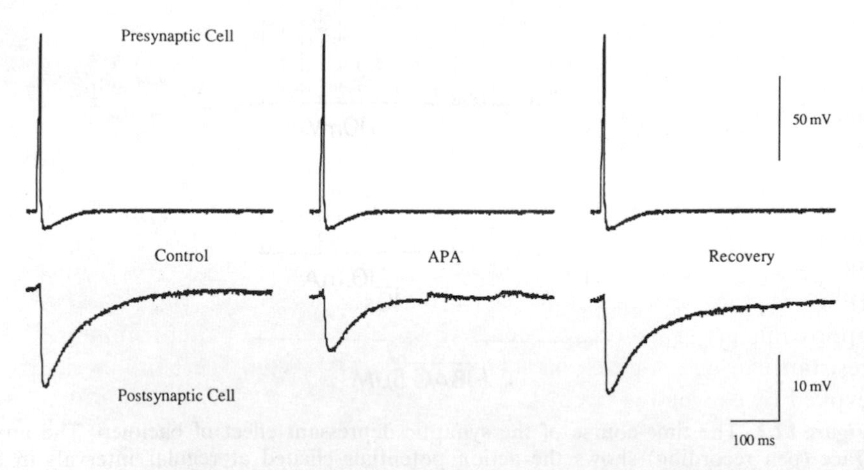

Figure 12.4. 3-Aminopropylphosphinic acid (3-APA) also decreases the amplitude of inhibitory postsynaptic potentials (IPSPs) recorded at synapses between cultured rat hippocampal neurons. In this experiment, 3-APA was applied from a micropipette containing 1 μM 3-APA. Note that the action potential in the presynaptic neuron is again unaltered

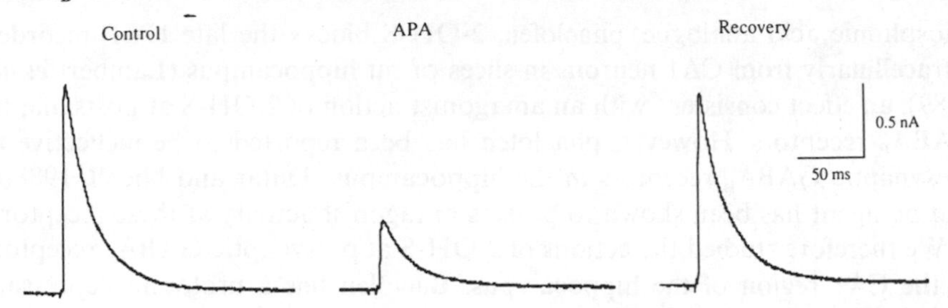

Figure 12.5. 3-Aminopropylphosphinic acid (3-APA) also decreases the amplitude of inhibitory postsynaptic currents (IPSCs). In this experiment, 3-APA was applied from a micropipette containing 1 μM 3-APA. Individual IPSCs recorded under voltage clamp at -40 mV are illustrated

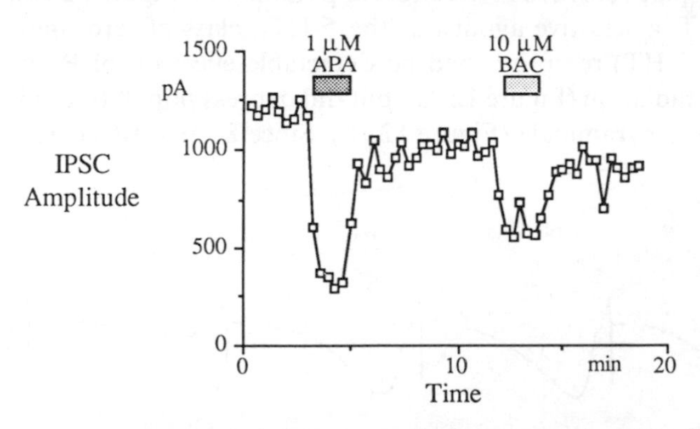

Figure 12.6. 3-Aminopropylphosphinic acid (3-APA) is approximately 10 times more potent than baclofen (BAC) in depressing synaptic transmission at inhibitory synapses between cultured rat hippocampal neurons. Each point represents the amplitude of a single inhibitory postsynaptic current (IPSC)

In seven experiments, application of 3-APA from a pipette containing 1 μM always reduced IPSC amplitude (e.g. Figure 12.5) by ~50% (mean ± SD: 58 ± 9%). The effect of 3-APA was dose-related; 10 μM 3-APA reduced the IPSC by 94%. It was found that 10 μM baclofen and 1 μM 3-APA were approximately equieffective (Figure 12.6). 3-APA did not alter membrane input resistance or action potential activity at 10 μM, nor did 3-APA reduce currents evoked by exogenous GABA.

Presynaptic Actions of Baclofen in the Hippocampal Slice

2-OH-S (3-amino-2-(*p*-chlorophenyl)-2-hydroxypropylsulphonic acid), an analogue of baclofen, was recently reported to be an antagonist at GABA$_B$ receptors in the guinea-pig ileum (Kerr *et al.*, 1988). Like the structurally related phosphonic acid analogue, phaclofen, 2-OH-S blocks the late IPSP recorded intracellularly from CA1 neurons in slices of rat hippocampus (Lambert *et al.*, 1989), an effect consistent with an antagonist action of 2-OH-S at postsynaptic $GABA_B$ receptors. However, phaclofen has been reported to be ineffective at presynaptic $GABA_B$ receptors in the hippocampus (Dutar and Nicoll, 1988b), and no agent has been shown to possess antagonist activity at these receptors.

We therefore studied the actions of 2-OH-S at presynaptic $GABA_B$ receptors in the CA1 region of the hippocampus. Baclofen has a profound depressant action on excitatory synaptic transmission, as assessed by recordings of pEPSPs in the stratum radiatum (Figure 12.7a). The effect of baclofen was dose-related; the threshold effective dose of baclofen was 0.5 μM, and the estimated median effective concentration (EC_{50}) was 3 μM. In addition, baclofen abolished the population spike recorded in the stratum pyramidale (Figure 12.7b). In contrast, 8-OH-DPAT, a selective agonist at the $5\text{-}HT_{1a}$ class of serotonin (5-hydroxytryptamine, 5-HT) receptors, had no detectable effect on pEPSPs recorded in the stratum radiatum (Figure 12.7a), but did depress population spikes recorded in the stratum pyramidale (Figure 12.7b). Since 8-OH-DPAT activates $5\text{-}HT_{1a}$

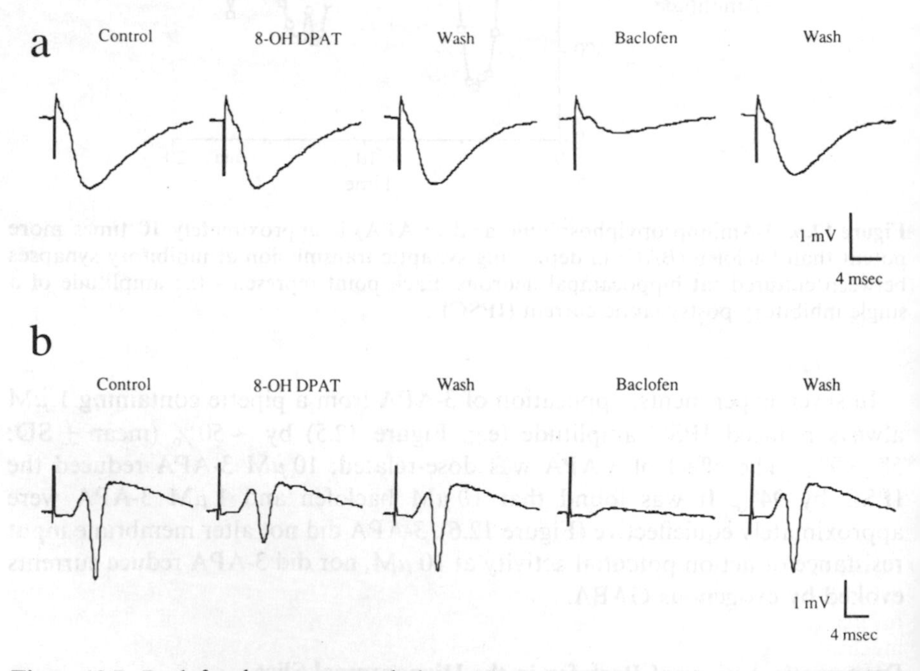

Figure 12.7. Baclofen depresses population excitatory postsynaptic potentials (pEPSPs) but 8-hydroxy-2-(di-*n*-propylamino)tetralin hydrogen bromide (8-OH-DPAT) does not. (a) Population EPSPs recorded in the stratum radiatum are depressed by 10 μM baclofen, but not by 100 nM 8-OH-DPAT. (b) Population spikes recorded in the stratum pyramidale are reduced by 8-OH-DPAT and abolished by baclofen

receptors located postsynaptically on CA1 pyramidal neurons (Andrade *et al.*, 1986; Colino and Halliwell, 1987), these findings argue strongly that depression of pEPSPs recorded in the stratum radiatum probably reflects the activation of *presynaptic* rather than *postsynaptic* GABA$_B$ receptors.

2-Hydroxysaclofen is an Antagonist of the Presynaptic Actions of Baclofen in the Hippocampal Slice

In our experiments, phaclofen was at best weak and inconsistent in antagonizing the synaptic depressant effect of 10 μM baclofen. Small effects (~30–50% reductions in response to baclofen) were seen at the highest dose of phaclofen used (1 mM). 2-OH-S itself, at 200 μM, was without effect on pEPSPs, but consistently attenuated the synaptic depressant effect of 10 μM baclofen (Figure 12.8a,b). In 15 experiments, 10 μM baclofen alone reduced pEPSP amplitude by

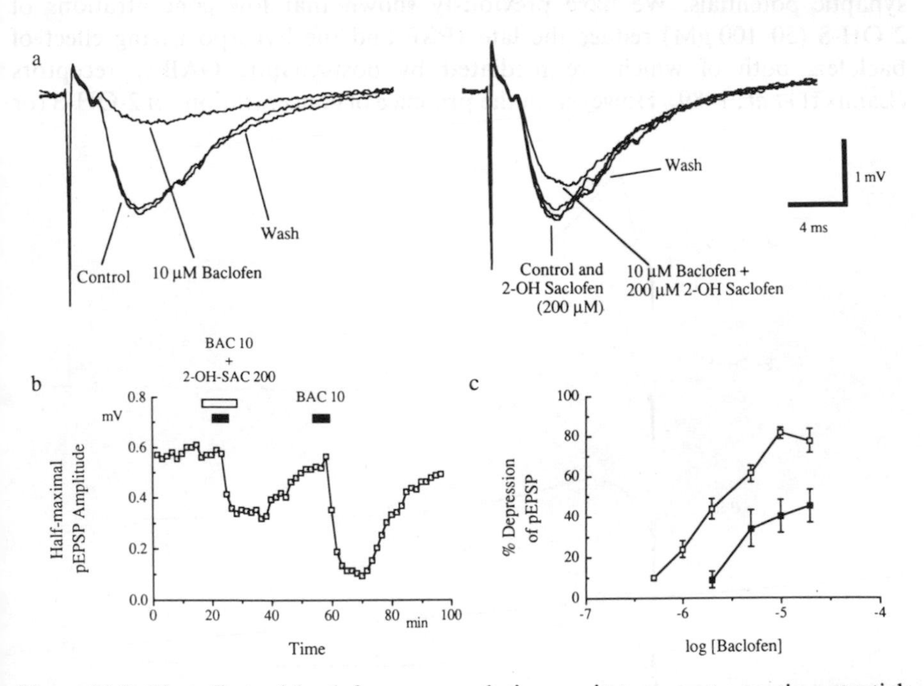

Figure 12.8. The effect of baclofen on population excitatory postsynaptic potentials (pEPSPs) is blocked by 2-hydroxysaclofen (2-OH-S). (a) Application of 10 μM baclofen reversibly reduced pEPSP amplitude. (b) The blockade of transmission by baclofen is much reduced in the presence of 200 μM 2-OH-S. To test the antagonist activity of 2-OH-S, the drug was applied for 2 min prior to the application of a combination of 2-OH-S and baclofen, after which the perfusion was returned to 2-OH-S alone for 2 min. (c) Dose–response curves for the action of baclofen on pEPSPs, obtained in the presence (■) and absence (□) of 200 μM 2-OH-S. The dose–response curve for baclofen is shifted to the right in the presence of 200 μM 2-OH-S

81 ± 12% (mean ± SD). In the presence of 200 μM 2-OH-S, 10 μM baclofen reduced pEPSP amplitude by only 40 ± 17% (n = 5). Statistical comparisons using a two-tailed t-test showed that the antagonistic effect of 2-OH-S was significant at the level $P < 0.01$. Further experiments with other doses of baclofen indicated that the dose–response curve for baclofen was shifted to the right in the presence of 200 μM 2-OH-S (Figure 12.8c).

These data, obtained using field potential recordings, strongly suggested an effect of 2-OH-S on presynaptic $GABA_B$ receptors. However, in order to confirm these observations, it was necessary to make intracellular recordings from CA1 pyramidal neurons. Stimulation of the Schaffer collateral/commissural pathways evokes a series of synaptic potentials in CA1 neurons: an initial fast EPSP is followed by a fast IPSP and a slow 'late' IPSP. Addition of baclofen (10–30 μM) to the perfusion medium normally has two effects: a membrane hyperpolarization of about 10 mV, and a decrease in the amplitude of all evoked synaptic potentials. We have previously shown that low concentrations of 2-OH-S (50–100 μM) reduce the late IPSP and the hyperpolarizing effect of baclofen, both of which are mediated by postsynaptic $GABA_B$ receptors (Lambert *et al.*, 1989). However, in the presence of concentrations of 2-OH-S (or

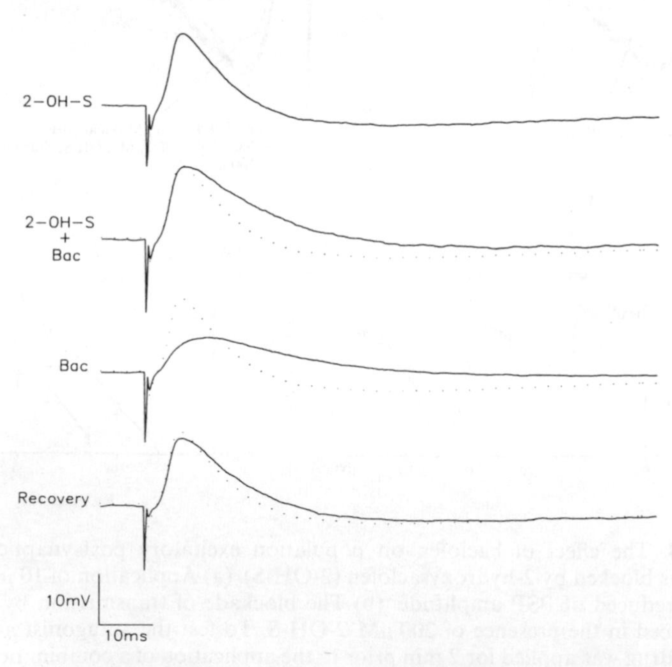

Figure 12.9. The effect of baclofen on intracellularly recorded excitatory postsynaptic potentials (EPSPs) is blocked by 2-hydroxysaclofen (2-OH-S). Application of 10 μM baclofen reversibly reduced the amplitude of intracellularly recorded EPSPs. The blockade of transmission by baclofen is abolished in the presence of 500 μM 2-OH-S

phaclofen) that block postsynaptic GABA$_B$ receptors, we found that baclofen still reduces the amplitude of evoked synaptic potentials, suggesting that the receptors mediating this effect are distinct from the postsynaptic receptors, as previously suggested (Dutar and Nicoll, 1988b). In control recordings, baclofen (10 μM) decreased the slope of intracellularly recorded EPSPs in CA1 neurons (Figure 12.9) by 69 $\pm$ 4% (mean $\pm$ SEM, $n = 7$). In 100 μM 2-OH-S, the synaptic depressant effect of baclofen was unaltered, although the late IPSP was largely blocked. In 500 μM 2-OH-S, baclofen depressed the slope of intracellularly recorded EPSPs (Figure 12.9) only by 11 $\pm$ 6% (mean $\pm$ SEM, $n = 7$).

2-Hydroxysaclofen is an Antagonist of the Presynaptic Actions of Baclofen in Cultured Rat Hippocampal Neurons

In cultures of rat hippocampal neurons, both inhibitory and excitatory synapses are depressed by baclofen, although postsynaptic GABA$_B$ receptors are absent. This presynaptic action of baclofen is not blocked by 200–500 μM phaclofen or by preincubation of the tissue with pertussis toxin (Harrison, 1990). Baclofen (10 μM) decreased the amplitude of EPSPs, and this effect of baclofen was largely blocked in the presence of 200 μM 2-OH-S (Figure 12.10), confirming

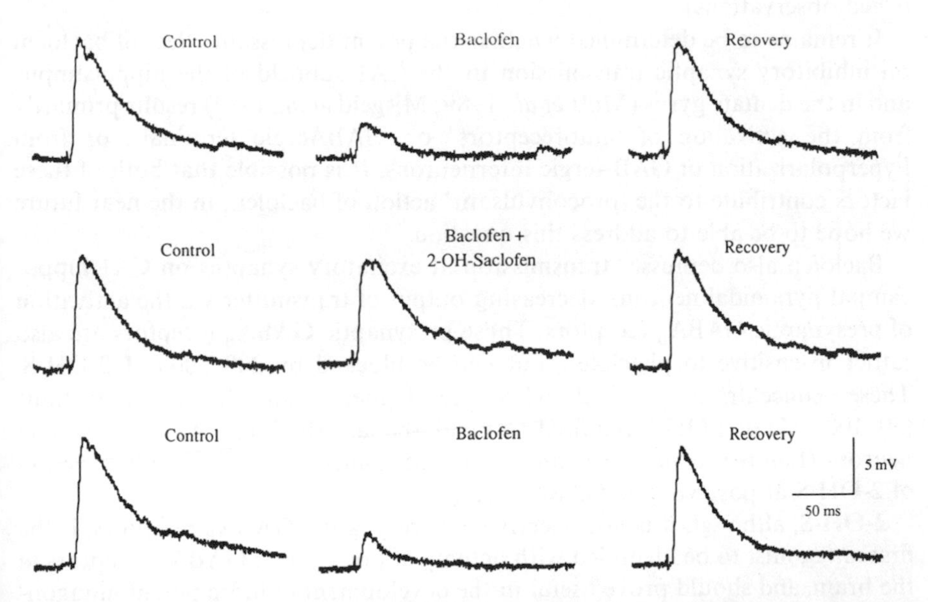

Figure 12.10. 2-Hydroxysaclofen (2-OH-S) blocks the presynaptic effect of baclofen at excitatory synapses between cultured neurons. Application of 10 μM baclofen reduced the amplitude of excitatory postsynaptic potentials (EPSPs) at an excitatory synapse in culture. The depressant action of baclofen is largely blocked in the presence of 200 μM 2-OH-S

the existence of presynaptic receptors that are sensitive to 2-OH-S in a system where postsynaptic $GABA_B$ receptors cannot complicate the interpretation of experimental results.

DISCUSSION

Baclofen depresses inhibitory synaptic transmission at inhibitory synapses between cultured rat hippocampal neurons. This effect in tissue culture is undoubtedly due to the action of baclofen at presynaptic $GABA_B$ receptors on or near the terminals of the GABAergic neurons. The existence of such receptors on GABAergic terminals in vivo is supported by biochemical studies in brain slices that show baclofen decreases GABA release (Bonanno *et al.*, 1989; Pittaluga *et al.*, 1987; Waldmeier *et al.*, 1988), and also by recent experiments with GABA uptake inhibitors in neocortical slices (Deisz and Prince, 1989). These putative GABA 'autoreceptors' are also activated by the GABA analogue 3-APA, which is approximately 10 times more potent than baclofen. The presynaptic $GABA_B$ receptors on GABAergic terminals are rather insensitive to phaclofen (but see Davies *et al.*, 1990); they are blocked by relatively high concentrations of 2-OH-S (Davies *et al.*, 1990; Harrison and Lambert, unpublished observations).

It remains to be determined whether the potent depressant effect of baclofen on inhibitory synaptic transmission in the CA1 subfield of the hippocampus and in the dentate gyrus (Mott *et al.*, 1989; Misgeld *et al.*, 1989) results primarily from the activation of 'autoreceptors' on GABAergic terminals, or from hyperpolarization of GABAergic interneurons. It is possible that both of these factors contribute to the 'proconvulsant' action of baclofen; in the near future we hope to be able to address this question.

Baclofen also depresses transmission at excitatory synapses on CA1 hippocampal pyramidal neurons, decreasing output of transmitter via the activation of presynaptic $GABA_B$ receptors. These presynaptic $GABA_B$ receptors are also rather insensitive to phaclofen, but can be blocked by 200–500 μM 2-OH-S. These concentrations of 2-OH-S are higher than the concentrations (50–100 μM) of 2-OH-S required to block the late IPSP in CA1 hippocampal neurons (Lambert *et al.*, 1989), an action that results from the antagonist action of 2-OH-S at postsynaptic $GABA_B$ receptors.

2-OH-S, although it is not selective for presynaptic $GABA_B$ receptors, is the first antagonist to be identified with activity at presynaptic $GABA_B$ receptors in the brain, and should prove useful in the development of more potent antagonists in the future. Furthermore, the identification of an antagonist for presynaptic $GABA_B$ receptors will be of great assistance in studying their possible physiological functions.

ACKNOWLEDGEMENTS

The authors wish to express their thanks to Drs Jeffery Barker (Bethesda), Forrest Weight (Rockville) and Tim Teyler (Rootstown) for their enthusiastic support of this work. We also thank our colleagues in Adelaide, Drs David Kerr, Jenny Ong and Rolf Prager for providing us with several samples of their baclofen analogues.

REFERENCES

Alger, B.E. and Nicoll, R.A. (1982) Feed-forward dendritic inhibition in rat hippocampal pyramidal cells studied in vitro. *J. Physiol.*, **328**, 105–123.

Andrade, R., Malenka, R.C. and Nicoll, R.A. (1986) A G protein couples serotonin and GABA$_B$ receptors to the same channels in hippocampus. *Science*, **234**, 1261–1265.

Ault, B. and Nadler, J.V. (1982) Baclofen selectively inhibits transmission at synapses made by axons of CA3 pyramidal cells in the rat hippocampal slice. *J. Pharmacol. Exp. Ther.*, **223**, 291–297.

Barker, J.L. and Harrison, N.L. (1988) Outward rectification of inhibitory postsynaptic currents in cultured rat hippocampal neurons. *J. Physiol.*, **403**, 41–55.

Bonanno, G., Cavazzani, P., Andrioli, G.C., Asaro, D., Pellegrini, G. and Raiteri, M. (1989) Release-regulating autoreceptors of the GABA$_B$ type in human cerebral cortex. *Br. J. Pharmacol.*, **96**, 341–346.

Bormann, J., Hamill, O.P. and Sakmann, B. (1987) Mechanism of anion permeation through channels gated by glycine and γ-aminobutyric acid in mouse cultured spinal neurons. *J. Physiol.*, **385**, 243–246.

Bowery, N.G., Hill, D.R., Hudson, A.L., Doble, A., Middlemiss, D.N., Shaw, J. and Turnbull, M. (1980) (−)Baclofen decreases neurotransmitter release in the mammalian CNS by an action at a novel GABA receptor. *Nature*, **283**, 92–94.

Bowery, N.G., Hudson, A.L. and Price, G.W. (1987) GABA$_A$ and GABA$_B$ receptor site distribution in the rat central nervous system. *Neuroscience*, **20**, 365–383.

Colino, A. and Halliwell, J.V. (1987) Differential modulation of three separate K^+ conductances in hippocampal CA1 neurons by serotonin. *Nature*, **328**, 73–77.

Colmers, W.F. and Williams, J.T. (1988) Pertussis toxin pretreatment discriminates between pre- and postsynaptic actions of baclofen in rat dorsal raphé nucleus in vitro. *Neurosci. Lett.*, **93**, 300–306.

Connors, B.W., Malenka, R.C. and Silva, L.R. (1988) Two inhibitory postsynaptic potentials, and GABA$_A$ and GABA$_B$ receptor-mediated responses in neocortex of rat and cat. *J. Physiol.*, **406**, 443–468.

Davies, J. (1981) Selective depression of synaptic excitation in cat spinal neurons by baclofen: an iontophoretic study. *Br. J. Pharmacol.*, **72**, 373–384.

Davies, C.H., Davies, S.N. and Collingridge, G.L. (1990) Paired pulse depression of monosynaptic GABA-mediated inhibitory postsynaptic responses in rat hippocampus. *J. Physiol.*, **424**, 513–531.

Deisz, R.A. and Prince, D.A. (1989) Frequency dependent depression of inhibition in the guinea-pig neocortex in vitro by GABA$_B$ receptor feedback on GABA release. *J. Physiol.*, **412**, 513–541.

Dutar, P. and Nicoll, R.A. (1988a) A physiological role for GABA$_B$ receptors in the central nervous system. *Nature*, **332**, 156–158.

Dutar, P. and Nicoll, R.A. (1988b) Pre- and postsynaptic $GABA_B$ receptors in the hippocampus have different pharmacological properties. *Neuron*, **1**, 585–591.

Gähwiler, B. and Brown, D.A. (1985) $GABA_B$ receptor-activated K^+ current in voltage-clamped CA3 pyramidal cells in rat hippocampal cultures. *Proc. Natl Acad. Sci. USA*, **82**, 1558–1562.

Gallagher, J.P., Stevens, D.R. and Shinnick-Gallagher, P. (1984) Actions of GABA and baclofen on neurons of the dorsolateral septal nucleus (DLSN) in vitro. *Neuropharmacology*, **23**, 825–826.

Hamill, O.P., Marty, A., Neher, E., Sakmann, B. and Sigworth, F.J. (1981) Improved patch-clamp techniques for high-resolution current recording from cells and cell-free membrane patches. *Pflügers Arch.*, **391**, 85–100.

Harrison, N.L. (1990) On the presynaptic action of baclofen at inhibitory synapses between cultured rat hippocampal neurons. *J. Physiol.*, **422**, 433–446.

Harrison, N.L., Lange, G.D. and Barker, J.L. (1988) Baclofen activates presynaptic $GABA_B$ receptors on GABAergic neurons from the embryonic rat hippocampus. *Neurosci. Lett.*, **85**, 105–109.

Hasuo, H. and Gallagher, J.P. (1988) Comparison of antagonism by phaclofen of baclofen induced hyperpolarizations and synaptically mediated late hyperpolarizing potentials recorded from rat dorsolateral septal neurons. *Neurosci. Lett.*, **86**, 77–81.

Hill, D.R. and Bowery, N.G. (1981) ^{3}H-Baclofen and ^{3}H-GABA bind to bicuculline-insensitive $GABA_B$ sites in rat brain. *Nature*, **290**, 149–152.

Hill, D.R., Bowery, N.G. and Hudson, A.L. (1984) Inhibition of $GABA_B$ receptor binding by guanine nucleotides. *J. Neurochem.*, **42**, 652–657.

Holz, G.G., Rane, S.G. and Dunlap, K. (1986) GTP-binding proteins mediate transmitter inhibition of voltage-dependent calcium channels. *Nature*, **319**, 670–672.

Howe, J.R., Sutor, B. and Zieglgansberger, W. (1987) Baclofen reduces postsynaptic potentials of rat neocortical neurons by an action other than its hyperpolarizing action. *J. Physiol.*, **384**, 539–570.

Huettner, J.E. and Baughmann, R.W. (1986) Primary culture of identified neurons from the visual cortex of postnatal rats. *J. Neurosci.*, **6**, 3044–3060.

Karbon, E.W. and Enna, S.J. (1985) Characterization of the relationship between γ-aminobutyric acid agonists and transmitter-coupled cyclic nucleotide-generating systems in rat brain. *Mol. Pharmacol.*, **27**, 53–59.

Karlsson, G. and Olpe, H.-R. (1989) Late inhibitory postsynaptic potentials in rat prefrontal cortex may be mediated by $GABA_B$ receptors. *Experientia*, **45**, 157–158.

Kerr, D.I.B., Ong, J., Prager, R.H., Gynther, B.D. and Curtis, D.R. (1987) Phaclofen: a peripheral and central baclofen antagonist. *Brain Res.*, **405**, 150–154.

Kerr, D.I.B., Ong, J., Johnston, G.A.R., Abbenante, J. and Prager, R.H. (1988) 2-Hydroxy-saclofen: an improved antagonist at central and peripheral $GABA_B$ receptors. *Neurosci. Lett.*, **92**, 92–96.

Lacey, M.G., Mercuri, N. and North, R.A. (1988) On the potassium conductance increase activated by $GABA_B$ and dopamine D_2 receptors in rat substantia nigra neurons. *J. Physiol.*, **401**, 437–454.

Lambert, N.A., Harrison, N.L., Prager, R.H., Ong, J., Kerr, D.I.B. and Teyler, T.J. (1989) Blockade of the late IPSP in rat CA1 hippocampal neurons by 2-hydroxy-saclofen. *Neurosci. Lett.*, **107**, 125–128.

McCormick, D. (1989) GABA as an inhibitory neurotransmitter in human cerebral cortex. *J. Neurophysiol.*, **62**, 1018–1027.

Misgeld, U., Müller, W. and Brunner, H. (1989) Effects of (−)baclofen on inhibitory neurons in the guinea pig hippocampal slice. *Pflügers Arch.*, **414**, 139–144.

Mott, D.D., Bragdon, A.C., Lewis, D.V. and Wilson, W.A. (1989) Baclofen has a proepileptic effect in the rat dentate gyrus. *J. Pharmacol. Exp. Ther.*, **249**, 721–725.

Newberry, N.R. and Nicoll, R.A. (1984) Direct hyperpolarizing action of baclofen on hippocampal pyramidal cells. *Nature*, **308**, 450–452.

Olsen, R.W. (1982) Drug interactions at the GABA receptor–ionophore complex. *Annu. Rev. Pharmacol.*, **22**, 245–277.

Peet, M.J. and McLennan, H. (1986) Pre- and postsynaptic actions of baclofen: blockade of the late synaptically-evoked hyperpolarization in CA1 hippocampal neurons. *Exp. Brain Res.*, **61**, 567–574.

Pittaluga, A., Asaro, D., Pellegrini, G. and Raiteri, M. (1987) Studies on [^{3}H]GABA and endogenous GABA release in rat cerebral cortex suggest the presence of autoreceptors of the GABA$_B$ type. *Eur. J. Pharmacol.*, **144**, 45–52.

Schofield, P.R., Darlison, M.G., Fujita, N., Burt, D.R., Stephenson, F.A., Rodriguez, H., Rhee, L.M., Ramachandran, J., Reale, V., Glencorse, T.A., Seeburg, P.H. and Barnard, E.A. (1987) Sequence and functional expression of the GABA$_A$ receptor shows a ligand-gated receptor super-family. *Nature*, **328**, 221–227.

Segal, M. and Barker, J.L. (1984) Rat hippocampal neurons in culture: voltage-clamp analysis of inhibitory synaptic connections. *J. Neurophysiol.*, **52**, 469–487.

Soltesz, I., Haby, M., Leresche, N. and Crunelli, V. (1988) The GABA$_B$ antagonist phaclofen inhibits the late K^+-dependent IPSP in cat and rat thalamic and hippocampal neurons. *Brain Res.*, **448**, 351–354.

Stevens, D.R., Gallagher, J.P. and Shinnick-Gallagher, P. (1985) Further studies on the action of baclofen on neurons of the dorsolateral septal nucleus of the rat. *Brain Res.*, **358**, 360–363.

Stirling, J.M., Cross, A.J., Robinson, T.N. and Green, A.R. (1989) The effects of GABA$_B$ receptor agonists and antagonists on potassium-stimulated $[Ca^{2+}]_1$ in rat brain synaptosomes. *Neuropharmacology*, **28**, 699–704.

Teyler, T.J. (1980) Brain slice preparation: hippocampus. *Brain. Res. Bull.*, **5**, 391–403.

Waldmeier, P.C., Wicki, P., Feldtrauer, J.-J. and Baumann, P.A. (1988) Potential involvement of a baclofen-sensitive autoreceptor in the modulation of the release of endogenous GABA from rat brain slices in vitro. *Naunyn-Schmiedeberg's Arch. Pharmacol.*, **337**, 289–295.

Wojcik, W.J. and Neff, N.H. (1984) γ-Aminobutyric acid B receptors are negatively coupled to adenylate cyclase in brain, and in the cerebellum these receptors may be associated with granule cells. *Mol. Pharmacol.*, **25**, 24–28.

13 $GABA_B$ Receptor Function in the Hippocampus: Development and Control of Excitability

P. A. SCHWARTZKROIN

INTRODUCTION

It has become clear that GABA release leads to two quite distinct effects, mediated by very different receptors. The $GABA_A$ receptor, now recognized as the receptor associated with the conventional chloride-mediated inhibitory postsynaptic potential (IPSP) in hippocampal neurons, has received much attention. Its chloride dependency, associated conductance increase, and various roles in control of excitability have been explored in a number of different experimental settings, not least of which have been studies of the hippocampus (e.g. Alger and Nicoll, 1982; Kandel *et al.*, 1961). Investigators generally agree that proper functioning of the $GABA_A$ receptor is critical in maintaining normal excitability in cortical regions such as the hippocampus. Blockade of $GABA_A$ receptors with specific antagonists such as bicuculline inevitably leads to hyperexcitability and often to the development of seizure discharge (e.g. Dingledine and Gjerstad, 1980; Meldrum, 1975). In fact, many of the most intensively studied models of epileptogenesis are based on blockade of the $GABA_A$ receptor.

Not so well studied, and certainly not so well understood, is the $GABA_B$ receptor. Over the past few years, investigators have come to recognize that this second type of GABA receptor is present and functioning in cortical neurons such as hippocampal pyramidal cells. Information has been obtained regarding the mechanism of action of this receptor (see other chapters) – for example, the potassium flux associated with opening of the $GABA_B$-mediated channels (Gähwiler and Brown, 1985; Inoue *et al.*, 1985a,c; Newberry and Nicoll, 1984; Peet and McLennan, 1986) and the second messenger pathways initiated by GABA binding to the $GABA_B$ receptor (Andrade *et al.*, 1986; Thalmann, 1987, 1988). There remain, however, many unanswered questions about the $GABA_B$ receptor and its function. In this chapter, I will examine some of those questions about $GABA_B$ function in the hippocampus, a region in which the balance of excitability is quite delicate, and where upset of that balance may have significance for the animal's behaviour (e.g. Swartzwelder *et al.*, 1987). Some of

$GABA_B$ Receptors in Mammalian Function. Edited by N.G. Bowery, H. Bittiger and H.-R. Olpe

the questions to be addressed are: (1) Where are the $GABA_B$ receptors located? (2) What is the relative importance of pre- versus postsynaptic $GABA_B$ receptors? (3) What is the role of $GABA_B$-mediated postsynaptic inhibition? (4) Do $GABA_B$ receptors play a significant role in the control of cell excitability?

To understand some of these questions, it may be useful to examine the development of $GABA_B$ function in hippocampal neurons. If there is an early period of development in which $GABA_B$ function is absent, hippocampal function during that time may give us some clues about the ultimate role of $GABA_B$ receptors in this region. As will become obvious, the issues are quite complex, and our understanding of this GABA receptor subtype is still rather limited.

$GABA_B$ RECEPTOR INVOLVEMENT IN SYNAPTIC ACTIVITY

Postsynaptic Localization

A number of studies have shown that there is a significant late component of the GABA-mediated IPSP in hippocampal pyramidal cells, a component that is not bicuculline-sensitive and apparently not chloride-dependent (Kehl and McLennan, 1983, 1985a,b; Knowles *et al.*, 1984; Newberry and Nicoll, 1984; Thalmann and Ayala, 1982). The characteristics of this late IPSP component are, in fact, similar to the characteristics of membrane changes produced by the $GABA_B$ agonist, (−)-baclofen (e.g. Inoue, *et al.*, 1985c); that is, the potential is long-lasting, dependent on the opening of potassium channels (and with a reversal potential close to that appropriate for a potassium-mediated ion flux), and associated with a relatively small conductance change. Recent studies have shown that both the late component of the IPSP and the baclofen-induced potential change are mediated by a G-protein (Andrade *et al.*, 1986) which is sensitive to pertussis toxin (Thalmann, 1987, 1988). Such a $GABA_B$-mediated response can be demonstrated even in the absence of intact synaptic transmission, indicating that at least some of these receptors are located on the pyramidal cell postsynaptic membrane. These $GABA_B$ potentials have been identified in both the CA1 and CA3 pyramidal cells of the hippocampus.

Some studies have suggested that the dendritic membrane is most likely to be involved in $GABA_B$-mediated activity, that this region is most sensitive to baclofen delivery (Inoue *et al.*, 1985a; Newberry and Nicoll, 1985). However, preferential localization of $GABA_B$ receptors to the dendrites has not been clearly established. It is certainly the case that baclofen application to somata as well as dendrites, at least in CA3 pyramidal cells, produces the typical $GABA_B$-mediated response (Janigro and Schwartzkroin, 1988a). Receptor binding studies show $GABA_B$ receptors are distributed over the entire axosomatic axis of the pyramidal neuron, with some suggestion of higher receptor density in the dendritic region (Bowery *et al.*, 1987). However, given the relatively low membrane density at the soma, these receptor localization studies could simply reflect the greater amount of postsynaptic membrane in dendritic regions, not a higher density of receptors on dendrites as opposed to somata.

There are, however, some physiological data that support the idea of preferential GABA$_B$ dendritic localization - or at least differential GABA$_A$ and GABA$_B$ localization. One such observation is that orthodromic stimulation in the hippocampus (of the Schaffer and commissural afferents) produces both early and late IPSP components (i.e. both GABA$_A$- and GABA$_B$-mediated events), whereas antidromic activation of pyramidal cells (i.e. alveus stimulation) produces only a GABA$_A$ IPSP (Alger and Nicoll, 1982). This finding suggests that different interneuron populations are involved in feedback versus feedforward inhibition, and are associated with different GABA receptor subtypes. Studies of local circuit neurons in the CA1 region of the hippocampus (Lacaille and Schwartzkroin, 1988) also support this view. Stimulation in the stratum radiatum of the hippocampus activates all three interneuronal cell types that have been characterized in CA1, whereas alveus stimulation activates only basket cells and oriens/alveus interneurons (both directly and via pyramidal cell collaterals). While alveus stimulation produces only a GABA$_A$ IPSP, the alveus-activated interneurons have GABAergic synapses not only on the pyramidal cell soma, but also on its dendrites (Lacaille *et al.*, 1987; Schwartzkroin and Kunkel, 1985). Thus, it is clear that GABA$_A$ synapses are not confined to the soma. Interestingly, the lacunosum-moleculare interneuron, which is activated only orthodromically, makes synapses preferentially on the dendrites of pyramidal cells (Kunkel *et al.*, 1988). Preliminary data, based on paired recordings between lacunosum-moleculare interneurons and pyramidal cells, indicate that the lacunosum-moleculare-induced IPSP is at least partially mediated by GABA$_B$ receptors (Lacaille and Schwartzkroin, 1988).

Presynaptic Localization

Studies employing baclofen application to the hippocampus have also identified a presynaptic locus for GABA$_B$ receptors (Ault and Nadler 1982, 1983b; Inoue *et al.*, 1985b; Lanthorn and Cotman, 1981; Olpe *et al.*, 1982). Results of early studies suggested that the presynaptic GABA$_B$ effect occurs preferentially in the afferent input to CA1, with a much smaller presynaptic GABA$_B$ effect on the mossy terminals onto CA3 pyramidal cells. Ault and Nadler (1982) and Inoue *et al.* (1985b) found no GABA$_B$ presynaptic effects in the mossy cell-to-CA3 synapse, as compared to a striking GABA$_B$ presynaptic influence in the Schaffer collateral-to-CA1 synapse. However, other studies have found that at least some afferents to the CA3 neurons do demonstrate GABA$_B$ (i.e. baclofen-mediated) effects (e.g. Lanthorn and Cotman, 1981). Experimental results also suggest that GABA receptors are located not only on excitatory but also on inhibitory terminals in CA1 and CA3 regions (Kehl and McLennan, 1985b; Misgeld *et al.*, 1984). $GABA_B$ presynaptic receptors at inhibitory synapses could account for baclofen-induced reduction of both early and late components of the IPSP. While this disinhibitory presynaptic (i.e. 'autoreceptor') $GABA_B$ function appears quite powerful, the relative paucity of $GABA_B$ receptors around pyramidal cell somata (Bowery *et al.*, 1987) is somewhat at odds with the dense concentration of GABA-positive boutons at this site (Kunkel *et al.*, 1986).

Presence in Immature Tissue

Only a few studies have explored $GABA_B$ receptor function in the immature hippocampus. Interpretation of $GABA_B$ function is complicated by the changing nature of the GABA-mediated IPSP during development. Inhibitory function develops late (relative to excitation), especially in the CA1 pyramidal cells of the hippocampus (Schwartzkroin, 1982; Schwartzkroin and Kunkel, 1982). $GABA_A$ effects on immature CA1 hippocampal pyramidal cells are depolarizing at both soma and dendrite, not the hyperpolarizing response normally associated with IPSPs in mature hippocampal neurons (Mueller *et al.*, 1984). However, as in the adult, the response is associated with a very large conductance increase, the membrane potential change is mediated by chloride flux, and the depolarization is bicuculline-sensitive. As the hippocampus matures, hyperpolarizing $GABA_A$ responses start to appear at the soma level, and the depolarizing responses are found primarily in the dendrites – the state typical of the CA1 region in the adult. The pattern of change is quite different in CA3 pyramidal neurons, where $GABA_A$-induced hyperpolarizations are seen much earlier (Schwartzkroin and Kunkel, 1982; Swann *et al.*, 1989). GABA application to immature CA3 dendrites produces a mixed response, with a clear hyperpolarizing response mediated by $GABA_B$ receptors as well as a depolarizing response (Janigro and Schwartzkroin, 1988a) (Figure 13.1). In mature CA3 dendrites, however, the GABA response is primarily hyperpolarizing, with little evidence for the significant $GABA_A$ dendritic response typical of CA1 cells (Janigro and Schwartzkroin, 1988b).

These observations illustrate the rapid changes that occur in the GABA profile of the developing hippocampus, and the clear distinction between CA1 and CA3 cells. As in the $GABA_A$ system, there appears to be a marked difference between CA1 and CA3 neurons in development of $GABA_B$-mediated responses. Whereas the late hyperpolarizing component of the IPSP is clear in immature CA3 pyramidal cells, and behaves like a $GABA_B$-mediated event (not chloride-dependent; reversal potential appropriate for potassium-mediated event; insensitive to bicuculline) (Swann *et al.*, 1989), no such late hyperpolarization is consistently apparent in CA1 (Schwartzkroin and Kunkel, 1982). At the time when CA3 can be driven to produce a biphasic IPSP with two hyperpolarizing components, an orthodromically driven response in CA1 is primarily depolariz-

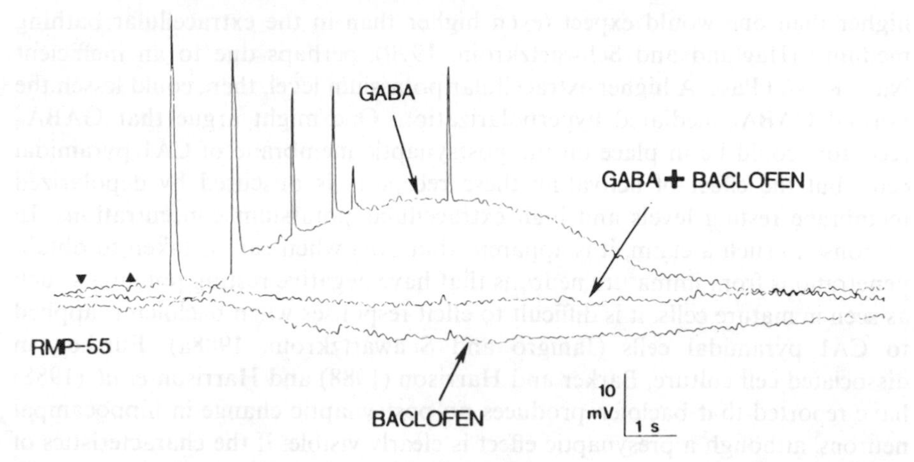

Figure 13.1. Responses of an immature CA3 pyramidal cell to applications of GABA alone, baclofen alone, and GABA plus baclofen. GABA (pipette concentration of 1 M, pH 4; +20 nA retaining current, −60 nA ejection current), when applied to the cell dendrite, produced a depolarization, mediated by GABA$_A$ receptors. Baclofen (pipette concentration of 10 mM, pH 3; + 10 nA retaining current, −70 nA ejection current) produced a long hyperpolarization, mediated by GABA$_B$ receptors. Simultaneous application of the two drugs produced a 'cancellation' of their respective effects; the directly activated GABA$_A$-mediated discharge was blocked by GABA$_B$ action. (From Janigro and Schwartzkroin, 1988a)

ing. Blockade of the GABA$_A$-mediated depolarization in these immature CA1 neurons reveals no clear underlying hyperpolarization that might be interpreted as a GABA$_B$-mediated component of the IPSP. What does this absence of GABA$_B$ responsiveness in immature CA1 cells mean for the overall function of the region? Is it really the case that GABA$_B$ receptors are absent? If so, does this absence significantly alter the balance of excitability?

There are a number of points to be made on both sides of these questions. First, it is worthwhile noting that the GABA$_B$-mediated response has significant voltage- and potassium-sensitivity (Inoue *et al*., 1985a; Knowles *et al*., 1984; Newberry and Nicoll, 1985); that is, as the cell is depolarized, the GABA$_B$ response decreases (anomalously, given that the reversal potential for potassium is then further away from the resting potential). Since recordings from CA1 pyramidal cells in immature tissue (particularly in animals less than a week old) are difficult to maintain without injury, many such recordings may monitor depolarized neurons. Cell depolarization (via electrode-induced injury) could then lead to 'inactivation' of the GABA$_B$-mediated response – as is seen when mature cells are depolarized. Also, the GABA$_B$-mediated response decreases as extracellular potassium increases (the driving force for the potassium influx is decreased). Recordings from immature tissue with potassium-sensitive electrodes suggest that, particularly in the CA1 region, the resting potassium level is higher than one would expect (even higher than in the extracellular bathing medium) (Haglund and Schwartzkroin, 1990), perhaps due to an inefficient Na^+, K^+-ATPase. A higher extracellular potassium level, then, could lessen the normal $GABA_B$-mediated hyperpolarization. One might argue that $GABA_B$ receptors could be in place on the postsynaptic membrane of CA1 pyramidal cells, but the effect of activating these receptors is obscured by depolarized membrane resting levels and high extracellular potassium concentrations. In response to such a claim, it is apparent that even when care is taken to obtain penetrations from immature neurons that have negative resting potentials such as seen in mature cells, it is difficult to elicit responses when baclofen is applied to CA1 pyramidal cells (Janigro and Schwartzkroin, 1988a). Further, in dissociated cell culture, Barker and Harrison (1988) and Harrison *et al.* (1988) have reported that baclofen produces no postsynaptic change in hippocampal neurons, although a presynaptic effect is clearly visible. If the characteristics of these cultured cells reflect immature pyramidal cell properties, then one might interpret these results as supporting the idea that $GABA_B$ receptors are late developing on CA1 pyramidal cells.

ROLE OF $GABA_B$ RECEPTORS IN CONTROL OF EXCITABILITY

Given the large size and long-lasting appearance of the late component of the IPSP - the $GABA_B$-mediated component - and an additional role for $GABA_B$ receptors in modulating transmitter release, it would seem that $GABA_B$ receptors must play a critical and major part in controlling cell excitability. To be sure, if one bathes the hippocampal tissue in baclofen, the level of excitability in the tissue is significantly depressed (Ault and Nadler, 1983a; Ault *et al.*, 1986; Janigro and Schwartzkroin, 1988a; Swartzwelder *et al.*, 1986b). In particular, synaptic potentials may be potently blocked so that baclofen can provide substantial relief from status epilepticus (Ault *et al.*, 1986). However, the physiological effects of GABA release onto $GABA_B$ receptors are not so easily understood, for GABA presumably is released discretely, at localized sites, during normal function. In analyzing $GABA_B$-mediated effects on tissue excitability, there are two separate issues to explore. First, what effects are mediated by the presynaptic $GABA_B$ receptors and how are those effects manifested in the known physiological activities of the cell? Second, what role do the postsynaptic $GABA_B$ receptors play - in particular, what is the effect of the late component of the IPSP?

Presynaptic Roles of $GABA_B$ Receptors

The first question appears to be complicated by the fact that both excitatory PSPs (EPSPs) and IPSPs can be reduced by GABA binding to $GABA_B$ receptors at presynaptic terminals. Investigators have shown that excitability can be reduced via GABA$_B$-mediated inhibition of transmitter release from excitatory terminals (e.g. Olpe *et al.*, 1982). However, a number of recent studies have also suggested that presynaptic GABA$_B$-mediated activity may actually be a potent inhibitor of inhibitory action (Ault and Nadler, 1983a; Misgeld *et al.*, 1984; Mott and Bragdon, 1987; Mott *et al.*, 1988). Thompson and Gähwiler (1989) have suggested that the habituation of GABA-mediated IPSPs often seen with repetitive activation of GABA synapses is due, at least in part, to the action of released GABA at the presynaptic terminal; binding of released GABA to GABA$_B$ receptors leads to a reduction in the amount of GABA subsequently released. This presynaptic effect of GABA on terminals of inhibitory interneurons may be especially potent, since GABA release is most directly associated with inhibitory synapses, especially at the soma. At the dendritic level, where excitatory terminals predominate, there are relatively few synapses that release GABA. Even if the excitatory terminals have GABA$_B$ receptors (as demonstrated with baclofen application), their physiological significance may be much less than the GABA$_B$ receptors on inhibitory terminals; there is a lower probability that there will be local GABA available to bind to those receptors on excitatory terminals. Thus, in considering the GABA$_B$ effects at presynaptic terminals, it is unclear whether one should think of them as being predominantly inhibitory of pyramidal cell output, or predominantly excitatory (via a disinhibition mechanism).

Postsynaptic GABA$_B$-mediated IPSP Effects

The postsynaptic receptors are equally confusing. The late component of the IPSP reflects a large potential change in many pyramidal cells, especially in CA3. This cell hyperpolarization should, in theory, be functionally inhibitory. However, the conductance change associated with the GABA$_B$ late IPSP is relatively small, and the inhibitory effects seem slight. Whereas GABA$_A$ inhibition is associated with a very large conductance shunt at the level of the cell body and initial segment (the conductance increase is often of 80 % or more), the GABA$_B$ conductance change is only of the order of 20 % or 30 %. Using a paired-pulse paradigm to assess inhibition, one typically sees strong inhibition with short interpulse intervals (e.g. 10–50 ms), but synaptic potentiation at longer intervals. The period of inhibition corresponds to the conductance change of the early GABA$_A$ IPSP; the period of potentiation is associated closely with the GABA$_B$-mediated hyperpolarization. To be sure, there may be processes other than those mediated by GABA$_B$ receptors that produce the late potentiation. The point here is simply that the postsynaptic hyperpolarization produced via GABA$_B$ receptors is not a very potent inhibitory event. To complicate matters further, Misgeld *et al.* (1989) have recently reported that baclofen inhibits inhibitory neurons in the dentate gyrus, thus showing that the postsynaptic membrane of inhibitory cells contains GABA$_B$ receptors; GABA released onto these cells, many of which are located near the cell body layer, would antagonize IPSP production, perhaps yielding relative hyperexcitability. Indeed, baclofen has a 'proconvulsive' effect in the dentate gyrus (Mott and Bragdon, 1987).

$GABA_B$ Control of Epileptiform Activity

Baclofen has been used in some models of epileptogenesis to explore the possible inhibitory role of $GABA_B$ receptors in reducing epileptiform discharge. In at least some cases, baclofen has been shown to suppress hippocampal epileptiform activity (at least in CA3) without suppressing synaptic activity (Swartzwelder *et al.*, 1986a). In such experiments, the postsynaptic $GABA_B$ receptors must be responsible for the reported change. In an in-vitro model of epileptiform bursting, Swartzwelder *et al.* (1986a) have explored the $GABA_B$ 'anticonvulsant' action on stimulus train-induced bursting (STIB). Baclofen reduced the late component of the burst response, as reflected in the loss of field potential multiple population spikes; the early component of the response was unaffected, indicating a normal synaptic input (Figure 13.2). When bursting activity was

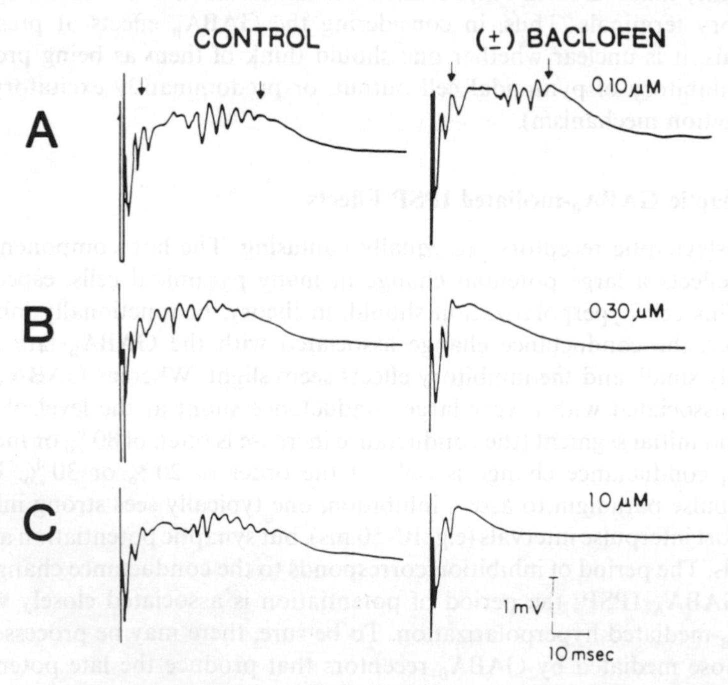

Figure 13.2. Effect of bath application of baclofen (at concentrations of 0.1, 0.3, and 1.0 μM) on burst discharge in the CA3 region of a hippocampal slice. Epileptiform field potential bursts, as shown in the left column, were produced by the stimulation paradigm developed by Stasheff et al. (1985). The late burst components were blocked by the higher concentrations of baclofen, leaving intact the initial, monosynaptically activated population spike (right column). (From Swartzwelder et al., 1986a)

induced by bathing the tissue in high potassium concentrations, the efficacy of baclofen in abolishing burst activity was much reduced, as compared to its effect against STIB- or bicuculline-induced discharge (Swartzwelder *et al.*, 1986b). This result would be predicted if the baclofen/GABA$_B$ mechanism was dependent on potassium efflux. Somewhat paradoxically, however, baclofen, when used in conjunction with a magnesium-free bathing medium, is a key element in the conversion of interictal-like burst discharge in hippocampal slices to seizure-like events; with baclofen in the bath, long periods of epileptiform discharge replace the brief and widely spaced interictal burst activity (Anderson *et al.*, 1988). In these studies, the GABA-mimetic effect of baclofen is only partially inhibitory – a major component of the baclofen effect was excitatory.

This confusion is seen, too, when baclofen is used to treat seizure-like events in immature tissue. As in mature cells, baclofen directly hyperpolarizes immature CA3 pyramidal cells, an effect associated with an increased membrane conductance and not dependent on functional synaptic activity (Brady and Swann, 1984); that is, this hyperpolarization is due to a direct postsynaptic effect. Since spontaneous epileptiform-like afterdischarges occur in immature rat CA3, Brady and Swann (1984) were able to assess the effect of baclofen on epileptiform activity that was spontaneously generated in the immature system. These authors reported that baclofen perfusion blocked all spontaneous epileptiform events, and also raised the stimulus intensity required to trigger epileptiform discharges with orthodromic activation (Figure 13.3). The duration of afterdischarge was not affected by the baclofen, and afterdischarges were eliminated only when the initial paroxysmal depolarization shift (PDS) was blocked. This baclofen effect on immature cells is in contrast to that seen on mature neurons, where the drug decreased the duration of afterdischarge in CA3 pyramidal cells (cf. Figure 13.2).

The inhibitory baclofen effect in the immature tissue has been attributed by Brady and Swann (1984) to a direct hyperpolarization of individual CA3 neurons, which results in a general depression of discharge within the population. A depression of cell discharge activity would discourage synchronization of the cell population, and therefore eliminate the PDS (which is dependent on the synchronized activation of the population). Recent experiments in cultured organotypic hippocampal slices give support to this hypothesis (Malouf *et al.*, 1990). In the cultured slices, a typical biphasic IPSP can be induced by stimulating the afferents to CA3 (or it may occur spontaneously). In these cultures, treating the tissue with phaclofen, a specific GABA$_B$ antagonist (Dutar and Nicoll, 1988a), has little effect on cell excitability, even though phaclofen blocks the late (GABA$_B$) component of the IPSP (Figure 13.4). Critically, this blockade of the late IPSP does not lead to hyperexcitability of the population or to epileptiform discharge – even though synaptic activation remains intact (presumably due to the fact that presynaptic GABA$_B$ receptors are unaffected by phaclofen; Dutar and Nicoll, 1988b). In contrast, bicuculline blockade of the

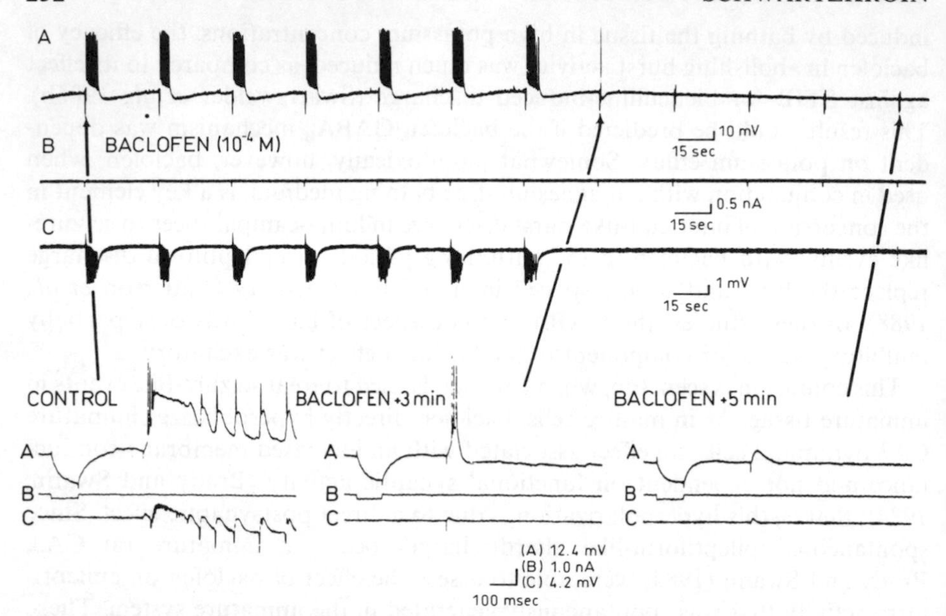

Figure 13.3. Antibursting effect of baclofen on immature (postnatal day 12) rat CA3 pyramidal cell discharge induced by 25 μM bicuculline added to the bathing medium. Orthodromic stimulation was used to elicit burst discharge as seen in both intracellular (A) and field potential (C) recordings. Upper traces, recorded at a slow speed, show the time-course of the baclofen effect. At the left are control burst discharges which are eventually blocked (10^{-4} M baclofen added to the bathing medium at the dot). Traces at faster sweep speed are shown below, corresponding to the times indicated by the arrows Baclofen led to a decrease in cell input resistance (as assessed with intracellular hyperpolarizing current pulses; current traces shown in B), and a clear blockade of the stimulus-induced discharge. (From Brady and Swann, 1984)

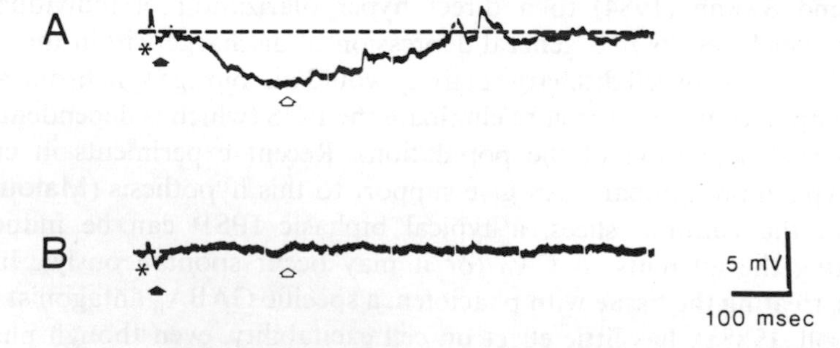

Figure 13.4. Intracellular recording from a CA3 pyramidal cell in an organotypic cultured hippocampal slice. Mossy-fiber stimulation (at the asterisk) was used to trigger a triphasic synaptic response consisting of a brief excitatory postsynaptic potential (EPSP), small early inhibitory PSP (IPSP) (solid arrow), and long hyperpolarizing IPSP (open arrow). A control response is shown in (A); (B) shows the response following tissue treatment with 1 mM phaclofen. (From Malouf *et al.*, 1990)

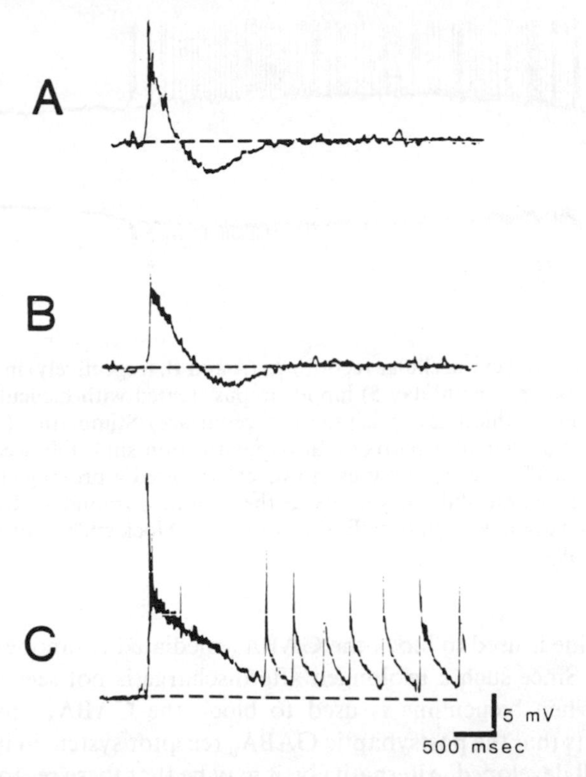

Figure 13.5. Spontaneous postsynaptic potentials (PSPs) in a CA3 cell from a cultured slice preparation untreated by drugs. The excitatory PSP (EPSP) was followed by a $GABA_B$ -mediated hyperpolarization which was quite variable in amplitude (cf. A and B). When this late inhibitory PSP (IPSP) component disappeared (C), the large initial EPSP was followed by a long train of afterdischarge activity, suggesting that the $GABA_B$-mediated component was an important factor in suppressing afterdischarge events. (From Malouf *et al.*, 1990)

$GABA_A$ component of the IPSP in these cultures results in PDS-like burst discharge, just as seen in many other preparations. What, then, is the function of this large, $GABA_B$-mediated, late IPSP component? In tissue in which the $GABA_A$ component of the IPSP is absent or reduced, phaclofen application or spontaneous variability in slow IPSP amplitude may lead to prolonged afterdischarge in the cultured hippocampal slices (Malouf *et al.*, 1990) (Figure 13.5). During periods of late IPSP fluctuation, it appears that synchronization of the population is facilitated when the late $GABA_B$ component is reduced. The result is long-term discharge of the CA3 pyramidal cell population. A similar long-term afterdischarge phenomenon is seen in immature CA3 pyramidal cells in acute slices (Swann *et al.*, 1989). These cells, in which postsynaptic effects of $GABA_B$ receptors have been demonstrated, show a prolonged afterdischarge

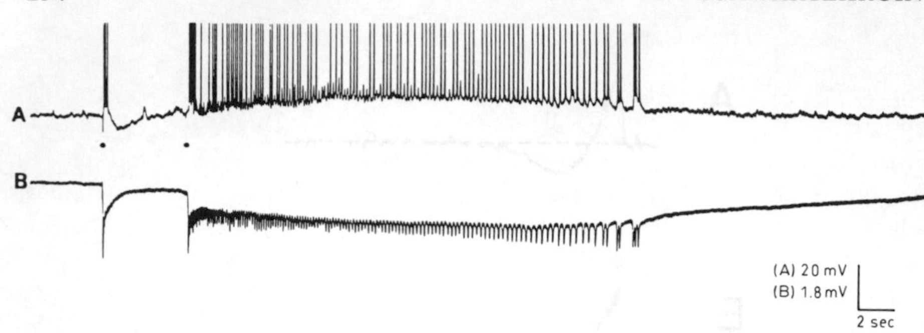

Figure 13.6. Intra- and extracellular recordings (A and B, respectively) in the CA3 region of an immature rat (postnatal day 5) hippocampus treated with bicuculline (droplet of 100 μM bicuculline methiodide applied to the slice surface). Stimulation (at dots) initially produced a burst similar to a paroxysmal depolarization shift, followed by a $GABA_B$ hyperpolarization. The second stimulus, however, triggered a prolonged afterdischarge, associated with an extracellular negativity at the stratum pyramidale. In this tissue, the $GABA_B$-mediated event was not of sufficient strength to block epileptiform events. (From Swann et al., 1989)

when bicuculline is used to block the $GABA_A$-mediated component of the IPSP (Figure 13.6). Since such a prolonged afterdischarge is not seen in the mature CA3 region when bicuculline is used to block the $GABA_A$ component, this finding suggests that the postsynaptic $GABA_B$ receptor system in immature CA3 cells is not well developed. Alternatively, it may be that these receptors normally see very little of the released GABA transmitter (i.e. their location does not correspond to presynaptic terminals in the immature animal).

DISCUSSION

$GABA_B$ receptors are present both pre- and postsynaptically on hippocampal neurons, and they are seen in immature neurons (at least in CA3) as well as in mature cells. The role of these receptors, however, is unclear. The presynaptic $GABA_B$ function appears to be a potent one, but it is difficult to predict the balance of effect given the presence of $GABA_B$ receptors on excitatory and inhibitory presynaptic elements. One might argue that presynaptic $GABA_B$ receptors are more likely to modulate inhibitory than excitatory terminals, thus producing net excitation. The postsynaptic effects, even in the mature animal, are also difficult to interpret. It is clear that a major component of the IPSP is mediated by GABA action at postsynaptic $GABA_B$ receptors. There are some data to indicate that these receptors are primarily localized to dendrites, and even some suggestions that particular interneuron populations preferentially (and maybe selectively) interact with these dendritic $GABA_B$ sites (Kehl and McLennan, 1985b; Lacaille and Schwartzkroin, 1988). A number of investigators have observed, however, that the late component of the IPSP is seen primarily in association with higher levels of stimulation, and its normal role in controlling cell excitability is unclear. Blockade of the late component of the IPSP with phaclofen appears to produce little change in normal cell responsiveness. An investigation of the interneuronal circuitry involved in mediating the late component of the IPSP, and a determination of those conditions under which the GABA$_B$ IPSP is activated, seems essential if we are to understand the role of this potential in normal cell function. One hypothesis which has received support from several studies is that the GABA$_B$ IPSP component hyperpolarizes the cell population, and therefore discourages synchronization. Such a mechanism is predicted by the modelling studies of Traub *et al.* (1989), and fits relatively well with the available data.

Developmentally, it appears that the GABA$_B$ system parallels the GABA$_A$ system in its relatively slow maturation in the CA1 population of the hippocampus. Even in CA3, where the GABA$_A$ IPSP is mature early in development, the efficacy of the late component of the IPSP in discouraging synchronized afterdischarge is much reduced compared to that seen in the adult. As in CA1, this functional immaturity may be attributed to lower receptor density or less well developed receptor binding properties. Perhaps more likely, the low potency of GABA$_B$-mediated inhibition may be due to a still immature interneuronal circuitry which does not appropriately match presynaptic transmitter release with available GABA$_B$ receptors on the postsynaptic membrane. Much work remains to unravel the relative development of GABA$_B$ and GABA$_A$ systems in the maturing hippocampus, and to understand the distinctive roles of these receptor subtypes.

ACKNOWLEDGEMENTS

Studies from the author's laboratory were supported by NIH, NINDS grant NS 15317. P.A.S. is a research affiliate of the Child Development and Mental Retardation Center, University of Washington.

REFERENCES

Alger, B.E. and Nicoll, R.A. (1982) Pharmacological evidence for two kinds of GABA receptor on rat hippocampal pyramidal cells studied in vitro. *J. Physiol.*, **328**, 125–141.

Anderson, W.W., Swartzwelder, H.S. and Wilson, W.A. (1988) Regenerative, all-or-none, electrographic seizures in the rat hippocampal slice in physiological magnesium medium. In: H.L. Haas and G. Buzsáki (eds) *Synaptic plasticity in the hippocampus*. Berlin: Springer-Verlag, pp. 180–183.

Andrade, R., Malenka, R.C. and Nicoll, R.A. (1986) A G protein couples serotonin and GABA$_B$ receptors to the same channels in hippocampus. *Science*, **234**, 1261–1265.

Ault, B. and Nadler, J.V. (1982) Baclofen selectively inhibits transmission at synapses made by axons of CA3 pyramidal cells in the hippocampal slice. *J. Pharmacol. Exp. Ther.*, **223**, 291–297.

Ault, B. and Nadler, J.V. (1983a) Anticonvulsant-like actions of baclofen in the rat hippocampal slice. *Br. J. Pharmacol.*, **78**, 701–708.

Ault, B. and Nadler, J.V. (1983b) Effects of baclofen on synaptically-induced cell firing in the rat hippocampal slice. *Br. J. Pharmacol.*, **80**, 211–219.

Ault, B., Gruenthal, M., Armstrong, D.R. and Nadler, J.V. (1986) Efficacy of baclofen and phenobarbital against the kainic acid limbic seizure–brain damage syndrome. *J. Pharmacol. Exp. Ther.*, **239**, 612–617.

Barker, J.L. and Harrison, N.L. (1988) Outward rectification of inhibitory postsynaptic currents in cultured rat hippocampal neurones. *J. Physiol.*, **403**, 41–55.

Bowery, N.G., Hudson, A.L. and Price, G.W. (1987) $GABA_A$ and $GABA_B$ receptor site distribution in the rat central nervous system. *Neuroscience*, **20**, 365–383.

Brady, R.J. and Swann, J.W. (1984) Postsynaptic actions of baclofen associated with its antagonism of bicuculline-induced epileptogenesis in hippocampus. *Cell. Mol. Neurobiol.*, **4**, 403–408.

Dingledine, R. and Gjerstad, L. (1980) Reduced inhibition during epileptiform activity in the in vitro hippocampal slice. *J. Physiol.*, **305**, 297–313.

Dutar, P. and Nicoll, R.A. (1988a) A physiological role for $GABA_B$ receptors in the central nervous system. *Nature*, **332**, 156–158.

Dutar, P. and Nicoll, R.A. (1988b) Pre- and postsynaptic $GABA_B$ receptors in the hippocampus have different pharmacological properties. *Neuron*, **1**, 585–591.

Gähwiler, B. and Brown, D.A. (1985) $GABA_B$ receptor-activated K^+ current in voltage-clamped CA3 pyramidal cells in hippocampal cultures. *Proc. Natl Acad. Sci. USA*, **82**, 1558–1562.

Haglund, M.M. and Schwartzkroin, P.A. (1990) Role of Na,K pump potassium regulation and IPSPs in seizures and spreading depression in immature rabbit hippocampal slices. *J. Neurophysiol.*, **63**, 225–239.

Harrison, N.L., Lange, G.D. and Barker, J.L. (1988) (−)-Baclofen activates presynaptic $GABA_B$ receptors on GABAergic inhibitory neurons from embryonic rat hippocampus. *Neurosci. Lett.*, **85**, 105–109.

Inoue, M., Matsuo, T. and Ogata, N. (1985a) Baclofen activates voltage-dependent and 4-aminopyridine sensitive K^+ conductance in guinea-pig hippocampal pyramidal cells maintained in vitro. *Br. J. Pharmacol.*, **84**, 833–841.

Inoue, M., Matsuo, T. and Ogata, N. (1985b) Characterization of pre- and postsynaptic action of (−)-baclofen in the guinea-pig hippocampus in vitro. *Br. J. Pharmacol.*, **84**, 843–851.

Inoue, M., Matsuo, T. and Ogata, N. (1985c) Possible involvement of K^+-conductance in the action of gamma-aminobutyric acid in the guinea-pig hippocampus. *Br. J. Pharmacol.*, **86**, 515–524.

Janigro, D. and Schwartzkroin, P.A. (1988a) Effects of GABA and baclofen on pyramidal cells in the developing rabbit hippocampus: an 'in vitro' study. *Dev. Brain Res.*, **41**, 171–184.

Janigro, D. and Schwartzkroin, P.A. (1988b) Effects of GABA on CA3 pyramidal cell dendrites in rabbit hippocampal slices. *Brain Res.*, **453**, 265–274.

Kandel, E.R., Spencer, W.A. and Brinley, F.J., Jr (1961) Electrophysiology of hippocampal neurons: I. Sequential invasion and synaptic organization. *J. Neurophysiol.*, **24**, 225–242.

Kehl, S.J. and McLennan, H. (1983) Evidence for a bicuculline-insensitive long-lasting inhibition in the CA3 region of the rat hippocampal slice. *Brain Res.*, **279**, 278–281.

Kehl, S.J. and McLennan, H. (1985a) An electrophysiological characterization of inhibitions and postsynaptic potentials in rat hippocampal CA3 neurones in vitro. *Exp. Brain Res.*, **60**, 299–308.

Kehl, S.J. and McLennan, H. (1985b) A pharmacological characterization of chloride- and potassium-dependent inhibitions in the CA3 region of the rat hippocampus in vitro. *Exp. Brain Res.*, **60**, 309–317.

Knowles, W.D., Schneiderman, J.H., Wheal, H.V., Stafstrom, C.E. and Schwartzkroin, P.A. (1984) Hyperpolarizing potentials in guinea pig hippocampal CA3 neurons. *Cell. Mol. Neurobiol.*, **4**, 207–230.

Kunkel, D.D., Hendrickson, A.E., Wu, J.-Y. and Schwartzkroin, P.A. (1986) Glutamic acid decarboxylase (GAD) immunocytochemistry of developing rabbit hippocampus. *J. Neurosci.*, **6**, 541–552.

Kunkel, D.D., Lacaille, J.-C. and Schwartzkroin, P.A. (1988) Ultrastructure of stratum lacunosum-moleculare interneurons of hippocampal CA1 region. *Synapse*, **2**, 382–394.

Lacaille, J.-C. and Schwartzkroin, P.A. (1988) Stratum lacunosum-moleculare interneurons of hippocampal CA1 region: II. Intrasomatic and intradendritic recordings of local circuit interactions. *J. Neurosci.*, **8**, 1411–1424.

Lacaille, J.-C., Mueller, A. L., Kunkel, D.D. and Schwartzkroin, P.A. (1987) Local circuit interactions between alveus/oriens interneurons and CA1 pyramidal cells in hippocampal slices: electrophysiology and morphology. *J. Neurosci.*, **7**, 1979–1993.

Lanthorn, T.H. and Cotman, C.W. (1981) Baclofen selectively inhibits excitatory synaptic transmission in the hippocampus. *Brain Res.*, **225**, 171–178.

Malouf, A.T., Robbins, C.A. and Schwartzkroin, P.A. (1990) Phaclofen inhibition of the slow IPSP in hippocampal slice cultures: a possible role for the GABA$_B$-mediated IPSP. *Neuroscience*, **35**, 53–61.

Meldrum, B.S. (1975) Epilepsy and gamma-aminobutyric acid-mediated inhibition. *Int. Rev. Neurobiol.*, **17**, 1–36.

Misgeld, U., Klee, M.R. and Zeise, M.L. (1984) Differences in baclofen-sensitivity between CA3 neurons and granule cells of the guinea-pig hippocampus in vitro. *Neurosci. Lett.*, **47**, 307–311.

Misgeld, U., Müller, W. and Brunner, H. (1989). Effects of (−)baclofen on inhibitory neurons in the guinea pig hippocampal slice. *Pflügers Arch.*, **414**, 139–144.

Mott, D.D. and Bragdon, A.C. (1987) Baclofen has a pro-epileptiform effect in the dentate gyrus of rats. *Neurosci. Abstr.*, **13**, 1162.

Mott, D.D., Lewis, D.V., Bragdon, A.C. and Wilson, W.A. (1988) Evidence that the depression of recurrent inhibition following tetanic stimulation in the rat dentate gyrus is mediated by GABA$_B$ receptors. *Neurosci. Abstr.*, **14**, 809.

Mueller, A.L., Taube, J.S. and Schwartzkroin, P.A. (1984) Development of hyperpolarizing inhibitory postsynaptic potentials and hyperpolarizing response to GABA in rabbit hippocampus in vitro. *J. Neurosci.*, **4**, 860–867.

Newberry, N.R. and Nicoll, R.A. (1984) A bicuculline-resistant inhibitory postsynaptic potential in rat hippocampal pyramidal cells in vitro. *J. Physiol.*, **348**, 239–254.

Newberry, N.R. and Nicoll, R.A. (1985) Comparison of the action of baclofen with gamma-aminobutyric acid on rat hippocampal pyramidal cells in vitro. *J. Physiol.*, **360**, 161–185.

Olpe, H.-R., Baudry, M., Fagni, L. and Lynch, G. (1982) The blocking action of baclofen on excitatory transmission in the rat hippocampal slice. *J. Neurosci.*, **2**, 698–703.

Peet, M.J. and McLennan, H. (1986) Pre- and postsynaptic actions of baclofen: blockade of the late synaptically-evoked hyperpolarization of CA1 hippocampal neurones. *Exp. Brain Res.*, **61**, 567–574.

Schwartzkroin, P.A. (1982) Development of rabbit hippocampus: physiology. *Dev. Brain Res.*, **2**, 469–486.
Schwartzkroin, P.A. and Kunkel, D.D. (1982) Electrophysiology and morphology of the developing hippocampus of fetal rabbits. *J. Neurosci.*, **2**, 448–462.
Schwartzkroin, P.A. and Kunkel, D.D. (1985) Morphology of identified interneurons in the CA1 region of guinea pig hippocampus. *J. Comp. Neurol.*, **232**, 205–218.
Stasheff, S.F., Bragdon, A.C. and Wilson, W.A. (1985) Induction of epileptiform activity in hippocampal slices by trains of electrical stimuli. *Brain Res.*, **344**, 296–302.
Swann, J.W., Brady, R.J. and Martin, D.L. (1989) Postsynaptic development of GABA-mediated synaptic inhibition in rat hippocampus. *Neuroscience*, **28**, 551–561.
Swartzwelder, H.S., Bragdon, A.C., Sutch, C.P., Ault, B. and Wilson, W.A. (1986a) Baclofen suppresses hippocampal epileptiform activity at low concentrations without suppressing synaptic transmission. *J. Pharmacol. Exp. Ther.*, **237**, 881–887.
Swartzwelder, H.S., Sutch, C.P. and Wilson, W.A. (1986b) Attenuation of epileptiform bursting by baclofen: reduced potency in elevated potassium. *Exp. Neurol.*, **94**, 726–734.
Swartzwelder, H.S., Tilson, H.A., McLamb, R.L. and Wilson, W.A. (1987) Baclofen disrupts passive avoidance retention in rats. *Psychopharmacology*, **92**, 398–401.
Thalmann, R.H. (1987) Pertussis toxin blocks a late inhibitory postsynaptic potential in hippocampal CA_3 neurons. *Neurosci. Lett.*, **82**, 41–46.
Thalmann, R.H. (1988) Evidence that guanosine triphosphate (GTP)-binding proteins control a synaptic response in brain: effects of pertussis toxin and GTP gamma S on the late inhibitory postsynaptic potential of hippocampal CA3 neurons. *J. Neurosci.*, **8**, 4589–4602.
Thalmann, R.H. and Ayala, G.F. (1982) A late increase in potassium conductance follows synaptic stimulation of granule neurons of the dentate gyrus. *Neurosci. Lett.*, **29**, 243–248.
Thompson, S.M. and Gähwiler, B.H. (1989) Activity-dependent disinhibition: III. Desensitization and $GABA_B$ receptor-mediated presynaptic inhibition in the hippocampus in vitro. *J. Neurophysiol.*, **61**, 524–533.
Traub, R.D., Miles, R. and Wong, R.K.S. (1989) Model of the origin of rhythmic population oscillations in the hippocampal slice. *Science*, **243**, 1319–1325.

14 GABA$_B$ Receptor-mediated Frequency Dependence of Inhibitory Postsynaptic Potentials of Neocortical Neurons In Vitro

R. A. DEISZ and W. ZIEGLGÄNSBERGER

INTRODUCTION

The classical GABA receptor, now termed the GABA$_A$ receptor, is found on most neurons of the central nervous system (for early reviews, see Krnjević, 1974; Nistri and Constanti, 1979). Activation of this receptor by presynaptically released GABA causes the early inhibitory postsynaptic potential (IPSP) in central neurons. The GABA$_A$ receptor is characterized by several key agonists (GABA, muscimol), antagonists (bicuculline, picrotoxin), modulators (benzodiazepines, barbiturates) and the selectivity of its associated ion channel for Cl^- (for a recent review, see Zieglgänsberger and Häuser, 1989). The receptor has been purified, and cloned cDNAs encoding the subunits have been isolated. The inferred amino-acid sequence of the subunits has homologies in primary sequence and transmembrane topology. The sequence of GABA receptor subunits indicates that they share structural features with other transmitter-gated ion channels. Functional expression in *Xenopus* oocytes revealed that the cloned cDNAs encode GABA receptors with appropriate pharmacological profile (Schofield *et al.*, 1987).

The second class of GABA receptors, termed GABA$_B$ receptors, were originally distinguished from the GABA$_A$ receptors by their resistance to blockade by the antagonist bicuculline and their activation by the agonist baclofen, which is ineffective at GABA$_A$ receptor sites (Bowery *et al.*, 1980; Hill and Bowery, 1981). Since then a wealth of evidence has indicated a distribution of B-type receptors almost as ubiquitous as that of the A type. GABA$_B$ receptors contribute to a comparatively slow IPSP in hippocampal (Dutar and Nicoll, 1988a,b; Misgeld *et al.*, 1984; Newberry and Nicoll, 1984, 1985; Thalmann, 1988) and cortical neurons (Howe *et al.*, 1987a). In addition to a small hyperpolarization and conductance increase (Connors *et al.*, 1988; Deisz and Prince, 1989; Howe *et al.*, 1987a), baclofen reduces excitatory PSPs (EPSPs) (Howe *et al.*, 1987a) and IPSPs (Connors *et al.*, 1988; Deisz and Prince, 1989; Howe *et al.*, 1987a; Scholfield, 1983) in the neocortex.

GABA$_B$ Receptors in Mammalian Function. Edited by N.G. Bowery, H. Bittiger and H.-R. Olpe

The mechanisms involved in the attenuation of central neurotransmission are not yet fully understood. On the one hand, a mechanism analogous to that found postsynaptically may reduce presynaptic excitability and also transmitter release. On the other hand, evidence from several laboratories indicates that transmitter release may be reduced by baclofen through a modulation of Ca^{2+} currents. Evidence for this mechanism was mainly obtained in spinal ganglion cells (Deisz and Lux, 1985; Robertson and Taylor, 1986; Scott and Dolphin, 1986) but the applicability of this concept to central neurons is limited. So far no direct evidence for the modulation of Ca^{2+} conductance by baclofen has been obtained in central neurons. In cultured hippocampal neurons, Gähwiler and Brown (1985) found a baclofen-induced K^+ conductance but no effect of baclofen on Ca^{2+} currents. In addition, presumably Ca^{2+}-dependent action potentials were not affected by baclofen in neocortical neurons (Howe *et al.*, 1987a). However, Harrison *et al.* (1988) reported the attenuation by baclofen of IPSPs of hippocampal neurons in tissue culture, without a detectable conductance change of the presynaptic neuron. This would indicate that baclofen may affect Ca^{2+} currents.

Here we review some data from neocortical neurons obtained in different laboratories. Taken together, the results suggest that $GABA_B$ receptors serve an important function as presynaptic autoreceptors regulating the release of GABA (Deisz and Prince, 1986, 1989). The findings are complemented by evidence from other neurons and we try to outline the similarities and discrepancies between different sets of observations.

CELLULAR ACTIONS OF BACLOFEN

Postsynaptic Effects of Baclofen

Several laboratories have recently studied the cellular actions and synaptic effects of baclofen in neocortical neurons. The data obtained in cat, guinea-pig and rat neocortical neurons are in fairly good agreement despite some minor differences that are probably due to experimental arrangements. Baclofen hyperpolarized the majority of neocortical neurons (Connors *et al.*, 1988; Deisz and Prince, 1989; Howe *et al.*, 1987a) and produced outward currents when the neurons were recorded in the voltage-clamp mode (Howe *et al.*, 1987a). The hyperpolarizations are fairly small, about 4 mV, and are associated with a small resistance decrease between 15 and 30%. Table 1 compares the mean values of some of the parameters of baclofen effects and synaptic inhibition obtained by different authors in various species.

The magnitude of baclofen-induced potential and conductance changes is comparatively small, yet these effects are sufficient to alter neuronal excitability. Although baclofen had no effect on the threshold voltage for action potential generation or action potential parameters of rat neocortical neurons, the amplitude of depolarizing current necessary to evoke action potentials was increased (Howe *et al.*, 1987a). Similar findings were also obtained in guinea-pig neocortical neurons. Depolarizations that normally were sufficient to just elicit a few action potentials failed to evoke action potentials during the application of baclofen. However, depolarizations with slightly increased current pulses indicated essentially unaltered excitability (Deisz and Prince, 1989). A more detailed analysis of the effects of baclofen on firing properties of neocortical neurons by frequency versus current intensity (F/I) plots substantiates this notion. Such F/I curves consist of two components: a very steep initial range for the first interspike interval (primary slope), which flattens fairly abruptly to a much smaller secondary slope. Baclofen shifted the onset of the primary slope by about 0.3 nA to the right, i.e. to higher current intensities, but the steepness of the primary slope was increased rather than decreased (Connors *et al.*, 1988). The secondary slope of the F/I relationship, however, in the presence of baclofen was not significantly different from control. As far as the output from a given neuron is concerned, the rather small conductance increase induced by baclofen would prevent the attainment of threshold only for those excitatory inputs that are just superthreshold, while inputs of higher intensity would be hardly affected. In essence, this action of baclofen, and perhaps also the late IPSP (see below), would tend to reduce the noise level, thereby improving the signal to noise ratio.

Baclofen Effects on EPSPs

The effects of baclofen on synaptic components evoked by orthodromic stimulation have been investigated by various means, including local application by ionophoresis or pressure and by superfusion. The bath application of baclofen used by Howe *et al.* (1987a) reduced EPSPs in a dose-dependent fashion (median effective concentration, EC_{50} 1 μM). These effects were shown to be stereoselective: D-baclofen at 10 μM had no significant effect on the amplitude of the EPSPs, whereas L-baclofen reduced the amplitudes by about 50% at a 10-fold lower concentration. Others failed to detect significant effects of baclofen on EPSPs. Connors *et al.* (1988) and Deisz and Prince (1989) applied baclofen with brief pressure pulses and found no consistent effects on EPSPs. In some instances baclofen even increased the amplitude of EPSPs. A previous study on olfactory cortex neurons also revealed an increase of EPSP amplitudes with baclofen, together with a pronounced depression of IPSPs (Scholfield, 1983). The failure of baclofen to depress EPSPs in some studies may be explained by the very localized application used by Connors *et al.* (1988) and Deisz and Prince (1989) and a spatial distribution of the excitatory inputs converging predominantly on remote dendritic localizations. In fact, higher doses of baclofen, which probably affect larger arrays of neurons, depressed EPSPs (Connors *et al.*, 1988). Nevertheless, EPSPs may be affected by baclofen to a different extent in different regions within the neocortex. At any rate, the attenuation of EPSPs has been demonstrated to be presynaptically mediated, since responses to local application of excitatory neurotransmitters, such as *N*-methyl-D-aspartate (NMDA) and L-glutamate, were unaffected by L-baclofen (Howe *et al.*, 1987a). Moreover, these observations are in line with neurochemical data suggesting a decreased release of glutamate and aspartate from neocortical slices in the presence of baclofen (Neal and Shah, 1989).

Properties of IPSPs

As shown in Table 1, most cortical neurons from different species exhibit with higher intensities of orthodromic stimulation two types of IPSPs that differ in conductance and reversal potential: an early IPSP with a reversal potential between -68 and -77 mV and a late component with a reversal potential between -81 and -92 mV. The early IPSP exhibits a five to eightfold higher conductance than the late IPSP. Changes in intracellular Cl^- and extracellular K^+ show that the early IPSP is Cl^- dependent and that the late IPSP is K^+ dependent (Thompson *et al.*, 1988). Hence, the early IPSP fulfils the criteria for a $GABA_A$ receptor-mediated Cl^- conductance, while the late component is mediated by a different receptor and ionic mechanism. Therefore we shall refer to the early and late components as $IPSP_A$ and $IPSP_B$, respectively.

The two components of inhibition can be quantified using current–voltage relationships to estimate the underlying conductances (see Deisz and Prince, 1989). The temporal overlap of EPSP, $IPSP_A$ and $IPSP_B$ inherently precludes the exact determination of the reversal potential. Therefore, the terms 'reversal potential' or 'synaptic conductance' refer to the 'apparent' value. The reversal potential of the $E_{IPSP,A}$ varied between -66 and -82 mV in guinea-pig neurons (-73.0 ± 4.4 mV), and was in the same range as that found in neocortical neurons from rats and cats (see Table 1). The synaptic conductance of the $IPSP_A$ ($g_{IPSA,A}$) determined at low frequencies (0.1–0.25 Hz) was, on the average, 193 ± 250 nS (depolarizing direction). At more negative membrane potentials (E_ms) $g_{IPSP,A}$ was always considerably smaller, on the average 33.5 ± 30 nS. The mean reversal potential of $IPSP_B$ ($E_{IPSP,B}$) was on the average -88.6 ± 6.1 mV in guinea-pig neurons (Deisz and Prince, 1989), similar to the values obtained by others (see Table 1). The conductance of $IPSP_B$ ($g_{IPSB,B}$) was higher in the depolarizing direction (24 ± 19 nS) compared with hyperpolarized levels (10 ± 11 nS). Changes in apparent R_m in the depolarizing direction (Connors *et al.*, 1982) were insufficient to account for the pronounced rectification of the synaptic events. For example, the average rectification ratio in guinea-pig neurons was 1.16 ± 0.27, while the mean rectification of conductance was 5.8 and 2.3 for $IPSP_A$ and $IPSP_B$, respectively (Deisz and Prince, 1989). Considering the differences in species and in experimental arrangements, for example stimulation conditions between these sets of data, the differences are surprisingly

Table 1. Mean values of membrane properties, baclofen responses and inhibitory components in different species

	Connors *et al.* (1988)		Deisz and Prince (1989)	Howe *et al.* (1987b)
	Cat	Rat	Guinea-pig	Rat
E_m (mV)	-74.5 ± 5	-79 ± 6.1	-75.8 ± 5.2	-80.4 ± 8.3
R_m (MΩ)	29 ± 5.9	30 ± 10.4	46 ± 27	21.6 ± 10
AP (mV)	n.g.	n.g.	107 ± 7.2	110 ± 11
$E_{baclofen}$ (mV)	n.g.	-90 ± 6.2	n.g.	-90.6
$g_{baclofen}$ (nS)	n.g.	12	10	n.g.
$E_{IPSP, A}$ (mV)	-77 ± 3.9	-75 ± 3.8	-73.0 ± 4.4	-68.7 ± 1.9
$g_{IPSP, A}$ (nS)	94 ± 51	72 ± 32	193 ± 250	n.g.
$E_{IPSP, B}$ (mV)	-92 ± 3	-91 ± 5.2	-88.6 ± 6.1	-81.4 ± 2.4
$g_{IPSP, B}$ (nS)	19 ± 8.7	12 ± 5.2	24 ± 19	n.g.
A/B	5	6	8	n.g.

Note the fairly good agreement of parameters such as reversal potentials for $IPSP_A$ and $IPSP_B$ between different species. AP, action potential amplitude; R_m, neuronal input resistance; n.g., not given in the paper; further abbreviations, see text.

small. The only marked differences are between the conductance estimates. This may be explained by the difference in data handling. The values given by Connors *et al.* (1988) refer to a total value, while Deisz and Prince (1989) determined the conductances in both the hyperpolarizing and the depolarizing directions. The values referred to are those determined in the depolarizing direction, where a higher conductance is obtained. Nevertheless, the concordance clearly indicates that the late IPSP, which has a more negative reversal potential and a smaller conductance than the early IPSP, is a common feature of neocortical neurons across species.

Frequency Dependence of IPSPs

The amplitudes of both IPSPs were usually stable at low frequencies of stimulation (below 0.2 Hz). At higher frequencies, however, the amplitudes of both IPSPs were attenuated with the second stimulus. The decline in synaptic response was largest from the first to the second stimulus, and after about 5–10 stimuli the depression reached a steady state (Deisz and Prince, 1989). In some neurons the amplitudes of either the $IPSP_A$ or the $IPSP_B$ were reduced even with the lowest stimulus frequency employed (0.1 Hz). At 1 Hz stimulation frequency, $IPSP_A$ and $IPSP_B$ were drastically reduced in amplitude. This attenuation at 1 Hz was accompanied by only marginal changes of E_{IPSP}. The $E_{IPSP, A}$ decreased about 5 mV, on the average, with a tenfold change in the stimulation frequency. For the same increase in stimulation frequency the change in $E_{IPSP, B}$ was even smaller, on the average, about 2 mV. By comparison, the changes in conductance of both synaptic components with a tenfold increase in stimulus frequency are quite large. On the average the conductance decreased 79 % and 61 % for the early and late component, respectively, without significant effects on E_m or R_m.

At these frequencies the effectiveness of the subcortical stimulus in evoking synaptic events was unaltered, judging from the virtual constancy of field potentials (Deisz and Prince, 1989). Moreover, the action potentials of pyramidal cells were not significantly affected in the same frequency range (Deisz and Prince, 1987) and interneurons were capable of firing at much higher rates (McCormick *et al.*, 1985). These experiments suggest that the frequency dependence may be located at the inhibitory synapse *per se*.

Effects of Nipecotic Acid

Given the conclusion that the depression of both IPSPs at higher frequencies is located at the inhibitory synapse, several conceivable pre- and postsynaptic mechanisms may contribute to the effect. In order to distinguish between the various possibilities, the effects of nipecotic acid, an established blocker of GABA reuptake (Krogsgaard-Larsen and Johnston, 1975) were investigated. Nipecotic acid produced no detectable effect on R_m, membrane rectification or action potential generation, but reduced the amplitudes of both $ISPS_A$ and $IPSP_B$. Quantitative evaluation revealed that application of nipecotic acid significantly decreased the conductance of the $IPSP_A$ by about 80% when evoked at low frequencies of stimulation (Deisz and Prince, 1989). The attenuation and even loss of IPSPs in the presence of nipecotic acid (Aickin and Deisz, 1981; see also Dingledine and Korn, 1985) was shown to be related to a decrease in GABA uptake, presumably leading to prolonged transients of the synaptically liberated GABA. This was mainly inferred from the change in frequency dependence of the IPSP in the presence of nipecotic acid. Unlike possible non-specific effects of nipecotic acid, which may cause a constant depression at all frequencies, nipecotic acid attenuated the $IPSP_A$ amplitude only at low frequencies. At 1 Hz stimulation, nipecotic acid caused no significant effect. Changes in postsynaptic sensitivity to local GABA application were also rendered unlikely. In fact, the GABA-induced conductance was slightly enhanced (20%) in the presence of nipecotic acid, judging from current–voltage relationships (Deisz and Prince, 1989).

Baclofen Effects on IPSP

The reduction of IPSPs may be brought about by various presynaptic effects of GABA, a shunting of the membrane by an increase in Cl^- conductance, or a more specific effect by alterations of Ca^{2+} spikes (Dunlap, 1981). We used baclofen to test these alternatives. This agonist for $GABA_B$ receptor sites, which reduces transmitter release (Bowery *et al.*, 1980; Holz *et al.*, 1989) and Ca^{2+} events (Bowery and Williams, 1986; Deisz and Lux, 1985; Holz *et al.*, 1986) should reveal which mechanism is involved. Figure 14.1A shows a pen-recording of part of an experiment to evaluate the effects of baclofen on IPSPs. Depolarizing and hyperpolarizing current pulses were applied at 0.2 Hz, while orthodromic stimulation evoked both IPSPs. From such families of traces resting and inhibitory conductances were determined. Application of a racemic mixture of 100 μM baclofen by brief pressure pulses caused a slow transient hyperpolarization of the postsynaptic membrane. Although the postsynaptic effects were comparatively small, IPSPs were markedly attenuated by baclofen (see Figure 14.1B). Despite the maintained generation of action potentials during depolarizing current pulses of sufficient intensity, the IPSP was virtually abolished within a few seconds of application (see Figure 14.1B). After the slight membrane hyperpolarization had dwindled, a second set of current pulses with orthodromic stimulation (interval C2 in Figure 14.1A) was applied. Comparison of the resting membrane and IPSP conductances, before and after the application of baclofen (see Figure 14.2), indicates that both IPSPs were still attenuated after about 2 min when postsynaptically detectable effects had dissipated (Deisz and Prince, 1989). The observation that the effect of baclofen outlasts postsynaptic effects has been made by all investigators in this field (Connors *et al.*, 1988;

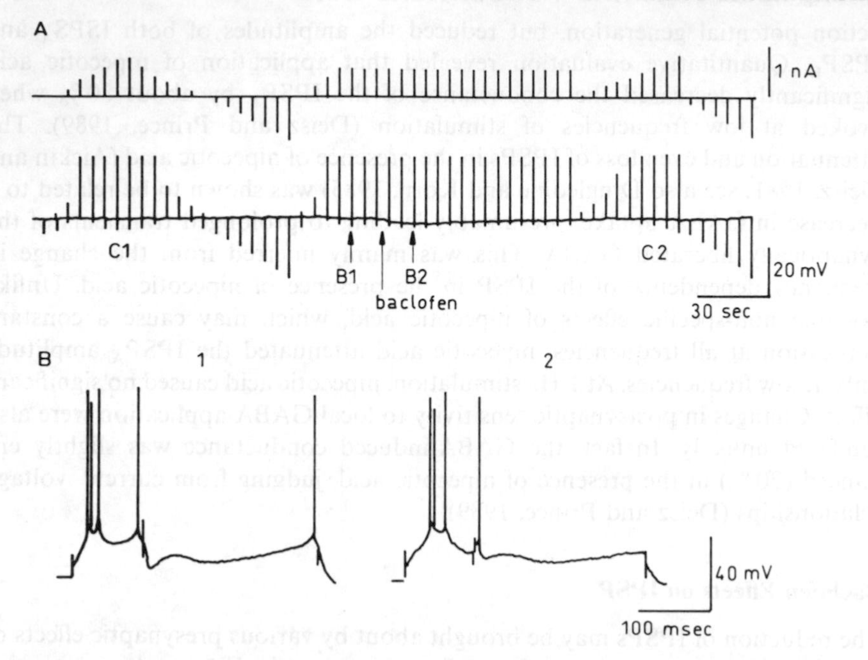

Figure 14.1. (A) Pen-recording showing part of an experiment. During the white-matter stimulation at low frequency (0.2 Hz), a series of depolarizing and hyperpolarizing current pulses were applied to determine the current–voltage characteristics. In order to monitor membrane excitability directly during the pressure application of baclofen (600 ms pulse from a 100 μM-containing electrode), constant suprathreshold depolarizing current pulses of 350 ms duration (about 0.6 nA) were applied at the same frequency, timed so that the orthodromic stimulus fell in the midst of each depolarizing current pulse. (B) Traces shown are from the same experiment as in (A), taken at the times indicated. Within 10 s of baclofen application the IPSP is abolished but the excitability is maintained (cf. B1 and B2). (From Deisz and Prince, 1989, with permission)

Deisz and Prince, 1989; Howe *et al.*, 1987a). As noted by several authors, the depression of IPSPs occurs in the face of unaltered postsynaptic responsiveness to GABA (Howe *et al.*, 1987a; Scholfield, 1983), indicating a decreased release of transmitter.

DISCUSSION

Postsynaptic Actions of Baclofen

In the early studies concerning the cellular actions of baclofen, Pierau and Zimmermann (1973) noted that the application of baclofen caused a slight hyperpolarization without noticeable changes in R_m. In the olfactory cortex

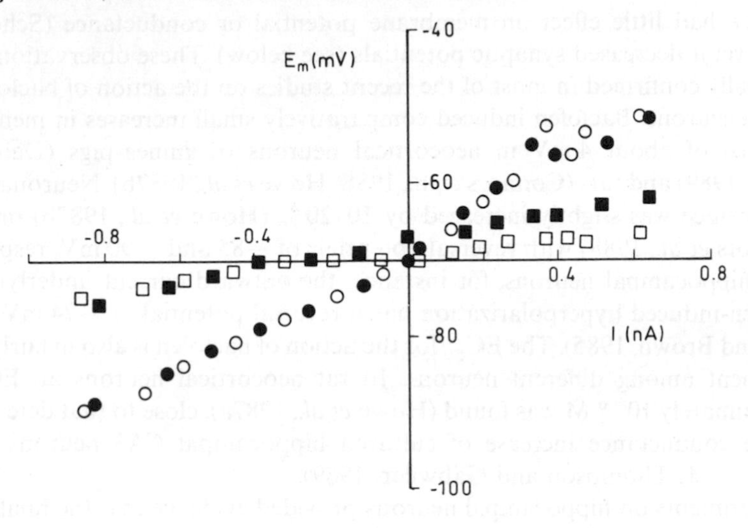

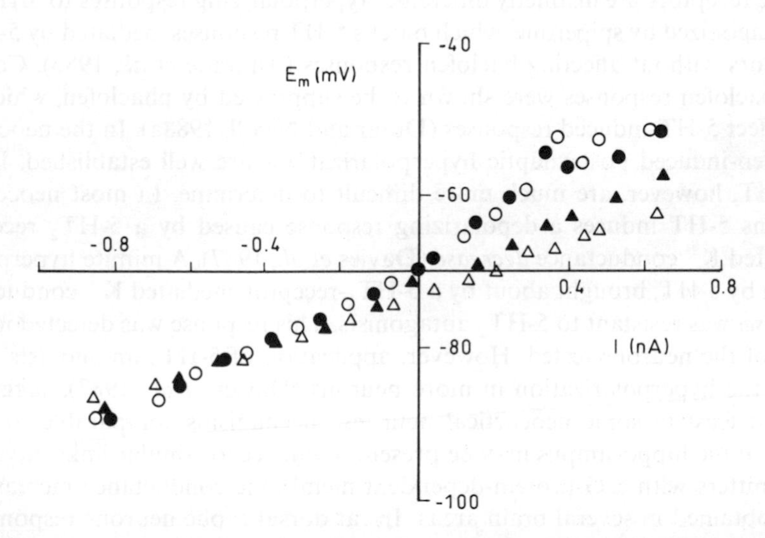

Figure 14.2. Current–voltage relationship before (open symbols) and about 2 min after (filled symbols) application of a minute quantity of baclofen. The membrane potential (E_m) attained has been plotted against the injected current amplitude both before orthodromic stimulation (circles) and at the peak of IPSP$_A$ (rectangles, upper plot) and IPSP$_B$ (triangles, lower plot), the early and late inhibitory postsynaptic potentials, respectively. (From Deisz and Prince, 1989, with permission)

baclofen had little effect on membrane potential or conductance (Scholfield, 1983), yet it decreased synaptic potentials (see below). These observations were essentially confirmed in most of the recent studies on the action of baclofen on central neurons. Baclofen induced comparatively small increases in membrane potential of about 4 mV in neocortical neurons of guinea-pigs (Deisz and Prince, 1989) and rats (Connors *et al.*, 1988; Howe *et al.*, 1987b). Neuronal input conductance was slightly increased by 10–20 % (Howe *et al.*, 1987b) or 12 nS (Connors *et al.*, 1988) with reversal potentials of −85 and −90 mV, respectively. In hippocampal neurons, for instance, the outward current underlying the baclofen-induced hyperpolarization has a reversal potential of −74 mV (Gähwiler and Brown, 1985). The EC_{50} for the action of baclofen is also in fairly good agreement among different neurons. In rat neocortical neurons an EC_{50} of approximately 10^{-6} M was found (Howe *et al.*, 1987a), close to that determined for the conductance increase of cultured hippocampal CA3 neurons (EC_{50} 5×10^{-7} M; Thompson and Gähwiler, 1989).

Experiments on hippocampal neurons provided evidence that the final target of postsynaptic $GABA_B$ receptors is a guanyl nucleotide binding protein (G-protein) -linked K^+ channel, a feature shared with various other receptors including 5-hydroxytryptamine (5-HT, serotonin) receptors. This was demonstrated by the non-additive responses of both transmitters (Andrade *et al.*, 1986), yet the receptors are distinctly different. Hyperpolarizing responses to 5-HT can be antagonized by spiperone, which blocks 5-HT responses mediated by 5-HT_{1A} receptors without affecting baclofen responses (Andrade *et al.*, 1986). Conversely, baclofen responses were shown to be suppressed by phaclofen, which did not affect 5-HT-induced responses (Dutar and Nicoll, 1988a). In the neocortex, baclofen-induced postsynaptic hyperpolarizations are well established. Effects of 5-HT, however, are much more difficult to determine. In most neocortical neurons 5-HT induces a depolarizing response caused by a 5-HT_2 receptor-mediated K^+ conductance *decrease* (Davies *et al.*, 1987). A minute hyperpolarization by 5-HT, brought about by a 5-HT_1-receptor-mediated K^+ conductance *increase*, was resistant to 5-HT_2 antagonists. This response was detected in only 25 % of the neurons tested. However, application of 5-HT_2 antagonists uncovered the hyperpolarization in more neurons (Davies *et al.*, 1987), suggesting that, at least in some neocortical neurons, mechanisms comparable to those found in the hippocampus may be present. Evidence for similar links of various transmitters with a G-protein-dependent membrane conductance increase has been obtained in several brain areas. In rat dorsal raphé neurons responses to both baclofen and 5-HT were reduced by pertussis toxin (Innis and Aghajanian, 1987). The receptors involved include those for, for example, α-adrenergic (Holz *et al.*, 1989), peptidergic and muscarinic agonists (Christie and North, 1988). In rat lateral parabrachial neurons opioid peptides, muscarine (M_2) and $GABA_B$ receptors were also shown to share the same effector system by an occlusion experiment (Christie and North, 1988).

Baclofen Effects on EPSPs

Baclofen exerts quite variable effects on EPSPs in different structures. In cat spinal motoneurons attenuation of EPSPs appears to be the prominent effect (Pierau and Zimmermann, 1973). There, baclofen almost completely abolished EPSPs without noticeable changes in R_m, although a slight hyperpolarization was noted. At supraspinal levels, however, EPSPs are less consistently affected. In neocortical neurons, fairly pure EPSPs, elicited by low intensities of orthodromic stimulation, are reduced by baclofen (Howe *et al.*, 1987a). EPSPs with superimposed IPSPs are increased by local application of small doses of baclofen (Connors *et al.*, 1988; Deisz and Prince, 1989). The concomitant depression of IPSPs by baclofen (see below) can readily account for the augmentation of the EPSP. The intimate temporal relationship of EPSP and IPSP as such would increase EPSPs when IPSPs are decreased. In addition, the shunting of the EPSP by the IPSP is reduced when inhibitory components are decreased. The more effective electrotonic spread of excitation could contribute to the enhancement of EPSPs. EPSPs are reduced only with a more widespread application of baclofen (Connors *et al.*, 1988) as with the bath application used by Howe *et al.* (1987a). In hippocampal neurons the effects of baclofen differ depending upon the region investigated. In granule cells the amplitude and duration of EPSPs evoked by perforant path stimulation increased with low doses of baclofen (Misgeld *et al.*, 1984). At higher concentrations the EPSPs were reduced concomitant with an increase in membrane conductance. In CA3 neurons, however, already low concentrations of baclofen increased the membrane conductance and virtually abolished synaptic responses (Misgeld *et al.*, 1982, 1984).

Baclofen Effects on IPSPs

High intensities of orthodromic stimulation elicit in addition to EPSPs two types of IPSPs with comparable reversal potentials and conductances in neocortical neurons of different species (see Table 1; Connors *et al.*, 1988; Deisz and Prince, 1989; Howe *et al.*, 1987a,b). The early component is probably brought about by a GABA$_A$-mediated increase in Cl^- conductance, as inferred from the change in reversal potential following Cl^- injection (Thompson *et al.*, 1988) and the sensitivity to antagonists of the GABA$_A$ receptor, picrotoxin or bicuculline (Connors *et al.*, 1988). The late component is mediated by an increase in K^+ conductance. This can be inferred from the value of the reversal potential (see Table 1) as well as from the dependence on extracellular K^+ (Howe *et al.*, 1987b), the slope of reversal potential changes approaching a Nernst response for K^+ ions (Thompson *et al.*, 1988). Further evidence for the involvement of GABA$_B$ receptors in the IPSP$_B$ stems from the observation that the IPSP$_B$ is reduced by the GABA$_B$ receptor antagonist phaclofen (Karlsson *et al.*, 1988; but see Müller and Misgeld, 1989). In hippocampal neurons a late, bicuculline-insensitive IPSP was described previously (Newberry and Nicoll, 1984) which shares several properties with baclofen responses (Newberry and Nicoll, 1985). The $IPSP_B$ is, like the response to baclofen, attenuated by pertussis toxin and intracellular application of GTPγS (Thalmann, 1987, 1988), suggesting a common conductance mechanism. Moreover, both the late synaptic component and the baclofen response were reduced by the $GABA_B$ receptor antagonists phaclofen (Dutar and Nicoll, 1988a,b) and 2-hydroxysaclofen (Lambert *et al.*, 1989). However, a possible contribution of other transmitters, activating the same G-protein-mediated K^+ conductance, to the $IPSP_B$ should be considered, since 2-hydroxysaclofen does not completely block the $IPSP_B$ (Lambert *et al.*, 1989).

Nevertheless, the similarity in ionic mechanisms (see Table 1) and pharmacology between $IPSP_B$ and baclofen responses suggests that the $IPSP_B$ is mainly mediated by GABA activating $GABA_B$ receptors. Yet, the prototype agonist for $GABA_B$ receptors markedly attenuates the $IPSP_B$ in most central neurons. Baclofen at 10–20 μM abolished IPSPs in CA3 neurons and in granule cells (Misgeld *et al.*, 1982). Also in hippocampal CA1 neurons it was noted that IPSPs are reduced prior to any effect on the EPSP (Peet and McLennan, 1986). In the neocortex the attenuation of both IPSPs outlasted the increase in postsynaptic conductance changes (Connors *et al.*, 1988; Deisz and Prince, 1986, 1989; Howe *et al.*, 1987a). But the EC_{50} values determined for the depression of synaptic potentials and postsynaptic conductance were found to be comparable (Howe *et al.*, 1987a). Likewise, the EC_{50} value for the reduction of inhibition (5×10^{-7} M) was similar to that for the postsynaptic conductance increase of cultured hippocampal CA3 neurons (Thompson and Gähwiler, 1989). The marked depression of inhibition was shown to occur in the face of unaltered postsynaptic GABA responses in hippocampal (Thompson and Gähwiler, 1989) and cortical neurons (Howe *et al.*, 1987a; Scholfield, 1983), indicating a presynaptic mechanism. Taken together, these data indicate that a prominent feature of the action of baclofen is the attenuation of synaptic inhibition by a presynaptic mechanism.

One major exception to the attenuation of IPSPs occurs in the spinal cord. As noted by Pierau and Zimmermann (1973), baclofen almost completely abolished EPSPs of cat spinal motoneurons, but had little or no effect on IPSPs. Since inhibition at the spinal level is mediated by glycine, it is tempting to speculate that glycinergic terminals may lack $GABA_B$ receptors. Alternatively, these receptors, if present, may assume a different physiological state under normal conditions.

Towards the Presynaptic Actions of Baclofen

The K^+ conductance increase by baclofen determined postsynaptically in neocortical and hippocampal neurons could decrease the spread of excitation, if present, in presynaptic terminals. The effect may be small though, considering the small effects on postsynaptic excitability (Connors *et al.*, 1988; Deisz and Prince, 1989). This raises the question of whether the postsynaptic effects outlined above contribute to the attenuation of synaptic potentials through presynaptic mechanisms.

At first sight the original observation by Dunlap (1981) that Ca^{2+}-dependent action potentials are shortened by baclofen might readily explain the reduced transmitter release. The established role of Ca^{2+} in transmitter release (Katz and Miledi, 1970; Llinás *et al.*, 1981) makes the neurotransmitter modulation of Ca^{2+} currents an interesting hypothesis for presynaptic inhibition (Dunlap and Fischbach, 1981). However, it is unclear whether the shortening of the duration of Ca^{2+} action potentials would have consequences on transmitter release depending probably on much briefer Ca^{2+} transients. Moreover, it was shown that the shortening of Ca^{2+} action potentials by baclofen was removed by injection of Cs^+ (Desarmenien *et al.*, 1984). This observation suggests that the interpretation of the shortening of Ca^{2+} action potentials by baclofen may have been complicated by baclofen activating a K^+ conductance (Gähwiler and Brown, 1985; Newberry and Nicoll, 1984). Patch-clamp measurements revealed that Ca^{2+} currents of cultured avian dorsal root ganglion neurons were indeed reduced by both baclofen and GABA in a bicuculline-insensitive manner (Deisz and Lux, 1985). This satisfies the operational definition of GABA$_B$ receptors. Since these measurements were also carried out with symmetrical CsCl across the membrane, a condition which should eliminate all other contaminating conductances, this observation indicates that net Ca^{2+} current had indeed been reduced. Reduction of Ca^{2+} currents by baclofen have been confirmed in dorsal root ganglion neurons of several species (Dolphin and Scott, 1986; Robertson and Taylor, 1986; Scott and Dolphin, 1986). Moreover it was shown that the modulation of Ca^{2+} currents by baclofen was removed by pertussis toxin (Holz *et al.*, 1986) and intracellular application of the GDP analogue GDPβS (Holz *et al.*, 1986; Scott and Dolphin, 1986), conditions interfering with the G-protein transduction mechanism. This G-protein-linked modulation of Ca^{2+} current is, like the postsynaptic K^+ conductance increase by baclofen, shared by several transmitters and neuromodulators including, for example, opioid peptides (e.g. Hescheler *et al.*, 1987) and noradrenaline (Bean, 1989; Holz *et al.*, 1986).

However, in supraspinal neurons baclofen has not yet been shown to affect Ca^{2+}-mediated events. In hippocampal CA3 neurons baclofen failed to affect Ca^{2+} inward currents (Gähwiler and Brown, 1985), and in neocortical neurons Ca^{2+} action potentials were unaltered by baclofen (Howe *et al.*, 1987a). Further evidence for a K^+ conductance mechanism underlying the presynaptic attenuation of transmitter release was provided by Misgeld *et al.* (1989). They showed that the condition which eliminates the postsynaptic K^+ conductance also removed the attenuation of the IPSP. However, in cultured hippocampal neurons paired recordings from pre- and postsynaptic neurons revealed a drastic reduction in inhibitory transmission (92 %) when spike threshold and input resistance of both neurons were unaffected (Harrison *et al.*, 1988), indicating that the action of baclofen may have been mediated by modulation of presynaptic Ca^{2+} conductances. The latter mechanism is also supported by measurements of synaptosomal Ca^{2+} levels. Baclofen was shown to attenuate the depolarization-induced increase in Ca^{2+} of synaptosomes measured using Ca^{2+}-sensitive dyes (Bowery and Williams, 1986).

Comparison of the effectiveness of baclofen at pre- and postsynaptic sites revealed the same EC_{50}, yet the presynaptic sensitivity appears much higher. The effect on transmitter release, however, may be potentiated by additional factors. As pointed out by Robertson and Taylor (1986), the steep dependence of transmitter release on extracellular Ca^{2+} (Dodge and Rahamimoff, 1967; Katz and Miledi, 1970) would enhance the effects of baclofen on transmitter output. Furthermore, the pronounced slowing of the Ca^{2+} current onset in the presence of GABA (Deisz and Lux, 1985) might effectively contribute to a reduction of the relevant current time integral. This slowing would render Ca^{2+} sequestration more effective. Hence during the short interval required for release, the total effect may thereby be larger than expected from the attenuation of peak currents. Along this line of reasoning it is noteworthy that, in the in-vivo spinal cord, baclofen mimics the effects of reduced presynaptic Ca^{2+} entry (Lev-Tov *et al.*, 1988).

Interestingly, there seems to be a marked difference between the mechanisms involved in the baclofen effects in different areas. In spinal ganglia, attenuation of Ca^{2+} currents is well documented (Deisz and Lux, 1985; Robertson and Taylor, 1986; Scott and Dolphin, 1986). Attenuation of Ca^{2+} currents has also been implicated in hippocampal neurons in dissociated culture (Harrison *et al.*, 1988) whereas effects through K^+ conductance increase are suggested in the hippocampal slice preparation (Gähwiler and Brown, 1985; Misgeld *et al.*, 1989). Most studies have shown that the modulation of Ca^{2+} currents depends upon the G-protein cascade (Hescheler *et al.*, 1987; Holz *et al.*, 1986; Scott and Dolphin, 1986). The same cascade has also been shown to couple α-adrenergic and $GABA_B$ receptors to inhibition of peptide secretion (Holz *et al.*, 1989). Additional second messenger systems are unlikely to be involved, suggesting a direct link between $GABA_B$ receptors and Ca^{2+} channels through G-proteins (Dolphin *et al.*, 1989). Neurochemical evidence from cerebral synaptosomes, however, suggests a coupling of $GABA_B$ receptors with adenylate cyclase through G-proteins (Nishikawa and Kuriyama, 1989). It remains to be seen whether these differences relate to the physiological state of the preparation or to the neuron type. Available data would support either view. Evidence supporting a K^+-conductance increase in hippocampal neurons (Gähwiler and Brown, 1985; Misgeld *et al.*, 1989) as opposed to a predominant modulation of Ca^{2+} currents in ganglion cells (Deisz and Lux, 1985; Robertson and Taylor,

1986; Scott and Dolphin, 1986) may indicate that different neurons couple GABA$_B$ receptors to different effector systems. On the other hand, evidence for a presynaptic K^+ conductance increase in adult neurons in a hippocampal slice (Misgeld *et al.*, 1989) as opposed to evidence towards a modulation of Ca^{2+} currents in cultured hippocampal neurons (Harrison *et al.*, 1988) suggests that the coupling of GABA$_B$ receptors to different effector systems may depend upon the physiological state or age of the neuron. In the CA1 region of the hippocampus, a distinct difference between pre- and postsynaptic GABA$_B$ responses was noted. Dutar and Nicoll (1988b) provided evidence that the postsynaptic responses to baclofen are sensitive to blockade by phaclofen and pertussis toxin. But the attenuation of EPSPs by baclofen was insensitive to phaclofen and pertussis toxin. Interestingly, K^+-induced increases in the level of Ca^{2+} in synaptosomes were shown to be reduced by baclofen; however, these effects were not antagonized by phaclofen (Stirling *et al.*, 1989). These data are difficult to equate with a single type of GABA$_B$ receptor–effector system and different GABA$_B$ receptor–effector systems have previously been proposed (Deisz and Lux, 1985).

Our view of a modulation of GABA release by GABA$_B$ receptors is corroborated by release experiments. Waldmeier *et al.* (1988) reported a pronounced reduction of evoked release from neocortical slice preparations by baclofen when low frequencies of stimulation were used. At higher frequencies of stimulation the effect of baclofen on stimulus-induced release was much weaker compared with that at low frequencies, analogous to the more pronounced reduction of inhibition at low frequencies compared with high frequencies observed with nipecotic acid.

In neocortical neurons baclofen increased the threshold for eliciting paroxysmal depolarization shifts (Howe *et al.*, 1987a), but in hippocampal granule cells low doses of baclofen caused a disinhibition resembling the action of bicuculline (Misgeld *et al.*, 1984). Furthermore, intracortical administration of baclofen in the rat causes stereoselective partial epilepsy with focal motor symptoms (Van Rijn *et al.*, 1987). Thus it appears as if the net effect of baclofen in different neurons may depend upon several factors including the relative density of different types of pre- and postsynaptic GABA$_B$ receptors. As far as neocortical neurons are concerned, the data obtained in different laboratories (Connors *et al.*, 1988; Deisz and Prince, 1986, 1989; Howe *et al.*, 1987; Scholfield, 1983) indicate that the predominant effect of baclofen is the attenuation of inhibition. The recent link between the well-established frequency-dependent depression of the IPSP (Ben-Ari *et al.*, 1979; McCarren and Alger, 1985) and GABA$_B$ receptors exerting negative feedback on GABA's own release (Deisz and Prince, 1989) makes GABA$_B$ receptor antagonists interesting tools. A GABA$_B$ receptor antagonist that interrupts selectively the negative feedback but spares the postsynaptic GABA$_B$ receptor may even become a therapeutically useful drug which prevents use-dependent depression of inhibition.

ACKNOWLEDGEMENTS

The original work forming the basis for this review was supported by grants from the SFB 220 (Howe *et al.*, 1987a,b) and NIH (Deisz and Prince, 1989).

REFERENCES

Aickin, C.C. and Deisz, R.A. (1981) Pentobarbitone interference with inhibitory synaptic transmission in crayfish stretch receptor neurones. *J. Physiol.*, **315**, 175–187.

Andrade, R., Malenka, R.C. and Nicoll, R.A. (1986) A G protein couples serotonin and $GABA_B$ receptors to the same channels in hippocampus. *Science*, **234**, 1261–1265.

Bean, B.P. (1989) Neurotransmitter inhibition of neuronal calcium currents by changes in channel voltage dependence. *Nature*, **340**, 153–156.

Ben-Ari, Y., Krnjević, K. and Reinhardt, W. (1979) Hippocampal seizures and failure of inhibition. *Can. J. Physiol. Pharmacol.*, **57**, 1462–1466.

Bowery, N.G. and Williams, L.C. (1986) $GABA_B$ receptor activation inhibits the increase in nerve terminal Ca^{++} induced by depolarization. *Br. J. Pharmacol.*, **87**, 37P.

Bowery, N.G., Hill, D.R., Hudson, A.L., Doble, A., Middlemiss, D.N., Shaw, J. and Turnbull, M. (1980) (−)Baclofen decreases neurotransmitter release in the mammalian CNS by an action at a novel GABA receptor. *Nature*, **283**, 92–94.

Christie, M.J. and North, R.A. (1988) Agonists at μ-opioid, M_2-muscarinic and $GABA_B$-receptors increase the same potassium conductance in rat lateral parabrachial neurones. *Br. J. Pharmacol.*, **95**, 896–902.

Connors, B.W., Gutnick, M.J. and Prince, D.A. (1982) Electrophysiological properties of neocortical neurons in vitro. *J. Neurophysiol.*, **48**, 1302–1320.

Connors, B.W., Malenka, R.C. and Silva, L.R. (1988) Two inhibitory postsynaptic potentials, and $GABA_A$ and $GABA_B$ receptor-mediated responses in neocortex of rat and cat. *J. Physiol.*, **406**, 443–468.

Davies, M.F., Deisz, R.A., Prince, D.A. and Peroutka, S.J. (1987) Two distinct effects of 5-hydroxytryptamine on single cortical neurones. *Brain Res.*, **423**, 347–352.

Deisz, R.A. and Lux, H.D. (1985) γ-aminobutyric acid-induced depression of calcium currents of chick sensory neurons. *Neurosci. Lett.*, **56**, 205–210.

Deisz, R.A. and Prince, D.A. (1986) Presynaptic GABA feedback causes frequency-dependent depression of ipsps in neocortical neurons. *Soc. Neurosci. Abstr.*, **12**, 19.

Deisz, R.A. and Prince, D.A. (1987) Effect of D890 on membrane properties of neocortical neurons. *Brain Res.*, **422**, 63–73.

Deisz, R.A. and Prince, D.A. (1989) Frequency-dependent depression of inhibition in guinea-pig neocortex in vitro by $GABA_B$ receptor feed-back on GABA release. *J. Physiol.*, **412**, 513–541.

Desarmenien, M., Feltz, P., Occhipinti, G., Santangelo, F. and Schlichter, R. (1984) Coexistence of $GABA_A$ and $GABA_B$ receptors on Aδ and C primary afferents. *Br. J. Pharmacol.*, **81**, 327–333.

Dingledine, R. and Korn, S.J. (1985) γ-Aminobutyric acid uptake and the termination of inhibitory synaptic potentials in the rat hippocampal slice. *J. Physiol.*, **366**, 387–409.

Dodge, F.A. and Rahamimoff, R. (1967) Co-operative action of calcium ions in transmitter release at the neuromuscular junction. *J. Physiol.*, **193**, 419–432.

Dolphin, A.C. and Scott, R.H. (1986) Inhibition of calcium currents in cultured rat dorsal root ganglion neurones by (−)-baclofen. *Br. J. Pharmacol.*, **88**, 213–220.

Dolphin, A.C., McGuirck, S.M. and Scott, R.H. (1989) An investigation into the mechanisms of inhibition of calcium channel currents in cultured sensory neurones of the rat by guanine nucleotide analogues and (−)-baclofen. *Br. J. Pharmacol.*, **97**, 263–273.

Dunlap, K. (1981) Two types of γ-aminobutyric acid receptor on embryonic sensory neurones. *Br. J. Pharmacol.*, **74**, 579–585.

Dunlap, K. and Fischbach, G.D. (1981) Neurotransmitters decrease the calcium conductance activated by depolarization of embryonic chick sensory neurones. *J. Physiol.*, **317**, 519–535.

Dutar, P. and Nicoll, R.A. (1988a) A physiological role for $GABA_B$ receptors in the central nervous system. *Nature*, **332**, 156–158.

Dutar, P. and Nicoll, R.A. (1988b) Pre- and postsynaptic $GABA_B$ receptors in the hippocampus have different pharmacological properties. *Neuron*, **1**, 585–591.

Gähwiler, B.H. and Brown, D.A. (1985) $GABA_B$-receptor-activated K^+ current in voltage-clamped CA_3 pyramidal cells in hippocampal cultures. *Proc. Natl Acad. Sci. USA.*, **82**, 1558–1562.

Harrison, N.L., Lange, G.D. and Barker, J.L. (1988) (−)-Baclofen activates presynaptic $GABA_B$ receptors on GABAergic inhibitory neurons from embryonic rat hippocampus. *Neurosci. Lett.*, **85**, 105–109.

Hescheler, J., Rosenthal, W., Trautwein, W. and Schultz, G. (1987) The GTP-binding protein, G_o, regulates neuronal calcium channels. *Nature*, **325**, 445–447.

Hill, D.R. and Bowery, N.G. (1981) ^{3}H-Baclofen and ^{3}H-GABA bind to bicuculline-insensitive $GABA_B$ sites in rat brain. *Nature*, **290**, 149–152.

Holz IV, G.G., Rane, S.G. and Dunlap, K. (1986) GTP-binding proteins mediate transmitter inhibition of voltage-dependent calcium channels. *Nature*, **319**, 670–672.

Holz IV, G.G., Kream, R.M., Spiegel, A. and Dunlap, K. (1989) G proteins couple α-adrenergic and $GABA_b$ receptors to inhibition of peptide secretion from peripheral sensory neurons. *J. Neurosci.*, **9**, 657–666.

Howe, J.R., Sutor, B. and Zieglgänsberger, W. (1987a) Baclofen reduces postsynaptic potentials of rat cortical neurones by an action other than its hyperpolarizing action. *J. Physiol.*, **384**, 539–569.

Howe, J.R., Sutor, B. and Zieglgänsberger, W. (1987b) Characteristics of long-duration inhibitory postsynaptic potentials in rat neocortical neurons in vitro. *Cell. Mol. Neurobiol.*, **7**, 1–18.

Innis, R.B. and Aghajanian, G.K. (1987) Pertussis toxin blocks 5-HT_{1A} and $GABA_B$ receptor-mediated inhibition of serotonergic neurons. *Eur. J. Pharmacol.*, **143**, 195–204.

Karlson, G., Pozza, M. and Olpe, H.-R. (1988) Phaclofen: a $GABA_B$ blocker reduces long-duration inhibition in the neocortex. *Eur. J. Pharmacol.*, **148**, 485–486.

Katz, B. and Miledi, R. (1970) Further study of the role of calcium in synaptic transmission. *J. Physiol.*, **207**, 789–801.

Krnjević, K. (1974) Chemical nature of synaptic transmission in vertebrates. *Physiol. Rev.*, **54**, 418–540.

Krogsgaard-Larsen, P. and Johnston, G.A.R. (1975) Inhibition of GABA uptake in brain slices by nipecotic acid, various isoxazoles and related compounds. *J. Neurochem.*, **25**, 797–802.

Lambert, N.A., Harrison, N.L., Kerr, D.I.B., Ong J., Prager, R.H. and Teyler, T.J. (1989) Blockade of the late IPSP in rat CA1 hippocampal neurons by 2-hydroxy-saclofen. *Neurosci. Lett.*, **107**, 125–128.

Lev-Tov, A., Meyers, D.E.R. and Burke, R.E. (1988) Activation of type B γ-aminobutyric acid receptors in the intact mammalian spinal cord mimics the effects of reduced presynaptic Ca^{2+} influx. *Proc. Natl Acad. Sci. USA.*, **85**, 5330–5334.

Llinás, R., Steinberg, I.Z. and Walton, K. (1981) Relationship between presynaptic calcium current and postsynaptic potential in squid giant synapse. *Biophys. J.*, **33**, 323-352.

McCarren, M. and Alger, B.E. (1985) Use-dependent depression of IPSPs in rat hippocampal pyramidal cells in vitro. *J. Neurophysiol.*, **53**, 557-571.

McCormick, D.A., Connors, B.W., Lighthall, J.W. and Prince, D.A. (1985) Comparative electrophysiology of pyramidal and sparsely spiny neurons of the neocortex. *J. Neurophysiol.*, **54**, 782-806.

Misgeld, U., Klee, M.R. and Zeise, M.L. (1982) Differences in burst characteristics and drug sensitivity between CA3 neurons and granule cells. In: M.R. Klee, H.D. Lux and E.J. Speckmann (eds) *Physiology and pharmacology of epileptogenic phenomena*. New York: Raven Press, pp. 131-139.

Misgeld, U., Klee, M.R. and Zeise, M.L. (1984) Differences in baclofen-sensitivity between CA3 neurons and granule cells of the guinea pig hippocampus in vitro. *Neurosci. Lett.*, **47**, 307-311.

Misgeld, U., Müller, W. and Brunner, H. (1989) Effects of (−)-baclofen on inhibitory neurons in the guinea pig hippocampal slice. *Pflügers Archiv*, **414**, 139-144.

Müller, W. and Misgeld, U. (1989) Carbachol reduces $I_{K,baclofen}$, but not $I_{K,GABA}$ in guinea pig hippocampal slices. *Neurosci. Lett.*, **102**, 229-234.

Neal, M.J. and Shah, M.A. (1989) Baclofen and phaclofen modulate GABA release from slices of rat cerebral cortex and spinal cord but not from retina. *Br. J. Pharmacol.*, **98**, 105-112.

Newberry, N.R. and Nicoll, R.A. (1984) A bicuculline-resistant inhibitory post-synaptic potential in rat hippocampal pyramidal cells in vitro *J. Physiol.*, **348**, 239-254.

Newberry, N.R. and Nicoll, R.A. (1985) Comparison of the action of baclofen with γ-aminobutyric acid on rat hippocampal pyramidal cells in vitro. *J. Physiol.*, **360**, 161-185.

Nishikawa, M. and Kuriyama, K. (1989) Functional coupling of cerebral γ-aminobutyric acid $(GABA)_B$ receptor with adenylate cyclase system: effect of phaclofen. *Neurochem. Int.*, **14**, 85-90.

Nistri, A. and Constanti, A. (1979) Pharmacological characterization of different types of GABA and glutamate receptors in vertebrates and invertebrates. *Progr. Neurobiol.*, **13**, 117-235.

Peet, M.J. and McLennan, H. (1986) Pre- and postsynaptic actions of baclofen: blockade of the late synaptically-evoked hyperpolarization of CA1 hippocampal neurones. *Exp. Brain Res.*, **61**, 567-574.

Pierau, F.K. and Zimmermann, P. (1973) Action of a GABA-derivative on postsynaptic potentials and membrane properties of cats' spinal motoneurones. *Brain Res.*, **54**, 376-380.

Robertson, B. and Taylor, W.R. (1986) Effects of γ-aminobutyric acid and (−)-baclofen on calcium and potassium currents in cat dorsal root ganglion neurones in vitro. *Br. J. Pharmacol.*, **89**, 661-672.

Schofield, P.R., Darlison, M.G., Fujita, N., Burt, D.R., Stephenson, F.A., Rodriguez, H., Rhee, L.M., Ramachandran, J., Reale, V., Glencorse, T.A., Seeburg, P.H. and Barnard, E.A. (1987) Sequence and functional expression of the $GABA_A$ receptor shows a ligand-gated receptor super-family. *Nature*, **328**, 221-227.

Scholfield, C.N. (1983) Baclofen blocks postsynaptic inhibition but not the effect of muscimol in the olfactory cortex. *Br. J. Pharmacol.*, **78**, 79-84.

Scott, R.H. and Dolphin, A.C. (1986) Regulation of calcium currents by a GTP analogue: potentiation of (−)-baclofen mediated inhibition. *Neurosci. Lett.*, **69**, 59-64.

Stirling, J.M., Cross, A.J., Robinson, T.N. and Green, A.R. (1989) The effects of $GABA_B$

receptor agonists and antagonists on potassium-stimulated $[Ca^{2+}]_i$ in rat brain synaptosomes. *Neuropharmacology*, **28**, 699–704.

Thalmann, R.H. (1987) Pertussis toxin blocks a late inhibitory post-synaptic potential in hippocampal CA_3 neurons. *Neurosci. Lett.*, **82**, 41–46.

Thalmann, R.H. (1988) Evidence that guanosine triphosphate (GTP)-binding proteins control a synaptic response in brain: effect of pertussis toxin and GTPγS on the late inhibitory postsynaptic potential of hippocampal CA3 neurons. *J. Neurosci.*, **8**, 4589–4602.

Thompson, S.M. and Gähwiler, B.H. (1989) Activity-dependent disinhibition: III. Desensitization and GABA$_B$ receptor mediated presynaptic inhibition in the hippocampus in vitro. *J. Neurophysiol.*, **61**, 524–533.

Thompson, S.M., Deisz, R.A. and Prince, D.A. (1988) Relative contributions of passive equilibrium and active transport to the distribution of chloride in mammalian cortical neurons. *J. Neurophysiol.*, **60**, 105–124.

Van Rijn, C.M., Van Berlo, M.J., Feenstra, M.G.P., Schoofs, M.L.F. and Hommes, O.R. (1987) R(−)-Baclofen: focal epilepsy after intracortical administration in the rat. *Epilepsy Res.*, **1**, 321–327.

Waldmeier, P.C., Wicki, P., Feldtrauer, J.-J. and Baumann, P.A. (1988) Potential involvement of a baclofen-sensitive autoreceptor in the modulation of the release of endogenous GABA from rat brain slices in vitro. *Naunyn-Schmiedeberg's Arch. Pharmacol.*, **337**, 289–295.

Zieglgänsberger, W. and Häuser, M. (1989) The electrophysiology and neuropharmacology of GABA. In: N.G. Bowery and G. Nistico (eds) *GABA: basic research and clinical applications*. Rome: Pythagora Press, pp. 59–101.

15 $GABA_B$-mediated Inhibition of Calcium Currents: A Possible Role in Presynaptic Inhibition

A. C. DOLPHIN, E. HUSTON and R. H. SCOTT

INTRODUCTION

In this paper we shall take as our hypothesis that $GABA_B$-mediated presynaptic inhibition is a result of inhibition of Ca^{2+} channels. We shall examine the properties of $GABA_B$-mediated inhibition of Ca^{2+} channel currents from rat dorsal root ganglion (DRG) neurons in primary culture, and compare them with the properties of $GABA_B$-mediated inhibition of glutamate release from cultured cerebellar neurons.

The examination of somatic events in DRG neurons as a model for presynaptic modulation was initially suggested by Dunlap and Fischbach (1981), who observed that the Ca^{2+}-dependent plateau phase of the DRG action potential was inhibited by several compounds that produced presynaptic inhibition, notably GABA, noradrenaline and opiates. Subsequently it was found that if the Ca^{2+} current in these neurons was isolated pharmacologically by blocking ions flowing through all other channels, these agents inhibited the Ca^{2+} current. The effect of GABA (Dunlap and Fischbach, 1981) was mimicked by baclofen (Scott and Dolphin, 1986). Much effort has subsequently been channelled into investigating the mechanism of this inhibition, and into determining whether it represents a means by which inhibition of transmitter release from presynaptic terminals may occur.

INHIBITION OF CALCIUM CHANNEL CURRENTS AND NEUROTRANSMITTER RELEASE

Involvement of a Pertussis Toxin-sensitive G-protein

The ability of baclofen to inhibit the Ca^{2+}-dependent phase of the DRG action potential is shown in Figure 15.1A. Ca^{2+} channel currents were examined as previously described, using Ba^{2+} rather than Ca^{2+} as the charge carrier (Dolphin and Scott, 1987). Baclofen inhibits Ca^{2+} channel currents in DRGs, and a current–voltage relationship is shown in Figure 15.1B for 100 μM

$GABA_B$ Receptors in Mammalian Function. Edited by N.G. Bowery, H. Bittiger and H.-R. Olpe

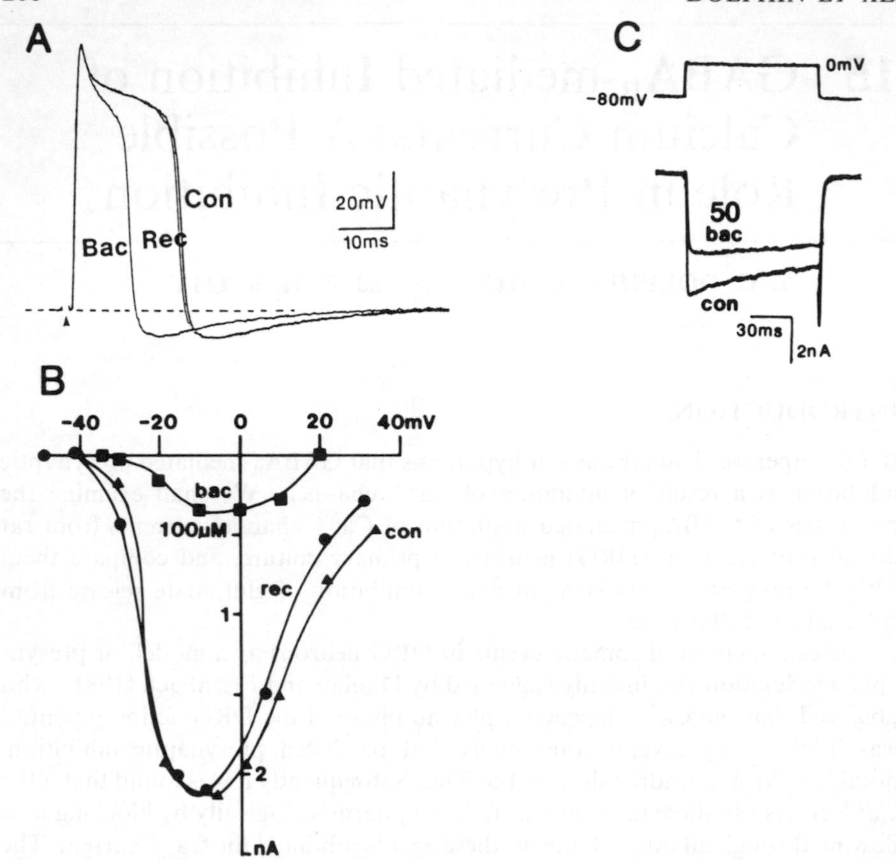

Figure 15.1. The effect of baclofen on dorsal root ganglion action potentials and Ca^{2+} channel current (I_{Ba}). (A) Reduction by 100 μM (−)-baclofen of the action potential duration, with no effect on resting membrane potential. The action potential was evoked by a brief depolarizing pulse (arrowhead), and was prolonged by 2.5 mM tetramethylammonium ions. (B) Current–voltage relationship, showing inhibition by 100 μM baclofen (■) of control Ca^{2+} channel current (▲), with complete recovery at 5 min (●). The current was carried by Ba^{2+} (2.5 mM). All currents are leak subtracted. The holding potential, V_H, was −80 mV. (C) Differential inhibition of the transient I_{Ba} by 50 μM baclofen. The Ca^{2+} channel current was recorded before (con), during and after (rec) the application of baclofen (bac)

baclofen. The median effective concentration (EC_{50}) for (−)-baclofen is 15 μM (Dolphin and Scott, 1987). At submaximal concentrations, baclofen is often, although not always, observed to inhibit differentially the initial transient component of the Ca^{2+} channel current (Figure 15.1C) but at a maximal concentration almost complete inhibition of the current is observed in some cells. The amount of inhibition is variable between cells and between cultures

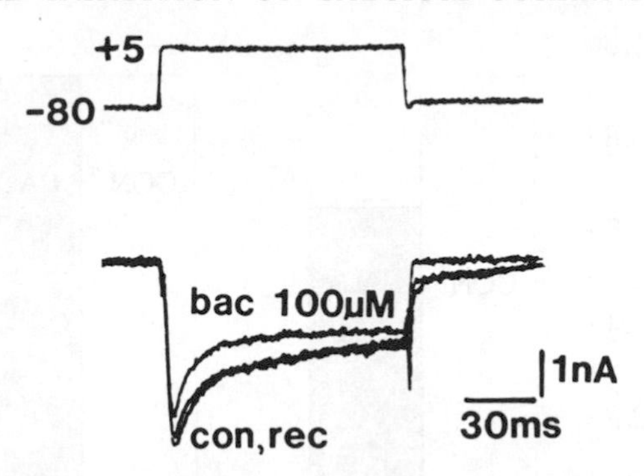

Figure 15.2. The lack of effect of baclofen on Ca^{2+} channel current (I_{Ba}) in dorsal root ganglion (DRG) neurons pretreated with pertussis toxin. DRGs were treated with 0.5 μg/ml pertussis toxin for 2 h at 37 °C. The maximum Ca^{2+} channel current was recorded before (con), during and after (rec) the application of 100 μM baclofen from a 'puffer' pipette. The holding potential, V_H, was -80 mV

and may well depend on culture conditions, and the presence or absence of non-neuronal cells. Some cultures were insensitive to baclofen, but the role of different components of culture media and growth substrates in the coupling of GABA$_B$ receptors to Ca^{2+} currents has yet to be investigated.

The involvement of a GTP binding protein (G-protein) in this process was first suggested from experiments in which non-hydrolysable GTP and GDP analogues were included in the patch pipette. GTPγS enhanced the response to baclofen (Scott and Dolphin, 1986), reducing the EC_{50} to 2 μM. In addition, GDPβS reduced the response by competing with endogenous GTP (Holz *et al.*, 1986). The finding that pertussis toxin prevented the response to baclofen (Figure 15.2) (Dolphin and Scott, 1987; Holz *et al.*, 1986) narrowed down the G-protein involved to one of the G_o- or G_i-subgroups of G-proteins.

Glutamate release from cerebellar granule neurons in culture was examined as previously described (Dophin and Prestwich, 1985). Two periods of 2-min stimulation (S_1, S_2) with 50 mM K^+ were used to depolarize the neurons and stimulate glutamate release. ($-$)-Baclofen (100 μM), when present during S_2, inhibited release by 20–30% (Figure 15.3). Pretreatment of neurons for 16 h with pertussis toxin (500 ng/ml) prevented the baclofen inhibition of glutamate release (Figure 15.3).

We have previously shown the adenosine agonist ($-$)-phenylisopropyladenosine (($-$)-PIA) to inhibit glutamate release in the same system, and this is converted by pertussis toxin to an enhancement of glutamate release by ($-$)-PIA (Dolphin and Prestwich, 1985). No parallel reversal of the effects of baclofen were observed. However, pertussis toxin alone enhanced K^+-stimulated glutamate release by 37 $\pm$ 5% (mean $\pm$ SEM, $n = 17$), confirming the initial report by Dolphin and Prestwich (1985). The mechanism of this

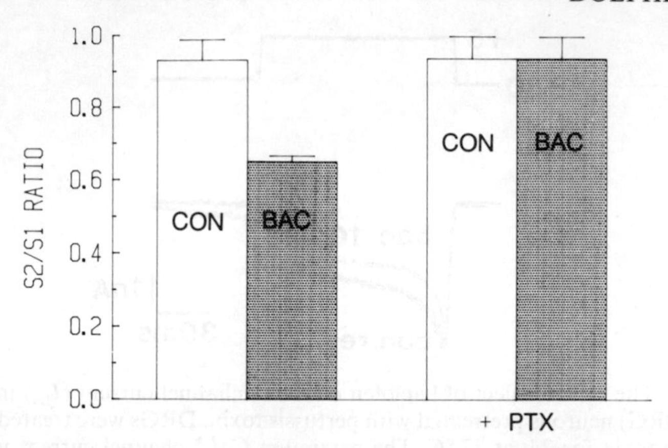

Figure 15.3. The prevention by pertussis toxin pretreatment of baclofen-mediated inhibition of transmitter release. Cerebellar neurons grown on coverslips were preincubated for 1 h with [^{3}H]glutamine (5 μCi) to prelabel a releasable pool of glutamate. [^{3}H]Glutamate release was stimulated by 50 mM K^+ (two 2-min stimulations, S_1 and S_2). Baclofen (BAC, 100 μM), present during S_2 only, produced a marked decrease of glutamate release (as determined by the S_2/S_1 ratio) in control (CON) neurons, but not in those pretreated with pertussis toxin (PTX, 500 ng/ml, for 16 h at 37 °C)

enhancement remains unclear, although it is possible that pertussis toxin is effecting relief of tonic presynaptic inhibition by an endogenous agent such as adenosine.

$GABA_B$ Inhibition of the Different Components of the Calcium Channel Current

Two components of the Ca^{2+} channel current can be recorded in isolation. The T current is a low-threshold transient current available only from hyperpolarized potentials (−90 mV) and activated by small depolarizations. It is present in 45% of our cultured DRGS. This current is partially inhibited by 100 μM (−)-baclofen by 31 ± 7% ($n = 5$; Figure 15.4B) and is enhanced by a lower concentration of (−)-baclofen (2 μM) by 25 ± 4% ($n = 14$; Figure 15.4A) (Scott *et al.*, submitted for publication). This current can be identified pharmacologically, since it is not sensitive to ω-conotoxin, but is inhibited by octanol (Dolphin *et al.*, 1989b; Llinas, 1988). The role of T currents has been suggested to include the control of repetitive and burst firing (Coulter *et al.*, 1989; Llinas, 1988), and a role in the supply of Ca^{2+} for transmitter release has been tentatively proposed (Seabrook and Adams, 1989). The $GABA_B$-mediated inhibition of T currents may represent a means by which neuronal excitability is reduced, but whether there is a physiological counterpart to the $GABA_B$-mediated enhancement of T currents remains to be elucidated.

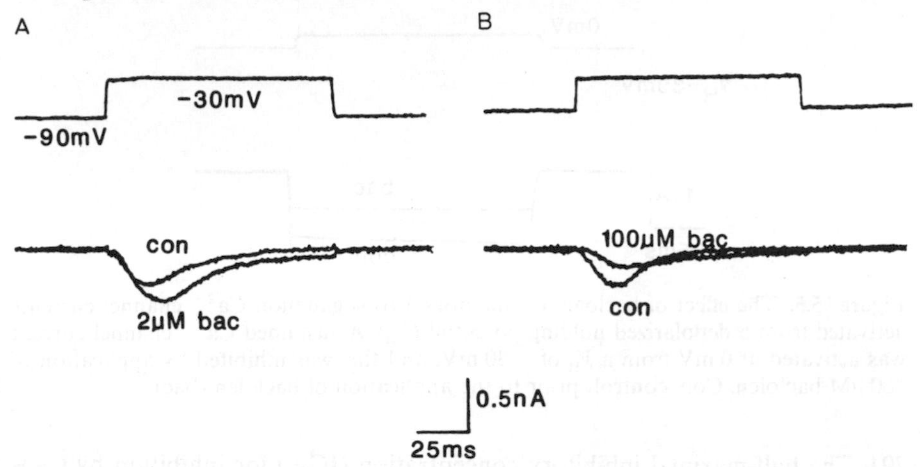

Figure 15.4. The dual effect of baclofen on T currents in dorsal root ganglion neurons. T currents were recorded from a holding potential, V_H, of -90 mV at a clamp potential of -30 mV. They were stable for 5 min, before (con) the application of baclofen (bac). Application of 2 μM baclofen by low-pressure 'puffer' pipette reversibly increased the T current ($n = 14$; A), whereas 100 μM baclofen reversibly decreased the T current ($n = 6$; B)

It is also possible to isolate the sustained current from the transient component of the high-threshold neuronal Ca^{2+} channel current. This may be done by depolarizing the holding potential (V_H) to -30 mV. At this potential a sustained current is still available, whereas the transient currents are inactivated. We have investigated the sensitivity of this current to dihydropyridine antagonists, and observed a $77 \pm 3\%$ inhibition ($n = 4$) by 5 μM (−)-202-791 over 10 min. The current available at -30 mV may represent L-type current, although it is not markedly more sensitive to (−)-202-791 than is the current available from a V_H of -80 mV ($44 \pm 9\%$ inhibition ($n = 6$) by 5 μM(−)-202-791 over 3.5 min; Dolphin and Scott, 1989). In addition, both components are similarly sensitive to ω-conotoxin (approximately 80% inhibition; Scott and Dolphin, unpublished results), indicating that in DRGs this toxin does not appear to distinguish between different components of the high-threshold current. (−)-Baclofen (100 μM) was still able to inhibit the Ca^{2+} channel current (I_{Ba}) activated at 0 mV from a V_H of -30 mV (Figure 15.5) (Dolphin and Scott, in press).

A parallel experiment on transmitter release has been performed by maintaining the cerebellar neurons under depolarized conditions, in Ca^{2+}-free medium containing 50 mM K^+. Glutamate release was then stimulated with 5 mM Ca^{2+}. Under these conditions the release of glutamate was entirely (>90%) sensitive to blockade by the dihydropyridine Ca^{2+} channel antagonist (−)-202-

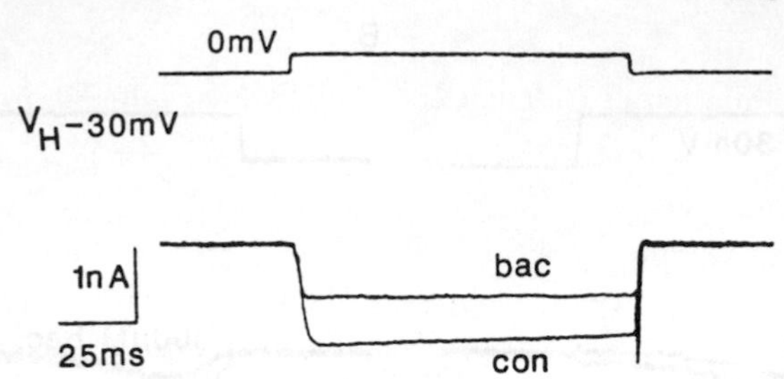

Figure 15.5. The effect of baclofen on the dorsal root ganglion Ca^{2+} channel currents activated from a depolarized holding potential (V_H). A sustained Ca^{2+} channel current was activated at 0 mV from a V_H of −30 mV, and this was inhibited by application of 100 μM baclofen. Con, control: prior to the application of baclofen (bac)

791. The half-maximal inhibitory concentration (IC_{50}) for inhibition by (−)-202-791 was 1 nM, a potency comparable with its effect on L-type Ca^{2+} currents in smooth muscle cells (Hof *et al.*, 1985). Baclofen (100 μM) inhibited glutamate release by a similar extent, under these conditions, to its effect in normal medium (27 ± 5 % inhibition, $n = 4$; e.g. Figure 15.6).

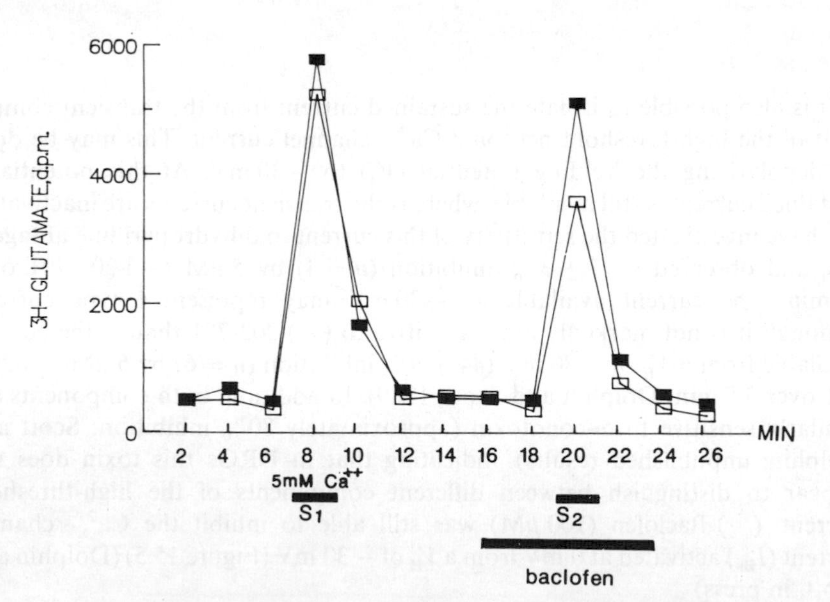

Figure 15.6. The effect of baclofen on the release of glutamate from depolarized cerebellar neurons. Cerebellar neurons were prelabelled with [^{3}H]glutamine (as in legend to Figure 15.3), and were maintained in Ca^{2+}-free Krebs medium containing 50 mM K^+ (replacing Na^+). Release of [^{3}H]glutamate was stimulated with two periods (S_1, S_2) of medium containing 5 mM Ca^{2+}. Baclofen (100 μM), present during S_2, markedly decreased release (□), compared to control (■)

Mechanism of Inhibition of Calcium Currents by Neuromodulators

We have examined the ability to prevent the effect of baclofen or GTPγS of different agents that either mimic or inhibit various second messenger systems.

Neither cyclic AMP nor forskolin prevented the effect of baclofen (Dolphin *et al.*, 1989a). However, some enhancement of neuronal Ca^{2+} channel currents by forskolin and cholera toxin was observed (Dolphin *et al.*, 1989a) although this was much less marked than that reported for cardiac cells (Kameyama *et al.*, 1986).

The evidence concerning the obligatory mediation of protein kinase C (PKC) in the inhibition of Ca^{2+} currents remains equivocal. Several studies have failed to show either an effect on Ca^{2+} currents of PKC activators or the ability of inhibitors of PKC to prevent the response of Ca^{2+} currents to neurotransmitters or GTP analogues (Brezina *et al.*, 1987; Dolphin *et al.*, 1989a; McFadzean and Docherty, 1989; Wanke *et al.*, 1987). In contrast, Rane and Dunlap (1986) observed phorbol esters and oleoylacetylglycerol to inhibit Ca^{2+} currents in chick DRGs. They have also shown specific peptide inhibitors of PKC to prevent inhibition of Ca^{2+} currents by noradrenaline (Rane *et al.*, 1987). The neurotransmitters and neuromodulators that inhibit Ca^{2+} currents are not normally considered to be in the class of 'Ca^{2+}-mobilizing' agonists that activate phospholipase C (PLC). However, we have found (−)-baclofen (100 μM) to produce a small but significant increase in total inositol phosphate production in 30 s (Dolphin *et al.*, 1989a). For comparison, the enhancement by bradykinin (1 μM) was about 70% (Figure 15.7). The effect of bradykinin on I_{Ba} was thus examined and compared to that of baclofen. It was found to have no effect on the majority of cells, but in 7/16 cells it produced a marked increase of

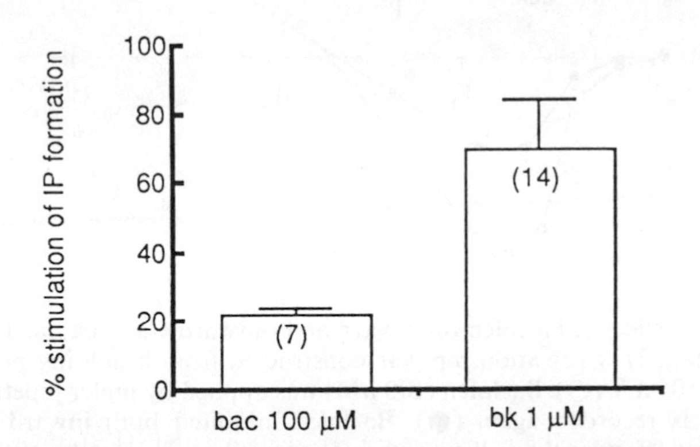

Figure 15.7. A comparison of the effect of 100 μM baclofen and 1 μM bradykinin on inositol phosphate production in cultured dorsal root ganglion neurons. Total inositol phosphate (IP) production was determined as described by Dolphin et al. (1989a). Stimulation with either baclofen (bac) or bradykinin (bk) was for 30 s

I_{Ba} by 72 ± 16% (Dolphin *et al.*, 1989a; McGuirk and Dolphin, unpublished results). In no case was bradykinin observed to mimic the effect of baclofen. These results do not provide evidence that activation of PLC is required for inhibition of Ca^{2+} channel currents by $GABA_B$ receptor activation.

Voltage Dependence of G-protein-mediated Inhibition of Calcium Channel Currents

It has recently been observed by Bean (1989) that noradrenaline did not inhibit frog DRG Ca^{2+} channel currents activated by large depolarizations. Under these conditions, the current flowing was outward due to Cs^+ passing out of the cell through Ca^{2+} channels.

In contrast we have observed that the outward Ca^{2+} channel current activated by large depolarizations from −80 mV to +100 or +120 mV for 100 ms is inhibited by baclofen although to a smaller extent than the inhibition of the maximum inward I_{Ba} (Figure 15.8). In these experiments the outward

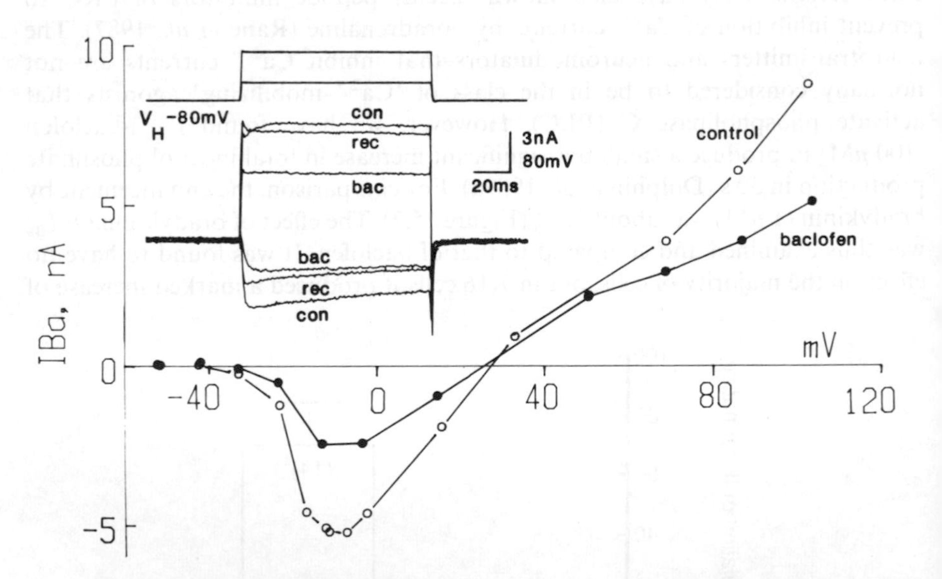

Figure 15.8. The effect of baclofen on inward and outward Ca^{2+} channel currents (I_{Ba}). A current–voltage (I–V) relationship was constructed from a holding potential, V_H, of −80 mV to +105 mV (○). Baclofen (100 μM) was applied by puffer pipette and the I–V relationship was recorded again (●). Baclofen inhibited both inward and outward currents without affecting the null potential. The inset traces show the maximum inward and outward currents before (con), and during, baclofen (bac) application, and following partial recovery (rec) after 5 min. Note that the activation of the inward current is slowed by baclofen, as well as being reduced in amplitude, whereas baclofen did not slow the outward current

current was inhibited by 33 ± 8% ($n = 9$) by 100 μM baclofen, and in the same cells I_{Ba} was inhibited by 45 ± 5% ($n = 9$). It is also clear from Figure 15.8 that, although the activation of the inward current was slowed by baclofen, the outward current, while reduced in amplitude, was rapidly activating. The effect of GTPγS to slow the activation of the Ca^{2+} channel current also persists during step depolarization to +100 mV although again the activation of the outward current is slowed to a lesser extent than the inward I_{Ba} (Dolphin and Scott, 1990). This finding suggests that a component of G-protein modulation of Ca^{2+} channel currents by GABA$_B$ receptor activation is able to persist upon extreme step depolarizations.

Another recent finding by Grassi and Lux (1989) is that a prepulse to +40 mV delivered before activation of the maximum inward Ca^{2+} current prevented inhibition of the DRG Ca^{2+} current by GABA. We have confirmed that the kinetic effect of GTPγS on the rate of activation of I_{Ba} (Dolphin *et al.*, 1989b) is abolished by such a prepulse (Figure 15.9) although the current amplitude remains reduced compared to the control current (Scott and Dolphin, 1990). This result suggests that agonists and GTPγS activate a G-protein

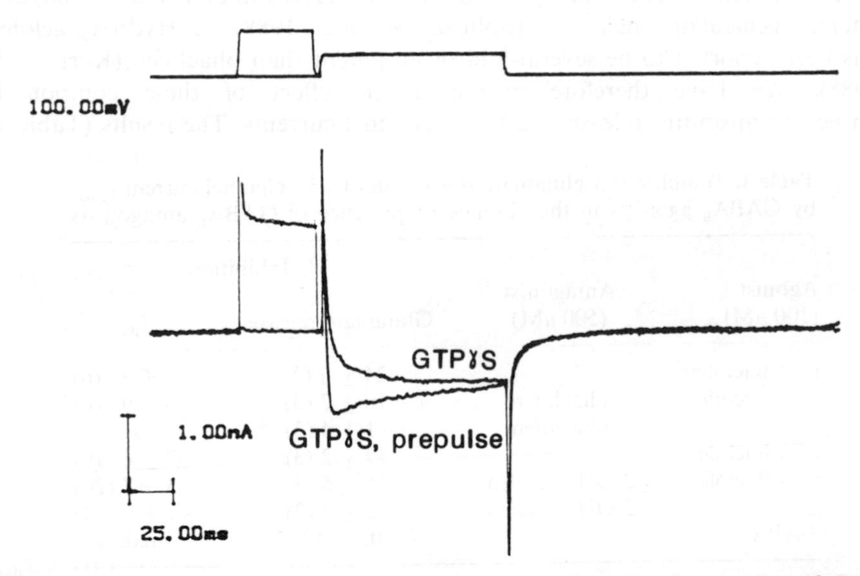

Figure 15.9. The effect of a prepulse to +40 mV on the rate of activation of Ca^{2+} channel current (I_{Ba}) in the presence of the GTP analogue GTPγS. The maximum Ca^{2+} channel current in the presence of 200 μM internal GTPγS was activated from a holding potential, V_H, of −80 mV by a 100 ms voltage step. A slowly activating current was recorded. A prepulse to +40 mV (near the null potential) for 40 ms, terminating 5 ms before activating the maximum inward current, caused expression of a rapidly activating I_{Ba}, by removal of a voltage-dependent block of the Ca^{2+} channels, presumably due to activated G-protein. Currents are not leak subtracted

which interacts with the Ca^{2+} channel, reducing its rate of activation upon depolarization. The slow rate of activation represents gradual voltage-dependent recovery from this blockade. Thus a prepulse to near the null potential causes prior relief of blockade, allowing the inward current to be activated immediately upon subsequent depolarization.

The residual current in the presence of agonist represents both current whose kinetics have been modified by G-protein activation and current which is not affected by agonist.

Pharmacology of the $GABA_B$ Receptors Associated with Inhibition of Calcium Channel Currents and Transmitter Release

Until recently there have been no agonists other than GABA and baclofen, and no antagonists available for the $GABA_B$ receptor. However, the agonist 3-aminopropylphosphinic acid (3-APA) has recently been reported to be a more potent agonist than baclofen at peripheral $GABA_B$ receptors (Hills *et al.*, 1989). The antagonist phaclofen (Kerr *et al.*, 1987) has been found to antagonize the baclofen-activated K^+ conductance and the $GABA_B$-mediated inhibitory postsynaptic potential (IPSP) in hippocampal neurons (Dutar and Nicoll, 1988) and lateral geniculate neurons (Soltesz *et al.*, 1988). 2-Hydroxysaclofen has been reported to be several-fold more potent than phaclofen (Kerr *et al.*, 1988). We have therefore examined the effect of these compounds on neurotransmitter release and Ca^{2+} channel currents. The results (Table 1)

Table 1. Inhibition of glutamate release and Ca^{2+} channel current (I_{Ba}) by $GABA_B$ agonists in the absence or presence of $GABA_B$ antagonists

Agonist (100 μM)	Antagonist (500 μM)	% Inhibition	
		Glutamate release	I_{Ba}
(−)-Baclofen	—	24 ± 6 (3)	36 ± 3 (6)
(−)-Baclofen	phaclofen	10 ± 7 (3)	6 ± 9 (6)
—	phaclofen	−1 ± 4 (3)	n.d.
(−)-Baclofen	—	19 ± 2 (3)	37 ± 6 (9)
(−)-Baclofen	2-OH-saclofen	25 ± 5 (3)	13 ± 7 (20)
—	2-OH-saclofen	−2 ± 7 (3)	−1 ± 9 (4)
3-APA	—	− 10, − 12 (2)	n.d.

Values are means ± SEM; *n* is given in parentheses.

For examination of the effect of agents on glutamate release from cerebellar neurons, either agonist alone or together with antagonist was present for 4 min prior to, and during, 2-min depolarization (S_2, see legend to Figure 15.3).

For examination of their effects on I_{Ba} in dorsal root ganglion neurons, agents were administered either alone or together by low-pressure application from a 'puffer' pipette.

The two sets of experiments were performed on the same batch of neurons.

3-APA, 3-aminopropylphosphinic acid.

show that 3-APA did not mimic the agonist action of (−)-baclofen and, although 500 μM phaclofen partially inhibited glutamate release and I_{Ba}, 2-hydroxysaclofen at the same concentration was less effective than phaclofen at preventing (−)-baclofen inhibition of I_{Ba}, and did not prevent (−)-baclofen from inhibiting glutamate release (Huston *et al.*, in press).

These results indicate that the GABA$_B$ receptors mediating inhibition of Ca^{2+} channel currents and inhibition of glutamate release show marked similarities, and may differ from those activating K^+ currents.

CONCLUSION

GABA$_B$ inhibition of Ca^{2+} currents and transmitter release is mediated in both cases by activation of a pertussis toxin-sensitive G-protein. We have no evidence that a second messenger is involved. It is likely that there is a voltage-dependent component to the inhibition of the current, which would have the physiological consequence that GABA$_B$-mediated inhibition could be partly overcome by depolarization. It also appears that GABA$_B$ receptors mediating inhibition of transmitter release and inhibition of Ca^{2+} channel currents in DRGs may have a pharmacology that is different from the GABA$_B$ receptors associated with activation of K^+ channels, and from peripheral GABA$_B$ receptors.

Identification of the type(s) of channel involved in GABA$_B$-mediated presynaptic inhibition remains equivocal. Because transmitter release is not normally sensitive to dihydropyridine antagonists, it has been said that N rather than L channels supply Ca^{2+} for release at the presynaptic terminal (for review, see Tsien *et al.*, 1988). However, because dihydropyridines interact with Ca^{2+} channels in a voltage-sensitive manner, this is not an ideal criterion. Indeed, the finding of the extreme dihydropyridine sensitivity of glutamate release in depolarized neurons suggests that L channels are present at release sites and are able to support transmitter release; and this remains sensitive to baclofen. Thus it may be the case that all types of Ca^{2+} channel are capable of coupling to GABA$_B$ receptors, possibly by different mechanisms, and that they may all support transmitter release, given suitable conditions.

Unfortunately there are insufficient tools available to distinguish between the different types of Ca^{2+} channels in neurons; and it is likely that in the many types of neuron making up the central and peripheral nervous system, the division into low-threshold, and high-threshold dihydropyridine-sensitive and -insensitive channels hides a further complexity.

REFERENCES

Bean, B.P. (1989) Neurotransmitter inhibition of neuronal calcium currents by changes in channel voltage-dependence. *Nature*, **340**, 153–155.

Brezina, V., Eckert, R. and Erxleben, C. (1987) Suppression of calcium current by an endogenous neuropeptide in neurones of *Aplysia californía. J. Physiol.*, **388**, 565–596.

Coulter, D.A., Huguenard, J.R. and Prince, D.A. (1989) Specific petit mal anticonvulsants reduce calcium currents in thalamic neurones. *Neurosci. Lett.*, **98**, 74–78.

Dolphin, A.C. and Prestwich, S.A. (1985) Pertussis toxin reverses adenosine inhibition of neuronal glutamate release. *Nature*, **316**, 148–150.

Dolphin, A.C. and Scott, R.H. (1987) Calcium channel currents and their inhibition by (−)-baclofen in rat sensory neurones: modulation by guanine nucleotides. *J. Physiol.*, **386**, 1–17.

Dolphin, A.C. and Scott, R.H. (1989) Interaction between calcium channel ligands and guanine nucleotides in cultured rat sensory and sympathetic neurones. *J. Physiol.*, **413**, 271–288.

Dolphin, A.C. and Scott, R.H. (1990) Activation of calcium channel currents in rat sensory neurones by large depolarizations: effect of guanine nucleotides and (−)-baclofen. *Eur. J. Neurosci.*, **2**, 104–108.

Dolphin, A.C., McGuirk, S.M. and Scott, R.H. (1989a) An investigation into the mechanisms of inhibition of calcium channel currents in cultured sensory neurones of the rat by guanine nucleotide analogues and (−)-baclofen. *Br. J. Pharmacol.*, **97**, 263–273.

Dolphin, A.C., Scott, R.H. and Wooton, J.H. (1989b) Photo-release of GTP-γ-S inhibits a low threshold calcium channel current in cultured rat DRG neurones. *J. Physiol.*, **410**, 16P.

Dunlap, K. and Fischbach, G.D. (1981) Neurotransmitters decrease the calcium conductance activated by depolarisation of embryonic chick sensory neurones. *J. Physiol.*, **317**, 519–535.

Dutar, P. and Nicoll, R.A. (1988) A physiological role for $GABA_B$ receptors in the central nervous system. *Nature*, **332**, 156–158.

Grassi, F. and Lux, H.D. (1989) Kinetic and voltage-dependence of GABA-induced modulation of Ca currents in chick DRG cells. *Eur. J. Neurosci.*, Suppl., 2.44.1.

Hills, J.M., Dingsdale, R.A., Parsons, M.E., Dolle, R.E. and Howson, W. (1989) 3-Amino propyl-phosphinic acid: a potent selective $GABA_B$ receptor agonist in the guinea pig ileum and rat anococcygeus muscle. *Br. J. Pharmacol*, **97**, 1292–1296.

Hof, R.P., Rüegg, U.T., Hof, A. and Vogel, A. (1985) Stereoselectivity at the calcium channel: opposite action of the enantiomers of a 1,4-dihyropyridine. *J. Cardiovasc. Pharmacol.*, **7**, 689–693.

Holz, G.G., Rane, S.G. and Dunlap, K. (1986) GTP binding proteins mediate transmitter inhibition of voltage-dependent calcium channels. *Nature*, **319**, 670–672.

Huston, E., Scott, R.H. and Dolphin, A.C. (1990) A comparison of the effect of calcium channel ligands and $GABA_B$ agonists and antagonists on transmitter release and somatic calcium channel currents in cultured neurones. *Neuroscience* (in press).

Kameyama, M., Hescheler, J., Hofmann, F. and Trauturein, W. (1986) Modulation of Ca current during the phosphorylation cycle in the guinea pig heart. *Pflüger's Arch.*, **407**, 123–128.

Kerr, D.I.B., Ong, J., Prager, R.H., Gynther, B.D. and Curtis, D.R. (1987) Phaclofen: a peripheral and central baclofen antagonist. *Brain Res.*, **405**, 150–154.

Kerr, D.I.B., Ong, J., Johnston, G.A.R., Abbenante, J. and Prager, R.H. (1988) 2-Hydroxy-saclofen: an improved antagonist at central and peripheral $GABA_B$ receptors. *Neurosci. Lett.*, **92**, 92–96.

Llinas, R. (1988) The intrinsic electrophysiological properties of mammalian neurons: insights into central nervous system function. *Science*, **242**, 1654–1664.

McFadzean, I. and Docherty, R. (1989) Noradrenaline and enkephalin-induced inhibition of voltage-sensitive calcium currents in NG 108-15 hybrid cells: transduction mechanisms. *Eur. J. Neurosci.*, **1**, 141–150.

Rane, S.G. and Dunlap, K. (1986) Kinase C activator 1,2-oleoylacetylglycerol attenuates voltage-dependent calcium current in sensory neurones. *Proc. Natl Acad. Sci. USA*, **83**, 184–188.

Rane, S.G., Walsh, M.P. and Dunlap, K. (1987) Norepinephrine inhibition of sensory neuron calcium current is blocked by a specific protein kinase C inhibitor. *Soc. Neurosci. Abstr.*, **13**, 557.

Scott, R.H. and Dolphin, A.C. (1986) Regulation of calcium currents by a GTP analogue: potentiation of (−)-baclofen mediated inhibition. *Neurosci. Lett.*, **56**, 59–64.

Scott, R.H. and Dolphin, A.C. (1990) Voltage-dependent modulation of rat sensory neurone calcium channel currents by G protein activation; effect of a dihydropyridine antagonist. *Br. J. Pharmacol.*, **99**, 629–630.

Seabrook, G.R. and Adams, D.J. (1989) Inhibition of neurally-evoked transmitter release by calcium channel antagonists in rat para-sympathetic ganglia. *Br. J. Pharmacol.*, **97**, 1125–1136.

Soltesz, I., Haby, M., Leresche, N. and Crunelli, V. (1988) The GABA$_B$ antagonist phaclofen inhibits the late K^+-dependent IPSP in cat and rat thalamic and hippocampal neurones. *Brain Res.*, **448**, 351–354.

Tsien, R.W., Lipscombe, D., Madison, D.V., Bley, K.R. and Fox, A.P. (1988) Multiple types of neuronal calcium channels and their selective modulation. *Trends Neurosci.*, **11**, 431–437.

Wanke, E., Ferroni, A., Malgaroli, A., Ambrosini, A., Pozzan, T. and Meldolesi, J. (1987) Activation of a muscarinic receptor selectively inhibits a rapidly inactivated Ca^{2+} current in rat sympathetic neurones. *Proc. Natl Acad. Sci. USA*, **884**, 4313–4317.

16 Physiological and Pharmacological Characterization of $GABA_B$ Receptor-mediated Potassium Conductance

N. OGATA

INTRODUCTION

Baclofen, a β-chlorophenyl derivative of GABA, depresses neuronal excitability in various parts of the central nervous system (CNS) (Pierau and Zimmermann, 1973). The site of action for this drug had once been considered to be distinct from GABA recognition sites (Johnston, 1978). In addition to the classical GABA recognition site ($GABA_A$ site), a new class of GABA receptor ($GABA_B$ site) has been characterized on the basis of pharmacological criteria (Bowery *et al.*, 1980). When $GABA_B$ sites on nerve terminals are activated, diminished transmitter release results, probably through a reduction in Ca^{2+} influx (Dunlap, 1981). Baclofen was shown to be a selective agonist for this novel $GABA_B$ recognition site (Bowery *et al.*, 1980).

It has been shown that baclofen directly hyperpolarizes the membrane of guinea-pig hypothalamic neurons (Ogata and Abe, 1982), in addition to its presynaptic action. When studied in hippocampal pyramidal neurons, this postsynaptic action of baclofen was shown to result from an increase in K^+ conductance (Inoue *et al.*, 1985a; Newberry and Nicoll, 1984a). GABA also activates K^+ conductance in hippocampal neurons (Gähwiler and Brown, 1985; Inoue *et al.*, 1985c; Newberry and Nicoll, 1985). This chapter will review our recent studies on the cellular mechanisms related to K^+ conductance induced by GABA and baclofen.

BACLOFEN-INDUCED HYPERPOLARIZATION

The first systematic study of the postsynaptic action of baclofen was made by Ogata and Abe (1982). Baclofen hyperpolarized the membrane and reduced the input resistance of neurons in the guinea-pig hypothalamus. Since the reversal potential was close to the equilibrium potential for K^+, it was suggested that baclofen activates postsynaptic K^+ conductance. Several years later, we reinvestigated the postsynaptic action of baclofen in more detail using slices of the

$GABA_B$ Receptors in Mammalian Function. Edited by N.G. Bowery, H. Bittiger and H.-R. Olpe

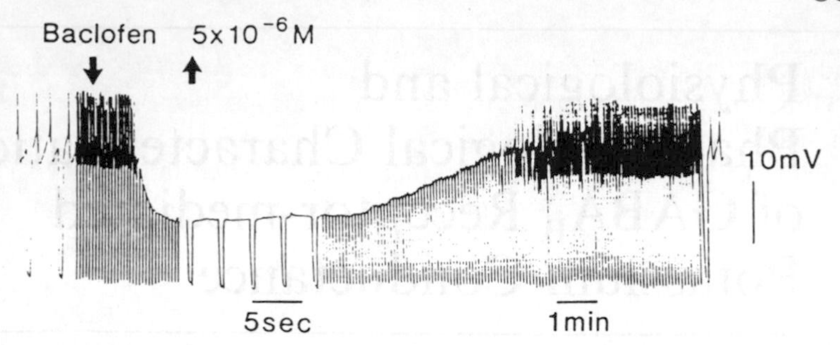

Figure 16.1. Effects of baclofen on the electrical activity of cells in slices of guinea-pig hippocampus recorded with an intracellular microelectrode. In this and subsequent figures recordings were made from CA3 pyramidal cells; repetitive negative deflections reflect the electrotonic potentials to inward current injections (0.3 Hz, 0.6-s pulse duration) of constant intensity for measurement of input resistance; downward and upward arrows represent the onset and offset of superfusion of test solution; upward deflection represents positive polarity; spikes were almost entirely eliminated from the traces due to the limit frequency bandwidth of the pen-recorder; the chart of the pen-recordings was intermittently run at a faster speed

guinea-pig hippocampus. Figure 16.1 shows a typical example of the hyperpolarizing action of baclofen in the hippocampus. Baclofen produces consistent hyperpolarization and a decrease in the input resistance as has been observed in the hypothalamus. The effects of baclofen persisted during a perfusion of Ca^{2+}-free medium, confirming its postsynaptic nature.

The concentration–response relations showed that the minimal effective concentration was approximately 0.1 μM, and ED_{50} was 1.1 μM. When the spatial responsiveness of the pyramidal cell to ionophoretically applied baclofen was studied, the dendritic region produced larger and more prolonged hyperpolarization than did the somatic region. This indicates that baclofen may act primarily in the dendrites.

IONIC MECHANISM UNDERLYING BACLOFEN-INDUCED HYPERPOLARIZATION

To examine a possible involvement of K^+ conductance, we studied the effect of external K^+ concentration on the baclofen response. The amplitude of hyperpolarization produced by baclofen in standard medium (K^+ concentration, 6.24 mM) was about 1.5 times increased in a low-K^+ (1.24 mM) medium. The reversal potential for baclofen-induced hyperpolarization shifted towards the hyperpolarizing direction, as expected from the Nernst equation. However, the amplitude of the hyperpolarization measured in high K^+ concentrations decreased much more steeply than expected from the Nernst equation. This was due not to changes in the equilibrium potential for K^+ but rather to an inward rectifying property of baclofen-induced K^+ conductance (see later). The baclofen-induced hyperpolarization was almost totally abolished in 25 mM external K^+. Thus, an ion-substitution experiment failed to demonstrate the ionic dependence of baclofen-induced hyperpolarization.

Conclusive evidence for the ionic nature of the baclofen-induced hyperpolarization came from an experiment in which the reversal potential for the hyperpolarization was measured in comparision with that for the Ca^{2+}-activated K^+ conductance, which was manifested as a slow afterhyperpolarization (AHP) (Alger and Nicoll, 1980; Hotson and Prince, 1980) subsequent to the burst discharges triggered by the afferent fibre stimulation. The value of the reversal potential for the baclofen-induced hyperpolarization approximated that of the reversal potential for the AHP, thus confirming that the baclofen-induced hyperpolarization is brought about by an increase in K^+ conductance.

The value of the reversal potential for the baclofen-induced hyperpolarization was considerably (about 20 mV) more negative than the equilibrium potential for K^+ (Inoue *et al.*, 1985a). The reason for this might be considered as follows. If the cell has no processes as shown in Figure 16.2A, the entire membrane can be clamped to the same potential levels, and thus, if V is held to the reversal

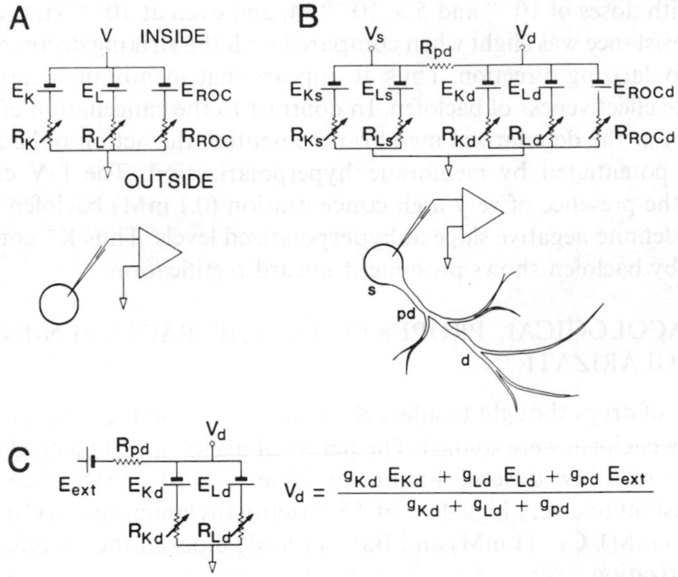

Figure 16.2. Leaky-condensor model representing the neuronal membrane. (A) A cell with no processes. (B) A cell with rich dendritic processes. (C) An equation for calculating the dendritic membrane potential. V_s and V_d represent somatic and dendritic membrane potentials, respectively, E_K and E_L represent equilibrium potentials for K^+ and other miscellaneous conductances, respectively. E_{ROC} represents the reversal potential for receptor-operated channels. E_{ext} represents the sum of batteries generated in the somatic region. Small characters 's', 'pd' and 'd' represent soma, proximal dendrite and distal dendrite, respectively

potential for receptor-operated channels (E_{ROC}), the membrane potential does not shift when ROC are activated, thus providing an accurate value of the reversal potential. However, if the cell has a large dendritic tree and if ROC are predominantly located on the remote dendrites as shown in Figure 16.2B, the situation becomes complicated. If the conductance increase induced by ROC in the dendrite (ROCd) is strong enough to clamp the dendritic potential, V_d, to its reversal potential (E_{ROCd}), the potential measured at the somatic membrane (V_s) gives the correct value for the reversal potential for ROC, because in such a case there is no current passing through the resistor R_{pd} when V_s is held to E_{ROCd}. However, if the contributions of E_{Kd} and E_{Ld} (E_K and E_L in the dendrite, respectively) are not negligible, V_d does not attain E_{ROCd} even when ROCd are fully activated, and thus to bring the V_d to the E_{ROCd}, V_s should be kept at potentials more negative than the E_{ROCd} (Figure 16.2C).

PHYSIOLOGICAL PROPERTY OF THE BACLOFEN-INDUCED HYPERPOLARIZATION

Measurements taken during applications of baclofen showed that the slope of the current–voltage (I–V) curve in the depolarizing direction apparently increased with doses of 10^{-6} and 5×10^{-6} M, and even at 10^{-4} M, the decrease in input resistance was slight when compared with the striking decrease noted in the hyperpolarizing direction. Thus, it appears that membrane depolarization reduces the effectiveness of baclofen. In contrast to the cancellation of the effect of baclofen at the depolarized membrane potential, the action of baclofen was strikingly potentiated by membrane hyperpolarization. The I–V curves obtained in the presence of very high concentration (0.1 mM) baclofen showed a small but definite negative slope at hyperpolarized levels. Thus K^+ conductance activated by baclofen shows prominent inward rectification.

PHARMACOLOGICAL PROPERTY OF THE BACLOFEN-INDUCED HYPERPOLARIZATION

The effects of drugs thought to affect K^+ conductance on the hyperpolarization induced by baclofen were studied. The action of baclofen was blocked during an application of a low dose of 4-aminopyridine (5×10^{-6} M), whereas it was totally resistant to a very high dose of TEA (tetraethylammonium) (10 mM) and procaine (5 mM). Cs^+ (1 mM) and Ba^{2+} (1 mM) blocked the baclofen-induced hyperpolarization.

The action of baclofen was unaffected in a medium containing picrotoxin (2×10^{-5} M). Bicuculline in relatively high doses over 10 μM antagonized the hyperpolarizing action of baclofen. This antagonism became progressively more apparent with time (full antagonism was usually attained after a perfusion of bicuculline of at least 5–10 min prior to the testing of the antagonism), and therefore did not appear to be due to specific antagonism at the receptor site.

Bicuculline methiodide, a water-soluble analogue of bicuculline, had essentially the same effect as bicuculline, although its potency was about one-tenth that of bicuculline. Therefore, the postsynaptic action of baclofen was 'resistant' rather than 'insensitive' to bicuculline.

EFFECTS OF BACLOFEN ON SYNAPTIC TRANSMISSION

As originally proposed, GABA$_B$ sites are present on nerve terminals and their activation results in diminished transmitter release, probably through a blockade of Ca^{2+} channels. Therefore, we tested whether the depressant action of baclofen on synaptic transmission was related to its postsynaptic action.

It has been reported that baclofen preferentially depresses transmission at synapses made by axons of CA3 pyramidal cells (Ault and Nadler, 1983). Therefore, we examined the effect of baclofen on evoked potentials in two synaptic connections within the hippocampus: one, between the Schaffer collateral/commissural fibre and the CA1 pyramidal cell; the other, between the mossy fibre and the CA3 pyramidal cell.

Stimulation of the hippocampal afferents generally evokes a sequence of events in both CA1 and CA3 pyramidal cells comprising (1) an excitatory postsynaptic potential (EPSP) which often provokes an action potential, (2) a fast-hyperpolarizing inhibitory postsynaptic potential (IPSP) which peaks at a latency of about 50 ms, and (3) a slow IPSP (the late hyperpolarizing potential) which peaks at about 200 ms. Baclofen in doses lower than 10 μM augmented the amplitude of the EPSP, whereas it totally suppressed the fast and slow IPSPs in both synaptic pathways. On the contrary, doses over 10 μm produced dose-dependent suppression of the EPSP only in the Schaffer collateral/commissural pathway. This was not due to membrane hyperpolarization or suppression of the action potential of the presynaptic fibres. Thus, baclofen preferentially depresses transmission at synapses made by axons of CA3 pyramidal cells in the hippocampus through a mechanism distinct from the action on the postsynaptic membrane.

Taken together, it appears that, in the hippocampus, baclofen at relatively low concentrations activates postsynaptic receptors which are linked to K^+ channels and are distributed on the pyramidal cells ubiquitously within the hippocampus, while baclofen at higher concentrations also activates presynaptic receptors which are preferentially distributed on the CA3 axon terminals.

FUNCTIONAL SIGNIFICANCE OF THE POSTSYNAPTIC ACTION OF BACLOFEN

Figure 16.3 shows the effects of baclofen on excitations induced by several neurotransmitters in the hypothalamus. The excitation induced by substance P was antagonized by a low concentration of baclofen, whereas excitations

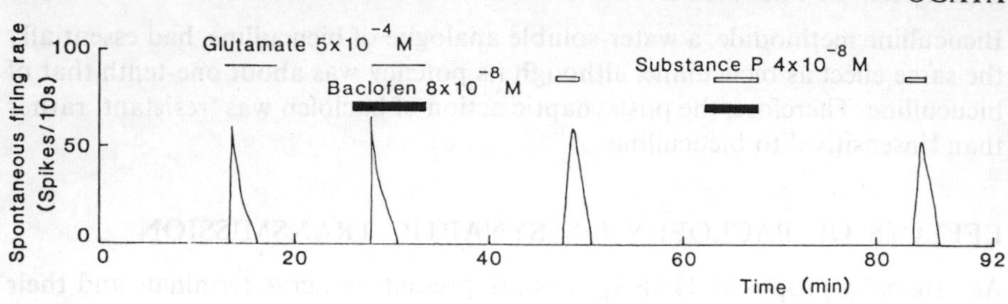

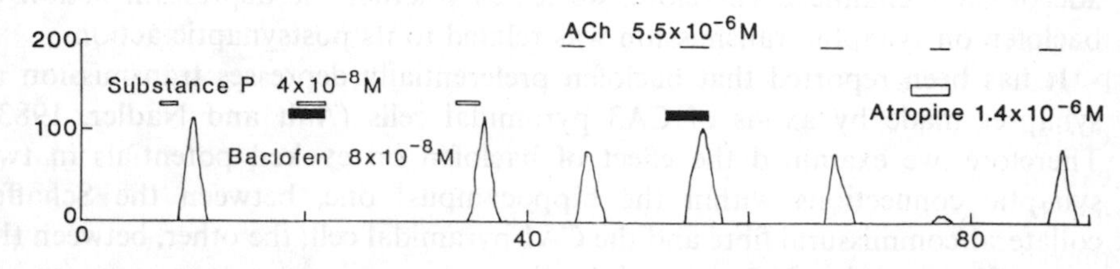

Figure 16.3. Baclofen selectively blocks the action of substance P but enhances the action of L-glutamate and acetylcholine (ACh) in the anterior hypothalamus. Excitation induced by ACh was readily blocked by atropine. Spontaneous extracellular unit discharges were recorded on magnetic tape, digitized by an A-D converter, and processed by a computer for compilation of the firing rate. Drugs were applied at periods indicated by bars. To exclude an involvement of synaptic events, the medium was Ca^{2+}-free and contained 12 mM Mg^{2+}

induced by acetylcholine or glutamate were not blocked (see Ogata and Abe, 1981). Baclofen was once reported by Saito *et al.* (1975) to be a 'specific' antagonist of substance P. Such a view was reasonable in the light of this remarkable selective action of baclofen on the substance P-induced excitation. However, several subsequent investigations have failed to confirm the specificity of antagonism between substance P and baclofen.

Recently, it has been reported that substance P blocks inward rectifying K^+ conductance through mechanism involving pertussis toxin-insensitive G-protein. In the hypothalamus, we have shown that baclofen increases inward rectifying K^+ conductance, whereas substance P suppresses K^+ conductance (Ogata and Abe, 1982). These mirror-image actions further suggest an important interaction between substance P and baclofen. Since such an interaction is not due to competitive antagonism at the receptor site (Hanley *et al.*, 1980), a possible explanation is that the same K^+ channel is reciprocally linked to these excitatory and inhibitory receptors mediated by two different G-proteins (Figure 16.4). Substance P blocks the inward rectifying K^+ channel through pertussis toxin-insensitive G-protein (Nakajima *et al.*, 1988), while baclofen increases the opening of this K^+ channel through pertussis toxin-sensitive G-protein (Andrade *et al.*, 1986).

Action potentials occurring spontaneously, evoked by direct membrane

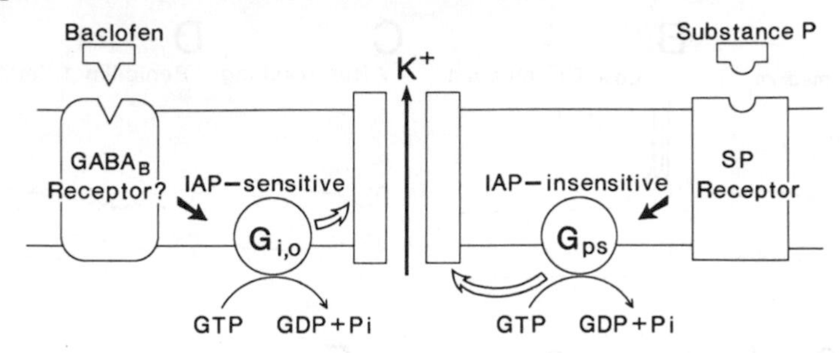

Figure 16.4. Possible reciprocal modulation of the inward rectifying K^+ channel by baclofen and substance P, mediated by separate G-proteins. IAP, islet-activating pertussis toxin. For explanation, see text

current injections or by anode break excitation, were blocked by baclofen at the extremely low concentration of 1 nM (Inoue *et al.*, 1985b), and this depressant action was due to the postsynaptic effect of baclofen. At synapses devoid of presynaptic GABA$_B$ receptors, baclofen augments the excitatory post-synaptic potential (Inoue *et al.*, 1985b; Ogata *et al.*, 1986). Even at synapses involving presynaptic GABA$_B$ receptors, baclofen in concentrations lower than 10 μM, which is sufficient to cause a postsynaptic effect, produces augmentation of synaptic transmission. Thus, baclofen may be particularly effective in improving the signal to noise ratio for excitatory synaptic transmission by reducing background (intrinsic) activity. Such an effect is evident also in cases of excitation induced by excitatory agonists (cf. responses to acetylcholine in the presence or absence of baclofen in Figure 16.3).

ANTIEPILEPTIC ACTION OF BACLOFEN

Baclofen was remarkably more potent than GABA and several antiepileptic drugs in suppressing epileptiform activity. The depressant action of baclofen was manifest at a concentration as low as 10 nM. An intriguing aspect is that a depressant action on epileptiform activity was obvious at concentrations much lower than those required to depress synaptic transmission. In contrast to baclofen, both GABA and all the anticonvulsant drugs used suppressed the EPSP at concentrations required to block the epileptiform activity (Ogata *et al.*, 1986). Thus, baclofen has a highly potent as well as preferential depressant action on hippocampal epileptiform activity. Amongst various types of seizure disorders, temporal lobe epilepsy (complex partial seizures) is characterized by temporal lobe focal electroencephalographic abnormalities, and its clinical signs are related to disturbance of functioning of the temporal cortex, amygdala and hippocampal formation. The treatment of temporal lobe epilepsy is generally less effective than that of generalized or absence seizures. Therefore, it is tempting to speculate that baclofen may be particularly effective against temporal lobe epilepsy with minimal side-effects.

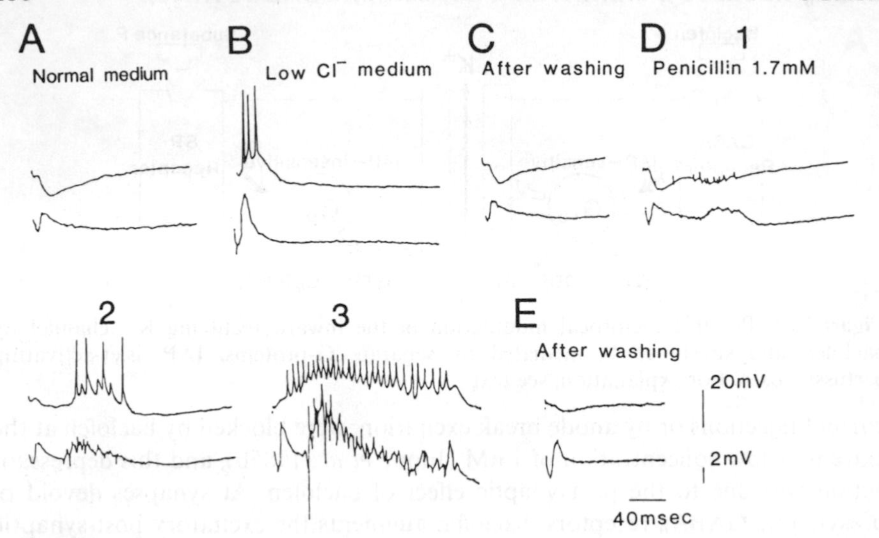

Figure 16.5. 'Epileptiform' activity and enhanced synaptic transmission. Continuous recordings of the intracellular potentials and concurrent field potentials were made in response to mossy-fibre stimulation. In (B), external, Cl^- concentration was reduced to 10.2 mM

Both antiepileptic (Ogata *et al.*, 1986; Swartzwelder *et al.*, 1986) and proepileptic (Mott *et al.*, 1989) effects of baclofen have been reported in experimental and clinical studies. A possible source of this discrepancy might be due to the concentration used. Terrence *et al.* (1983) reported that baclofen in normal therapeutic doses does not seem to be epileptogenic or to have a deleterious effect on seizure control. Mott *et al.* (1989) appear to have studied two separate electrical phenomena, i.e. potentiation of excitatory synaptic transmission and generation of epileptiform activity. As illustrated in Figure 16.5, these two phenomena are independent of each other: the former is augmented by baclofen, whereas the latter is readily blocked by a low concentration of baclofen.

MULTIPHASIC MEMBRANE EFFECT OF GABA

As shown in Figure 16.6, in the hippocampus GABA exerts a complex multiphasic action comprising hyperpolarizing and depolarizing components, whereas baclofen consistently hyperpolarizes the membrane. Relatively high concentrations of GABA (over 0.1 mM) were required to manifest the response in the bath application. This was due to a powerful uptake mechanism operating within the hippocampus, since procedures which suppress the uptake, such as depletion of external Na^+ or lowering the temperature of the medium, markedly reduced the effective concentration of GABA. As shown in Figure 16.7, the response to GABA was enhanced by a reduction of external Na^+ from 137 mM

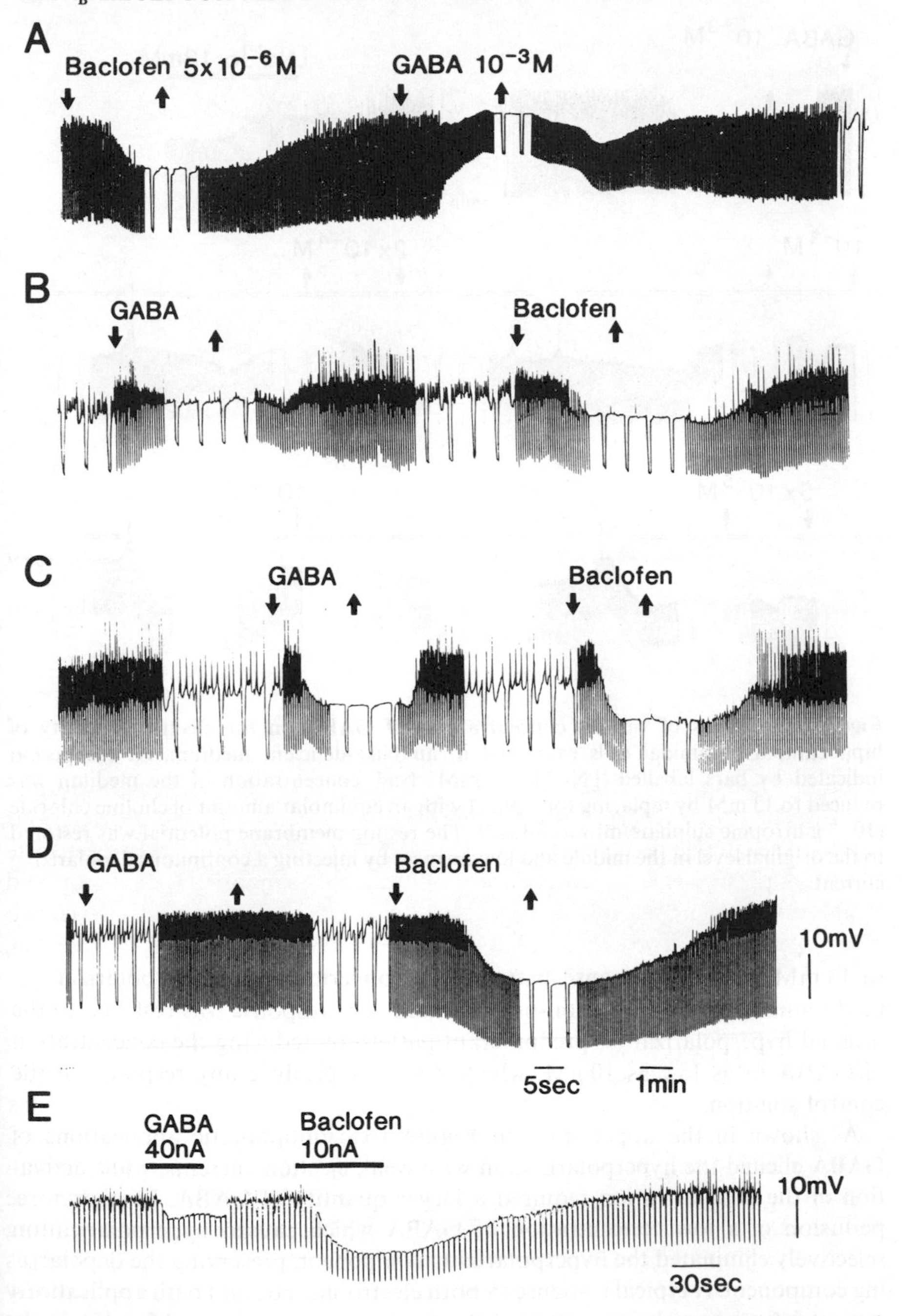

Figure 16.6. The membrane effects of GABA in the hippocampus are shown in comparison with those of baclofen. (A–D) Bath application. (E) Ionophoretic application

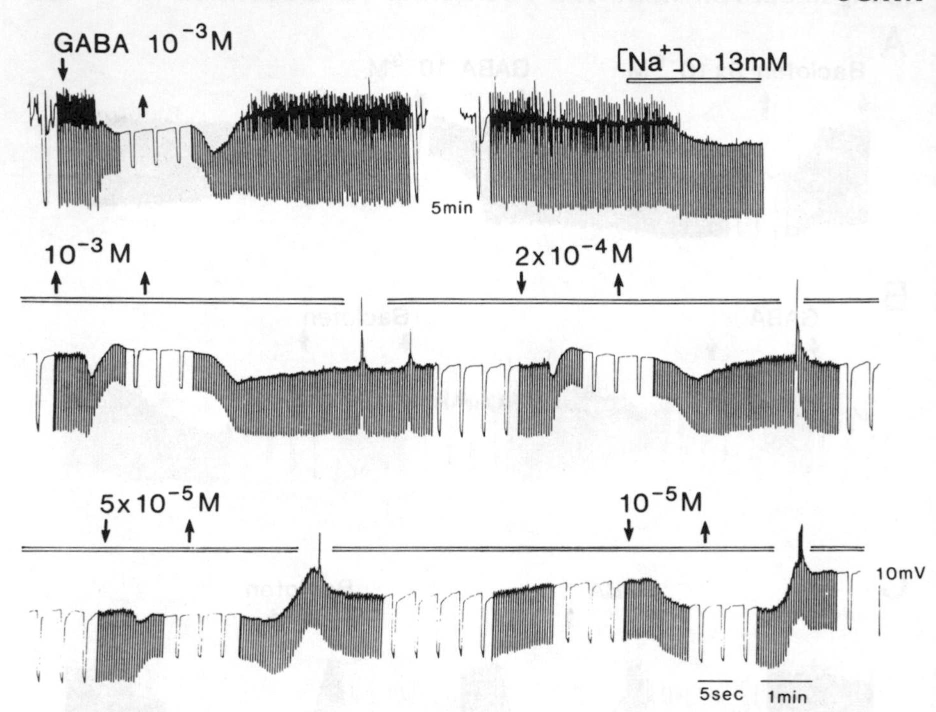

Figure 16.7. Effects of various concentrations of GABA on the electrical activity of hippocampal pyramidal cells examined in an Na^+-deficient medium. At the period indicated by bars labelled '$[Na^+]_o$ 13 mM' Na^+ concentration of the medium was reduced to 13 mM by replacing total NaCl with an equimolar amount of choline chloride (10^{-6} g atropine sulphate/ml was added). The resting membrane potential was restored to the original level in the middle and lower traces by injecting a continuous depolarizing current

to 13 mM, and the response pattern was converted from hyperpolarization-predominant to depolarization-predominant. The response was restored to the original hyperpolarization-predominant pattern by reducing the concentration of GABA to as low as 10 μM, which does not produce any response in the control solution.

As shown in the upper trace in Figure 16.8 ionophoretic applications of GABA elicited the hyperpolarization with weak ejection currents, while activation of the depolarization required a larger quantity of GABA. Furthermore, perfusion of a low concentration of GABA which caused hyperpolarization selectively eliminated the hyperpolarizing component, preserving the depolarizing component. A typical response in both electrophoretic and bath applications was a triphasic membrane potential change, comprising an initial fast hyperpolarization, sustained depolarization, and late hyperpolarization (Ogata *et al.*,

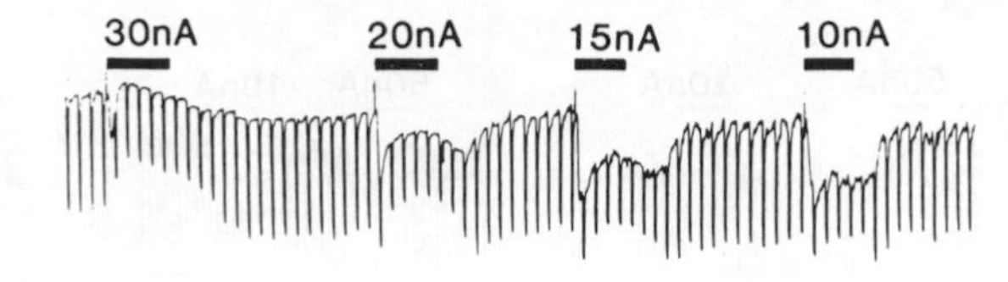

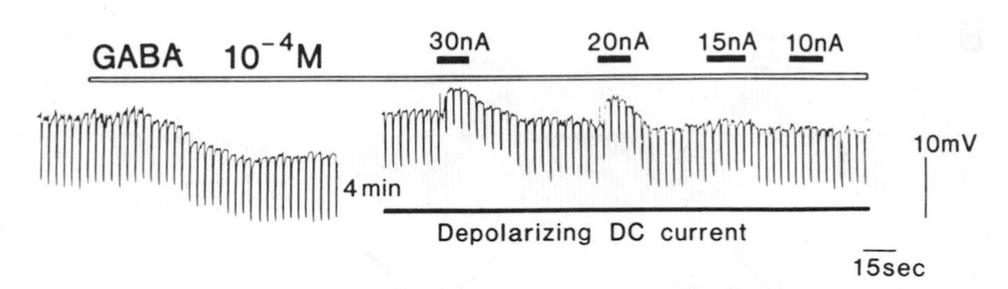

Figure 16.8. Membrane effects of ionophoresed GABA. GABA was applied with various ejection current intensities to the soma. Upper and lower traces were continuous records. In the lower trace, ionophoresis of GABA was made during a superfusion of GABA. A hyperpolarization produced by superfused GABA was compensated by injecting a continuous depolarizing current

1987b). The late hyperpolarization always appeared as an 'off' response subsequent to termination of GABA application.

Figure 16.9A illustrates the spatial responsiveness of the pyramidal cell to ionophoresed GABA. GABA applied to the somatic region provoked a depolarization-predominant response at 50 nA, or a hyperpolarization-predominant response at 10 nA (left trace). When the ionophoretic electrode was moved to the middle portion of the apical dendrites (approximately 150–200 μm from the pyramidal layer), the depolarizing component was augmented at both ejection current intensities (middle trace), confirming that GABA has a predominantly depolarizing effect on dendrites (e.g. Anderson *et al.*, 1980). Slight lateral displacement of the ionophoretic electrode from the soma-dendritic axis (right trace), which would reduce the effective concentration of dendritic GABA, altered the response so that it became hyperpolarization-predominant. Thus, both depolarization and hyperpolarization of comparable amplitudes to those induced by the somatic application of GABA were induced when a low concentration was administered in the dendritic region. As shown in Figure 16.9B, GABA responses of comparable amplitude were induced by somatic and dendritic applications of different ejection intensity. The time required to wash out the GABA effect was much longer for dendritic applications, suggesting that the uptake system for GABA may be more active in the somatic region.

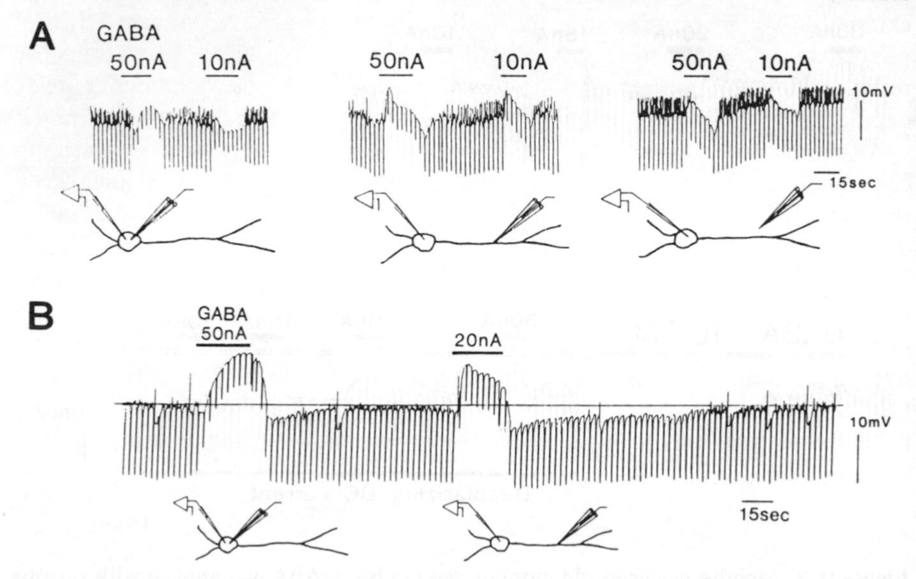

Figure 16.9. The spatial responsiveness of the pyramidal cell to ionophoretically applied GABA. (A) GABA was applied using two ejection current intensities near the soma (left), middle portion of the apical dendrites (middle), and a site slightly lateral to the middle (right), as shown under each trace. All traces were recorded from the same neuron. In this and subsequent figures, the duration of the electrophoretic application of the drug is indicated by solid lines. (B) The electrophoretic electrode was first positioned at the soma and subsequently moved to the dendrites

These observations might indicate that the GABA receptor responsible for the hyperpolarization represents the high-affinity site, while the receptor for the depolarization represents the low-affinity site in both soma and dendrites. The triphasic nature appears to be best explained in terms of the change in effective GABA concentration at the receptor sites. At the initial stage of application, the concentration of GABA, either ionophoresed or perfused, might be low, and as a consequence, the high-affinity GABA recognition sites would be predominantly activated, thus producing a hyperpolarization. Meanwhile, the concentration of GABA reaches a maximum, and the low-affinity GABA recognition sites responsible for the depolarization (see above) would be increasingly activated. On termination of application of ionophoresed or perfused GABA, the concentration of GABA would slowly return to zero, and consequently the low-affinity recognition sites would immediately become unsaturated, whereas the high-affinity recognition sites would remain active for some time.

IONIC MECHANISMS UNDERLYING THE ACTION OF GABA

The depolarizing component of the response to GABA was dependent on the external Cl^- concentration, whereas it was relatively insensitive to changes in external K^+ concentration, was markedly depressed by a low concentration of picrotoxin, and had a reversal potential about 15 mV more positive than the resting membrane potential (Inoue *et al.*, 1985c). Thus, it was concluded that the depolarizing component results from an activation of Cl^- conductance.

An application of picrotoxin eliminated the major part of the depolarizing component of the multiphasic response to GABA and consequently uncovered the hyperpolarizing component. Likewise, a low concentration of bicuculline depressed the depolarizing component of the multiphasic response to GABA. Thus, we could 'isolate' and maintain the relatively pure hyperpolarizing response by superfusing the slice continuously with both picrotoxin and bicuculline. The reversal potential for the above pharmacologically isolated hyperpolarizing response coincided with that of the AHP triggered by mossy-fibre stimulation, as in the case of baclofen. Thus, it was concluded that the hyperpolarization induced by GABA is mediated by an increase in K^+ conductance.

IONIC MECHANISMS UNDERLYING THE FAST INHIBITORY POSTSYNAPTIC POTENTIAL

Ionic mechanisms related to the fast IPSP in the mammalian CNS have long been a matter of controversy. Eccles and co-workers originally proposed that the IPSP of spinal motoneurons (Eccles *et al.*, 1964) and the hippocampus (Allen *et al.*, 1977) is associated with an increased permeability of the membrane to both K^+ and Cl^- ions. In contrast, several investigators claimed that the increased permeability to Cl^- is sufficient explanation for the generation of the IPSP (e.g. Misgeld *et al.*, 1986), and this single ionic dependence theory is now accepted by many investigators.

Since the IPSP in the hippocampus is probably mediated by GABA (Storm-Mathisen, 1977), the above observation that the response to GABA in the hippocampus is mediated by at least two separate receptor–ionophore complexes again raises the possibility that the IPSP in the hippocampus may be attributed to concomitant activation of K^+ and Cl^- channels.

It has been suggested that the fast hyperpolarization induced by GABA when ionophoresed onto the soma may primarily result from a rapid increase in Cl^- permeability, on the basis of findings that the fast hyperpolarization was highly sensitive to bicuculline, reversed its polarity at the potential positive to the equilibrium potential for K^+, and was dependent on the Cl^- gradient across the cell membrane (Newberry and Nicoll, 1985). However, the fast hyperpolarizations were fused to a late hyperpolarization of the triphasic response by reducing electrophoretic current (Figure 16.8) or by a combined application of bicuculline and picrotoxin (Ogata *et al.*, 1987a,b). Furthermore, it is unlikely that the fast and late hyperpolarizations of the typical triphasic GABA response are attributed to different kinetics of activation, since the triphasic nature was reproducible in both superfusion and ionophoresis of GABA. These results suggest that these two hyperpolarizations may be identical.

Although the reversal point of the fast hyperpolarization was less negative than the equilibrium potential for K^+ (Figure 4 of Inoue *et al.*, 1985c), it should be noted that this reversal was due to the contamination of a Cl^- conductance. Thus, the fast hyperpolarization examined in the presence of bicuculline and picrotoxin had a reversal potential corresponding to the equilibrium potential for K^+ (Inoue *et al.*, 1985c). Furthermore, as shown in Figure 16.10, the reversal point of the fast hyperpolarization was often more negative than that of the IPSP.

The observation by Newberry and Nicoll (1985) that the fast hyperpolarization was blocked by bicuculline methiodide in concentrations lower than those required to block the late hyperpolarization may be interpreted on the basis of concentration–response curves, since the agonist (GABA) concentrations used

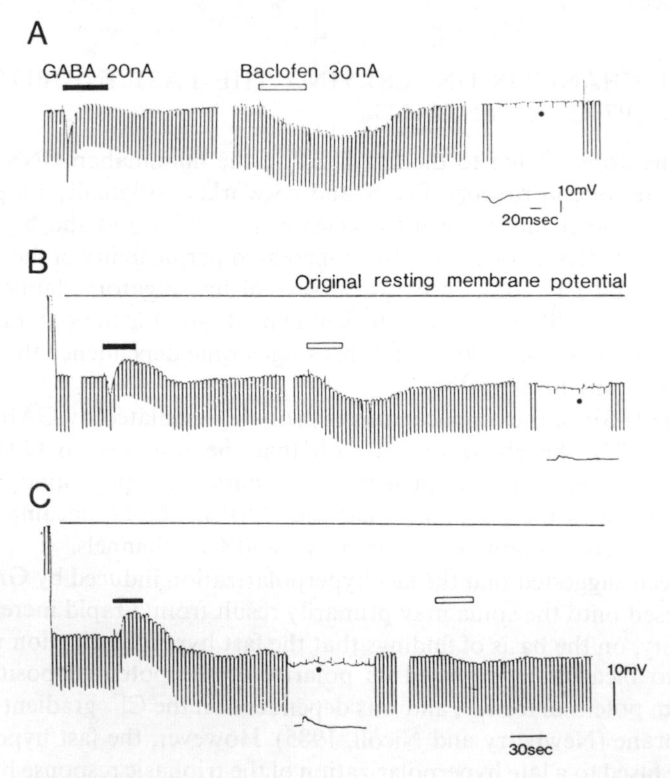

Figure 16.10. Voltage dependence of the fast hyperpolarization induced by ionophoresed GABA and the fast inhibitory postsynaptic potential (IPSP). The voltage dependence of the action of baclofen is also illustrated for reference. The potentials were examined at (A) −58 mV, (B) −83 mV and (C) −93 mV. The fast IPSP was triggered by mossy-fibre stimulation (dot). Inset traces illustrate the evoked response to mossy-fibre stimulation

in the respective responses differed. In the experiments of Newberry and Nicoll (1985), the quantity of ionophoresed GABA used to induce a 'pure' hyperpolarization would have been fairly small, as they themselves noted. We observed that the late hyperpolarization was actually antagonized by a relatively high concentration of bicuculline (Inoue *et al.*, 1985c). Conversely, the response induced by a fairly low dose of GABA would readily be blocked even by a relatively low concentration of the antagonist which does not affect the late hyperpolarization. These findings might warrant reinvestigation of the original hypothesis of a dual ionic mechanism for the generation of the IPSP. The same mechanism appears to operate in some synaptic potentials in *aplysia* (Gerschenfeld, 1973).

GABA$_B$ RECEPTORS AND THE SLOW INHIBITORY POSTSYNAPTIC POTENTIAL

It appears to have been established that the postsynaptic action of baclofen is mediated by postsynaptic GABA$_B$ receptors. In fact, as shown in Figure 16.11,

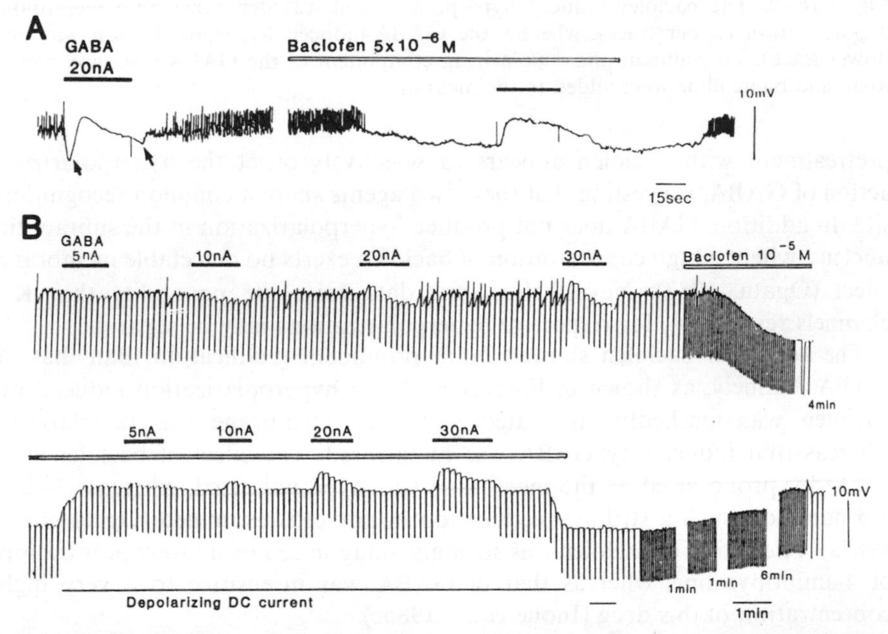

Figure 16.11. Baclofen selectively blocks the hyperpolarizing component of the multiphasic GABA response. Ionophoretic applications of GABA were repeated in the control and baclofen-containing media. GABA, ionophoresed with currents identical to those used before, failed to elicit the inital fast hyperpolarization and the late hyperpolarization, producing only a depolarization. In (B), the hyperpolarization induced by baclofen was compensated by passing a continuous hyperpolarizing current through the recording electrode

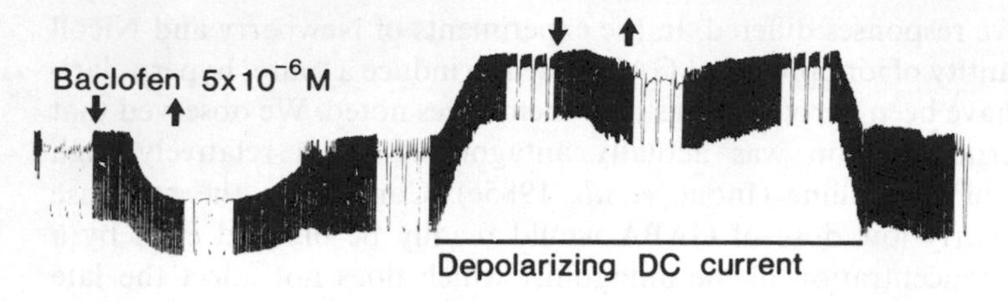

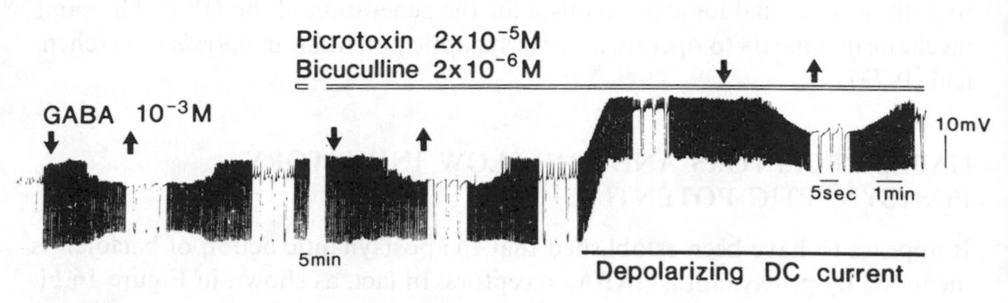

Figure 16.12. The baclofen-induced hyperpolarization was depressed by a membrane depolarization (upper trace), whereas the GABA-induced hyperpolarization was not (lower trace). To eliminate the depolarizing component of the GABA response, picrotoxin and bicuculline were added to the medium

pretreatment with baclofen appears to selectively offset the hyperpolarizing action of GABA, suggesting that these two agents share a common recognition site. In addition, GABA does not produce hyperpolarization in the supraoptic nucleus, where a high concentration of baclofen exerts no detectable membrane effect (Ogata, 1987). Nevertheless, our data raise the possibility that K^+ channels responsible for respective hyperpolarizations may be separate.

The action of baclofen showed more prominent rectification than that of GABA. Namely, as shown in Figure 16.12, the hyperpolarization induced by baclofen was markedly attenuated when the membrane was depolarized, whereas that induced by GABA was augmented. The effect of baclofen was markedly pronounced as the membrane was hyperpolarized, whereas GABA did not show such a striking rectification in any cells examined (Ogata *et al.*, 1987a). The action of baclofen was strongly antagonized by a low concentration of 4-aminopyridine, whereas that of GABA was insensitive to a very high concentration of this drug (Inoue *et al.*, 1985c).

The slow IPSP in the mammalian brain has been shown to be due to activation of K^+ conductance (Newberry and Nicoll, 1984b). Its highly voltage-sensitive nature (Satou *et al.*, 1982) and relative resistance to bicuculline (Newberry and Nicoll, 1984b) are reminiscent of the baclofen-induced hyperpolarization. Recent studies have demonstrated that the slow IPSP is blocked by phaclofen, reportedly an antagonist of baclofen, suggesting that baclofen

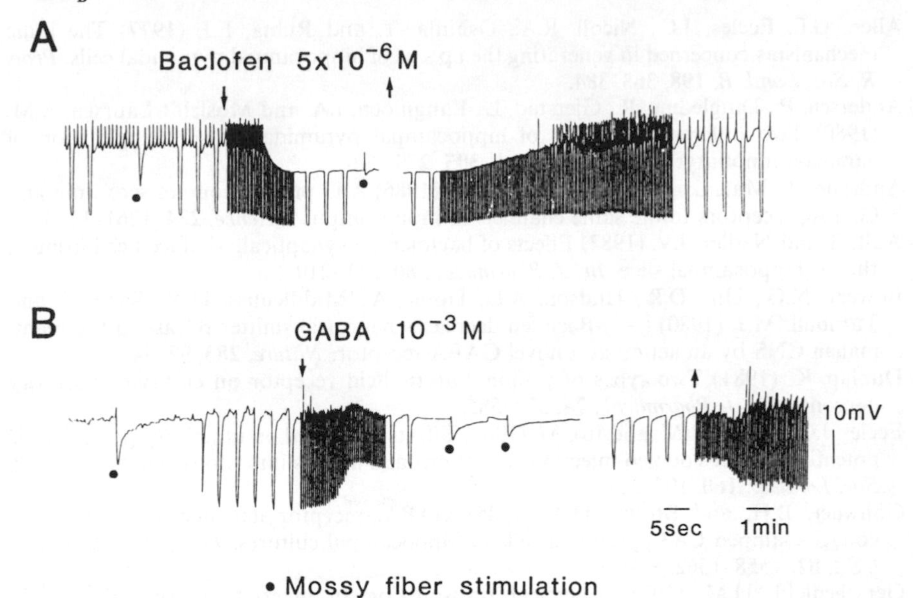

Figure 16.13. Effects of (A) baclofen and (B) GABA on membrane potential and evoked potentials in response to mossy-fibre stimulation (dot). Due to limited frequency bandwidth of pen-recordings, only the slow afterhyperpolarization (AHP) was manifest in the traces

activates receptors responsible for the slow IPSP. Consistent with this, baclofen caused a total suppression of the slow IPSP (Figure 16.13A). An important point is that GABA itself had no appreciable effect on the slow IPSP (Figure 16.13B). The reduction of the slow IPSP was proportional to the decrease in input resistance. The recognition sites for GABA, whether somatic or dendritic, would have been exposed to GABA, since GABA had been applied by superfusion. Thus, it remains to be elucidated whether or not the slow IPSP is mediated by a subclass of GABA recognition sites.

ACKNOWLEDGEMENTS

The author thanks Professor H. Kuriyama for help and advice, Drs M. Yoshii and D. Allsop, Psychiatric Research Institute of Tokyo, for their critical reading of the manuscript, and CIBA-Geigy (Japan) for the gift of (−)-baclofen.

REFERENCES

Alger, B.E. and Nicoll, R.A. (1980) Epileptiform burst afterhyperpolarization: calcium-dependent potassium potential in hippocampal CA1 pyramidal cells. *Science*, **210**, 1122–1124.

Allen, G.I., Eccles, J.C., Nicoll, R.A., Oshima, T. and Rubia, F.J. (1977) The ionic mechanisms concerned in generating the i.p.s.ps of hippocampal pyramidal cells. *Proc. R. Soc. Lond. B*, **198**, 363–384.

Andersen, P., Dingledine, R., Gjerstad, L., Langmoen, I.A. and Mosfeldt Laursen, A.M. (1980) Two different responses of hippocampal pyramidal cells to application of gamma-aminobutyric acid. *J. Physiol.*, **305**, 279–296.

Andrade, R., Malenka, R.C. and Nicoll, R.A. (1986) A G protein couples serotonin and $GABA_B$ receptors to the same channels in hippocampus. *Science*, **234**, 1261–1265.

Ault, B. and Nadler, J.V. (1983) Effects of baclofen on synaptically-induced cell firing in the rat hippocampal slice. *Br. J. Pharmacol.*, **80**, 211–219.

Bowery, N.G., Hill, D.R., Hudson, A.L., Doble, A., Middlemiss, D.N., Shaw, J. and Turnbull, M.J. (1980) (−)-Baclofen decreases neurotransmitter release in the mammalian CNS by an action at a novel GABA receptor. *Nature*, **283**, 92–94.

Dunlap, K. (1981) Two types of γ-aminobutyric acid receptor on embryonic sensory neurones. *Br. J. Pharmacol.*, **74,** 579–585.

Eccles, J.C., Eccles, R.M. and Ito, M. (1964) Effects produced on inhibitory postsynaptic potentials by the coupled injections of cations and anions into motoneurons. *Proc. R. Soc. Lond. B*, **160**, 197–211.

Gähwiler, B.H. and Brown, D.A. (1985) $GABA_B$-receptor-activated K^+ current in voltage-clamped CA3 pyramidal cells in hippocampal cultures. *Proc. Natl Acad. Sci. USA*, **82**, 1558–1562.

Gerschenfeld, H.M. (1973) Chemical transmission in invertebrate central nervous systems and neuromuscular junctions. *Physiol. Rev.*, **53**, 1–119.

Hanley, R.R., Sandberg, B.E.B., Lee, C.M., Iversen, L.L., Brundish, D.E. and Wade, R. (1980) Specific binding of ^{3}H-substance P to rat brain membranes. *Nature*, **286**, 810–812.

Hotson, J.R. and Prince, D.A. (1980) A calcium-activated hyperpolarization follows repetitive firings in hippocampal neurones. *J. Neurophysiol.*, **43,** 409–419.

Inoue, M., Matsuo, T. and Ogata, N. (1985a) Baclofen activates voltage-dependent and 4-aminopyridine sensitive K^+ conductance in guinea-pig hippocampal pyramidal cells maintained in vitro. *Br. J. Pharmacol.*, **84**, 833–841.

Inoue, M., Matsuo, T. and Ogata, N. (1985b) Characterization of pre- and postsynaptic actions of (−)-baclofen in the guinea-pig hippocampus in vitro. *Br. J. Pharmacol.*, **84**, 833–841.

Inoue, M., Matuo, T. and Ogata, N. (1985c) Possible involvement of K^+-conductance in the action of gamma-aminobutyric acid in the guinea-pig hippocampus. *Br. J. Pharmacol.*, **86**, 515–524.

Johnston, G.A.R. (1978) Neuropharmacology of amino acid inhibitory transmitters. *Annu. Rev. Pharmacol. Toxicol.*, **18**, 269–289.

Misgeld, U., Deisz, R.A., Dodt, H.U. and Lux, H.D. (1986) The role of chloride transport in postsynaptic inhibition of hippocampal neurons. *Science*, **232**, 1413–1415.

Mott, D.D., Bragdon, A.C., Lewis, D.V. and Wilson, W.A. (1989) Baclofen has a proepileptic effect in the rat dentrate gyrus. *J. Pharmacol. Exp. Ther.*, **249**, 721–725.

Nakajima, Y., Nakajima, S. and Inoue, M. (1988) Pertussis toxin-insensitive G protein mediates substance P-induced inhibition of potassium channels in brain neurons. *Proc. Natl Acad. Sci. USA*, **85**, 3643–3647.

Newberry, N.R. and Nicoll, R.A. (1984a) Direct hyperpolarizing action of baclofen on hippocampal pyramidal cells. *Nature*, **308**, 450–452.

Newberry, N.R. and Nicoll, R.A. (1984b) A bicuculline-resistant inhibitory post-synaptic potential in rat hippocampal pyramidal cells in vitro. *J. Physiol.*, **348**, 239–254.

Newberry, N.R. and Nicoll, R.A. (1985) Comparison of the action of baclofen with gamma-aminobutyric acid on rat hippocampal pyramidal cells in vitro. *J. Physiol.*, **360**, 161–185.

Ogata, N. (1987) Gamma-aminobutyric acid (GABA) causes consistent depolarization of neurons in the guinea pig supraoptic nucleus due to an absence of GABA$_B$ recognition sites. *Brain Res.*, **403,** 225–233.

Ogata, N. and Abe, H. (1981) Further support for the postsynaptic action of substance P and its blockade with baclofen in neurons of the guinea-pig hypothalamus in vitro. *Experientia*, **37**, 759–761.

Ogata, N. and Abe, H. (1982) Neuropharmacology in the brain slice: effects of substance P on neurons in the guinea-pig hypothalmus. *Comp. Biochem. Physiol.*, **72c**, 171–178.

Ogata, N., Matsuo, T. and Inoue, M. (1986) Potent depressant action of baclofen on hippocampal epileptiform activity in vitro: possible use in the treatment of epilepsy. *Brain Res.*, **337**, 362–367.

Ogata, N., Inoue, M. and Matsuo, T. (1987a) Contrasting properties of K^+ conductances induced by baclofen and gamma-aminobutyric acid in slices of the guinea pig hippocampus. *Synapse*, **1**, 62–69.

Ogata, N., Inoue, M. and Matsuo, T. (1987b) Possible mediation of hippocampal IPSP by two receptor–ionophore complexes. In: N. Charazonitis and M. Gola (eds) *Inactivation of hypersensitive neurons.* New York: Alan R. Liss, pp. 121–128.

Pierau, F.K. and Zimmermann, P. (1973) Action of a GABA-derivative on postsynaptic potentials and membrane properties of cat spinal motoneurons. *Brain Res.*, **54**, 376–380.

Saito, K., Konishi, S. and Otsuka, M. (1975) Antagonism between Lioresal and substance P in rat spinal cord. *Brain Res.*, **97**, 177–180.

Satou, M., Mori, K., Tazawa, Y. and Takagi, S.F. (1982) Two types of postsynaptic inhibition in pyriform cortex of the rabbit: fast and slow inhibitory postsynaptic potentials. *J. Neurophysiol.*, **48**, 1142–1156.

Storm-Mathisen, J. (1977) Localization of transmitter candidates in the brain: the hippocampal formation as a model. *Prog. Neurobiol.*, **8**, 119–181.

Swartzwelder, H.S., Bragdon, A.C., Sutch, C.P., Ault, B. and Wilson, W.A. (1986) Baclofen suppresses hippocampal epileptiform activity at low concentrations without suppressing synaptic transmission. *J. Pharmacol. Exp. Ther.*, **237**, 881–886.

Terrence, C.F., Fromm, G.H. and Roussan, M.S. (1983) Baclofen: its effect on seizure frequency. *Arch. Neurol.*, **40**, 28–29.

Part VI

APPLIED ASPECTS

Antidepressants

17 Antidepressants and $GABA_B$ Site Upregulation

K. G. LLOYD

INTRODUCTION

Clinical efficacy in depression is well documented for compounds which increase the availability of monoamines (most notably serotonin (5-hydroxytryptamine), 5-HT, and noradrenaline, NA) at their central nervous system (CNS) receptors. The mechanisms of action include inhibition of the specific nerve terminal monoamine uptake system and the inhibition of monoamine metabolism by monoamine oxidase (MAO) inhibitors. Although the most recent generation of antidepressant drugs includes highly specific inhibitors of 5-HT or NA uptake or MAO-B, considerable disadvantages remain in the use of such compounds. Examples of these are cardiovascular effects, parasympathetic effects and a considerable delay until the onset of a significant antidepressant action.

The neuronal circuitry involved in the perception, regulation and expression of mood is not understood; however, it is certain that these are complex processes implicating multiple synapses and different neurotransmitters. In this regard it is thus highly probable that manipulation of neurotransmitters other than 5-HT and NA will lead to mood alteration. GABA is a neurotransmitter which occurs in at least 25% of all CNS synapses (see Iversen, 1978), and is known to be involved in diverse neuropsychiatric functions (see Lloyd and Morselli, 1987). GABA thus appears to be a likely candidate for involvement in the regulation of mood disorders and/or the action of antidepressant drugs.

ACTIVITY OF ANTIDEPRESSANT DRUGS ON GABA SYNAPSES IN NORMAL RATS

When administered acutely to rats at behaviourally meaningful doses, antidepressant drugs are virtually devoid of activity on GABA synapses (levels, uptake, synthesis and $GABA_A$ and $GABA_B$ receptors; Pilc and Lloyd, 1984; see Lloyd, 1989). However, upon repeated (18 days daily s.c. or i.p.) administration, antidepressants (e.g. amitriptyline, desipramine, viloxazine, fluoxetine, fengabine) have a specific action within CNS GABA synapses. From the data presented in Figure 17.1 it can be seen that biochemical indicators of GABA terminal function are unaltered (glutamic acid decarboxylase activity, GABA

$GABA_B$ Receptors in Mammalian Function. Edited by N.G. Bowery, H. Bittiger and H.-R. Olpe

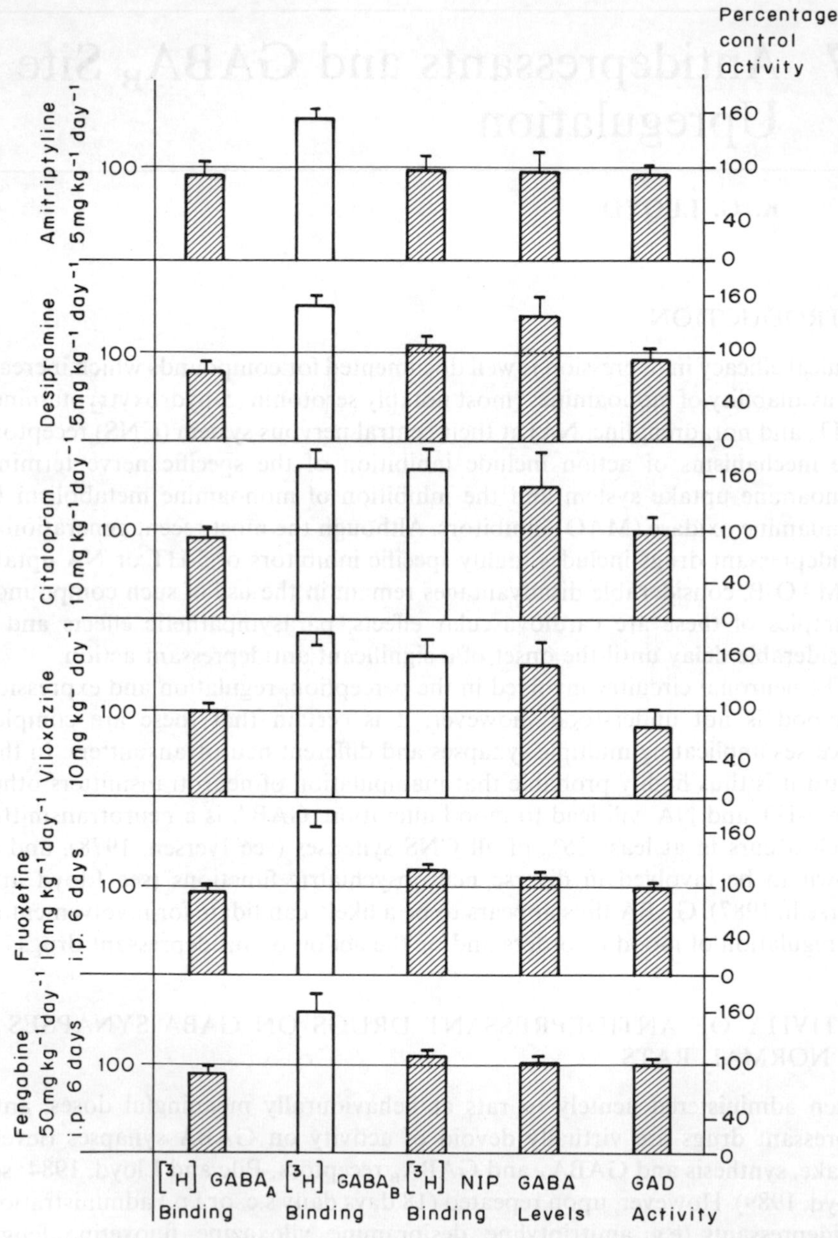

Figure 17.1. Effect of antidepressant drugs on biochemical indicators of GABA terminal function in normal rats. Hatched bars: $p > 0.05$ vs saline control, open bars: $p < 0.01$ vs saline. Data from Lloyd and Pilc (1984, unpublished)

levels, [^{3}H]nipecotic acid binding to GABA uptake sites) as is the number of $GABA_A$ receptors. In contrast, these compounds consistently enhance the number of $GABA_B$ receptors in the frontal cortex, in spite of their very diverse actions on monoamine neurotransmitter systems. The effects are dependent on time (greater at 18 days than 6 days) and dose (desipramine: 5.0 mg/kg > 1.25 mg/kg; $P < 0.05$), and have their earliest onset in the frontal cortex, although the number of $GABA_B$ sites in the hippocampus is also enhanced after longer periods of drug administration ($GABA_B$ sites in the parietal cortex, occipital cortex and olfactory cortex are not increased by 18 days of antidepressant treatment; Lloyd *et al.*, 1985).

This effect of antidepressant drugs on $GABA_B$ receptors has been investigated in some detail (Lloyd *et al.*, 1989; also unpublished data). A possible relation between the most potent action of different antidepressants on various brain monoamine receptors and the enhancement of $GABA_B$ binding was considered (Table 1). Although those compounds with actions on brain 5-HT synapses are on average the most effective, there do not appear to be any significant differences between the biochemical groups. The greatest increase observed occurs with idazoxan, an α_2 antagonist. All other clinically effective antidepressant drugs acting via adrenergic synapses also enhance $GABA_B$ binding in the rat frontal cortex, including uptake inhibitors and phosphodiesterase inhibitors. DL-threo-dihydrophenylserine, a direct precursor of NA also increases $GABA_B$ binding (to 135% control; $P < 0.05$). It is unknown to the author as to whether or not this compound presents a clinically effective antidepressant activity. It is worthy of note that, at doses of 1 or 2 mg/kg s.c. per day, amphetamine administration does not significantly enhance [^{3}H]$GABA_B$ binding to rat frontal cortex membranes.

In a like manner, antidepressant drugs inhibiting 5-HT uptake consistently induce a large increase in $GABA_B$ binding (mean $\pm$ SEM 173 $\pm$ 5%) to rat frontal cortex membranes. Compounds acting directly at 5-HT_{1A} receptors appear to have a somewhat weaker effect. 8-Hydroxy-2-(di-*n*-propylamino)tetralin (8-OH-DPAT) also induces a significant 23% increase in [^{3}H]$GABA_B$ binding, but clinical data on its antidepressant activity are not available. Buspirone (3 mg/kg s.c.), an anxiolytic 5-HT_{1A} agonist, did not alter [^{3}H]$GABA_B$ binding (107% control).

Interestingly, some compounds with GABA mimetic activity also enhance $GABA_B$ binding to rat frontal cortex membranes. This appears to be related to their antidepressant potential, as clinically effective compounds enhancing $GABA_A$ function only (adinazolam, alprazolam) are as effective as those directly acting at both $GABA_A$ and $GABA_B$ receptors (fengabine, progabide), whereas diazepam does not significantly increase $GABA_A$ binding (113 % control) and is not known for clinical antidepressant activity (Schatzberg and Cole, 1978).

Table 1. $[^3H]GABA_B$ binding enhancement by antidepressants: relation to biochemical specificity

Noradrenergic	Serotoninergic	GABAergic	Dopaminergic	Others
Idazoxan (5) 207%	Fluoxetine (10) 183%	Adinazolam (0.5) 161%	Bupropion (5) 154%	ECS (5 over 10 days) 177%
Viloxazine (10) 188%	Citalopram (10) 172%	Fengabine (50) 159%	Nomifensine (5) 146%	Trazodone (10) 155%
Desipramine (5) 151%	Zimeldine (10) 165%	Alprazolam (1) 137%		Iprindole (5) 129%
Levoprotiline (2) 147%	Gepirone (3) 123%	Progabide (100) 127%		Amitryptiline (10) 154%
Maprotiline (10) 135%				Pargyline (20) 142%
Rolipram (2) 135%				
Mianserine (10) 133%				
—	—	—	—	
Mean 156	161	146	150	
±SEM ±11%	±13%	±7%		

Data are expressed as the mean % of the respective control (saline or solvent) group and are significantly different ($P \leqslant 0.05$). Antidepressants were administered daily by the s.c. site to male albino rats at the doses (mg/kg) indicated in parentheses. (Fengabine was administered i.p.) Groups of 5 rats per dose were used and the animals sacrificed and the frontal cortices processed for $GABA_B$ binding as described by Lloyd *et al.* (1985).
Data from Lloyd (1989), Lloyd *et al.* (1985), Lloyd and Pichat (1987), Lloyd (unpublished data).

There are limited data available which suggest that decreasing GABAergic function has a depressant effect. To the author's knowledge cycloserine is the only compound which decreases GABAergic functions and has been used therapeutically in man (in the treatment of tuberculosis). A reversible depression is one of the more frequent adverse effects of cycloserine (e.g. Lewis *et al.*, 1957; Ruiz, 1964).

The two dopamine uptake inhibitors which are clinically proven antidepressants (buproprion and nomifensine) also induce a marked (50%) enhancement of GABA$_B$ binding. Two antagonists of dopamine receptors are without significant effect (haloperidol, 107% control; chlorpromazine, 82% control). In contrast, depletion of the three major monoamines by reserpine significantly decreases GABA$_B$ binding to rat frontal cortex membranes (70% control; $P < 0.01$). It is of interest that reserpine is reputed to be associated with the induction of depression in some people (e.g. Goodwin *et al.*, 1972).

Antidepressants which do not have a specific mechanism of action at a given monoaminergic synapse (e.g. mixed uptake or MAO inhibitors; trazodone, iprindole) also enhance GABA$_B$ binding to rat frontal cortex membranes. In addition, and most importantly (as it is a non-chemotherapeutic treatment), a series of mild electroshocks also enhances GABA$_B$ binding to rat frontal cortex membranes.

ANIMAL MODELS OF DEPRESSION AND GABA$_B$ RECEPTORS

Although it is impossible to attest to the occurrence of a mood disorder in rodents, several experimental situations exist in which behavioural modifications respond consistently and selectively to administration of antidepressant drugs. The removal of the olfactory bulbs in the rat results in multiple behavioural changes, including open field activity and a deficit in passive avoidance learning. These deficits respond to antidepressant treatment, but not to neuroleptics, anxiolytics or other psychotropics. Interestingly, like antidepressants, GABA mimetics also reverse these behavioural abnormalities in the olfactory-bulbectomized (OLB) rat (Broekkamp *et al.*, 1980; Delina-Stula and Vassout, 1978; Leonard 1984).

GABA$_B$ receptors are decreased (by 50–60 %) in the frontal cortex, but not in the occipital cortex, amygdala, cerebellum or hippocampus of the OLB rat. This is evidenced both by a decrease in [^{3}H]GABA$_B$ binding sites (Lloyd and Pichat, 1986) and by a decrease in GABA$_B$ receptor function (GABA$_B$ enhancement of isoproterenol-induced cyclic-AMP production; Lloyd and Pichat, 1987). Antidepressant drugs (e.g. desipramine) reverse the passive avoidance deficit in OLB rats, but not all of the animals respond to treatment. This allows the grouping of the data into that from animals 'responding to antidepressant treatment' (i.e. normalized to the response by sham-operated rats) and that from 'response

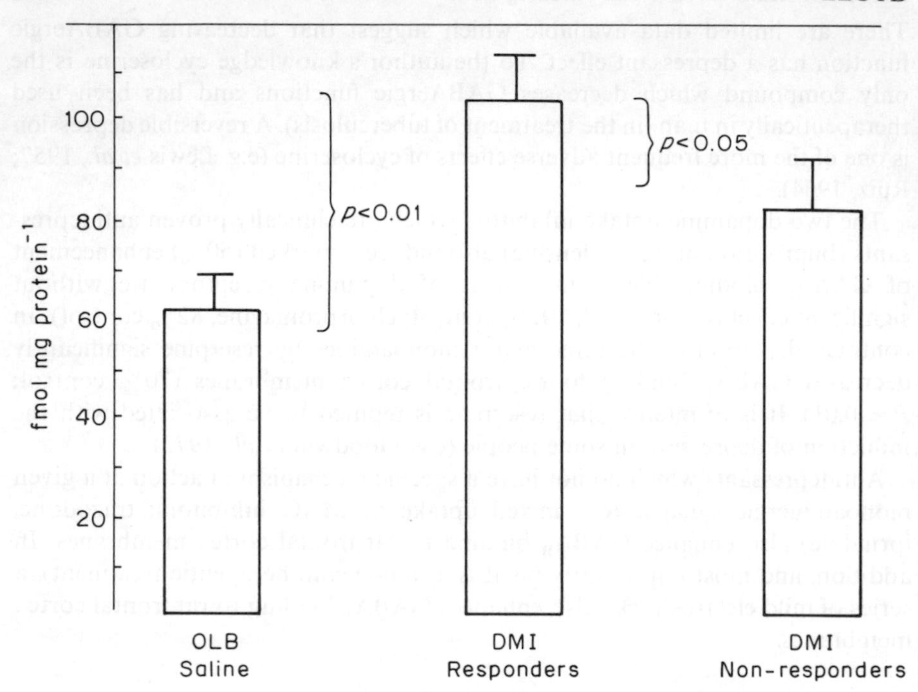

Figure 17.2. $GABA_B$ receptor binding in frontal cortex of OLB rats responding, or failing to respond, to repeated desipramine (DMI) treatment (14 days, 5 mg kg^{-1} day^{-1}). Data from Joly *et al.* (1987)

failures' (animals responding at the same rate as untreated OLB rats). As can be seen from Figure 17.2, $GABA_B$ receptor binding is upregulated in desipramine responders ($P < 0.01$), whereas the non-responders do not show a significant elevation of $GABA_B$ sites (which is significantly lower than for the desipramine responders) (Joly *et al.*, 1987; Lloyd and Pichat, 1987).

A similar situation occurs in the learned helplessness model, where $GABA_B$ receptors are decreased in the frontal cortex and $GABA_B$ sites are upregulated only in those animals which exhibit a behavioural response to antidepressant treatment (Martin *et al.*, in press; Pichat and Lloyd, unpublished observations). These experiments strongly suggest that the level of $GABA_B$ receptors in the frontal cortex is related to the behavioural state, at least in the rat.

The above results present an apparently homogeneous set of data; however, most have been obtained in the same laboratory, albeit by different individuals and by various methodologies. An important aspect is the question of inter-laboratory reproducibility.

CONFIRMATION OF THE UPREGULATION OF $GABA_B$ RECEPTORS BY ANTIDEPRESSANT DRUGS

As shown in Table 2 there is a broad confirmation of the upregulation of $GABA_B$ receptors by antidepressants in the rat and mouse. This is indicated by binding to $GABA_B$ receptors in membrane preparations and also using autoradiographic methods in situ (cf. Pratt and Bowery, Chapter 19). Furthermore the data available suggest that not all $GABA_B$ receptors are involved: the $GABA_B$ sites in the β-adrenergic receptor complex as well as those regulating 5-HT release respond to repeated antidepressant treatment (Gray and Green, 1987; Suzdak and Gianutsos, 1986), whereas the non-specific stimulation of cyclic AMP production induced by forskolin is not altered by repeated antidepressant administration (Szekely *et al.*, 1987).

The upregulation of $GABA_B$ receptors is demonstrated in vivo as well as ex vivo, and those $GABA_B$ receptors modulating hypothermia are upregulated by antidepressant drugs (Gray *et al.*, 1987), whereas those (possibly peripheral) $GABA_B$ receptors modulating nociception are apparently not increased by repeated antidepressant administration (Borsini *et al.*, 1986).

The team from one laboratory has consistently been unable to observe an upregulation of $GABA_B$ receptors following antidepressant treatment (see Cross *et al.*, Chapter 18). The reasons for this are unclear, but methodological aspects may play an important role. Thus, as shown in Table 3, the kinetic characteristics (B_{max}) of the [^{3}H]$GABA_B$ binding site in control animals in the studies of Cross and Horton (1986, 1987) are similar to or greater than those enhanced levels observed by researchers from other laboratories *after* antidepressant treatment.

CONCLUSIONS

The above data indicate that, although there are still several points to ponder, it is evident that $GABA_B$ receptors do play a role in the behavioural responses to repeated antidepressant administration in the rat. The translation to man is one of the interesting challenges of this hypothesis.

REFERENCES

Borsini, F., Giuliani, S. and Meli, A. (1986) Functional evidence for altered activity of GABAergic receptors following chronic desipramine treatment in rats. *J. Pharm. Pharmacol.*, **38**, 934–935.

Broekkamp, C.L., Garrigou, D. and Lloyd, K.G. (1980) Serotonin-mimetic antidepressant drugs on passive avoidance learning by olfactory bulbectomized rats. *Pharmacol. Biochem. Behav.*, **13**, 643–646.

Cross, J.A. and Horton, R.W. (1986) Cortical $GABA_B$ binding is unaltered following chronic oral administration of desmethylimipramine and zimelidine in the rat. *Br. J. Pharmacol.*, **89**, 521P.

Table 2. Effects of repeated antidepressant administration on $GABA_B$ receptor function

Investigating group	Effect of antidepressants on $GABA_B$ receptors	Parameter studied	Species	Brain region	Antidepressants*
Lloyd and collaborators[a]	Enhanced	[^{3}H]$GABA_B$ binding	Rat	Frontal cortex	ECS + 20 ADs
		$GABA_B$-β-adrenoceptor cAMP	Rat	Frontal cortex	DMI
Martin and collaborators	Enhanced	[^{3}H]$GABA_B$ binding	Rat	Frontal cortex (LH)	IMI
Suzdak and Gianutsos (1986)	Enhanced	[^{3}H]$GABA_B$ binding	Mouse	Cortex	IMI
		$GABA_B$-β-adrenoceptor cAMP	Mouse	Cortex	IMI
Bowery and collaborators	Enhanced	[^{3}H]$GABA_B$ autoradiography	Rat	Frontal cortex	DMI
Green and collaborators[b]	Enhanced	$GABA_B$-5-HT release	Mouse	Cortex	ECS + 4 ADs
		Baclofen hypothermia	Mouse	—	ECS + 4 ADs
Szekely *et al.* (1987)	Enhanced	[^{3}H]$GABA_B$ binding	Rat	Frontal cortex	DMI, IMI
	Unchanged	[^{3}H]$GABA_B$ binding	Rat	Frontal cortex	MAP
		[^{3}H]-(−)baclofen binding	Rat	Frontal cortex	DMI, IMI, MAP
		$GABA_B$-forskolin cAMP	Rat	Frontal cortex	DMI, IMI, MAP
Motohashi *et al.* (1989)	Enhanced	[^{3}H]-(−)baclofen binding	Rat	Hippocampus	Li, CBZ
	Unchanged	[^{3}H]-(−)baclofen binding	Rat	Frontal cortex	Li, CBZ
Cross and Horton (1986/87)	Unchanged	[^{3}H]$GABA_B$ binding	Rat	Frontal cortex	DMI
Borsini *et al.* (1986)	Decreased	Baclofen nociception	Rat	—	DMI

[a]See text for references. [b]Gray and Green (1987); Gray *et al.* (1987)

Abbreviations: ECS, electroconvulsive shock; ADs, antidepressant drugs; DMI, desipramine; IMI, imapramine; MAP, maprotoline; Li, lithium; CBZ, carbamazepine.

Table 3. Comparison of [^{3}H]GABA$_B$ binding in four studies following repeated imipramine or desipramine (active IMI metabolite) administration

		[^{3}H]GABA$_B$ binding					
		Assay conditions			Results		
Authors	Species region	GABA$_A$ exclusion	GABA$_B$ definition	Drug (dose)	K$_D$ (nM)	B$_{MAX}$ (fmol mg protein^{-1})	% saline
Lloyd *et al.* (1985)	Rat frontal cortex 18 days s.c.	Isoguvacine (40 μM)	Baclofen	Saline	35	870	—
				DMI (1.25)	37	1230	141
				DMI (5.0)	22, 144	2050	236
Suzdak and Gianutsos (1986)	Mouse cerebral cortex 14 days i.p.	Isoguvacine (100 μM)	Baclofen	Saline	30	605	—
				IMI (32)	45	750	124
Szekely *et al.* (1987)	Rat frontal cortex	Bicuculline (100 μM)	Baclofen	Saline	82	616	—
	21 days, BID, i.p.			DMI (10)	77	790	128
				IMI (7.5)	88	761	124
Cross and Horton (1987)	Rat frontal cortex	Isoguvacine (40 μM)	Baclofen	Saline	17	1240	—
	21 days, i.p.			DMI (10)	19	1330	107

Abbreviations: see footnote to Table 2.

Cross, J.A. and Horton, R.W. (1987) Are increases in $GABA_B$ receptors consistent findings following chronic antidepressant administration? *Eur. J. Pharmacol.*, **141**, 159–162.

Delina-Stula, A. and Vassout, A. (1978) Influence of baclofen and GABA-mimetic agents on spontaneous and olfactory-bulb-ablation-induced muricidal behaviour in the rat. *Arzneim-Forsch.*, **28**, 1508–1509.

Goodwin, F.K., Ebert, M.H. and Bunney, W.E., Jr (1972) Mental effects of reserpine in man: a review. In: R.I. Shader (ed.) *Psychiatric complications of medicinal drugs.* New York: Raven Press, pp. 73–101.

Gray, J.A. and Green, A.R. (1987) Increased $GABA_B$ receptor function in mouse frontal cortex after repeated administration of antidepressant drugs of electroconvulsant shocks. *Br. J. Pharmacol.*, **92**, 357–362.

Gray, J.A., Goodwin, G.M., Heal, D.J. and Green, A.R. (1987) Hypothermia induced by baclofen, a possible index of $GABA_B$ receptor function in mice, is enhanced by antidepressant drugs and ECS. *Br. J. Pharmacol.*, **92**, 863–870.

Iversen, L.L. (1978) Biochemical psychopharmacology of GABA. In: M.A. Lipton, A. Di Mascio and K.F. Killim (eds) *Psychopharmacology: a generation of progress.* New York: Raven Press, pp. 25–38.

Joly, D., Lloyd, K.G., Pichat, P. and Sanger, D.J. (1987) Correlation between the behavioural effect of desipramine and $GABA_B$ receptor regulation in the olfactory bulbectomized rat. *Br. J. Pharmacol.*, **90**, 125P.

Leonard, B.E. (1984) The olfactory bulbectomized rat as a model of depression. *Pol. J. Pharmacol.*, **36**, 561–569.

Lewis, W.C., Calden, G., Thurston, J.R. and Gelson, W.F. (1957) Psychiatric and neurological reactions to cycloserine in the treatment of tuberculosis. *Dis. Chest*, **32**, 172–182.

Lloyd, K.G. (1989) GABA and depression. In: G. Nistico and N.G. Bowery (eds) *GABA: basic research and clinical applications.* Rome: Pythagora Press, pp. 301–343.

Lloyd, K.G. and Morselli, P.L. (1987) Psychopharmacology of GABAergic drugs: In: H. Meltzer (ed.) *Psychopharmacology: the third generation of progress.* New York: Raven Press, pp. 183–195.

Lloyd, K.G. and Pichat, P. (1986) Decrease in $GABA_B$ binding to the frontal cortex of olfactory bulbectomized rats. *Br. J. Pharmacol.*, **87**, 36P.

Lloyd, K.G. and Pichat, P. (1987) GABA synapses, depression and antidepressant drugs. In: S.G. Dahl, L.F. Gram, S.M. Paul and W.Z. Potter (eds) *Clinical pharmacology in psychiatry.* Berlin: Springer-Verlag, pp. 113–126.

Lloyd, K.G., Thuret, F. and Pilc, A. (1985) Upregulation of γ-aminobutyric acid (GABA) B binding sites in rat frontal cortex: a common action of repeated administration of different classes of antidepressants and electroshock. *J. Pharmacol. Exp. Ther.*, **235**, 191–199.

Lloyd, K.G., Zivkovic, B., Scatton, B., Morselli, P.L. and Bartholini, G. (1989) The GABAergic hypothesis of depression. *Progr. Neuro-Psycho-Pharmacol., Biol. Psychiatr.*, **13**, 341–351.

Martin, P., Pichat, P., Messol, J., Soubrie, P., Lloyd, K.G. and Puech, A.J. (in press) Decreased $GABA_B$ receptors in helpless rats: reversal by tricyclic antidepressants. *Neuropsychology.*

Motohashi, N., Ikawa, K. and Kariya, T. (1989) $GABA_B$ receptors are up-regulated by chronic treatment with lithium or carbamazepine: GABA hypothesis of affective disorder? *Eur. J. Pharmacol.*, **166**, 95–99.

Pilc, A. and Lloyd, K.G. (1984) Chronic antidepressants and GABA ‘B’ receptors: a GABA hypothesis of antidepressant drug action. *Life Sci.*, **35**, 2149–2154.

Ruiz, R.C. (1964) D-Cycloserine in the treatment of pulmonary tuberculosis resistant to the standard drugs. *Dis. Chest*, **45**, 181–186.

Schatzberg, A.F. and Cole, J.O. (1978) Benzodiazepines in depressive disorders. *Arch. Gen. Psychiatr.*, **35**, 1359–1365.

Suzdak, P.D. and Gianutsos, G. (1986) Effect of chronic imipramine or baclofen on GABA-B binding and cyclic AMP production in cerebral cortex. *Eur. J. Pharmacol.*, **131**, 129–133.

Szekely, A.M., Barbaccia, M.L. and Costa, E. (1987) Effect of a protracted antidepressant treatment on signal transduction and [^{3}H](−)-baclofen binding at $GABA_B$ receptors. *J. Pharmacol. Exp. Ther.*, **243**, 155–159.

18 $GABA_B$ Binding Sites in Depression and Antidepressant Drug Action

J. A. CROSS, S. C. CHEETHAM, M. R. CROMPTON,
C. L. E. KATONA and R. W. HORTON

INTRODUCTION

Research into the mechanism of action of antidepressant treatments has been dominated for the last 30 years by hypotheses involving noradrenaline (NA) and serotonin (5-hydroxytryptamine, 5-HT). Although antidepressant drugs differ widely in their acute pharmacological interactions with these and other neurotransmitter systems, rather similar adaptive changes in several classes of central NA and 5-HT receptors have been reported following repeated administration of antidepressants in rodents. The most consistent finding is a reduction in the number of cortical β-adrenoceptor binding sites and in the activity of NA-stimulated adenylate cyclase following most classes of antidepressant drugs (including classical tricyclics, monoamine oxidase inhibitors and novel antidepressants that lack NA/5-HT uptake blocking properties) and following repeated electroconvulsive shock (ECS) (for review, see Sugrue, 1983). Cortical 5-HT_2 receptor binding sites are also decreased by most classes of antidepressant drugs (Peroutka and Snyder, 1980). These similarities have led to the suggestion that antidepressant treatments may act through a common mechanism. However, exceptions to these findings would seem to preclude reduction in both β-adrenoceptor binding and 5-HT_2 receptor binding as a final common pathway in antidepressant action. The more potent and selective 5-HT reuptake inhibitors (such as paroxetine, fluvoxamine and citalopram) do not appear to reduce β-adrenoceptor binding (Nelson *et al.*, 1989) and ECS increases 5-HT_2 receptor binding (Green and Nutt, 1987).

$GABA_B$ RECEPTORS AND ANTIDEPRESSANT ACTION

The involvement of other neurotransmitter systems in antidepressant action had received much less attention until the proposal was made that a GABA mechanism, mediated via $GABA_B$ synapses, represents a mode of action that is shared by all forms of antidepressant treatment (Lloyd *et al.*, 1985).

$GABA_B$ Receptors in Mammalian Function. Edited by N.G. Bowery, H. Bittiger and H.-R. Olpe

Initial experiments demonstrated a marked increase in the number of rat frontal cortical $GABA_B$ binding sites (by 38–66%, with no significant change in binding affinity) following 18 days s.c. infusion of amitriptyline, desmethylimipramine (DMI), citalopram, viloxazine or pargyline (Pilc and Lloyd, 1984). Subsequently, similar findings were reported for all of 14 established or putative antidepressants, representing the entire chemical spectrum of antidepressant drugs, and for repeated ECS, but not for other psychotropic drugs including neuroleptics, anxiolytics and anticonvulsants (Lloyd *et al.*, 1985). The effect of DMI was time- and dose-dependent, with a significant increase in $GABA_B$ binding following 18 days s.c. infusion at 1.25 mg/kg/day. This dose is lower than that which has generally been demonstrated to reduce β-adrenoceptor and $5\text{-}HT_2$ binding, although administration by s.c. infusion, rather than the more widely used i.p. or oral routes, makes such comparisons problematical.

To compare the effects of antidepressants on $GABA_B$ receptors with the well-documented effects on monoamine receptors, we have measured $GABA_B$ and $5\text{-}HT_2$ binding in rat frontal cortex following 21 days administration of DMI or zimelidine (ZIM), once daily at 10 mg/kg i.p. or twice daily orally at 5 and 10 mg/kg. Rats treated with vehicle or a single dose of drug (at the higher dose) were studied for comparison. Animals were killed 24 h after the final treatment. For $5\text{-}HT_2$ binding, tissue was stored at −20 °C and saturation binding of $[^3H]$ketanserin (8 concentrations 0.1–5 nM; specific binding defined with 1 μM methysergide) was subsequently performed according to Leysen *et al.* (1982). For $GABA_B$ binding, membranes were prepared according to the method of Bowery *et al.* (1983) and stored in 20 volumes 50 mM Tris–HCl, pH 7.4 at −20 °C. On the day of assay, membranes were allowed to thaw at room temperature and were washed three times with 50 mM Tris–HCl, pH 7.4, containing 2 mM $CaCl_2$. Saturation binding to $GABA_B$ sites was determined with $[^3H]$GABA (1 nM alone or in the presence of 5–150 nM unlabelled GABA) in 50 mM Tris–HCl, pH 7.4, containing 2 mM $CaCl_2$ and 40 μM isoguvacine. Specific binding was defined with 100 μM (±)-baclofen. Equilibrium dissociation constants (K_D) and maximal number of binding sites (B_{max}) were determined by computerized non-linear regression analysis.

Repeated i.p. administration of DMI (10 mg/kg) reduced $5\text{-}HT_2$ binding site B_{max} by 16%, with no change in affinity, whereas the same dose of ZIM did not significantly alter the affinity or number of $5\text{-}HT_2$ sites. Neither drug significantly altered the number or affinity of $GABA_B$ binding sites (Cross and Horton, 1987).

Twice daily oral administration of DMI significantly reduced $5\text{-}HT_2$ binding B_{max} by 24% (5 mg/kg) and 37% (10 mg/kg); the effect of ZIM was significant only at 10 mg/kg: $5\text{-}HT_2$ binding B_{max} was reduced by 22% (Table 1). Again, neither drug significantly altered the number or affinity of $GABA_B$ binding sites (Table 1).

Our results clearly demonstrate that chronic administration of DMI and

Table 1. Effect of oral administration of desmethylimipramine (DMI) and zimelidine (ZIM) on rat frontal cortical 5-HT$_2$ and GABA$_B$ binding

	5-HT$_2$		GABA$_B$	
	K_D (nM)	B_{max} (fmol/mg protein)	K_D (nM)	B_{max} (pmol/mg protein)
Control	0.59 ± 0.02	290 ± 13	17 ± 3	1.06 ± 0.08
DMI, 5 mg/kg	0.60 ± 0.06	219 ± 6*	14 ± 1	1.06 ± 0.13
DMI, 10 mg/kg	0.67 ± 0.05	183 ± 10**	14 ± 3	0.99 ± 0.14
ZIM, 5 mg/kg	0.57 ± 0.04	242 ± 13	16 ± 4	1.16 ± 0.18
ZIM, 10 mg/kg	0.62 ± 0.02	225 ± 13*	18 ± 2	1.19 ± 0.13

DMI and ZIM at the doses specified were administered twice daily for 21 days. K_D, equilibrium dissociation constant; B_{max}, maximal number of binding sites. Values are means ± SEM ($n = 6$–8). Significant differences from control are denoted by $^*P < 0.05$, $^{**}P < 0.01$ (Student's t-test, unpaired, two-tailed).

ZIM, under conditions where the number of 5-HT$_2$ binding sites were reduced, produced no significant alteration in GABA$_B$ binding in homogenates of frontal cortex. Our results do not preclude localized alterations in GABA$_B$ binding such as reported by Pratt and Bowery (Chapter 19).

A number of methodological differences may account for the discrepancies between our results and those of Lloyd *et al.* (1985). Subcutaneous infusion of drugs with osmotic minipumps is likely to produce more stable brain drug concentrations than the single daily i.p. or twice daily oral administration we have used. However, this is unlikely to be of major importance since significant increases in GABA$_B$ binding were found following once daily i.p. administration of fluoxetine, progabide, fengabine and sodium valproate (Lloyd *et al.*, 1985).

A further consideration is the time interval between drug discontinuation and sacrifice of the animals. Lloyd *et al.* (1985) killed their animals 72 h after drug discontinuation, whereas we chose 24 h. Since neither DMI nor ZIM interacts directly with GABA$_B$ binding sites in vitro, our results are unlikely to be influenced by residual drug effects. Indeed we found no significant difference in GABA$_B$ binding in a separate series of rats killed 72 h after twice daily oral administration of DMI (5 mg/kg) for 21 days (Cross and Horton, 1988).

Tissue preparation and storage also needs to be considered. Lloyd *et al.* (1985) froze their tissue at −80 °C for up to 72 h and stored their crude membranes at −20 °C for at least 24 h prior to assay. We prepared crude membranes immediately from fresh tissue and stored them at −20 °C in 20 volumes Tris–HCl, pH 7.4. We have examined the effects of different storage conditions and times on GABA$_B$ binding (Figure 18.1). Storage in 20 volumes of buffer at −20 °C gave the lowest K_D values. B_{max} values were little influenced by storage conditions, except that lower values were found when pellets were stored

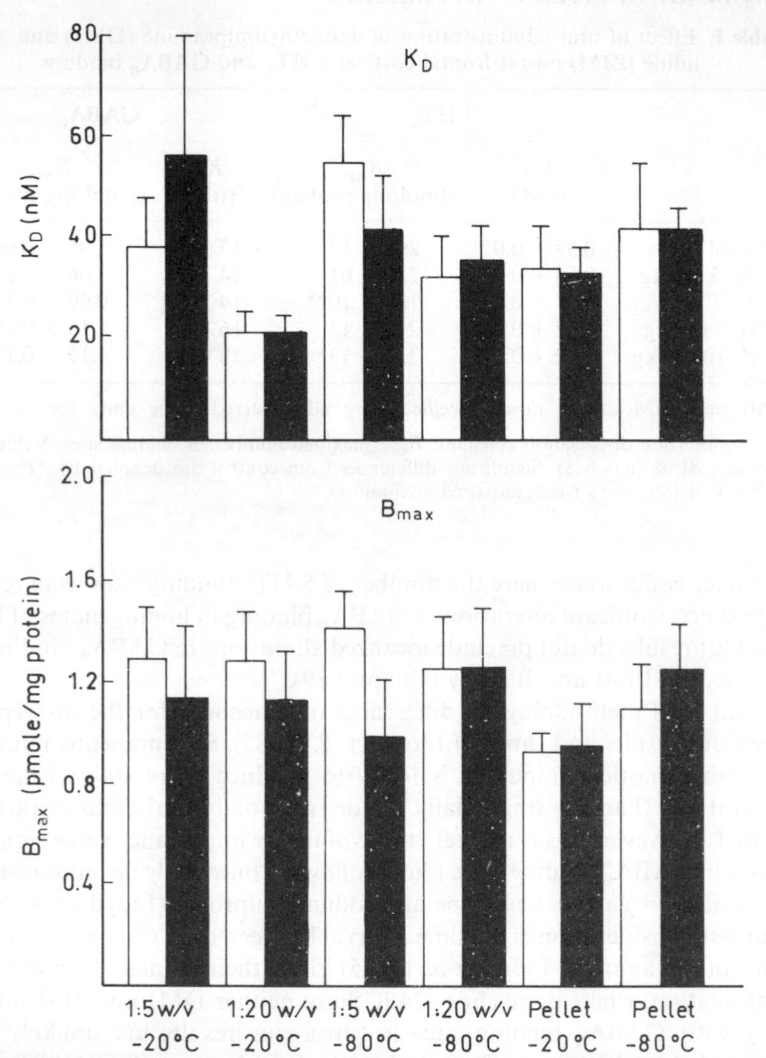

Figure 18.1. Rat cortical membranes were prepared by the method of Bowery *et al.* (1983) and stored in 5 or 20 volumes (w/v) Tris buffer or as pellets without buffer at − 20 °C or − 80 °C for 48 h (□) or 1 month (■) prior to assay. Values are means ± SEM for 4 determinations. K_D, equilibrium dissociation constant; B_{max}, maximal number of binding sites

in the absence of buffer at − 20 °C. Duration of tissue storage (48 h or 1 month) did not significantly alter K_D or B_{max} values under any of the conditions studied.

Several authors have reported the presence of high- and low-affinity $GABA_B$ binding sites, using [^{3}H]GABA and [^{3}H]baclofen as ligands. Lloyd *et al.* (1985) reported only a single high-affinity component in membranes prepared from control rats and rats treated with nomifensine and DMI (1.25 mg/kg/day). However, following DMI (5 mg/kg/day) and ZIM (10 mg/kg/day), an additional lower affinity site, with a B_{max} of 2–3 pmol/mg protein, was evident. We have, by using a restricted range of ligand concentrations (up to 150 nM), limited our studies to the high-affinity GABA$_B$ binding sites. An increase in GABA$_B$ binding following antidepressants which predominantly affected the lower affinity sites could provide an explanation for the discrepancies between our results and those of Lloyd *et al.* (1985). However, the findings of other studies are not entirely compatible with this interpretation. Suzdak and Gianutsos (1986) observed high- (K_D 30 nM) and low- (K_D 500 nM) affinity GABA$_B$ binding sites in control mouse cerebral cortex and the number of both high- and low-affinity sites was higher (by 24% and 15%, respectively) following once daily i.p. administration of imipramine for 14 days, albeit at a relatively high dose of 32 mg/kg. However, these findings are not directly comparable with those of Lloyd *et al.* (1985) or our own because the membranes were treated with 0.03% Triton X-100. A slightly higher concentration of Triton X-100 (0.05%) has previously been shown to increase the affinity of high-affinity GABA$_B$ binding but to reduce the number of sites by 50% compared to that in frozen and thawed membranes (Bowery *et al.*, 1983).

Szekely *et al.* (1987) reported higher B_{max} of frontal cortical GABA$_B$ binding sites, labelled with [^{3}H]GABA in the presence of bicuculline, following twice daily i.p. administration of DMI and imipramine but not following maprotiline. A single binding component was reported and inspection of Scatchard plots gave no evidence of deviation from linearity with ligand concentrations up to 400 nM. However, this study provides some evidence that perhaps only a subpopulation of GABA$_B$ binding sites are influenced by antidepressants, since sites labelled with [^{3}H](−)-baclofen, whose density was less than half of those sites labelled with [^{3}H]-GABA in the presence of bicuculline, were not significantly altered by repeat antidepressant administration.

GABA AND DEPRESSION

Although not extensively studied, there is some evidence to suggest that GABA function may be altered in drug-free depressed patients (reviewed by Cross *et al.*, 1988). Cerebrospinal fluid GABA concentration has been reported to be lower in depressed patients than in patients with other psychiatric or neurological disorders and healthy controls. Plasma GABA concentration is also low in unipolar depressed patients. Indices of GABA function in brain tissue from depressed patients have been little studied, although a decrease in glutamic acid decarboxylase (GAD) activity has been reported (Perry *et al.*, 1977).

We have attempted to evaluate the role of GABA in clinical depression by studies in brain tissue, obtained at autopsy from subjects dying by suicide. Suicide is associated most frequently with depression but also with other diagnoses such as schizophrenia, personality disorder and alcoholism (Barraclough *et al.*, 1974). In order to relate our findings more specifically to depression, we have selected from a large group of suicides only those subjects in whom a firm retrospective diagnosis of depression, based on documentary medical evidence, could be established. The depressed suicides studied died by hanging ($n = 7$), carbon monoxide poisoning ($n = 5$), drug overdose ($n = 5$), self-inflicted wounding ($n = 2$), and drowning ($n = 1$). Postmortem blood samples were assayed for psychoactive drugs and evidence on previous and current drug treatment was sought. Thirteen of the suicides had not been prescribed antidepressants recently and none were detected in their blood at autopsy. Six suicides were receiving antidepressants: three were receiving an antidepressant alone, one an antidepressant and a neuroleptic, and two an antidepressant and a benzodiazepine. Controls were subjects who had died suddenly from causes not involving the central nervous system, mainly myocardial infarction, and were without documentary evidence of mental illness. Control subjects were individually matched to the suicides for age and sex. An upper age limit of 60 years was imposed to minimize the likelihood of morphological abnormalities. Saturation binding of [^{3}H]GABA was performed in frontal (Brodmann area 11) and temporal cortical (Brodmann area 21/22) samples using the same membrane preparation and assay as we used for rat tissue. Hippocampal $GABA_B$ binding was measured at two ligand concentrations (1 and 51 nM). Assays were performed on coded samples, without knowledge of subject classification. Samples from suicides and controls were assayed concurrently. Tissue availability did not allow us to assay each region in all subjects.

Controls and suicides were well matched for age, sex, postmortem delay (the time from death to storage of tissue at -80 °C) and the duration of tissue storage prior to assay (Table 2).

Mean B_{max} values for the drug-free suicides did not differ significantly from those for the controls in frontal or temporal cortex (Table 3), nor did specific binding in the hippocampus (Table 4). K_D values did not differ in frontal cortex but were significantly higher in the drug-free suicides than in the controls in temporal cortex (Table 3). $GABA_B$ binding did not differ from that in the controls in those suicides who were receiving antidepressants (alone or in combination with other drugs) prior to death (Tables 3 and 4).

The affinity and number of $GABA_B$ binding sites in postmortem human brain are similar to values reported for rat brain. The lack of significant correlations between postmortem delay and tissue storage and $GABA_B$ binding demonstrates the stability of these sites after death and on tissue storage, and thus measurements in postmortem brain are likely to reflect the in-vivo situation immediately before death.

The results from the drug-free suicides suggest that depression is not associated with an abnormality in the number of high-affinity $GABA_B$ recogni-

Table 2. Demographic details of depressed suicides and controls

	n	Age (years)	Sex	Postmortem delay (h)	Storage time (weeks)
Controls	16	39 ± 3 (18–54)	12M, 4F	39.0 ± 3.7 (15–71)	85.2 ± 4.9 (50–116)
Depressed suicides	16	39 ± 3 (16–57)	12M, 4F	33.9 ± 4.5 (15–69)	83.1 ± 11.0 (15–156)

Data are expressed as means ± SEM; the range is shown in parentheses.

tion sites within the brain areas studied. In contrast, we have some evidence to suggest that abnormalities in GABA$_A$ receptors may be associated with depression. We found higher numbers of benzodiazepine binding sites (a component of the GABA$_A$ receptor–ionophore complex) in frontal cortex, but not in temporal cortex, hippocampus or amygdala, of depressed suicides (Cheetham *et al.*, 1988b; Stocks *et al.*, 1990). GAD activity was markedly reduced in those suicides who had died by carbon monoxide poisoning. However, when these subjects were excluded, there were no differences in GAD activity between suicides and controls. A previous report of markedly reduced GAD activity in depressed subjects (Perry *et al.*, 1977) may be related more to hypoxia associated with terminal illness than to depression.

Table 3. GABA$_B$ binding sites in the cortex of depressed suicides and controls

	Controls		Depressed suicides	
	K_D (nM)	B_{max} (pmol/mg protein)	K_D (nM)	B_{max} (pmol/mg protein)
Frontal cortex				
All (16)	28 ± 2	1.02 ± 0.06	29 ± 2	0.99 ± 0.06
Drug-free (11)	30 ± 3	1.06 ± 0.08	30 ± 3	1.01 ± 0.08
Drug-treated (5)	22 ± 4	0.94 ± 0.08	26 ± 5	0.95 ± 0.12
Antidepressant-treated (4)	22 ± 5	0.96 ± 0.10	28 ± 5	1.00 ± 0.15
Temporal cortex				
All (16)	22 ± 3	0.68 ± 0.06	29 ± 3	0.76 ± 0.06
Drug-free (11)	19 ± 1	0.62 ± 0.07	29 ± 3*	0.72 ± 0.07
Drug-treated (5)	30 ± 8	0.81 ± 0.11	28 ± 4	0.85 ± 0.12
Antidepressant-treated (4)	31 ± 10	0.73 ± 0.10	25 ± 4	0.83 ± 0.16

K_D, equilibrium dissociation constant; B_{max}, maximal numbers of binding sites. Values are means ± SEM for the number of subjects shown in parentheses. Significant differences between control and suicide groups are denoted by *$P < 0.05$ (Student's t-test, unpaired, two-tailed). The drug-treated subject not included in the antidepressant-treated group was receiving diazepam alone.

Table 4. $[^3H]GABA_B$ binding in the hippocampus of depressed suicides and controls

	$[^3H]GABA_B$ binding (fmol/mg protein)			
	Controls		Depressed suicides	
	1 nM	51 nM	1 nM	51 nM
All (17)	16.8 ± 1.3	280 ± 30	16.6 ± 1.5	279 ± 31
Drug-free (13)	16.1 ± 1.3	258 ± 31	15.4 ± 1.6	270 ± 38
Drug-treated (4)	19.0 ± 3.8	351 ± 74	20.5 ± 3.8	309 ± 59
Antidepressant-treated (3)	22.3 ± 2.4	376 ± 98	16.7 ± 0.5	253 ± 23

Specific $[^3H]GABA_B$ binding was measured at two ligand concentrations (1 and 51 nM). Values are means ± SEM for the number of subjects shown in parentheses. The drug-treated subject not included in the antidepressant-treated group was receiving diazepam alone.

Other studies from our laboratory have also identified differences in monoamine neurotransmitter systems in comparable groups of subjects. For example, the number of β-adrenoceptor binding sites was significantly lower in several cortical areas (De Paermentier *et al.*, 1989) and the number of $5\text{-}HT_1$ and $5\text{-}HT_2$ binding sites was lower in the hippocampus (Cheetham *et al.*, 1988a, 1989) of drug-free depressed suicides than in controls.

Although no differences in $GABA_B$ binding were apparent in suicides who had received antidepressants prior to death, it would be premature to draw any conclusions from these data. The number of subjects is small, they were receiving different antidepressants (some in combination with other psychoactive drugs) and we have no evidence on compliance other than that the concentrations of drugs in the blood at autopsy were within the therapeutic range.

An unexpected finding from the present study was a striking negative correlation between age and B_{max} of $GABA_B$ binding which was restricted to the temporal cortex. This was apparent within the total control and suicide groups ($r = -0.66$, $P = 0.001$) and when drug-treated subjects were excluded ($r = -0.57$, $P = 0.006$), and is thus likely to be a genuine age-related effect rather than the result of some other variable. Age-related decreases in other neurotransmitter receptors have been reported in the human brain, although such effects are most prominent in the elderly. The present effect is surprising since the oldest subject was 57 years of age. A more detailed study of $GABA_B$ binding in a larger group including elderly subjects would seem to be justified.

ACKNOWLEDGEMENT

We thank the Wellcome Trust for financial support.

REFERENCES

Barraclough, B., Bunch, J., Nelson, B. and Sainsbury, P. (1974) A hundred cases of suicide: clinical aspects. *Br. J. Psychiatr.*, **125**, 355–373.

Bowery, N.G., Hill, D.R. and Hudson, A.L. (1983) Characteristics of GABA$_B$ receptor binding sites on rat whole brain synaptic membranes. *Br. J. Pharmacol.*, **78**, 191–206.

Cheetham, S.C., Crompton, M.R., Katona, C.L.E. and Horton, R.W. (1988a) Brain 5-HT$_2$ receptor sites in depressed suicide victims. *Brain Res.*, **443**, 272–280.

Cheetham, S.C., Crompton, M.R., Katona, C.L.E., Parker, S.J. and Horton, R.W. (1988b) Brain GABA$_A$/benzodiazepine binding sites and glutamic acid decarboxylase activity in depressed suicide victims. *Brain Res.*, **460**, 114–123.

Cheetham, S.C., Crompton, M.R., Katona, C.L.E. and Horton, R.W. (1989) 5-HT$_1$ and 5-HT$_{1A}$ binding sites in depressed suicides. *Br. J. Pharmacol.*, **96**, 1P.

Cross, J.A. and Horton, R.W. (1987) Are increases in GABA$_B$ receptors consistent findings following chronic antidepressant administration? *Eur. J. Pharmacol.*, **141**, 159–162.

Cross, J.A. and Horton, R.W. (1988) Effects of chronic oral administration of the antidepressants, desmethylimipramine and zimelidine on rat cortical GABA$_B$ binding sites: a comparison with 5-HT$_2$ binding site changes. *Br. J. Pharmacol.*, **93**, 331–336.

Cross, J.A., Cheetham, S.C., Crompton, M.R., Katona, C.L.E. and Horton, R.W. (1988) Brain GABA$_B$ binding sites in depressed suicide victims. *Psychiatr. Res.*, **26**, 119–129.

De Paermentier, F., Cheetham, S.C., Crompton, M.R., Katona, C.L.E. and Horton, R.W. (1989) Lower cortical β-adrenoceptor binding sites in post-mortem samples from depressed suicides. *Br. J. Pharmacol.*, **98**, 812P.

Green, A.R. and Nutt, D.J. (1987) Psychopharmacology of repeated seizures: possible relevance to the mechanism of action of electroconvulsive therapy. In: L.L. Iversen, S.D. Iversen and S.H. Snyder (eds) *Handbook of psychopharmacology*, vol. 19, New York: Plenum Press. pp. 375–419.

Leysen, J.E., Niemegeers, C.J.E., Van Neuten, J.M. and Laduron, P.M. (1982) [^{3}H] Ketanserin (R 41 468), a selective ^{3}H-ligand for serotonin$_2$ receptor binding sites: binding properties, brain distribution and functional role. *Mol. Pharmacol.*, **21**, 301–314.

Lloyd, K.G., Thuret, F. and Pilc, A. (1985) Upregulation of γ-aminobutyric acid (GABA)$_B$ binding sites in rat frontal cortex: a common action of repeated administration of different classes of antidepressants and electroshock. *J. Pharmacol. Exp. Ther.*, **235**, 191–199.

Nelson, D.R., Thomas, D.R. and Johnson, A.M. (1989) Pharmacological effects of paroxetine after repeated administration to animals. *Acta Psychiatr. Scand.*, **80** (Suppl. 350), 21–23.

Peroutka, S.J. and Snyder, S.H. (1980) Long-term antidepressant treatment decreases spiroperidol-labelled serotonin receptor binding. *Science*, **210**, 88–90.

Perry, E.K., Gibson, P.H., Blessed, G., Perry, R.H. and Tomlinson, B.E. (1977) Neurotransmitter enzyme abnormalities in senile dementia: choline acetyltransferase and glutamic acid decarboxylase in necropsy brain tissue. *J. Neurol. Sci.*, **34**, 247–265.

Pilc, A. and Lloyd, K.G. (1984) Chronic antidepressants and GABA$_B$ receptors: a GABA hypothesis of antidepressant drug action. *Life Sci.*, **35**, 2149–2154.

Stocks, G.M., Cheetham, S.C., Crompton, M.R., Katona, C.L.E. and Horton, R.W. (1990) Benzodiazepine binding sites in the brains of depressed suicides. *J. Affect. Disord.*, **18**, 11–15.

Sugrue, M.F. (1983) Chronic antidepressant therapy and associated changes in central monoaminergic functioning. *Pharmacol. Ther.*, **21**, 1–33.

Suzdak, P.D. and Gianutsos, G. (1986) Effects of chronic imipramine or baclofen on GABA-B binding and cyclic AMP production in cerebral cortex. *Eur. J. Pharmacol.*, **131**, 129–133.

Szekely, A.M., Barbaccia, M.L. and Costa, E. (1987) Effect of a protracted antidepressant treatment on signal transduction and [^{3}H](−)baclofen binding at $GABA_B$ receptors. *J. Pharm. Exp. Ther.*, **243**, 155–159.

19 Autoradiographic Analysis of $GABA_B$ Receptors in Rat Frontal Cortex Following Chronic Antidepressant Administration

G. D. PRATT and N. G. BOWERY

INTRODUCTION

Adaptations of the populations of central cortical monoamine receptors of rodents following chronic antidepressant administration are now well documented; downregulation of the numbers of β-adrenoceptors and 5-hydroxytryptamine (serotonin, 5-HT_2) receptors has been consistently obtained with a variety of antidepressant classes.

Recently, however, focus has turned to the GABAergic system following the observation that levels of GABA are reduced in the cerebrospinal fluid and plasma of depressed patients (Berrettini *et al.*, 1982). This evidence, together with the fact that the GABA agonists, progabide and fengabine, have antidepressant action in animal models as well as clinically (Lloyd *et al.*, 1983, 1987; Musch and Garreau, 1986), has directed attention to the GABA receptor and in particular to the possible alterations in the $GABA_B$ receptor subtype following chronic antidepressant administration. This stemmed from the observations of Lloyd *et al.* (1985), who demonstrated increases in $GABA_B$ binding to rat cerebral cortical membranes following 18-day s.c. infusions of a variety of antidepressants. These effects have been substantiated using i.p. administration (Szekely *et al.*, 1987) and also in mouse cortical membranes by Suzdak and Gianutsos (1986), who reported that the potentiation of noradrenaline (NA) -stimulated cyclic AMP production by the $GABA_B$ agonist, baclofen, was enhanced by chronic imipramine, suggesting an increased functionality of $GABA_B$ receptors.

Not all researchers, however, have been able to demonstrate an upregulation of $GABA_B$ receptors after such chronic dosing regimes. Indeed, Cross and Horton (1988) failed to observe any significant changes in $GABA_B$ binding sites following twice daily oral administration for 21 days of desipramine or zimelidine, in spite of the fact that the 5-HT_2 binding site density was significantly reduced in the same population of rat cortical membranes.

$GABA_B$ Receptors in Mammalian Function. Edited by N.G. Bowery, H. Bittiger and H.-R. Olpe

The aim of this study, therefore, was to employ receptor autoradiography in an attempt to resolve these discrepancies. If indeed $GABA_B$ receptor densities are increased following chronic antidepressant treatment, it could be that these changes are discretely localized within certain cerebral cortical laminae which would be considerably diluted in a membrane preparation. Autoradiography offers the advantage of retaining the morphology of the brain intact, thus enabling the localization of discrete regions. If changes in $GABA_B$ binding sites are discretely localized, then an autoradiographic approach should be able to detect the change. The antidepressants employed in this study were selected on the basis of their monoamine uptake properties; thus while amitriptyline has mixed NA/5-HT uptake blocking activity, desipramine is more selective for NA, and paroxetine is a specific and potent inhibitor of 5-HT uptake. In addition to the possible modulation of $GABA_B$ receptors, the effects of the above antidepressants on β-adrenoceptor binding were also examined using the ligand $[^{125}I]$iodopindolol to correlate the possible upregulation of $GABA_B$ receptors with downregulated β-adrenoceptors.

METHODS

Drug Treatment and Tissue Preparation

Male CFY rats (140–180 g) received amitriptyline (26.2 mg/kg/day), desipramine (17.1 mg/kg/day) or paroxetine (8.3 mg/kg/day) via the drinking water for 21 days (these drug doses were calculated on the basis of the amount of drinking water consumed by each rat per day). Following a 24-h wash-out period, rats were anaesthetized with sodium pentobarbitone (40 mg/kg i.p.) and perfused-fixed with 250 ml 0.1% paraformaldehyde in 0.01 M phosphate-buffered saline (pH 7.4) via intracardiac administration. Brains were then frozen in isopentane cooled to -40 °C in liquid nitrogen and stored at -80 °C until use. Sagittal and coronal sections (10 μm) were cut at -20 °C and thaw-mounted onto glass slides.

$GABA_B$ Receptor Autoradiography

Brain sections were preincubated for 60 min at room temperature in 50 mM Tris–HCl buffer (pH 7.4) containing 2.5 mM $CaCl_2$. After drying, sections were incubated for 20 min with 150 μl buffer containing $[^3H]$GABA (25–400 nM). Labelling of $GABA_B$ receptors was achieved in the presence of 40 μM isoguvacine and non-specific binding defined by 100 μM ($-$)-baclofen. Following incubation, sections were washed for two 3-s periods in ice-cold buffer and dried rapidly in air before apposition to tritium-sensitive film.

β-Adrenoceptor Autoradiography

Sections were incubated for 60 min at room temperature in Tris–saline buffer (20 mM Tris, 135 mM NaCl; pH 7.4) containing (−)-[^{125}I]iodopindolol (37.5–600 pM). Non-specific binding was defined using 200 μM (−)-isoprenaline whilst the selective labelling of β_1-adrenoceptors was achieved in the presence of the β_2 antagonist ICI 118551 (50 nM) as described by Ordway *et al.* (1985). Binding to β_2-adrenoceptors was taken as the difference between total and β_1-adrenoceptor binding. Following incubation, sections were washed for two 5-min periods in buffer at 4 °C, given a rapid 3-s rinse in ice-cold distilled water to remove buffer salts and then quickly dried in a stream of cold air. Sections and calibrated brain paste standards were apposed to LKB [^{3}H] 'Ultrofilm' to generate autoradiograms.

Optical densities of both [^{3}H]GABA and (−)-[^{125}I]iodopindolol binding to the individual laminae of the frontal cortex were measured in nCi/mg tissue using a Quantimet 970 image analysis system (Cambridge Instruments) and converted to fmol/mg tissue. The kinetic parameters K_D (equilibrium dissociation constant) and B_{max} (maximal number of binding sites) of the binding were derived by Scatchard analysis and statistical comparisons between control and drug-treated animals made using Student's *t*-test (two-tailed).

Autoradiographic density measurements were restricted to the frontal cortex, which was taken to be represented by the anterior third of the cerebral cortex. Both $GABA_B$ and β-adrenoceptor binding was found to be clearly laminated; for $GABA_B$ binding the laminations were resolved into four distinct regions represented as laminae I, II–III, V and VI, whilst β-adrenoceptor binding fell into three bands denoted by laminae I, II–III and V–VI.

RESULTS AND DISCUSSION

$GABA_B$ Receptor Modulation by Antidepressants

The effects of amitriptyline, desipramine and paroxetine on $GABA_B$ binding in rat frontal cortex are visualized autoradiographically in Figure 19.1, whilst the kinetic parameters of the binding are summarized in Table 1. From the Scatchard plots for $GABA_B$ binding in lamina I of the frontal cortex of control and antidepressant-treated rats (Figure 19.2), it was apparent that $GABA_B$ binding was upregulated by each of the three treatments; however, only desipramine produced a significant increase and this corresponded to 32% ($P < 0.05$) above control. No significant changes in K_D values were obtained, indicating a lack of change in affinity. In laminae II–III (possessing the highest density of $GABA_B$ sites), desipramine and paroxetine produced smaller but non-significant increases in B_{max}, whereas amitriptyline was without effect. From the localized increase in $GABA_B$ binding produced by desipramine in

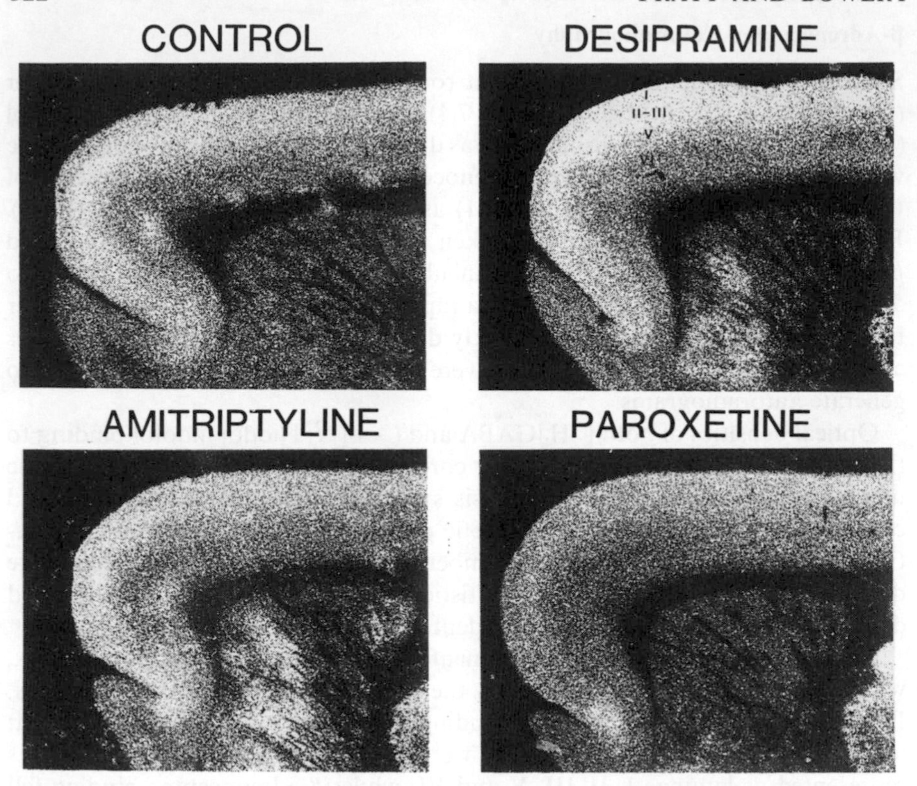

Figure 19.1. Autoradiograms showing [^{3}H]GABA binding to $GABA_B$ receptors in the frontal cortex of control and antidepressant-treated rats

lamina I of the frontal cortex, it could be inferred that the modulation of $GABA_B$ receptors by this particular antidepressant may have occurred as a consequence of the selective inhibition of the uptake of NA since amitriptyline and paroxetine were without effect under the conditions employed in this study. This hypothesis would contradict the findings of Lloyd *et al.* (1985) since both amitriptyline and fluoxetine (also a selective 5-HT uptake inhibitor) in addition to desipramine produced increases in the $GABA_B$ receptor density of rat cortical membranes of between 51% and 83% after chronic s.c. administration. Chronic administration of amitriptyline (and iprindole) prenatally, however, has been shown by Knott *et al.* (Chapter 20) to upregulate $GABA_B$ binding in the outer frontal cortex of neonatal rats.

β-Adrenoceptor Modulation by Antidepressants

Autoradiographic representation of (−)-[^{125}I]iodopindolol binding to β-adrenoceptors in sagittal and coronal sections of rat brain is shown in Figure 19.3.

Table 1. The effect of antidepressants on [^{3}H]GABA binding to $GABA_B$ receptors in rat frontal cortex

	Frontal cortex lamina			
	I	II–III	V	VI
	B_{max} (fmol/mg tissue)			
Control	232.8	280.4	213.9	194.1
Amitriptyline	264.0	284.3	214.7	189.1
Desipramine	307.2*	312.7	245.8	198.3
Paroxetine	275.5	304.3	241.2	205.3
	K_D (nM)			
Control	54.5	54.6	62.3	98.6
Amitriptyline	49.0	45.9	43.3	87.6
Desipramine	79.9	68.2	79.3	105.6
Paroxetine	50.5	58.4	62.4	98.3

B_{max}, maximal number of binding sites; K_D, equilibrium dissociation constant.
*$P < 0.05$ (Student's t-test).

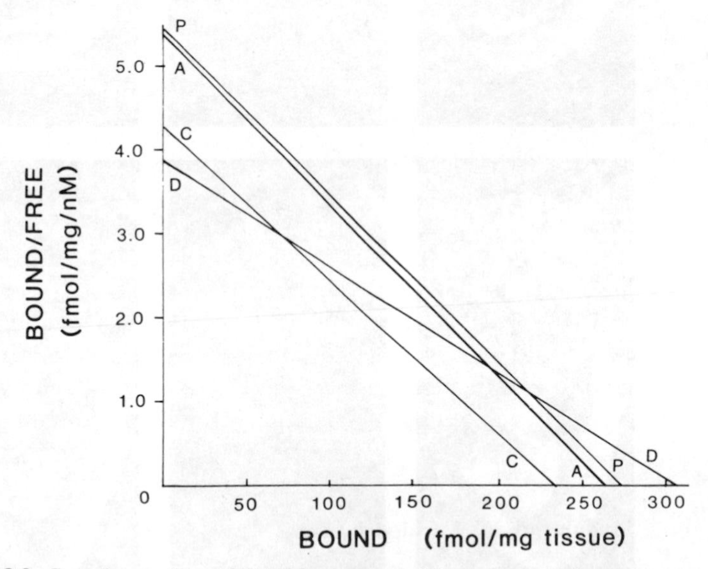

Figure 19.2. Scatchard plots of $GABA_B$ receptor binding to the frontal cortex (lamina I) of control (C) and amitriptyline (A), desipramine (D) or paroxetine (P) -treated rats

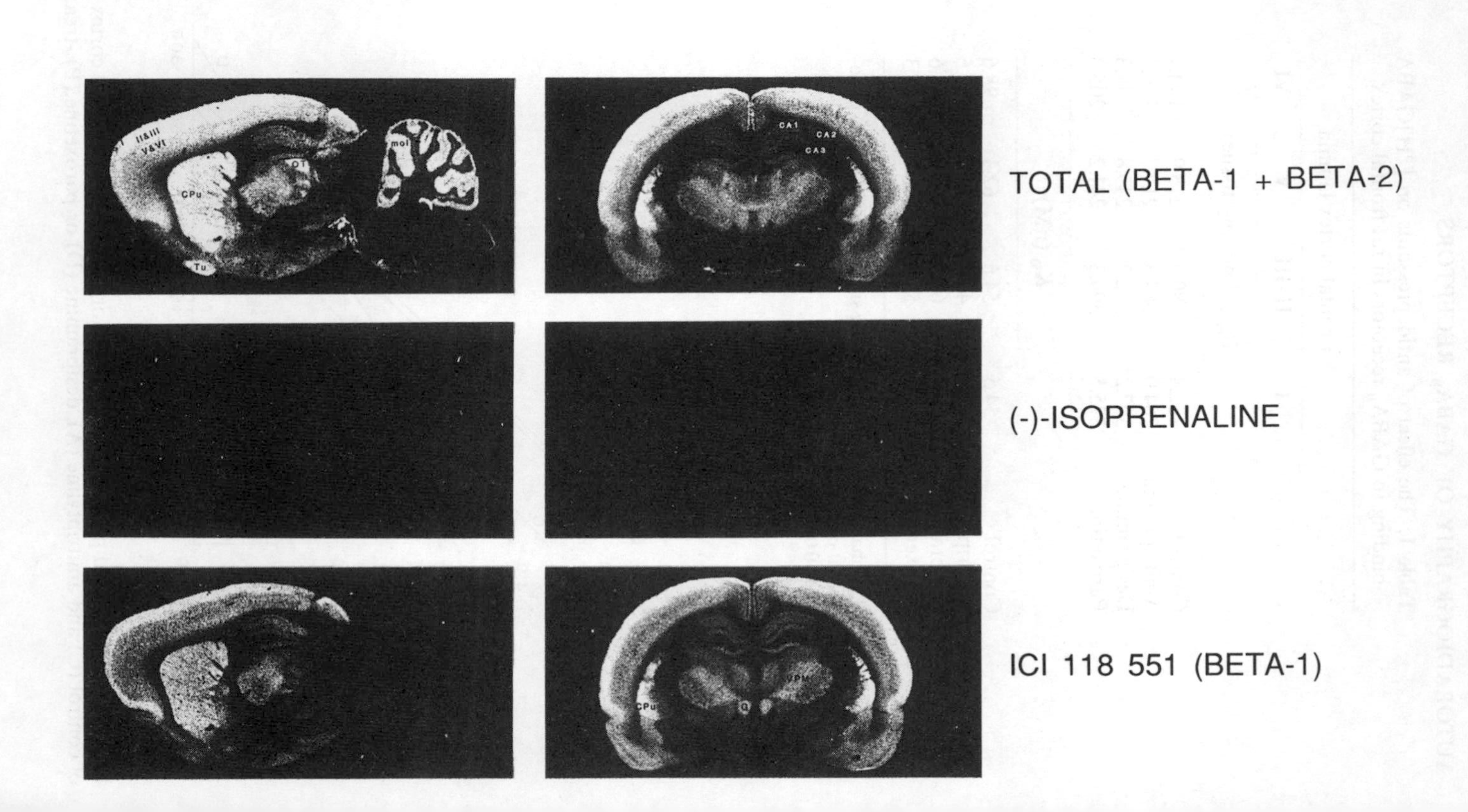
TOTAL (BETA-1 + BETA-2)
(-)-ISOPRENALINE
ICI 118 551 (BETA-1)
II&III
V&VI
CPu
Tu
mol
CA1
CA2
CA3
VPM

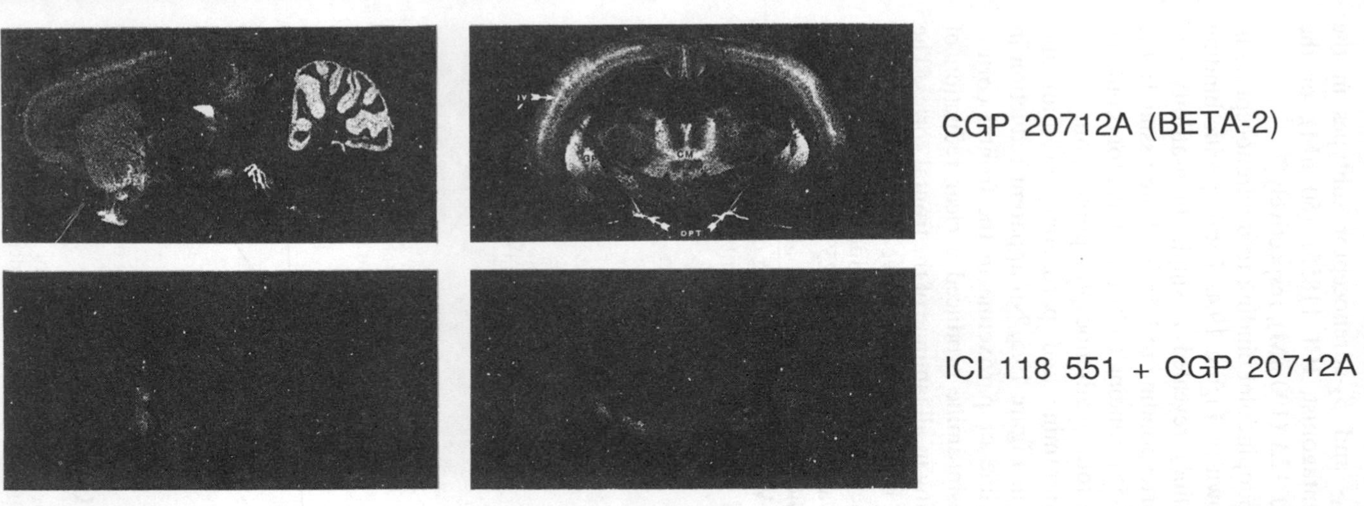

Figure 19.3. Autoradiograms of (−)-[^{125}I]iodopindolol binding to β-adrenoceptors in sagittal and coronal sections of rat brain. β_1-Adrenoceptors (defined in the presence of ICI 118551) predominate in the cerebral cortex, hippocampus (CA1 region), caudate putamen (CPu), ventral posteromedial thalamic nucleus (VPM) and gelatinosis thalamic nucleus (G). β_2-Adrenoceptors (defined in the presence of CGP 20712A) predominate in the molecular layer of the cerebellum (mol), nucleus of the optic tract (OT), optic tract (OPT), lamina IV of the parietal cortex (IV), olfactory tubercle (Tu), central medial thalamic nucleus (CM) and globus pallidus (GP)

This binding was resolved into β_1- and β_2-adrenoceptor subtypes in the presence of the β_2-adrenoceptor antagonist, ICI 118551 (50 nM), or the β_1-adrenoceptor antagonist, CGP 20712A (100 nM), respectively.

Scatchard analysis of (−)-[^{125}I]iodopindolol binding to β-adrenoceptors in lamina I of rat frontal cortex is shown in Figure 19.4. Total specific binding (defined by 200 μM (−)-isoprenaline) revealed a single high-affinity site (K_D = 114 pM) with a B_{max} value corresponding to 2.56 fmol/mg tissue. In the presence of the β_2 antagonist, ICI 118551, binding to β_1-adrenoceptors could be resolved and represented 73% of the total β-adrenoceptor population.

The effects of amitriptyline, desipramine and paroxetine on total and β_1-adrenoceptor binding are shown in Figure 19.5. No apparent reduction in binding was produced by amitriptyline or paroxetine in the frontal cortex; however, chronic treatment with desipramine produced a clear reduction of both total and β_1-adrenoceptor binding in all laminae of the frontal cortex. The kinetics of this binding are shown by the representative Scatchard plot of binding to lamina I (Figure 19.6). The downregulation induced by desipramine determined by comparison of B_{max} values amounted to 25% below control in both cases. This reduction was significant ($P < 0.01$) and also occurred to the same extent in laminae II–III and V–VI. Although binding of β_1-adrenoceptors appeared to be downregulated in all laminae, a significant decrease was only

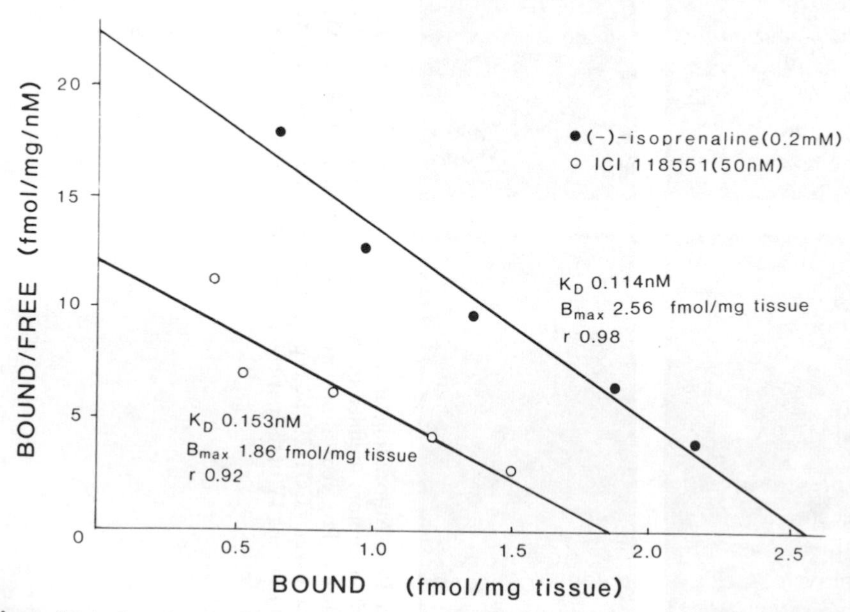

Figure 19.4. Scatchard plots of total (β_1 and β_2) adrenoceptor binding (defined by isoprenaline) and β_1-adrenoceptor binding (defined in the presence of ICI 118551). K_D, equilibrium dissociation constant; B_{max}, maximal number of binding sites

obtained in laminae V–VI. Neither amitriptyline nor paroxetine reduced the density of β-adrenoceptors; indeed, with the exception of amitriptyline in laminae V–VI, they produced small but non-significant increases in total and β_1-adrenoceptor binding. No significant alterations in affinities of the ligand for either of the sites were observed (Table 2).

From the analysis of β_2-adrenoceptor binding, it is apparent that these sites are also downregulated by desipramine. Binding densities for these sites were calculated as the difference between total and β_1-adrenoceptor binding. However, the B_{max} and K_D figures quoted are derived from the combined binding density values of five rats in triplicate (i.e. 15 determinations for each point of the saturation curve), since calculating the kinetic parameters for individual rats resulted in too much variation to allow accurate construction of Scatchard plots. For this reason, therefore, statistical comparisons were unable to be made.

Interestingly, the anticipated downregulation of β-adrenoceptors was only observed after treatment with desipramine; amitriptyline and paroxetine were without significant effect. This reduction, observed in the total β-adrenoceptor population of the three laminal areas examined, was attributed to the β_1-adrenoceptor upon resolution of the two subtypes. The involvement of β_2-adrenoceptors cannot be negated for reasons already mentioned. The downregulation of β-adrenoceptors by desipramine is attributable to the β_1 subtype (Beer *et al.*, 1987; Heal *et al.*, 1989; Ordway *et al.*, 1988). This might suggest that in our study possibly not all of the β_2-adrenoceptors were blocked by the β_2 antagonist, ICI 118551. However, the concentration used (50 nM) is in agreement with the methodology of Rainbow *et al.* (1984) and Ordway *et al.* (1988) in the rat, and of De Paermentier *et al.* (1989) in the human brain, so that this would seem an unlikely cause for concern.

The correlation produced by ranking the individual $GABA_B$ B_{max} values for desipramine in descending order of increase from the mean control value tended towards a negative correlation ($r = 0.67$) when the corresponding B_{max} values for β-adrenoceptor binding were paired with those for $GABA_B$ binding. So, for example, the rat, which produced the greatest decrease in β-adrenoceptor binding, exhibited the smallest increase in $GABA_B$ binding (Table 3).

SUMMARY

The present investigation is, to our knowledge, the first to employ the technique of receptor autoradiography to examine the effects of chronic antidepressant treatment on central $GABA_B$ binding using saturation and Scatchard analysis. Moreover, since the initial autoradiographic study visualizing $GABA_B$ binding sites in rat cerebral cortex (Bowery *et al.*, 1984), it has now become evident that $GABA_B$ sites are clearly localized within the individual laminae of the frontal cortex. This has proved invaluable to the more detailed assessment of the

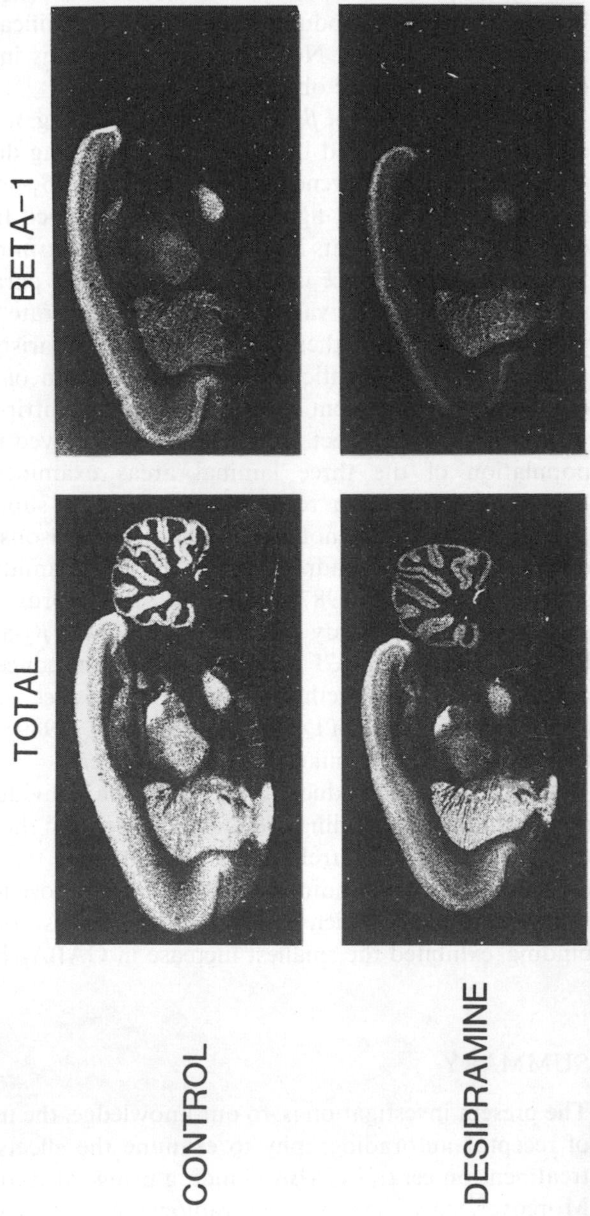

BETA−1

TOTAL

CONTROL

DESIPRAMINE

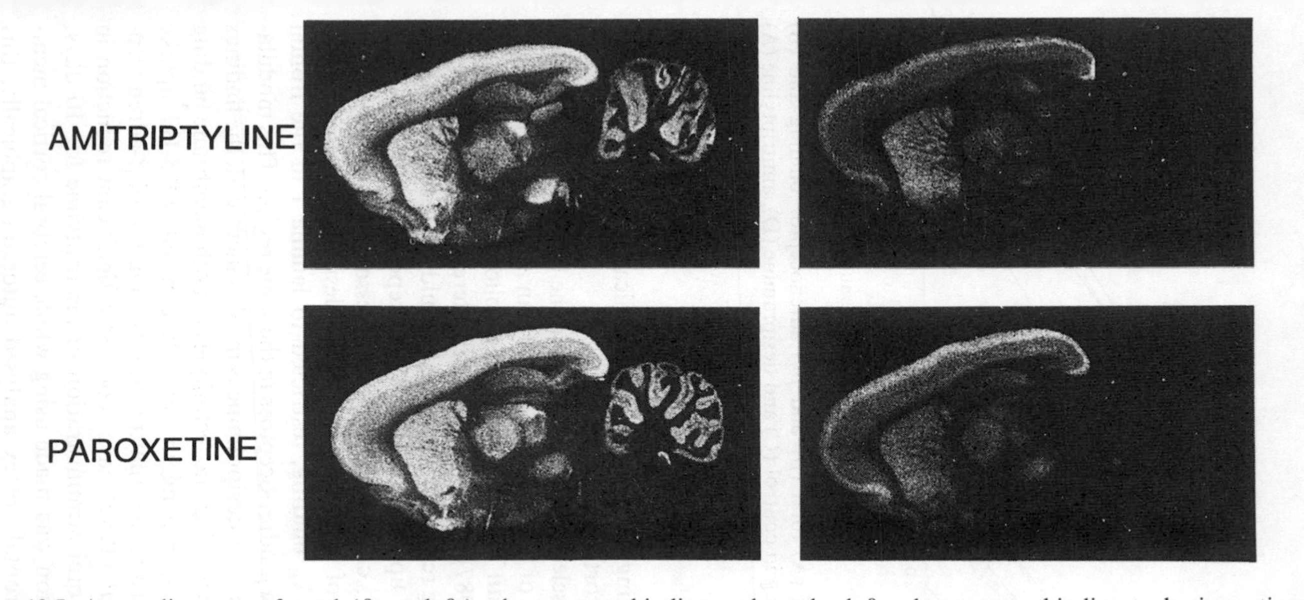

Figure 19.5. Autoradiograms of total (β_1 and β_2) adrenoceptor binding and resolved β_1-adrenoceptor binding to brain sections of control and antidepressant-treated rats

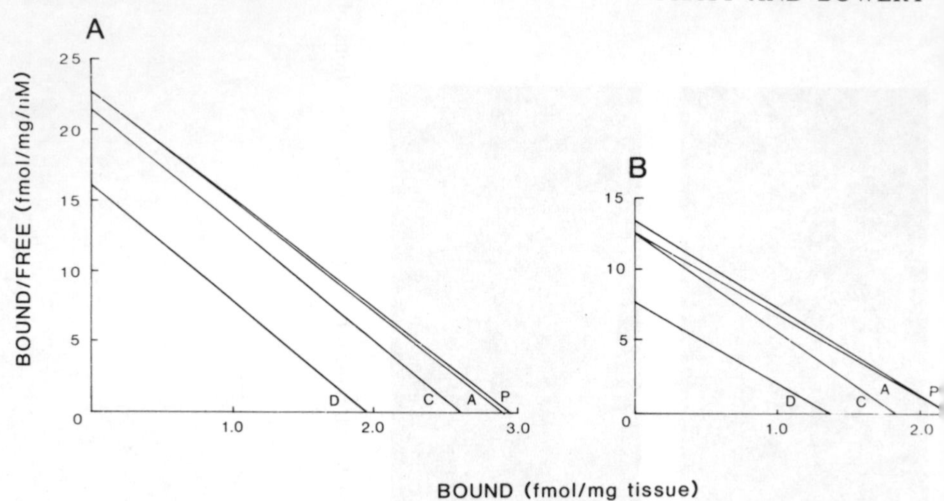

Figure 19.6. Scatchard plots of (A) total (β_1 and β_2) adrenoceptor binding and (B) resolved β_1-adrenoceptor binding in control (C) and desipramine (D), amitriptyline (A) or paroxetine (P) -treated rats

modulation of $GABA_B$ binding sites by certain chronically administered antidepressants within discrete cortical areas.

The antidepressants were selected for this study on the basis of their ability to selectively inhibit the uptake of NA, 5-HT or both. In an attempt to attain more consistent plasma levels than would be expected following twice daily i.p. injections (Szekely *et al.*, 1987), drugs were administered chronically via the drinking water. Such a dosing regime, it was hoped, would allow the observation of similar $GABA_B$ receptor upregulations to those reported by Lloyd *et al.* (1985) following continuous s.c. infusions of antidepressants via osmotic minipumps. As a consequence of this, however, only desipramine produced a significant increase in $GABA_B$ binding, restricted to lamina I in the frontal cortex, whilst downregulating β-adrenoceptors in the same region. This modulation of the two receptor systems by desipramine is in agreement with the findings of a number of reported studies using the techniques of both membrane binding and autoradiography. It was surprising, in view of the mixed NA/5-HT uptake inhibiting properties of amitriptyline, that no downregulation of β-adrenoceptors was observed. Heal *et al.* (1989) have reported a significant reduction in β_1-adrenoceptors following oral administration of amitriptyline for 10 days; however, since this observation was made using whole cerebral cortical membranes in contrast to the frontal cortex analysed autoradiographically, this could possibly explain the lack of effect of amitriptyline on both β-adrenoceptors and $GABA_B$ receptors in this study.

Table 2. The effect of antidepressants on [^{125}I]iodopindolol binding to β_1- and β_2-adrenoceptors in rat frontal cortex

	Frontal cortex lamina								
	I	II–III	V–VI	I	II–III	V–VI	I	II–III	V–VI
	Total			β_1			β_2		
	B_{max} (fmol/mg tissue)								
Control	2.61	3.34	2.69	1.83	2.16	1.68	0.85	1.20	1.04
Amitriptyline	2.90	3.40	2.56	2.17	2.50	1.79	0.88	1.03	0.91
Desipramine	1.97**	2.50**	1.91**	1.38	1.82	1.29*	0.62	0.77	0.73
Paroxetine	2.96	3.49	2.56	2.16	2.46	1.70	0.76	0.97	0.91
	K_D(pM)								
Control	122	134	128	146	134	151	84	124	97
Amitriptyline	127	126	121	172	163	181	82	92	78
Desipramine	122	119	117	178	163	182	62	71	68
Paroxetine	130	130	109	161	148	149	56	71	68

B_{max}, maximal number of binding sites; K_D, equilibrium dissociation constant.
*$P < 0.05$, **$P < 0.01$ (Student's t-test).

Table 3. Comparison of the effect of desipramine on $GABA_B$ and β_1- and β_2-adrenoceptor binding by pairing respective B_{max} values

	B_{max} (fmol/mg tissue)			
	$GABA_B$			β
Control (mean value)	232.8			2.61
Desipramine (ranked values)	373.1	*1*	*3*	1.95
	363.5	*2*	*5*	2.29
	299.7	*3*	*4*	2.11
	256.0	*4*	*2*	1.94
	243.9	*5*	*1*	1.56

B_{max}, maximal number of binding sites.

The influence of antidepressants on $GABA_B$ binding was examined in parallel with binding to β-adrenoceptors in an attempt to establish a possible correlation between the two receptor systems. We had hypothesized that a rat showing the largest increase in $GABA_B$ binding would also exhibit the greatest downregulation of β-adrenoceptors. However, in the case of desipramine, it was found that instead the trend was towards that of a negative correlation. This suggests that the desipramine-induced modulation of the two systems may indicate that the GABAergic and the noradrenergic systems are finely balanced so that a strong upregulatory signal for $GABA_B$ receptors may in some way suppress β-adrenoceptor downregulation and vice versa. This interesting new finding will require further experimentation to establish whether or not this biochemical marker of depression is valid.

REFERENCES

Beer, M., Hacker, S., Poat, J. and Stahl, S.M. (1987) Independent regulation of β_1- and β_2-adrenoceptors. *Br. J. Pharmacol.*, **92**, 827–834.

Berrettini, W.H., Nurnberger, J.I., Hare, T., Gershon, E.S. and Post, R.M. (1982) Plasma and CSF GABA in affective illness. *Br. J. Psychiatr.*, **141**, 483–487.

Bowery, N.G., Price, G.W., Hudson, A.L., Hill, D.R., Wilkin, G.P. and Turnbull, M.J. (1984) GABA receptor multiplicity: visualisation of different receptor types in the mammalian CNS. *Neuropharmacology*, **23**, 219–231.

Cross, J.A. and Horton, R.W. (1988) Effects of chronic oral administration of the antidepressants, desmethylimipramine and zimelidine on rat cortical $GABA_B$ binding sites: a comparison with $5\text{-}HT_2$ binding site changes. *Br. J. Pharmacol.*, **93**, 331–336.

De Paermentier, F., Cheetham, S.C., Crompton, M.R. and Horton, R.W. (1989) Beta-adrenoceptors in human brain labelled with [^{3}H]dihydroalprenolol and [^{3}H]CGP 12177. *Eur. J. Pharmacol.*, **167**, 397–405.

Heal, D.J., Butler, S.A., Hurst, E.M. and Buckett, W.R. (1989) Antidepressant treatments, including sibutramine hydrochloride and electroconvulsive shock, decrease β_1- but not β_2-adrenoceptors in rat cortex. *J. Neurochem.*, **53**, 1019–1025.

Lloyd, K.G., Morselli, P.L., Depoortere, H., Fournier, V., Zivkovic, R., Scatton, B., Broekkamp, C.L., Worms, P. and Bartholini, G. (1983) The potential use of GABA agonists in psychiatric disorders: evidence from studies with progabide in animal models and clinical trials. *Pharmacol. Biochem. Behav.*, **18**, 957–966.

Lloyd, K.G., Thuret, F. and Pilc, A. (1985) Upregulation of γ-aminobutyric acid $(GABA)_B$ binding sites in rat frontal cortex: a common action of repeated administration of different classes of antidepressants and electroshock. *J. Pharmacol. Exp. Ther.*, **235**, 191–199.

Lloyd, K.G., Zivkovic, R., Sanger, D., Joly, D., Depoortere, H. and Bartholini, G. (1987) Fengabine, a novel antidepressant GABAergic agent: I. Activity in models for antidepressant drugs and psychopharmacological profile. *J. Pharmacol. Exp. Ther.*, **241**, 245–250.

Musch, B. and Garreau, M. (1986) An overview of the antidepressant activity of fengabine in open clinical trials. In: C. Shagass *et al.* (eds) *Biological psychiatry*. New York: Elsevier, pp. 920–922.

Ordway, G.A., Gambarana, C. and Frazer, A. (1988) Quantitative autoradiography of central beta adrenoceptor subtypes: comparison of the effects of chronic treatment with desipramine or centrally administered *l*-isoproterenol. *J. Pharmacol. Exp. Ther.*, **247**, 379–389.

Rainbow, T.C., Parsons, B. and Wolfe, B.B. (1984) Quantitative autoradiography of β_1- and β_2-adrenergic receptors in rat brain. *Proc. Natl Acad. Sci. USA*, **81**, 1585–1589.

Suzdak, P.D. and Gianutsos, G. (1986) Effect of chronic imipramine or baclofen on $GABA_B$ binding and cyclic AMP production in cerebral cortex. *Eur. J. Pharmacol.*, **131**, 129–133.

Szekely, A.M., Barbaccia, M.L. and Costa, E. (1987) Effect of a protracted antidepressant treatment on signal transduction and $[^3H](-)$-baclofen binding at $GABA_B$ receptors. *J. Pharmacol. Exp. Ther.*, **243**, 155–159.

20 Neurochemical Effects of Prenatal Antidepressant Administration on Cortical $GABA_B$ Receptor Binding and Striatal [^{3}H]Dopamine Release

C. KNOTT, N. G. BOWERY, C. DEFILIPE, D. MONTERO and J. DEL RIO

INTRODUCTION

Approaches to the treatment of depression have been largely based on neurochemical differences observed in clinically depressed patients, and on the subjective effects of drugs with known mechanisms of action, such as propranolol and reserpine. Cerebrospinal fluid metabolites of 5-hydroxytryptamine (serotonin, 5-HT), dopamine (DA) and noradrenaline (NA) were found to be low in some forms of depression (Asberg *et al.*, 1984), and reduced 5-HT uptake and binding have been found in blood platelets from unmedicated depressed patients, which may be of some predictive value (Langer *et al.*, 1982). Treatment strategems geared towards correcting such imbalances generally involve chronically inhibiting the uptake mechanisms for 5-HT (assessed by [^{3}H]imipramine or [^{3}H]paroxetine binding to mammalian brain tissue; Cortes *et al.*, 1988; Gleiter and Nutt, 1988) or NA (which correlates with [^{3}H]desipramine binding; Biegon, 1978; Langer *et al.*, 1984), or the inhibition of neurotransmitter metabolism. Such manipulations inevitably tend to raise synaptic monoamine concentrations, and this may be partly responsible for the downregulation of β-adrenoceptors which often occurs in the frontal cortex (Beer *et al.*, 1987; Ordway *et al.*, 1988; Pandey and Davis, 1983; Sugrue, 1983), where the β_1 subtype predominates (Rainbow *et al.*, 1984). This effect is time and dose related (Sethy *et al.*, 1988) and appears to be associated with a desensitization of receptor-coupled (Suzdak and Gianutsos, 1988) but not of forskolin-stimulated (Szekely *et al.*, 1987) adenylyl cyclase, and is perhaps the strongest biochemical predictor of a drug's antidepressant potential. Many antidepressants, however, also bring about a dose-dependent downregulation of 5-HT_2 binding sites in the brain (Blackshear and Sanders-Bush, 1982; Cross and Horton, 1988), as occurs with repeated electroconvulsive shock (Gleiter and Nutt, 1988), while rather

$GABA_B$ Receptors in Mammalian Function. Edited by N.G. Bowery, H. Bittiger and H.-R. Olpe

better correlations have been found for α_1-adrenoceptor, muscarinic or DA receptor binding with antidepressant side-effects than with effects on mood (Enna, 1978; Pugsley and Lippmann, 1981). Selectivity for different receptors may of course be concentration related, and not confined to the monoamines.

A possible role for GABAergic dysfunction in the aetiology of affective disorders has been suggested. Evidence favouring a reduction in GABA synaptic function in depression comes from the findings that cortical $GABA_B$ binding (Lloyd *et al.*, 1985) and GABA release (Green and Vincent, 1987; Sherman and Petty, 1982) are reduced in animal models of depression (Petty, 1986) and that some therapeutic benefit is afforded by GABA agonists such as progabide (Bartholini and Morselli, 1983). However, the effects of chronic treatment with antidepressant drugs on GABAergic transmission are less clear-cut. Both downregulation (Suzdak and Gianutsos, 1985) and no change (Pilc and Lloyd, 1984) in $GABA_A$ receptor binding and downregulation of benzodiazepine receptors (Barbaccia *et al.*, 1986; Suranyi-Cadotte *et al.*, 1985) have been reported. However, this last finding may be less related to drug treatment and more to disease than previously thought (Cheetham *et al.*, 1988). Similarly, $GABA_B$ binding is found to be enhanced (Lloyd, Chapter 17; Suzdak and Gianutsos, 1986), unaltered (Horton *et al.*, Chapter 18) or dependent upon the assay methods employed (Szekely *et al.*, 1987), although we have not confirmed this last explanation. These studies have been re-examined in adult rats and the results are presented in previous chapters.

Indeed all of the studies mentioned hitherto have been performed on adult animals. However, the ontogenic development of 5-HT, NA and DA systems begins around gestation day 14, and these systems steadily mature, with adult receptor levels being reached by the eighth to tenth postnatal week (Bruinink and Lichtensteiger, 1984), while $GABA_B$ receptor maturation may occur earlier (Skerritt and Johnston, 1982). It is logical therefore to speculate that if antidepressant drugs are able to cross the placenta and modify brain neurochemistry this may be facilitated by their administration during gestation, when disruption of the normal patterns of neuronal organization would be possible. Antidepressant drugs are low molecular weight, lipid-soluble bases and thus they are able to pass from mother to fetus with ease (DeVane and Simpkins, 1985). During chronic administration, fetal exposure is determined by maternal drug concentrations and the equilibrium distribution between mother and fetus. The normal pH gradient across the placenta of 0.1 pH unit usually results in higher fetal compared with maternal plasma concentrations for these drugs and their active metabolites (DeVane and Simpkins, 1985), a phenomenon called ion trapping. The majority of antidepressants, however, are not recognized to be dysmorphogens in rats and humans, but they may induce subtle neurochemical changes in the developing brain with concomitant behavioural manifestations.

Del Rio *et al.* (1988) have investigated this possibility in the rat. Their results indicate that after chronic prenatal exposure to iprindole (IPR), chlorimipramine, mianserin or nomifensine both β-adrenoceptor and $5\text{-}HT_2$ receptor binding sites in neonatal rat brain membranes were downregulated, but no corresponding changes in the receptor affinities were found (DeCeballos *et al.*, 1985). These neurochemical changes were of a greater magnitude and more enduring than those usually reported in adults, persisting for at least 25 days postnatally in contrast to one week after drug withdrawal in adults (Pandey and Davis, 1983). In these same animals spontaneous locomotor activity was significantly reduced compared with that in control animals, and high doses of apomorphine (200 μg/kg) greatly enhanced locomotor activity, confirming earlier electrophysiological (Chiodo and Antelman, 1980) and behavioural (Serra *et al.*, 1979) studies in adult rats, which suggested that antidepressants induced a subsensitivity in DA autoreceptors. As previously documented in adult animals no corresponding changes were found in DA receptor binding to account for these behavioural observations when the unlabelled antagonists spiperone and haloperidol were used to displace [^{3}H]DA binding from neonatal striatal membranes (DeCeballos *et al.*, 1985). Interestingly, however, when the natural transmitter was used to displace [^{3}H]spiperone from these sites, significantly lower K_1 values were measured in the animals exposed to antidepressants (DeCeballos *et al.*, 1985). Hence the present studies were undertaken firstly to examine the regional effects of prenatal administration of amitriptyline (AMI) and IPR on DA and $GABA_B$ receptors using the technique of receptor autoradiography, and secondly, since agonist rather than antagonist binding data correlate better with functionally relevant receptor changes, to examine the effects of antidepressants on [^{3}H]DA release, and the ability of the $GABA_B$ agonist, (−)-baclofen, to modify this.

The study design was as follows. Pregnant female rats were injected s.c. with AMI (10 mg/kg/day), IPR (10 mg/kg/day) or saline (equal volumes; control group), from gestation day 7 until term. The pups were studied 35 days after delivery but received no antidepressants during this period.

PRENATAL ANTIDEPRESSANT EFFECTS ON $GABA_B$ AND DOPAMINE RECEPTORS

Pups from each treatment group were sacrificed on day 35, and the brains rapidly frozen at −80 °C. Cryostat sections (10 μm) were prepared from each brain for $GABA_B$ and DA receptor binding. $GABA_B$ receptor autoradiography was performed as described by Bowery *et al.* (1987) using [^{3}H]GABA (50 nM) in the presence of isoguvacine (40 μM), and unlabelled GABA (0–100 μM) as displacer. DA receptor binding was performed using [^{3}H]spiperone (0.2 nM), and unlabelled DA (1 mM) as displacer. Sections were incubated for 45 min, and then rinsed briefly (two 1-min periods) before air drying.

Dry, labelled sections were apposed to LKB ‘Ultrofilm’ and the resulting photographic images were analysed with a Quantimet 970 image analyser. In

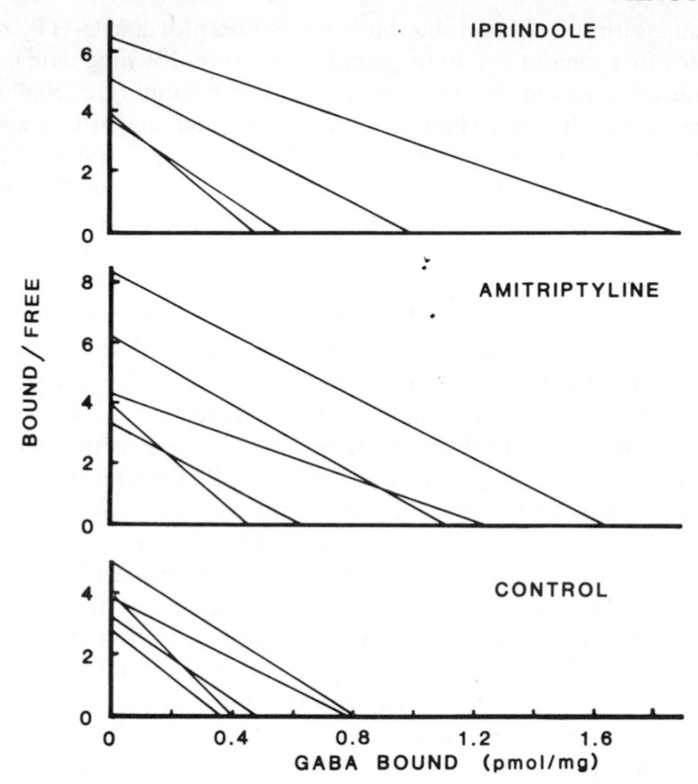

Figure 20.1. Individual Scatchard plots for [^{3}H]GABA binding to $GABA_B$ sites in the frontal cortex of rats injected with saline (control), amitriptyline or iprindole. [^{3}H]GABA bound expressed in pmol/mg tissue protein is plotted against bound/free values expressed as fmol/mg tissue protein/nM

addition to the regional autoradiographic analysis for DA we performed Scatchard analysis on striatal membranes using [^{3}H]spiperone (0.2–5 nM) binding and unlabelled DA as displacer in groups of rats.

Binding to $GABA_B$ receptor sites in the outer laminae (I and II) of the frontal cerebral cortex was enhanced by prenatal AMI and IPR in some, but not all, animals. Preliminary data from individual animals, illustrated in Figure 20.1, show that this results from enhanced binding capacity (B_{max}) with no significant changes in receptor affinity (K_D). In agreement with the findings of Pratt and Bowery (Chapter 19) these differences are confined to the outer layers of the frontal cortex. There were no significant differences in [^{3}H]GABA (50 nM) binding to $GABA_B$ sites in the striata of those rats exposed to AMI or IPR compared with that in the control group (Table 1). Striatal [^{3}H]spiperone binding measured by receptor autoradiography in animals exposed to AMI or IPR was significantly reduced compared with that in the control group (Figure

Table 1. Mean values for [^{3}H]GABA binding to $GABA_B$ sites

	Frontal cortex		Striatum	n
	B_{max}(fmol/mg)	K_D(nM)	Bound (fmol/mg)	
Control	565 ± 95	149 ± 16	326 ± 62	5
Amitriptyline	986 ± 230	193 ± 28	284 ± 42	5
Iprindole	979 ± 470	191 ± 37	365 ± 73	4

[^{3}H]GABA binding to $GABA_B$ receptors in the frontal cortex and in the corpus striatum of rats injected with saline (control), amitriptyline or iprindole. n represents the number of animals per group. Values are means ± SEM.
B_{max}, binding capacity; K_D, equilibrium dissociation constant.

Table 2. Mean values for striatal [^{3}H]spiperone binding

	Membrane homogenates				Brain slices
	K_D (nM)	B_{max} (fmol/mg)	K_D (nM)	B_{max} (fmol/mg)	Total bound (fmol/mg)
Control	0.06 ± 0.01	17.2 ± 2.1	13.8 ± 5.8	1165 ± 443	77.04 ± 3.6
Amitriptyline	—	—	—	—	60.30 ± 2.2
Iprindole	0.04 ± 0.00	9.3 ± 1.5	7.2 ± 2.6	832 ± 241	48.41 ± 2.2
Chlorimipramine	0.04 ± 0.01	11.0 ± 2.6	5.3 ± 2.2	1009 ± 432	—

Mean striatal [^{3}H]spiperone binding in rat brain membranes and in rat brain slices in animals prenatally exposed to saline (control), amitriptyline, iprindole or cholorimipramine. Values are means ± SEM.
B_{max}, binding capacity; K_D, equilibrium dissociation constant.

20.2). This is the result of reduced receptor number in the high-affinity binding sites, with no significant changes in receptor affinity, as shown from Scatchard analysis (Table 2). These data support our previous findings of enhanced agonist affinity at striatal DA receptors in rats exposed prenatally to antidepressants (DeCeballos *et al.*, 1985). However, changes in receptor binding need not reflect changes in receptor function and so we have examined the effect of antidepressants on striatal DA and $GABA_B$ receptor function by means of [^{3}H]DA release.

STRIATAL [^{3}H]DOPAMINE RELEASE AFTER PRENATAL ANTIDEPRESSANTS

Slices of rat striatum (260 μm) were incubated in oxygenated Krebs–bicarbonate buffer containing [^{3}H]DA (50 nM; 40 Ci/mmol; New England Nuclear) in the presence of pargyline (40 nM) and ascorbic acid (0.2 mM) at 37 °C for

CONTROL

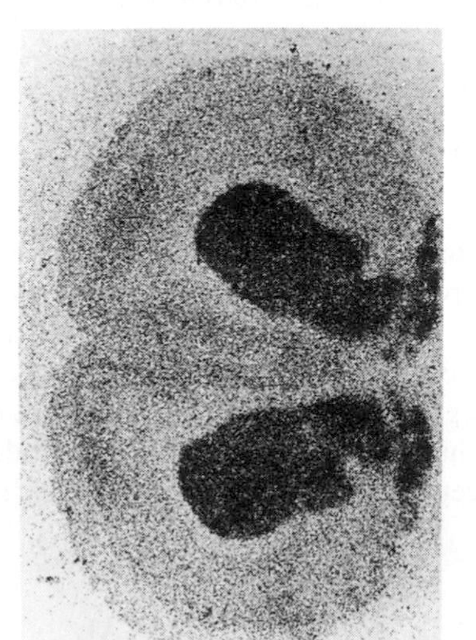

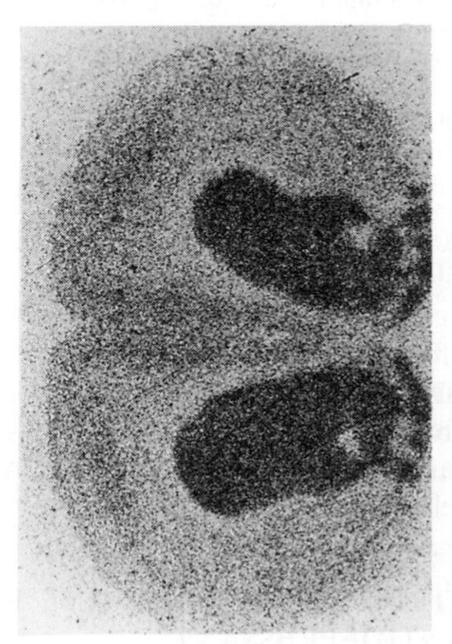

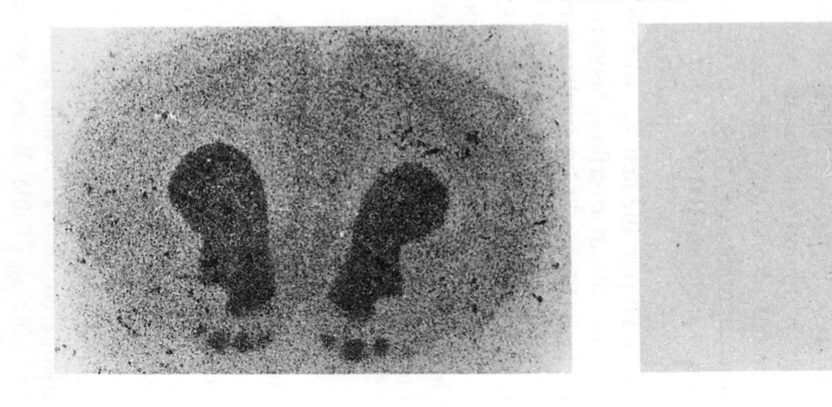

Figure 20.2. Autoradiograms for [^{3}H]spiperone binding to rat brain sections in animals from control, amitriptyline and iprindole groups. On the left are total binding sections and on the right the corresponding non-specific binding sections incubated in the presence of 1 mM unlabelled dopamine (DA)

20 min. Slices were then rinsed briefly, transferred to individual (500 μl) chambers and superfused with buffer at a rate of 1.8 ml/min. After 20 min 3.6-ml fractions of perfusate were collected and total tritium efflux was measured by liquid scintillation spectrometry. In each experiment, potassium (K^+; 20 mM) was added to the superfusion fluid on three occasions (2-min pulses separated by 14-min intervals). During the second K^+ pulse, (–)-baclofen (100 μM) was present. (–)-Baclofen had no effect on basal tritium efflux in any group.

There was no statistically significant difference in tissue accumulation of [^{3}H]DA between the three groups and the [^{3}H]DA release was approximately 80% calcium-dependent. Basal tritium efflux was also similar between groups; the total tritium accumulated in a 2-min period was as follows: control, 0.24 ± 0.02% ($n = 11$); AMI, 0.24 ± 0.03% ($n = 8$); and IPR, 0.2 ± 0.03% ($n = 10$). The addition of K^+ (20 mM) to the superfusate resulted in a mean increase in tritium overflow (Figure 20.3a) by 0.25–0.45% above basal release. This increase was transient and returned toward the basal values by the next collection period. Successive K^+ pulses resulted in a progressively diminished response such that the third responses were 63–73% of the initial responses. In the absence of calcium the K^+-evoked tritium release was approximately 0.06% of the total accumulation for all groups. The effect of (–)-baclofen (100 μM) was to inhibit K^+-evoked [^{3}H]DA release while having no effect on basal tritium release, as previously described by Bowery *et al.* (1980). During the first K^+ stimulation period, there was no significant difference in tritium overflow between the groups, although intergroup variability was high (Figure 20.3a). In all treatment groups, the K^+-evoked DA release during the second stimulation period was suppressed to a similar extent by (–)-baclofen: control, 57.6%; AMI, 55.1%; and IPR, 52.9%. These data are shown in Figure 20.3b, in which the response obtained in the first stimulation period in each case is taken as the maximal response, and the effect of (–)-baclofen is compared with this.

COMMENT

The present studies confirm the traditionally held view that changes are produced in monoamine systems after chronic antidepressant treatment (Willner, 1983), and emphasize the usefulness of prenatal drug administration as a tool for studying their effects on the more plastic neonatal circuitry. We have found that the effects of prenatal exposure to antidepressants are greater and more persistent than the corresponding effects seen in adults. This characteristic has also been stressed after prenatal exposure to benzodiazepines (Kellogg *et al.*, 1983) and haloperidol (Moon, 1984; Rosengarten and Friedhoff, 1979). Our findings of reduced striatal [^{3}H]spiperone binding (Table 2) and associated enhanced sensitivity to DA and to DA agonists (DeCeballos *et al.*, 1985) support the suggestion of reduced sensitivity of the DA autoreceptors. These biochemical changes were not, however, associated with significant effects on

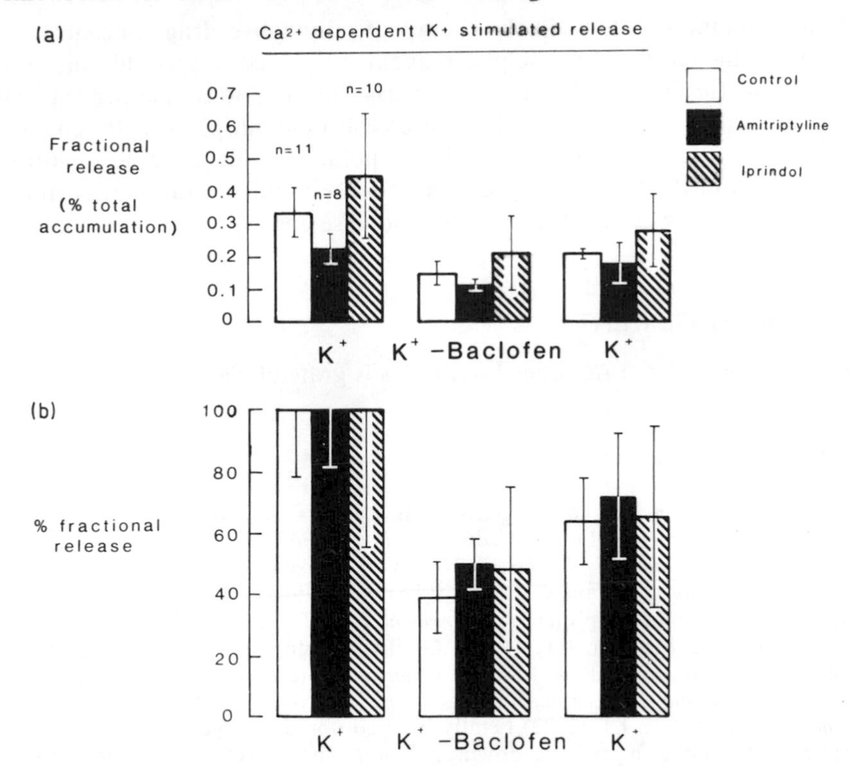

Figure 20.3. (a) The increase in K^+-evoked [^{3}H]dopamine (DA) release over basal release is shown as a percentage of total accumulation in control, amitriptyline and iprindole groups. The numbers of animals in each group are given above histogram bar, and each bar represents the mean $\pm$ SEM [^{3}H]DA released for those animals. ($-$)-Baclofen (100 μM) was included in the second stimulation period. (b) The same data plotted as a percentage of K^+-evoked release in the first stimulation period

K^+-evoked [^{3}H]DA release or on the response of striatal dopaminergic terminals to ($-$)-baclofen stimulation.

In the rat frontal cortex, β-adrenoceptor binding is reduced, which may reflect a response to enhanced noradrenergic innervation of these areas (Beaulieu and Coyle, 1982), and in some animals GABA$_B$ binding is enhanced (Figure 20.1). GABA$_B$ receptors are present on approximately 80% of the noradrenergic terminals in rat forebrain (Karbon *et al.*, 1983), but their upregulation appears not to be highly correlated with the β-adrenergic changes after chronic antidepressant treatment, indicating that these two effects may occur independently. The wide interindividual variation in biochemical responses to antidepressants, particularly the inconsistent findings of upregulation of GABA$_B$ receptors, may be partly related to individual variations in phenotype for hydroxylation (Bertilsson *et al.*, 1980; Sjoqvist and Bertilsson, 1984) and concomitant variations in the therapeutically active drug concentrations available at the receptors. These pharmacokinetic variations would only serve to accentuate the individual biochemical variations in time-course and extent of second messenger responses – the integrated effect of multiple receptor changes.

The clinical relevance of our findings points to the need for cautious consideration of the benefits of prescribing these antidepressant drugs during human pregnancies against their possible adverse consequences in neonates.

ACKNOWLEDGEMENT

Financial support from Acciones Integradas is gratefully acknowledged.

REFERENCES

Asberg, M., Bertilsson, L. and Martensson, B. (1984) CSF monoamine metabolites, depression and suicide. In: E. Usdin *et al.* (eds) *Frontiers in biochemical and pharmacological research in depression.* New York: Raven Press, pp. 87–97.

Barbaccia, M.L., Ravizza, L. and Costa, E. (1986) Maprotiline: an antidepressant with an unusual pharmacological profile. *J. Pharmacol. Exp. Ther.*, **236**, 307–312.

Bartholini, G. and Morselli, P.L. (1983) GABA receptor agonists: pharmacological spectrum and clinical actions. In: L.F. Gram *et al.* (eds) *Clinical pharmacology in psychiatry: bridging the experimental–therapeutic gap*, pp. 386–394.

Beaulieu, M. and Coyle, J.T. (1982) Fetally-induced noradrenergic hyperinnervation of central cortex results in persistent downregulation of beta-receptors. *Dev. Brain Res.*, **4**, 491–494.

Beer, M., Hacker, S., Poat, J. and Stahl, S.M. (1987) Independent regulation of β_1- and β_2-adrenoceptors. *Br. J. Pharmacol.*, **92**, 827–834.

Bertilsson, L., Eichelbaum, M., Mellstrom, B., Sawe, J., Schulz, H.-U. and Sjoqvist, F. (1980) Nortriptyline and antipyrine clearance in relation to debrisoquine hydroxylation in man. *Life Sci.*, **27**, 1673–1677.

Biegon, A. (1978) Quantitative autoradiography in drug research: tricyclic antidepressants. In: H.I. Yamamura, S.J. Enna and M.J. Kuhar (eds) *Neurotransmitter receptor binding.* New York: Raven Press, pp. 477–486.

Blackshear, M.A. and Sanders-Bush, E. (1982) Serotonin receptor sensitivity after acute and chronic treatment with mianserin. *J. Pharmacol. Exp. Ther.*, **221**, 303–308.

Bowery, N.G., Hill, D.R., Hudson, A.L., Doble, A., Middlemiss, D.N., Shaw, J. and Turnbull, M. (1980) (–)Baclofen decreases neurotransmitter release in the mammalian CNS by an action at a novel GABA receptor. *Nature*, **283**, 92–94.

Bowery, N.G., Hudson, A.L. and Price, G.W. (1987) $GABA_A$ and $GABA_B$ receptor site distribution in the rat central nervous system. *Neuroscience*, **20**, 365–383.

Bruinink, A. and Lichtensteiger, W. (1984) β-Adrenergic binding sites in fetal rat brain. *J. Neurochem.*, **43**, 578–581.

Cheetham, S.C., Crompton, M.R., Katona, C.L.E., Parker, S.J. and Horton, R.W. (1988) Brain $GABA_A$/benzodiazepine binding sites and glutamic acid decarboxylase activity in depressed suicide victims. *Brain Res.*, **460**, 114–123.

Chiodo, L.A. and Antelman, S.M. (1980) Repeated tricyclics induce a progressive dopamine autoreceptor subsensitivity independent of daily drug treatment. *Nature*, **287**, 451–454.

Cortes, R., Soriano, E., Pazos, A., Probst, A. and Palacios, J.M. (1988) Autoradiography of antidepressant binding sites in the human brain: localisation using [^{3}H]imipramine and [^{3}H]paroxetine. *Neuroscience*, **27**, 473–496.

Cross, J.A. and Horton, R.W. (1988) Effects of chronic oral administration of the antidepressants, desmethylimipramine and zimelidine on rat cortical $GABA_B$ binding sites: a comparison with $5HT_2$ binding site changes. *Br. J. Pharmacol.*, **93**, 331–336.

DeCeballos, M.L., Benedi, A., DeFilipe, C. and Del Rio, J. (1985) Prenatal exposure of rats to antidepressants enhances agonist affinity of brain dopamine receptors and dopamine-mediated behaviour. *Eur. J. Pharmacol.*, **116**, 257–262.

Del Rio, J., Montero, D. and DeCeballos, M.L. (1988) Long-lasting changes after perinatal exposure to antidepressants. *Prog. Brain Res.*, **78**, 173–187.

DeVane, C.L. and Simpkins, J.W. (1985) Pharmacokinetics of imipramine and its major metabolities in pregnant rats and their fetuses following a single high dose. *Drug Metab. Dispos.*, **13**, 438–442.

Enna, S.J. (1978) In vivo receptor modifications of antidepressant potential. In: H.I. Yamamura, S.J. Enna and M.J. Kuhar (eds) *Neurotransmitter receptor binding*. New York: Raven Press, pp. 409–427.

Gleiter, C.H. and Nutt, D.J. (1988) Measuring the serotonin uptake site using [^{3}H]paroxetine: a new serotonin uptake inhibitor. *Adv. Alcohol Subst. Abuse*, **7**, 107–111.

Green, A.R. and Vincent, N.D. (1987) The effect of repeated electroconvulsive shock on GABA synthesis and release in regions of rat brain. *Br. J. Pharmacol.*, **92**, 19–24.

Karbon, E.W., Duman, R. and Enna, S.J. (1983) Biochemical identification of multiple $GABA_B$ binding sites: association with noradrenergic terminals in rat forebrain. *Brain Res.*, **274**, 393–396.

Kellogg, C.K., Chisholm, J., Simmons, R.D., Ison, J.R. and Miller, R.K. (1983) Neural and behavioural consequences of prenatal exposure to diazepam. *Monogr. Neural Sci.*, **9**, 119–129.

Langer, S.Z., Zarifian, E., Briley, M., Raisman, R. and Sechter, D. (1982) High affinity ^{3}H-imipramine binding: a new biological marker in depression. *Pharmacopsychiatry*, **15**, 4–10.

Langer, S.Z., Raisman, R., Sechter, D., Gay, C. Loo, H. and Zarifian, E. (1984) ^{3}H-Imipramine and ^{3}H-desipramine binding sites in depression. In: E. Usdin *et al.* (eds) *Frontiers in biochemical and pharmacological research in depression*. New York: Raven Press, pp. 113–125.

Lloyd, K.G., Thuret, F. and Pilc, A. (1985) Upregulation of γ-aminobutyric acid $(GABA)_B$ binding sites in rat frontal cortex: a common action of repeated administration of different classes of antidepressants and electroshock. *J. Pharmacol. Exp. Ther.*, **235**, 191–199.

Moon, S.L. (1984) Prenatal haloperidol alters striatal dopamine and opiate receptors. *Brian Res.*, **323**, 109–113.

Ordway, G.A., Gambarana, C. and Frazer, A. (1988) Quantitative autoradiography of central beta adrenoceptor subtypes: comparison of the effects of chronic treatment with desipramine or centrally administered *l*-isoproterenol. *J. Pharmacol. Exp. Ther.*, **247**, 379–389.

Pandy, G.N. and Davis, J.M. (1983) Treatment with antidepressants and down-regulation of β-adrenergic receptors. *Drug Dev. Res.*, **3**, 393–406.

Petty, F. (1986) GABA mechanisms in learned helplessness. In: G. Bartholini (ed.) GABA and mood disorders. [*Experimental and clinical research*, vol. 4]. New York: Raven Press, pp. 61–66.

Pilc, A. and Lloyd, K.G. (1984) Chronic antidepressants and $GABA_B$ receptors: a GABA hypothesis of antidepressant drug action. *Life Sci.*, **35**, 2149–2154.

Pugsley, T.A. and Lippmann, W. (1981) Affinity of butriptyline and other tricyclic antidepressants for α-adrenoceptor binding sites in rat brain. *J. Pharm. Pharmacol.*, **33**, 113–115.

Rainbow, T.C., Parsons, B. and Wolfe, B.B. (1984) Quantitative autoradiography of β_1 and β_2-adrenergic receptors in rat brain. *Proc. Natl Acad. Sci. USA*, **81**, 1585–1589.

Rosengarten, H. and Friedhoff, A.J. (1979) Enduring changes in dopamine receptor cells of pups from drug administration to pregnant and nursing rats. *Science*, **203**, 1133–1135.

Serra, G., Argiolas, A., Klimek, V., Fadda, F. and Gessa, G. L. (1979) Chronic treatment with antidepressants prevents the inhibitory effect of small doses of apomorphine on dopamine synthesis and motor activity. *Life Sci.*, **25**, 415–424.

Sethy, V.H., Day, J.S. and Cooper, M.M. (1988) Dose-dependent downregulation of beta-adrenergic receptors after chronic intravenous infusion of antidepressants. *Prog. Neuropsychopharmacol. Biol. Psychiatr.*, **12**, 673–682.

Sherman, A.D. and Petty, F. (1982) Additivity of neurochemical changes in learned helplessness and imipramine. *Behav. Neural Biol.*, **35**, 344–353.

Sjoqvist, F. and Bertilsson, L. (1984) Clinical pharmacology of antidepressant drugs: pharmacogenetics. In: E. Usdin *et al.* (eds) *Frontiers in biochemical and pharmacological research in depression.* New York: Raven Press, pp. 359–372.

Skerritt, J.H. and Johnston, G.A.R. (1982) Postnatal development of GABA binding sites and their endogenous inhibitors in rat brain. *Dev. Neurosci.*, **5**, 189–197.

Sugrue, M.F. (1983) Do antidepressants possess a common mechanism of action? *Biochem. Pharmacol.*, **32**, 1811–1817.

Suranyi-Cadotte, B.E., Dam, T.V. and Quirion, R. (1985) Antidepressant-anxiolytic interaction: decreased density of benzodiazepine receptors in rat brain following chronic administration of antidepressants. *Eur. J. Pharmacol.*, **106**, 673–675.

Suzdak, P.D. and Gianutsos, G. (1985) Parallel changes in GABAergic and noradrenergic receptor sensitivity following administration of antidepressant and GABAergic drugs: a possible role for GABA in affective disorders. *Neuropharmacology*, **24**, 217–222.

Suzdak, P.D. and Gianutsos, G. (1986) Effect of chronic imipramine or baclofen on $GABA_B$ binding and cyclic AMP production in cerebral cortex. *Eur. J. Pharmacol.*, **131**, 129–133.

Szekely, A.M., Barbaccia, M.L. and Costa, E. (1987) Effect of a protracted antidepressant treatment on signal transduction and $[^3H](-)$baclofen binding at $GABA_B$ receptors. *J. Pharmacol. Exp. Ther.*, **243**, 155–159.

Willner, P. (1983) Dopamine and depression: a review of recent evidence. III. The effects of antidepressant treatments. *Brain Res. Rev.*, **6**, 237–246.

Epilepsy

21 $GABA_B$ Receptors and Experimental Models of Epilepsy

G. KARLSSON, M. SCHMUTZ, C. KOLB, H. BITTIGER and H.-R. OLPE

INTRODUCTION

It is well established that GABA plays an important role in epilepsy (Meldrum, 1975). Pharmacological interventions at the GABAergic synapse which lead to a reduction of transmission induce epileptic phenomena both in vivo and in vitro. Conversely, drugs which facilitate GABAergic transmission have anticonvulsant properties in man and animals. GABA acts via two main receptors in the brain, $GABA_A$ and $GABA_B$ (Hill and Bowery, 1981). Whereas pharmacological blockade of $GABA_A$ receptors by bicuculline leads to convulsions, the consequences of blocking $GABA_B$ receptors have remained unclear, largely because of the lack of suitable selective blockers that penetrate the blood–brain barrier. Some information on the possible function of $GABA_B$ receptors in this context is available from studies with the $GABA_B$ agonist baclofen (Lioresal®), which is an antispastic agent. However, no systematic investigations on the effect of baclofen in epileptic patients have been performed and the preclinical experimental data on baclofen in various in-vivo and in-vitro models are not conclusive (see Tables 1 and 2).

Very recently, the first parenterally active $GABA_B$ receptor blocker, CGP 35348, was described (see Bittiger *et al.*, Chapter 3). We were thus in the position to investigate the effects of $GABA_B$ receptor blockade in various experimental models of epilepsy in vivo (convulsion tests in mice) and in vitro (rat hippocampal slices). A brief description of the main pharmacological features of this compound is in order. CGP 35348 was shown to bind to the $GABA_B$ receptor with an affinity of 39 μM (half-maximal inhibitory concentration (IC_{50}) when [^{3}H]CGP 27492 was used as a ligand. Up to a concentration of 1 mM, CGP 35348 did not bind to $GABA_A$, *N*-methyl-D-aspartate (NMDA), quisqualate or kainate receptors. Moreover, no effect was found at biogenic amine and cholinergic receptors. In hippocampal slice preparations made epileptic by penicillin (1.2 mM), the depressant effect of baclofen (6 μM) on epileptic-like discharges was attenuated by 10 and 100 μM CGP 35348. In the same preparation, the depressant effects of adenosine (30 μM) and of the $GABA_A$ agonist THIP (4,5,6,7-tetrahydroisooxazolo[5,4-*c*]pyridin-3-ol, 30 μM) were not reduced, indicating that the drug selectively blocks $GABA_B$

$GABA_B$ Receptors in Mammalian Function. Edited by N.G. Bowery, H. Bittiger and H.-R. Olpe

receptors. Ionophoretic experiments performed on anaesthetized rats showed that CGP 35348 crosses the blood-brain barrier. This was shown as the blockade of the inhibitory effects of locally applied baclofen onto cortical neurons by parenteral administration of CGP 35348 (see Bittiger *et al.*, Chapter 3).

The present study attempts to elucidate the role of $GABA_B$ receptors in epileptic phenomena by including recent new data from our laboratory as well as published information. First, new in-vitro and in-vivo data on the action of CGP 35348 in various experimental models of epilepsy will be presented. This will be followed by a presentation of published and new results on the action of the $GABA_B$ agonist baclofen in various in-vitro and in-vivo models of epilepsy. Finally, we will review the literature on clinical baclofen data with special emphasis on its potential anti- or proconvulsive properties.

METHODS

Preparation of Hippocampal Slices

Male Sprague-Dawley rats (Tif: RA 25, Sisseln, Switzerland) weighing 100-150 g were decapitated under light ether anaesthesia. The brain was rapidly dissected out and immersed in a cooled artificial cerebrospinal fluid (ACSF) solution (3-5 °C, pregassed with 95% O_2/5% CO_2). The hippocampi were isolated carefully and 400-μm-thick transverse slices were cut using a McIllwain tissue chopper. The slices were placed on a nylon net located in a superfusion chamber (Spencer *et al.*, 1976) and were maintained at room temperature for the first hour in a static condition with the ACSF level adjusted to their upper surface. Warm humidified air was continuously circulated over the slices. The slices were subsequently superfused at 32 °C with gassed ACSF at a rate of 3 ml/min (1 chamber volume/min). The standard ACSF had the following composition in mM: NaCl, 124.0; KCl, 2.5; KH_2PO_4, 1.2; $MgSO_4$, 2.0; $CaCl_2$, 2.5; $NaHCO_3$, 26.0; D-glucose, 10.0.

Epileptic-like activity was induced in three different sets of experiments. In the first set, penicillin (1.2 mM) was added to the ACSF and the KCl concentration was raised to 10 mM. In the second set, bicuculline methiodide (5 μM) was present in the standard ACSF throughout the experiments. In the third set of experiments, the Mg^{2+} concentration was lowered to 0.1 mM (all other ions present as in standard ACSF). Extracellular recordings of spontaneous or stimulation-induced field potentials were performed by means of glass electrodes (2-5 MΩ) filled with 4 M NaCl. Spontaneous field potentials which consisted of multiple population spikes appeared in the presence of penicillin (recorded in CA3) and bicuculline (recorded in CA1 and CA3). In the experiments performed with low Mg^{2+} levels (0.1 mM), the synaptic input to the CA1 region was activated by electrical stimulation (0.1–0.3 ms pulse width, 3–15 V) of the Schaffer collateral/commissural fibres. Single pulse stimuli were given at a rate of 0.125/s during 5–20 min until responses of constant amplitude were obtained. The evoked responses were averaged ($n = 4$) by means of a Hitachi Digital oscilloscope VC-6041 and the averaged population spikes were plotted on a W + W 320 recorder.

Convulsion Tests in Mice

Animals

Male mice (Tif: MAGf) weighing 19–25 g were used. The animals were maintained under standard laboratory conditions with free access to food and tap water.

Drug administration

CGP 35348 and baclofen were dissolved in saline and administered 30 and 60 min respectively prior to the injection of a convulsant agent. CGP 35348 was tested at 100 and 300 mg/kg and baclofen at 6 and 12 mg/kg. The NMDA antagonist CGP 37849 was dissolved in saline and injected at 3 mg/kg 60 min before NMDA. In the NMDA test, baclofen and CGP 35348 were both injected 30 min prior to NMDA. The drugs were injected at the volume of 0.1 ml/10 g body weight. The control animals received saline (0.1 ml/10 g). All drugs and saline were administered i.p.

Induction of seizures

To detect the possibility of subtle pro- and anticonvulsive effects of CGP 35348 and baclofen respectively, the convulsants were in some of the tests injected at doses which were clearly submaximal (40–70% of control animals showing convulsions) in the experiments with CGP 35348, or near-maximal or maximal (70–100% of control animals showing convulsions) in the experiments with baclofen.

(+)-Bicuculline was administered i.p. at a dose of 3.5 mg/kg in the experiments with CGP 35348 (40% convulsions in control animals). In the experiments with baclofen, bicuculline was injected i.p. at 4.0 mg/kg (70% convulsions in control animals). The number of animals presenting clonic or tonic seizures within 15 min after the bicuculline injection was recorded.

Picrotoxin was administered i.p. at 7.5 mg/kg (80–100% convulsions in control animals) in all experiments. The number of animals presenting clonic seizures within 15 min after the injection was recorded.

Pentylenetetrazole was administered i.p. at 50 (80% convulsions in control animals) and 70 mg/kg (100% convulsions in control animals) in tests with CGP 35348 and at 70 mg/kg in the experiments with baclofen. The number of animals presenting clonic seizures within 10 min after the pentylenetetrazole injection was recorded.

Strychnine nitrate was administered i.p. at 0.85 mg/kg (50% convulsions in control animals) in the experiments with CGP 35348 and at 1.75 mg/kg (60% convulsions in control animals) and 2.5 mg/kg (100% convulsions in control animals) in the experiments with baclofen. The latency to the onset of tonic seizures and the number of animals surviving the seizures for longer than 10 min after the strychnine injection was recorded.

NMDA was injected i.p. at a dose of 75 mg/kg. The number of animals exhibiting circling behaviour within 30 min after the injection was recorded.

Maximal electroshock convulsions were induced by applying current (18 mA, 0.2 s, 50 Hz) through corneal electrodes. The number of animals with tonic hind-limb extension and the duration of the limb extension was recorded.

In all models pilot studies were performed. In the subsequent experiments the number of animals was increased to make a total of at least $n = 10$ in each group. The values are given as mean $\pm$ SD.

Statistics

Fisher's exact test (convulsions, mortality) and Student's *t*-test (convulsion latency or duration) were used for the statistical analyses.

RESULTS

The Effects of CGP 35348 and Baclofen on Epileptic-like Discharges In Vitro

The aim of these investigations was to investigate how the $GABA_B$ blocker CGP 35348 and baclofen influence epileptic-like conditions in vitro in three different models. In the penicillin model, spontaneous discharge activity of CA3 neurons occurred at a higher frequency than in the bicuculline model (see Figure 21.1). This may be due to the higher K^+ concentration (10 mM compared to 3.7 mM) used in the penicillin model and/or to the fact that a threshold concentration of bicuculline (5 μM) was chosen. In the penicillin model, baclofen (3 and 6 μM), as previously shown (Karlsson and Olpe, 1989), consistently and potently depressed the epileptic-like discharges ($n > 12$ experiments). CGP 35348 slightly but reversibly increased the spontaneous frequency in this model. At a concentration of 100 μM, the activity was increased in 7 out of 15 hippocampal slices (Figure 21.1). The mean activation never exceeded the spontaneous activity by more than 50%. A weak increase was also observed at 10 μM in 4 out of 5 slices (data not shown). In the bicuculline model, baclofen (6 μM) also invariably suppressed spontaneous epileptiform activity ($n = 3$). The action of CGP 35348 (100 μM) was investigated in four experiments in the CA1 region and four experiments in the CA3 region. In half ($n = 9$) of the slices investigated ($n = 18$), the compound weakly increased the spontaneous discharge frequency induced by bicuculline in CA3 neurons (Figure 21.1) but only 5 out of 21 slices showed a weak activation in the CA1 region (not shown).

In the experiments in which the Mg^{2+} level was lowered (0.1 mM), submaximal stimulation of the Schaffer collateral/commissural fibres was adjusted to evoke two population spikes of which the second is indicative of epileptic-like activity (Coan and Collingridge, 1985). Baclofen administered at concentrations of 2 ($n = 1$), 3 ($n = 1$) and 6 μM ($n = 5$) invariably induced the appearance of a third population spike. In 3 slices baclofen (6 μM) reduced the amplitude of the first population spikes. In 2 slices the action of baclofen (12 μM) was compared with that of the GABA$_A$ agonist THIP. Whereas baclofen elicited the above described effects, THIP (60 μM) reduced the amplitude of the second population spikes in both instances (Figure 21.2).

The Effects of CGP 35348 and Baclofen in Various Convulsion Models In Vivo

The aim of this study was to investigate the effects of GABA$_B$ receptor modulation in induced convulsions in mice.

The GABA$_B$ blocker CGP 35348 (100 and 300 mg/kg) and baclofen (6 and 12 mg/kg) had no significant effects on the number of animals presenting

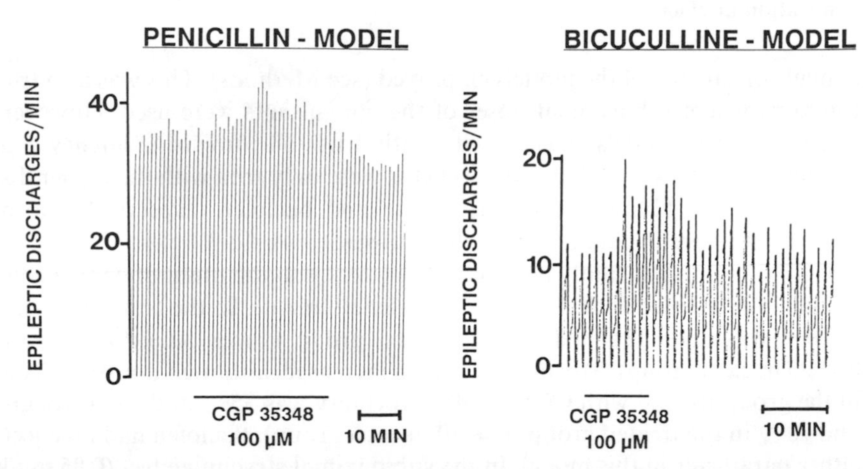

Figure 21.1. The effect of CGP 35348 (100 μM) on epileptic-like activity recorded in the CA3 region of hippocampal slices is depicted. Epileptic-like bursting was induced by penicillin (1.2 mM) or bicuculline (5 μM) and consisted of spontaneous field potentials with multiple population spikes. CGP 35348 weakly increased the bursting frequency in both models

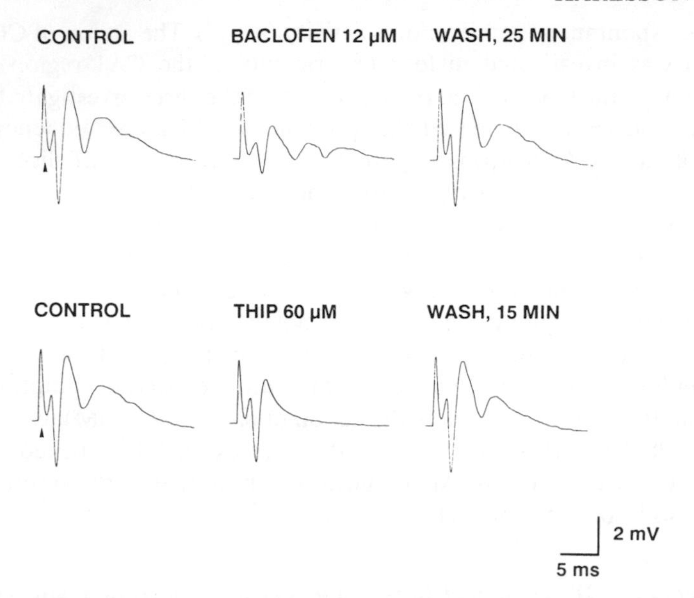

Figure 21.2. Baclofen (12 μM) and the $GABA_A$ receptor agonist THIP (60 μM) exert opposite effects on evoked responses in the CA1 region in low-Mg^{2+} (0.1 mM) medium. Control responses in low Mg^{2+} consist of more than one population spike. Baclofen depressed the first and the second population spikes but invariably induced the appearance of a third spike. THIP reduced the amplitude of the first spike, abolished the second and did not induce any additional population spikes. The effects of both baclofen and THIP were reversible within 20 min after starting the washout. Arrows indicate the stimulation artefact

convulsions in any of the models employed (see Methods). This was also true in the cases where submaximal doses of the convulsants were used. However, in some of the test models an effect with both drugs on convulsion latency and/or mortality was observed. Since these effects were only seen with the higher doses, 300 mg/kg and 12 mg/kg for CGP 35348 and baclofen respectively, we only describe the results obtained with these doses.

CGP 35348 significantly decreased the latency to the onset of convulsions in the animals injected with 70 mg/kg pentylenetetrazole. In the control group the latency was 88 ± 37 s and in the treated group it was 60 ± 40 s ($P < 0.05$, $n = 20$ in each group). Mortality was also increased, although non-significantly, in the group treated with CGP 35348. Mortality was 35 % in the control group and 70 % in the treated group ($n = 20$ in each group). Baclofen had no effect on either parameter in this model. In the submaximal strychnine test (0.85 mg/kg), CGP 35348 significantly decreased the latency to the onset of convulsions from 312 ± 113 s (in the control group) to 200 ± 49 s (in the treated group) ($P < 0.05$, $n = 10$ in each group). Baclofen, on the other hand, markedly increased the latency to convulsions induced by both a submaximal (1.75 mg/

kg) and a maximal (2.5 mg/kg) dose of strychnine. The increases were from 118 ± 77 to 309 ± 77 s ($P < 0.01$, $n = 10$) and from 130 ± 68 to 270 ± 78 s ($P < 0.01$, $n = 10$) respectively. Baclofen also significantly reduced the mortality in the maximal strychnine test from 100 % to 40 % ($P < 0.01$, $n = 10$ in each group). Whereas CGP 35348 was inactive, baclofen significantly increased the latency to picrotoxin-induced convulsions. The convulsion latency was 392 ± 52 s in the control group and 717 ± 110 s in the baclofen-treated group ($P < 0.01$, $n = 10$). NMDA (75 mg/kg) induced hyperactivity with grooming and biting of the tail in addition to circling behaviour in 9 out of the 10 animals tested. No convulsions were observed with this dose of NMDA. These behavioural effects were abolished in all animals ($n = 10$) by the NMDA antagonist CGP 37849 (3 mg/kg). CGP 35348 (300 mg/kg) and baclofen (3 mg/kg) did not modify the NMDA-induced behavioural symptoms or the latency to the onset of the circling. However, in the baclofen-treated group one animal presented clonic convulsions and died. In the maximal electroshock test CGP 35348 was without effect. On the other hand, baclofen tended (non-significantly) to reduce the number of animals presenting tonic hind-leg extension (from 100 % to 70 %, $n = 10$ in each group). The duration of the tonic hind-leg extension was significantly reduced from 12 ± 1.4 to 4.3 ± 3.7 s ($P < 0.0005$).

DISCUSSION

The most interesting but also unexpected aspect of this study concerns the results obtained with the new $GABA_B$ receptor blocker CGP 35348. Taken together, the in-vitro and in-vivo findings demonstrate that this compound has only subtle, if any, proconvulsant features. Its action differs in this respect quite distinctly from that of a $GABA_A$ blocker such as bicuculline.

The present results show that CGP 35348 has a weak potentiating effect on epileptic-like activity induced by bicuculline and penicillin in the hippocampal slice preparation (Figure 21.1). It has recently been reported that the $GABA_B$ blocker phaclofen does not induce any signs of proconvulsant activity in these slice models (Karlsson and Olpe, 1989). However, since CGP 35348 has been shown to be considerably more potent than phaclofen in blocking the $GABA_B$-mediated late inhibitory postsynaptic potential (IPSP) (see Karlsson *et al.*, Abstracts section, this volume), the negative findings with phaclofen could reflect the difference in potency of the two blockers. Similarly to the findings in vitro, CGP 35348 did not have any clear proconvulsive effects in chemically or electrically induced seizures in mice. However, in two tests where convulsions were induced by either a high dose of pentylenetetrazole or a submaximal dose of strychnine, CGP 35348 significantly decreased the latency to the convulsions.

On the basis of classical pharmacological principles, the $GABA_B$ agonist baclofen should give an activity pattern opposite to the one obtained with a $GABA_B$ blocker. Rather unexpectedly, this does not seem to be the case. A review of published preclinical in-vivo data (Table 2) shows that the activity profile of baclofen in animal models of epilepsy is complex. Thus, depending on the epilepsy model used, baclofen has been reported to be either anticonvulsive or ineffective (see Table 2). There are some indications that it may even be proconvulsive in animals at high doses (see Table 2) (Cottrell and Robertson, 1987; Meldrum, 1984; Vergnes *et al.*, 1984).

Our findings show that baclofen was active in suppressing epileptic-like activity induced by bicuculline and penicillin in vitro and this is in agreement with numerous other studies. Table 1 shows that baclofen is, in fact, effective in suppressing epileptic-like activity in a variety of in-vitro models. However, in one study performed using Mg^{2+}-free medium an epileptogenic action of baclofen was indicated. It was shown that baclofen suppressed the interictal bursting activity in this model and this allowed ictal activity to emerge (Swartzwelder *et al.*, 1987). On the other hand, in a more recent study performed using Mg^{2+}-free medium but in another brain area, Jones (1989) showed that baclofen was effective in blocking also ictal activity.

The in-vivo data presented in this study demonstrate that baclofen was inactive in protecting the animals from convulsions in all models tested. These findings are consistent with those reported by others that baclofen is not active in protecting animals from convulsions induced by picrotoxin (Bernard *et al.*, 1980; Worms and Lloyd, 1981), bicuculline (Worms and Lloyd, 1981), pentylenetetrazole (Benedito and Leite, 1981; Bernard *et al.*, 1980; Ulloque *et al.*, 1986), strychnine (Lembeck and Beubler, 1977) and electroshock (Bernard *et al.*, 1980; Mehta and Ticku, 1986; Ulloque *et al.*, 1986). As indicated in Table 2, there appear to be discrepancies beween findings with baclofen in the same convulsion models (no effect versus anticonvulsant activity) reported by different authors. However, these differences may in most cases be accounted for by the fact that different parameters were recorded (i.e. not protection from convulsions) or different convulsant doses or current intensities were used to induce convulsions. For example Naik *et al.* (1976) reported that baclofen possessed anticonvulsant properties in that it delayed the onset of convulsions in the strychnine test (observed also in the present study). Furthermore, baclofen was not active against the clonic convulsions induced by pentylenetetrazole in the present study and those mentioned in Table 2. However, Petersen (1983) reported that baclofen did reduce tonic convulsions induced by high doses of pentylenetetrazole. In the present study and those by others (see Table 2), baclofen was not active in protecting the animals from convulsions induced by maximal electroshock. Baclofen did, however, as shown in this study, significantly reduce the duration of tonic hind-leg extension. Similar findings were reported by Czuczwar *et al.* (1984) and Benedito and Leite (1981), who also found baclofen to be active in that it increased threshold intensities and convulsion latencies respectively. On the other hand, in certain kinds of in-vivo models of epilepsy, baclofen has been reported to be proconvulsive (see Table 2). It was reported that a high

Table 1. Effect of baclofen on epileptic-like activity induced chemically or electrically in slices

Convulsant agent	Baclofen (μM)*	Type of discharges and effect		Slice preparations (area)	Reference
Penicillin (1.2 mM)	6	spontaneous	↓	hippocampus (CA3)	Karlsson and Olpe (1989)
Bicuculline-methiodide					
(100 μM)	5	evoked	↓	hippocampus (CA1, CA3)	Ault and Nadler (1983)
(25 μM)	5	spontaneous	↓	hippocampus (CA3)	Brady and Swann (1984)
(50 μM)	10	spontaneous	↓	neocortex	Horne *et al.* (1986)
(30 μM)	0.3–1	spontaneous	↓	hippocampus (CA3)	Swartzwelder *et al.* (1986b)
(50 μM)	0.1	spontaneous	↓	hippocampus (CA3)	Ault *et al.* (1986b)
Zero Mg^{2+}	0.1–1	spontaneous (interictal)	↓	entorhinal cortex	Jones (1989)
	0.1–1	spontaneous (ictal)	↓	entorhinal cortex	Jones (1989)
	2	spontaneous (interictal)	↓	hippocampus (CA3)	Swartzwelder *et al.* (1987)
	2	(ictal)	↑	hippocampus (CA3)	Swartzwelder *et al.* (1987)
	10	spontaneous	↓	neocortex	Horne *et al.* (1986)
Elevated K^+					
(7 mM)	1–10	spontaneous	↓	hippocampus (CA3)	Swartzwelder *et al.* (1986b)
(3.5 mM)	27–500 nM	spontaneous	↓	hippocampus (CA3)	Ault *et al.* (1986b)
(5 mM)	27–500 nM	spontaneous	↓	hippocampus (CA3)	Ault *et al.* (1986b)
(7 mM)	27–500 nM	spontaneous	↓	hippocampus (CA3)	Ault *et al.* (1986b)
Kainic acid (50 nM)	27–500 nM	spontaneous	↓	hippocampus (CA3)	Ault *et al.* (1986b)
Stimulation train induced bursts	0.03–0.3	spontaneous	↓	hippocampus (CA3)	Swartzwelder *et al.* (1986a)
	0.3–1	evoked	↓	hippocampus (CA3)	Swartzwelder *et al.* (1986a)

*Except where indicated.

Table 2. Effect of baclofen in various seizure models in vivo

	Rat	Mouse	Reference
Chemically induced seizures			
GABA-related models			
Bicuculline		∅	Worms and Lloyd (1981)
Picrotoxin		∅	Bernard *et al.* (1980)
	∅		Worms and Lloyd (1981)
Isoniazid		↓	Lembeck and Beubler (1977)
	↓		Naik *et al.* (1976)
3-Mercaptopropionic acid	↓		Benedito and Leite (1981)
Allylglycine	↓		Ashton and Wauquier (1979)
DMCM		↓	Petersen (1983)
Non-GABA-related models			
Pentylenetetrazole	∅		Bernard *et al.* (1980)
		∅	Benedito and Leite (1981)
		∅	Ulloque *et al.* (1986)
		↓	Petersen (1983)
Strychnine		∅	Lembeck and Beubler (1977)
	↓		Naik *et al.* (1976)
Kainic acid	↓		Ault *et al.* (1986a)
	↓		Bernard *et al.* (1980)
NMDLA		↑	Czuczwar *et al.* (1985)
Electrically induced seizures			
Electroshock		↓	Czuczwar *et al.* (1984)
	∅		Bernard *et al.* (1980)
		↓	Benedito and Leite (1981)
	∅		Mehta and Ticku (1986)
		∅	Ulloque *et al.* (1986)

Amygdala kindled	∅		Morimoto *et al.* (1987)
Hippocampus kindled	∅		Vartanian *et al.* (1986)
Local electrical stimulation (inferior collicular cortex)	↓		McCown *et al.* (1987)
Genetic models			
Photosensitive epilepsy in baboons	(low dose) (high dose)	↓ ↑	Meldrum (1984)
Spontaneous petit mal-like epilepsy in rats		↑	Vergnes *et al.* (1984)
Audiogenic seizure-susceptible mice		↓	Benedito and Leite (1981)
Other models			
Epileptic myoclonus in rats kindled in amygdala or hippocampus		↑	Cottrell and Robertson (1987)
Muscimol-induced myoclonic jerks in mice		↓	Menon and Vivonia (1981)
Sound-induced seizures in ethanol-dependent withdrawal in rats		↓	Frye *et al.* (1986)
Focal eilepsy after intracortical administration of baclofen in rats		↑	Van Rijn *et al.* (1987)

∅, Baclofen inactive in protecting animals from convulsions.
DMCM, methyl-6,7-dimethoxy-4-ethyl-β-carboline-3-carboxylate; NMDLA, N-methyl-D,L-aspartate.

dose of baclofen induces myoclonus in rats kindled in the amygdala or hippocampus (Cottrell and Robertson, 1987). Meldrum (1984) reported that, in the photosensitive baboon, a toxic syndrome with myoclonias and electroencephalographic changes appears with high doses of baclofen. Furthermore, in rats with spontaneous petit-mal-like epilepsy, baclofen was shown to dose-dependently enhance the duration of the seizures (Vergnes *et al.*, 1984). However, in the last two models, this effect was not baclofen-specific since other GABAergic compounds (including $GABA_A$ agonists) also induced similar proconvulsive effects.

Whereas an early clinical study suggested that baclofen was ineffective as an antiepileptic agent and in fact had a deleterious effect in some patients with epilepsy (Pinto *et al.*, 1972), more recent studies have found that baclofen did not seem to have an effect on the intensity or frequency of epileptic seizures (Milla and Jackson, 1977; Terrence *et al.*, 1983). However, a few rare cases have been reported where seizures occurred during baclofen therapy (see Table 3). Furthermore, there have been reports of seizures provoked by either baclofen overdose (Haubenstock *et al.*, 1983) or withdrawal (Barker and Grant, 1982; Hyser and Drake, 1984; Terrence and Fromm, 1981). These phenomena do not, however, exclude the use of baclofen as an antiepileptic drug as they are known to occur also with established antiepileptics (Troupin and Ojemann, 1976). The overall picture of the antiepileptic profile of baclofen in various animal models of epilepsy does not appear to justify any further testing of this compound as a broadly active antiepileptic agent. It is conceivable, however, that baclofen might have beneficial therapeutic effects in certain types of epilepsy.

In the case of baclofen we are faced with the contrast between its pronounced antiepileptic features in vitro and its heterogeneous activity pattern in vivo. It appears that the intact brain responds differently to baclofen compared to epileptic slice preparations. This is particularly interesting in the case of bicuculline-induced convulsions since baclofen is very potent in suppressing epileptic-like activity in vitro but is inactive in vivo in this model. The reasons for this discrepancy are not clear. It may be speculated that afferent and/or efferent connections of the hippocampus have an important impact on the reactivity of hippocampal tissue to the action of epileptic agents and/or the $GABA_B$ agonist baclofen. It has recently been shown that the stability of the hippocampus is ensured by the feed-forward inhibitory action of subcortical afferents (Buzsaki *et al.*, 1989).

In view of the potent inhibitory effects seen with baclofen in various in-vitro models it is particularly interesting that a few authors observed proconvulsive effects in vivo (see above). We found some actions of baclofen in the hippocampal slice which could explain why the drug may facilitate epilepsy in models where epilepsy is already established, for example in genetic models of epilepsy. In the hippocampal slice, maintained in regular ACSF, the overall effect of baclofen is a depression of the excitatory PSP (EPSP) and the population spike

Table 3. Incidence/frequency of seizures in man during baclofen therapy: published data

	No. of patients with seizures/no. of patients studied	No. of studies	References
Patients with epilepsy (receiving antiepileptics)	5/137	6	Young (1980); Minford *et al.* (1980); Terrence *et al.* (1983); Zalman (1976); Hattab (1978); Primrose (1980)
Patients with a history of seizures (not receiving antiepileptics)	0/4	2	Milla and Jackson (1977); Minford *et al.* (1980)
Patients with no history of seizures	3/80	3	Primrose (1980); Hudgson and Weightman (1972); Seyfert and Straschill (1982)

Published reports on the frequency of epilepsy in man during baclofen therapy are summarized and classified into three categories. Most of the patients were undergoing baclofen treatment for their spasticity.

(Olpe *et al.*, 1982). These effects are in keeping with the marked depressant effects seen in the different in-vitro epilepsy models. However, in regular ACSF, baclofen has been shown to potently impair GABA-mediated paired-pulse inhibition. This effect was observed at concentrations which only moderately depressed the amplitude of the population spikes (Karlsson and Olpe, 1989). It has been suggested by several investigators that baclofen suppresses GABA transmission by inhibition at the level of the GABAergic interneurons (Harrison *et al.*, 1988; Karlsson and Olpe, 1989; Misgeld *et al.*, 1989). We furthermore observed that baclofen induced a third population spike in the low Mg^{2+} model (see Figure 21.2). Finally, we found that a second population spike may be evoked by baclofen in slices kept in regular ACSF but stimulated supramaximally (unpublished data). The appearance of multiple population spikes is indicative of epileptic-like activity. Thus, the findings that baclofen appears to be proconvulsive in situations of strong stimulation or lowered Mg^{2+} concentrations (Mg^{2+} blocks the NMDA receptor channel) indicates an interaction with the NMDA system. This could be explained by the above-mentioned disinhibitory action of baclofen, which would lead to the removal of the GABA-mediated hyperpolarization-block of the NMDA receptors. In favour of this suggestion are our recent findings that the proconvulsive effects of baclofen in vitro could be blocked by NMDA antagonists (unpublished data). It has furthermore been reported by Czuczwar *et al.* (1985) that baclofen decreased the convulsion latency and increased mortality in NMDA-induced convulsions in mice. In the present study we were not able to demonstrate a proconvulsant effect in this model; however, we used a threshold dose of NMDA and a much lower dose of baclofen (3 mg/kg compared to 20 mg/kg).

Taken together, these findings clearly show that $GABA_B$ receptors are less important than $GABA_A$ receptors in the control of epileptic processes. However, a modulatory role cannot be excluded. Intracellular recordings from epileptic tissue might help to discover more subtle effects of $GABA_B$ agonists and antagonists.

REFERENCES

Ashton, D. and Wauquier, A. (1979) Effects of some anti-epileptic, neuroleptic and GABAminergic drugs on convulsions induced by D,L-allylglycine. *Pharmacol. Biochem. Behav.*, **11**, 221–226.

Ault, B. and Nadler, J.V. (1983) Anticonvulsant-like actions of baclofen in the rat hippocampal slice. *Br. J. Pharmacol.*, **78**, 701–708.

Ault, B., Gruenthal, M., Armstrong, D.R. and Nadler, J.V. (1986a) Efficacy of baclofen and phenobarbital against the kainic acid limbic seizure-brain damage syndrome. *J. Pharmacol. Exp. Ther.*, **239**, 612–617.

Ault, B., Gruenthal, M., Armstrong, D.R., Nadler, J.V. and Wang, C.M. (1986b) Baclofen suppresses bursting activity induced in hippocampal slices by differing convulsant treatments. *Eur. J. Pharmacol.*, **126**, 289–292.

Barker, I. and Grant, I.S. (1982) Convulsions after abrupt withdrawal of baclofen. *Lancet*, **iv**, 556–557.

Benedito, M.A.C. and Leite, J.R. (1981) Baclofen as an anticonvulsant in experimental models of convulsions. *Exp. Neurol.*, **72**, 346–351.

Bernard, P.S., Sobiski, R.E. and Dawson, K.M. (1980) Antagonism of a kainic acid syndrome by baclofen and other putative GABAmimetics. *Brain Res. Bull.*, **5**, 519–523.

Brady, R.J. and Swann, J.W. (1984) Postsynaptic actions of baclofen associated with its antagonism of bicuculline-induced epileptogenesis in hippocampus. *Cell. Mol. Neurobiol.*, **4**, 403–408.

Buzsaki, G., Ponomareff, G.L., Bayardo, F., Ruiz, R. and Gage, F.H. (1989) Neuronal activity in the subcortically denervated hippocampus: a chronic model for epilepsy. *Neuroscience*, **28**, 527–538.

Coan, E.J. and Collingridge, G.L. (1985) Magnesium ions block an *N*-methyl-D-aspartate receptor-mediated component of synaptic transmission in rat hippocampus. *Neurosci. Lett.*, **53**, 21–26.

Cottrell, G.A. and Robertson, H.A. (1987) Baclofen exacerbates epileptic myoclonus in kindled rats. *Neuropharmacology*, **26**, 645–648.

Czuczwar, S.J., Chmielewska, B., Turski, W.A. and Kleinrok, Z. (1984) Differential effects of baclofen, gamma-hydroxybutyric acid and muscimol on the protective action of phenobarbital and diphenylhydantoin against maximal electroshock-induced seizures in mice. *Neuropharmacology*, **23**, 159–163.

Czuczwar, S.J., Frey, H.-H, and Löscher, W. (1985) Antagonism of *N*-methyl-D,L-aspartic acid-induced convulsions by antiepileptic drugs and other agents. *Eur. J. Pharmacol.*, **108**, 273–280.

Frye, G.D., McCown, T.J., Breese, G.R. and Peterson, S.L. (1986) GABAergic modulation of inferior colliculus excitability: role in the ethanol withdrawal audiogenic seizures. *J. Pharmacol. Exp. Ther.*, **237**, 478–485.

Harrison, N.L., Lange, G.D. and Barker, J.L. (1988) (–)-Baclofen activates presynaptic $GABA_B$ receptors on GABAergic inhibitory neurons from embryonic rat hippocampus. *Neurosci. Lett.*, **85**, 105–109.

Hattab, J.R. (1978) Lioresal in the treatment of spasticity of cerebral origin. In: A.M. Jukes (ed.) *Baclofen: spasticity and cerebral pathology*, pp. 60–67. Northampton: Cambridge Medical Publications Ltd.

Haubenstock, A., Hruby, K., Jäger, U. and Lenz, K. (1983) Baclofen (Lioresal®): intoxication report of 4 cases and review of the literature. *Clin. Toxicol.*, **20**, 59–68.

Hill, D.R. and Bowery, N.G. (1981) [^{3}H] Baclofen and [^{3}H] GABA bind to bicuculline-insensitive GABA-B sites in rat brain. *Nature*, **290**, 149–152.

Horne, A.L., Harrison, N.L., Turner, J.P. and Simmonds, M.A. (1986) Spontaneous paroxysmal activity induced by zero magnesium and bicuculline: suppression by NMDA antagonists and GABA mimetics. *Eur. J. Pharmacol.*, **122**, 231–238.

Hudgson, P. and Weightman, D. (1972) Eine Doppelblindstudie mit Lioresal. In: W. Birkmayer (ed.) *Aspekte der Muskelspastik*, pp. 138–140. Vienna: Hand Huber.

Hyser, C.L. and Drake, M.E. (1984) Status epilepticus after baclofen withdrawal. *J. Natl Med. Assoc.*, **76**, 533–538.

Jones, S.G. (1989) Ictal epileptiform events induced by removal of extracellular magnesium in slices of entorhinal cortex are blocked by baclofen. *Exp. Neurol.*, **104**, 155–161.

Karlsson, G. and Olpe, H.-R. (1989) Inhibitory processes in normal and epileptic-like rat hippocampal slices: the role of $GABA_B$ receptors. *Eur. J. Pharmacol.*, **163**, 285–290.

Lembeck, F. and Beubler, E. (1977) Convulsions induced by hyperbaric oxygen:

inhibition by phenobarbital, diazepam and baclofen. *Naunyn-Schmiedeberg's Arch. Pharmacol.*, **297**, 47–51.

McCown, T.J., Givens, B.S. and Breese, G.R. (1987) Amino acid influences on seizures elicited within the inferior colliculus. *J. Pharmacol. Exp. Ther.*, **243**, 603–608.

Mehta, A.K. and Ticku, M.K. (1986) Comparison of anticonvulsant effect of pentobarbital and phenobarbital against seizures induced by maximal electroshock and picrotoxin in rats. *Pharmacol. Biochem. Behav.*, **25**, 1059–1065.

Meldrum, B.S. (1975) Epilepsy and gamma-aminobutyric acid-mediated inhibition. *Int. Rev. Neurobiol.*, **17**, 1–37.

Meldrum, B. (1984) GABAergic agents as anticonvulsants in baboons with photosensitive epilepsy. *Neurosci. Lett.*, **47**, 345–349.

Menon, M.K. and Vivonia, C.A. (1981) Serotonergic drugs, benzodiazepines and baclofen block muscimol-induced myoclonic jerks in a strain of mice. *Eur. J. Pharmacol.*, **73**, 155–161.

Milla, P.J. and Jackson, A.D.M. (1977) A controlled trial of baclofen in children with cerebral palsy. *J. Int. Med. Res.*, **5**, 398–404.

Minford, A.M.B., Brown, J.K., Minns, R.A., Frazer, P., Hollway, L., Gibb, N., Campbell, L. and Neijerink, I. (1980) The effect of baclofen on the gait of hemiplegic children assessed by means of polarised light goniometry. *Scot. Med. J.*, **25** (suppl.), 29–35.

Misgeld, U., Müller, W. and Brunner, H. (1989) Effects of (−)-baclofen on inhibitory neurons in the guinea pig hippocampal slice. *Pflügers Arch.*, **414**, 139–144.

Morimoto, K., Holmes, K.H. and Goddard, G.V. (1987) Kindling-induced changes in EEG recorded during stimulation from the site of stimulation: III. Direct pharmacological manipulations of the kindled amygdala. *Exp. Neurol.*, **97**, 17–34.

Naik, S.R., Guidotti, A. and Costa, E. (1976) Central GABA receptor agonists: comparison of muscimol and baclofen. *Neuropharmacology*, **15**, 479–484.

Olpe, H.-R., Baudry, M., Fagni, L. and Lynch, G. (1982) The blocking action of baclofen on excitatory transmission in the rat hippocampal slice. *J. Neurosci.*, **2**, 698–703.

Petersen, E.N. (1983) DMCM: a potent convulsive benzodiazepine receptor ligand. *Eur. J. Pharmacol.*, **94**, 117–124.

Pinto, O., Polikar, M. and Loustalot, P. (1972) Die klinische Prüfung von Lioresal in der Übersicht. In: W. Birkmayer (ed.) *Aspekte der Muskelspastik*, pp. 192–207. Vienna: Hans Huber.

Primrose, D.A. (1980) Self injury in subnormal patients. *Scot. Med. J.*, **25** (suppl.), 44–49.

Seyfert, S. and Straschill, M. (1982) Elektroenzephalographische Veränderungen unter Baclofen. *Z. EEG–EMG*, **13**, 161–166.

Spencer, H., Gribkoff, V.K., Cotman, C.W. and Lynch, G.S. (1976) GDEE antagonism of iontophoretic amino acid excitations in the intact hippocampus and in the hippocampal slice preparation. *Brain Res.*, **105**, 471–480.

Swartzwelder, H.S., Bragdon, A.W., Sutch, C.P., Ault, B. and Wilson, W.A. (1986a) Baclofen suppresses hippocampal epileptiform activity at low concentrations without suppressing transmission. *J. Pharmacol. Exp. Ther.*, **237**, 881–887.

Swartzwelder, H.S., Sutch, C.P. and Wilson, W.A. (1986b) Attenuation of epileptiform bursting by baclofen: reduced potency in elevated potassium. *Exp. Neurol.*, **94**, 726–734.

Swartzwelder, H.S., Lewis, D.V., Anderson, W.W. and Wilson, W.A. (1987) Seizure-like events in brain slices: suppression by interictal activity. *Brain Res.*, **410**, 362–366.

Terrence, C.F. and Fromm, G.H. (1981) Complications of baclofen withdrawal. *Arch. Neurol.*, **38**, 588–589.

Terrence, C.F., Fromm, G.H. and Roussan, M.S. (1983) Baclofen: its effects on seizure frequency. *Arch. Neurol.* **40**, 28–29.

Troupin, A.S. and Ojemann, L.M. (1976) Paradoxical intoxication: a complication of anticonvulsant administration. *Epilepsia*, **16**, 753–758.

Ulloque, R.A., Chweh, A.Y. and Swinyard, E.A. (1986) Effects of gamma-aminobutyric acid (GABA) receptor agonists on the neurotoxicity and anticonvulsant activity of barbiturates in mice. *J. Pharmacol. Exp. Ther.*, **237**, 468–472.

Van Rijn, C.M., Van Berlo, M.J., Feenstra, M.G.P., Schoofs, M.L.F. and Hommes, O.R. (1987) *R*(−)-Baclofen: focal epilepsy after intracortical administration in the rat. *Epilepsy Res.*, **1**, 321–327.

Vartanian, M.G., Mickevicius, P.J., Stieber, B.Z. and Taylor, C.P. (1986) Effects of anticonvulsant drugs on afterdischarge threshold in kindled rats: a model of temporal lobe epilepsy. *Epilepsia*, **27**, 638.

Vergnes, M., Marescaux, C., Micheletti, G., Depaulis, A., Rumbach, L. and Warter, J.M. (1984) Enhancement of spike and wave discharges by GABAmimetic drugs in rats with spontaneous petit-mal-like epilepsy. *Neurosci. Lett.*, **44**, 91–94.

Worms, P. and Lloyd, K.G. (1981) Functional alterations of GABA synapses in relation to seizures. In: P.L. Morselli *et al.* (eds) *Neurotransmitters, seizures and epilepsy*, pp. 37–46. New York: Raven Press.

Young, J.A. (1980) Clinical experience in the use of baclofen in children with spastic cerebral palsy: a further report. *Scot. Med. J.*, **25** (suppl.), 23–25.

Zalman, E. (1976) Motorische Störungen im Kindes- und Jugendalter. *Ärztiche Praxis*, **64**, 3–7.

Thomas, [illegible] and Ohmann, [illegible] (1979) [illegible]
Ultsch, A., [illegible] (1993) [illegible] GABA_A [illegible] activity of [illegible]
Van Rijn, [illegible] [illegible]
Vermeulen, [illegible] [illegible]
Wieland, [illegible] (1992) [illegible] GABA_A [illegible] 44 [illegible]
Wong, [illegible] GABA_A [illegible] New York [illegible]
Yeh, [illegible] [illegible]
Zhang, [illegible]

Analgesia

22 $GABA_B$ Receptors and Analgesia

J. SAWYNOK

INTRODUCTION

The systemic administration of baclofen, a prototype $GABA_B$ receptor agonist (Bowery, 1982), produces antinociception in a variety of test systems commonly used to detect the antinociceptive and analgesic activity of centrally active drugs. These include the hot-plate test (Bartolini *et al.*, 1982; Cutting and Jordan, 1975; Hill *et al.*, 1981; Levy and Proudfit, 1977; Sawynok and LaBella, 1982; Vaught *et al.*, 1985), the tail-flick test (Bartolini *et al.*, 1982; Levy and Proudfit, 1977; Proudfit and Levy, 1978; Saelens *et al.*, 1980; Sawynok, 1983), writhing (Hill *et al.*, 1981; Levy and Proudfit, 1977; Saelens *et al.*, 1980; Sivam and Ho, 1983), tooth pulp (Foong and Satoh, 1983) and shock titration tests (Hill *et al.*, 1981). Systemic administration of baclofen also produces sedation and impairment of motor activity, but antinociception occurs at doses lower than those which produce motor effects (Cutting and Jordan, 1975; Sawynok and LaBella, 1982; Tamayo *et al.*, 1988; Vaught *et al.*, 1985). Antinociception also results from intracerebroventricular (i.c.v.) injection of baclofen, as well as microinjection into discrete brain sites (Table 1). Motor impairment (ataxia) as well as other motor effects are observed following i.c.v. administration and microinjection into the periaqueductal grey and nucleus gigantocellularis. Following systemic administration, the antinociceptive action of baclofen is markedly reduced by spinal transection (Proudfit and Levy, 1978; Zorn and Enna, 1985), suggesting a predominant supraspinal action.

$GABA_B$ binding sites are widely distributed throughout the brain, including regions normally associated with pain transmission and suppression (e.g. the central grey, thalamus, and brainstem nuclei such as the nucleus raphé magnus) (Bowery *et al.*, 1987). Following systemic administration, the antinociceptive action of baclofen is stereoselective for the L-isomer (Bartolini *et al.*, 1982; Sawynok and LaBella, 1982), and it has generally been assumed that this action results from activation of central $GABA_B$ receptors. The lack of availability of specific $GABA_B$ antagonists made this hypothesis difficult to examine directly for some time. However, recently it was demonstrated that the antinociceptive action of baclofen administered s.c. was blocked by i.c.v. administration of the $GABA_B$ antagonist phaclofen supporting this concept (Guiliani *et al.*, 1988).

Baclofen also produces antinociception following local actions within the spinal cord. Thus, intrathecal (i.t.) administration to rats (Table 1) and primates

$GABA_B$ Receptors in Mammalian Function. Edited by N.G. Bowery, H. Bittiger and H.-R. Olpe

Table 1. Central nervous regions in which microinjection of baclofen produces antinociception in rats

Region	Dose (μg)	Test	Reference
i.c.v.	1	pinch	Liebman and Pastor (1980)
PAG	0.75, 1.5	tail flick	Levy and Proudfit (1979)
NGC	1.5	tail flick	Levy and Proudfit (1979)
SN	0.3	hot plate	Frye *et al.* (1986)
LPOA	2	hot plate	Lim *et al.* (1985)
Spinal cord	0.1, 1	tail flick hot plate	Wilson and Yaksh (1978)
Spinal cord	1	tail flick hot plate	Hammond and Drower (1984)
Spinal cord	0.25–25	tail flick	Smith (1984)
Spinal cord	0.03–0.3	tail flick	Sawynok and Dickson (1985a)

i.c.v., intracerebroventricular; PAG, periaqueductal grey; NGC, nucleus gigantocellularis; SN, substantia nigra; LPOA, lateral preoptic area.

(Yaksh and Reddy, 1981) produces antinociception. Intrathecal doses which are effective in rats are similar to those administered directly to supraspinal sites (Table 1). The potential for a significant interaction between baclofen administered simultaneously to spinal and supraspinal sites, as demonstrated for morphine (Yeung and Rudy, 1980), has not been examined. Impairment of motor activity (hind-limb flaccidity) also occurs following spinal administration but, as with systemic administration, antinociception occurs in doses which are lower than those which produce alterations in motor function (0.1–0.5 μg compared to 1–10 μg). $GABA_B$ receptors in the spinal cord are concentrated in the substantia gelatinosa of the dorsal horn, which is the predominant area of termination of small-diameter fibres conveying noxious sensory information into the spinal cord (Bowery *et al.*, 1987; Price *et al.*, 1987). Following i.t. administration, antinociception is stereoselective for the L-isomer (Sawynok and Dickson, 1985a; Smith 1984; Wilson and Yaksh, 1978). Again, it has been assumed that baclofen activates $GABA_B$ receptors in the spinal cord to produce antinociception. However, the results of some recent studies suggest that baclofen may interact with a receptor variant in the spinal cord to produce this effect (see below).

NEUROTRANSMITTER SYSTEMS INVOLVED IN THE ANTINOCICEPTIVE ACTION OF BACLOFEN

The role of a variety of neurotransmitters in the antinociceptive action of systemically administered baclofen has been examined by using antagonists for specific receptor types. Endogenous opioids are not involved in the action of baclofen because antinociception is not modified by a wide range of doses of naloxone (2–50 mg/kg) (Bartolini *et al.*, 1982; Levy and Proudfit, 1977; Sawynok and LaBella, 1982; Sivam and Ho, 1983). GABA$_A$ receptors also are not involved because antinociception is generally not modified by bicuculline or picrotoxin (Bartolini *et al.*, 1982; Sawynok and LaBella, 1982; Vaught *et al.*, 1985). However, cholinergic mechanisms do appear to be involved as antinociception is reduced by the muscarinic antagonists atropine and scopolamine (Bartolini *et al.*, 1982; Kendall *et al.*, 1982; Tamayo *et al.*, 1988; Vaught *et al.*, 1985) and increased by the choline esterase inhibitor physostigmine (Tamayo *et al.*, 1988).

The role of amines in antinociception has been examined using both antagonists for specific receptors and agents to deplete endogenous amine levels. The dopamine (DA) antagonists haloperidol and chlorpromazine potentiate the action of baclofen (Sawynok, 1983; Vaught *et al.*, 1985), as do the α-adrenergic antagonists ergotamine, phentolamine and tolazoline (Sawynok, 1983; Tamayo *et al.*, 1988; Vaught *et al.*, 1985) and the β-adrenergic antagonists propranolol and naldol (Tamayo *et al.*, 1988). α-Methyl-*p*-tyrosine, a tyrosine hydroxylase inhibitor which depletes brain and spinal cord levels of noradrenaline (NA) and DA, also enhances the action of baclofen (Sawynok, 1983), which is consistent with observations with the respective antagonists. Finally, the 5-hydroxytryptamine (serotonin, 5-HT) antagonists methysergide (Bartolini *et al.*, 1982) and cinanserin (Sawynok, 1987) also increase the antinociceptive action of baclofen, although a lack of significant effect with methysergide and *p*-chlorophenylalanine, a tryptophan hydroxylase inhibitor which depletes central 5-HT levels, also has been noted (Sawynok, 1987). The involvement of DA, NA and 5-HT mechanisms in pharmacological effects of baclofen also is supported by the results of biochemical studies which have reported significant effects of baclofen on the content, release and turnover of these agents in various brain regions (for review, see Sawynok, 1989).

EFFECTS OF NEUROTOXIN-INDUCED LESIONS TO SPECIFIC AMINERGIC PATHWAYS ON THE ANTINOCICEPTIVE ACTION OF BACLOFEN

The use of receptor antagonists and depleting agents can provide information on the overall involvement of neurotransmitter systems in antinociception, but it cannot provide information on the role of particular pathways in antinociception. Central NA and 5-HT pathways consist of distinct ascending and descending projections (Dahlström and Fuxe, 1964, 1965), and both aspects of these projections have been implicated in antinociception produced by morphine (for review, see Sawynok, 1989). Recently, the role of specific NA and 5-HT pathways in the antinociceptive action of baclofen has been examined following microinjection of the neurotoxins 6-hydroxydopamine (6-OHDA) and 5,7-dihydroxytryptamine (5,7-DHT) into specific brain regions. Antinociception generally was examined 3–14 days following induction of lesions and, at the end of the experiment, brain and spinal cord regions were analysed for NA or 5-HT. In addition to amine depletion, neurotoxins also can produce receptor supersensitivity, and the potential for both effects to contribute to observed changes must be considered. Nevertheless, this method can implicate specific pathways, and in some cases mechanisms, in the action of baclofen.

The effects of neurotoxin-induced lesions to ascending and descending NA and 5-HT pathways on the antinociceptive action of baclofen are summarized in Table 2. Lesions to ascending NA pathways induced by microinjection of 6-OHDA into the dorsal bundle deplete NA levels in a number of brain regions but not in the spinal cord, and inhibit the antinociceptive action of baclofen in the tail-flick test. Lesions to descending NA pathways induced by microinjection of 6-OHDA into the medullary A1 region, and by i.t. injection of 6-OHDA, produce a marked depletion of spinal cord NA levels without substantially altering brain levels, and inhibit the antinociceptive action of baclofen (Table 2). These results indicate that both ascending and descending NA pathways are critical to the antinociceptive action of baclofen. The observation that i.t. administration of phentolamine reverses the antinociceptive action of systemically administered baclofen (Sawynok and Dickson, 1985b) further suggests that the action of baclofen is mediated by *activation* of descending NA pathways. This conclusion is supported by the biochemical observation that systemically administered baclofen increases the turnover of NA in the spinal cord (Sawynok and Reid, 1986b). Paradoxically, when NA levels are depleted simultaneously in ascending and descending pathways by 6-OHDA microinjected into the locus coeruleus, the antinociceptive action of baclofen is *increased* (Table 2). This indicates that there are significant interactions between ascending and descending aspects of NA pathways as expressed in mechanisms of antinociception.

The microinjection of 5,7-DHT into the ventromedial tegmentum, nucleus raphé medianus and nucleus raphé dorsalis produces a selective depletion of 5-HT levels in brain structures without substantially altering levels in the spinal cord (Sawynok and Reid, 1988). Each of these lesions inhibits the antinociceptive action of baclofen (Table 2), although the patterns of reduction differ in that the first two lesions transiently inhibit the action of baclofen, while raphé dorsalis lesions inhibit baclofen at a later time interval (Sawynok and Reid, 1988). These results suggest that ascending 5-HT pathways are involved in the antinociceptive action of baclofen, but mechanisms of compensation (perhaps receptor supersensitivity) can develop. The effects of lesions to descending 5-HT pathways depend on the method used to induce the lesion (Table 2). Thus, i.t. injection of 5,7-DHT increases the antinociceptive action of baclofen, but 5,7-DHT microinjected into the nucleus raphé magnus or into the ventral raphé region does not. The effect of simultaneously depleting 5-HT in the brain and spinal cord by i.c.v. injections of 5,7-DHT is to increase the action of baclofen

Table 2. Summary of effects of lesions to ascending and descending noradrenergic and serotonergic pathways on antinociception produced by systemically administered baclofen

	Ascending	Descending	Ascending and descending
NA	↓ dorsal bundle*	↓ medullary A1†	↑ locus coeruleus*
		↓ i.t. 6-OHDA‡	
5-HT	↓ ventromedial tegmentum§	↑ i.t. 5,7-DHT‡	↑ i.c.v. 5,7-DHT§
		↔ raphé magnus§	
	↓ raphé dorsalis§	↔ ventral raphé§	
	↓ raphé medianus§		

Lesions to noradrenergic and serotonergic pathways were produced by the stereotaxic microinjection (intrathecal, i.t., or intracerebroventricular, i.c.v.) of the neurotoxins 6-hydroxydopamine (6-OHDA) and 5,7-dihydroxytryptamine (5,7-DHT), respectively, into specific brain regions. The antinociceptive action of i.p. baclofen was determined in the rat tail-flick test 3–14 days following microinjection of the toxin or the corresponding vehicle. At the end of the experiment, noradrenaline (NA) and 5-hydroxytryptamine (5-HT) levels were determined in selected brain regions and the spinal cord by high-pressure liquid chromatography. Data from: *Sawynok and Reid (1986a); †Sawynok (1987); ‡Sawynok and Dickson (1985b); §Sawynok and Reid (1988).

(Table 2), suggesting that the influence of the descending pathways is predominant over that of the ascending pathways. The increase seen with both i.t. and i.c.v. 5,7-DHT appears to correspond to the presence of supersensitivity to 5-HT receptors, as these manipulations induce a supersensitive response to i.t. 5-HT in the tail-flick test, while 5,7-DHT microinjected into the nucleus raphé magnus and ventral raphé region does not (Sawynok and Reid, 1990). This result suggests that 5-HT mechanisms are indeed involved in the antinociceptive action of systemically administered baclofen, but the nature of this involvement is not clear. It appears that baclofen does not directly activate descending 5-HT pathways because there is no reduction in effect following depletion of spinal cord 5-HT levels, and because i.t. injection of methysergide in doses which selectively block 5-HT receptors does not reverse the action of baclofen (Sawynok and Dickson, 1985b).

ROLE OF G-PROTEINS AND SECOND MESSENGERS IN THE SPINAL ANTINOCICEPTIVE ACTION OF BACLOFEN

Baclofen inhibits adenylyl cyclase in subcellular fractions from a variety of brain regions and from the spinal cord (Wojcik and Neff, 1984). Inhibition of adenylyl cyclase also has been demonstrated in cultured cells, where this effect is blocked by pertussis toxin (Xu and Wojcik, 1986). Pertussis toxin ADP-ribosylates and inactivates the GTP-binding protein G_i, which mediates inhibition of adenylyl cyclase as well as interactions with other effector systems, and

G_o, linked directly to ion channels (Rosenthal *et al.*, 1988). The effects of baclofen on adenylyl cyclase are complex, however, because in brain slice preparations baclofen does not alter basal activity, enhances stimulation induced by noradrenaline, histamine, adenosine and vasoactive intestinal polypeptide (Hill, 1985; Karbon and Enna, 1985; Watling and Bristow, 1986), but inhibits stimulation of adenylate cyclase by forskolin (Hill, 1985; Karbon and Enna, 1985). Recently, it was demonstrated that 3-aminopropylphosphonic acid (3-APPA), a structural analogue of GABA, could bind to $GABA_B$ receptors and mimic the effect of baclofen on forskolin-stimulated activity, but not on isoproterenol-stimulated activity (Scherer *et al.*, 1988). It was suggested that the receptors activated by baclofen in producing these responses may represent different recognition sites.

The effects of pertussis toxin, phosphodiesterase inhibitors and forskolin on the spinal antinociceptive action of baclofen have been examined in order to determine whether G-proteins and changes in adenylate cyclase are involved in this pharmacological response to baclofen. In-vivo pretreatment with pertussis toxin inhibited the antinociceptive action of i.t. baclofen in both the tail-flick test

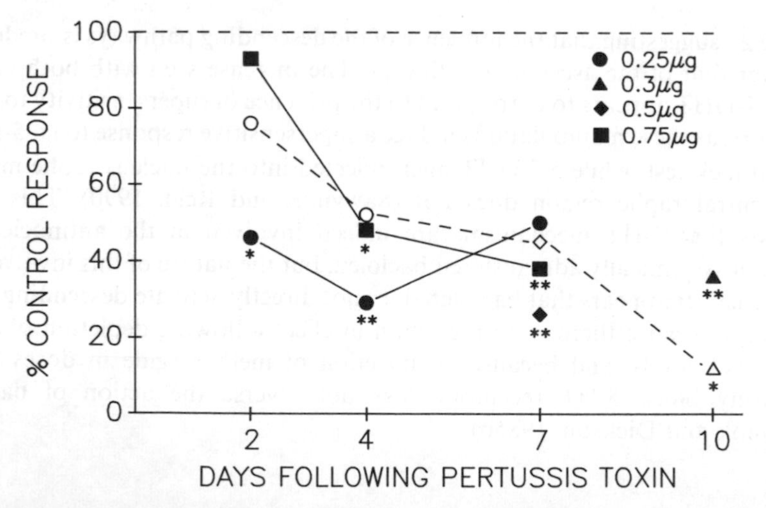

Figure 22.1. Effect of intrathecal (i.t.) pretreatment with 0.25–0.75 μg pertussis toxin on the spinal antinociceptive action of baclofen. On indicated days, 0.3–0.8 μg L-baclofen was injected i.t. into vehicle- or toxin-pretreated rats following baseline determinations, and tail-flick and hot-plate latencies were determined at 15-min intervals for 60 min. Data are expressed as the mean percentage (SEM 4–27%) of the control area under the time–response curve for the tail-flick (solid symbols) and the hot-plate (open symbols) test. Hot-plate data are omitted where control responses were low or where the characteristic end-point was absent. $^{*}P < 0.05$, $^{**}P < 0.01$, by Student's *t*-test on raw data for corresponding vehicle- and toxin-pretreated groups. Data derived from Hoehn *et al.* (1988) (0.25, 0.5, 0.75 μg) or unpublished experiment (0.3 μg)

and the hot-plate test, an effect seen most clearly 4–10 days following administration of the toxin (Hoehn *et al.*, 1988; Figure 22.1). Intrathecal pretreatment with aminophylline, a phosphodiesterase inhibitor and an adenosine receptor antagonist, did not affect the spinal antinociceptive action of baclofen (Figure 22.1). The lack of effect with an adenosine receptor antagonist indicates that the action of baclofen is not due to spinal release of adenosine as has been proposed for morphine (Sweeney *et al.*, 1987). The non-xanthine phosphodiesterase inhibitor Ro 201724 did not affect baclofen at 15 μg but reduced its action at 30 μg (Figure 22.2). These doses of Ro 201724 respectively decrease and increase the action of morphine, presumably due to phosphodiesterase inhibition (Nicholson *et al.*, submitted for publication). Finally, forskolin did not alter the

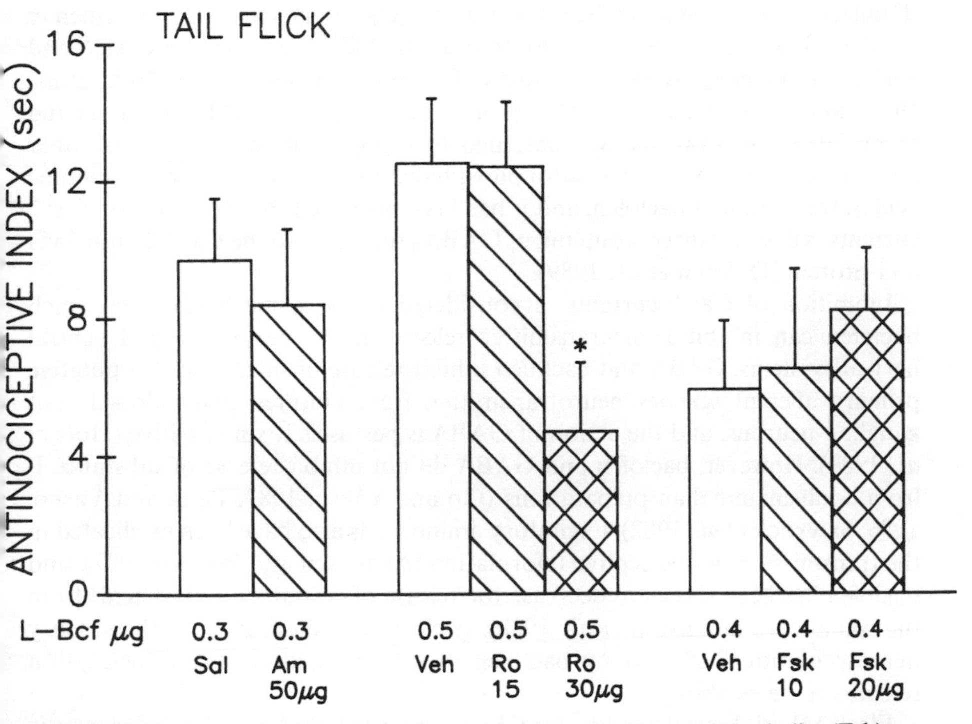

Figure 22.2. Effect of aminophylline (Am), Ro 201724 (Ro) and forskolin (Fsk) on antinociception produced by intrathecal (i.t.) injection of L-baclofen (L-Bcf). The Antinociceptive Index represents the area under the time–response curve for determinations at 15-min intervals for 60 min following injection of baclofen. Intrathecal pretreatments with saline (Sal), vehicle (Veh, 50 % ethanol/saline) or the various drugs indicated in the figure were administered 15 min prior to injection of L-baclofen. The data depict means + SEM for $n = 6$–8 per group. *$P < 0.05$ compared to control by Student's *t*-test. See Hoehn *et al.* (1988) for chronic i.t. cannulation methodology. Data are from unpublished experiments

action of baclofen, although it inhibits the spinal antinociceptive action of morphine and NA (Nicholson *et al.*, submitted for publication, and unpublished data). Collectively, these results indicate that the spinal antinociceptive action of baclofen is mediated by a mechanism involving a pertussis toxin-sensitive G-protein, but do not provide support for inhibition of adenylate cyclase in this action.

Within the spinal cord, $GABA_B$ receptors are selectively concentrated in the substantia gelatinosa (Bowery *et al.*, 1987; Price *et al.*, 1987). The level of these receptors is reduced 40–50% by dorsal rhizotomy and neonatal capsaicin pretreatment, indicating a population of receptors on small-diameter primary afferent neurons (Price *et al.*, 1984, 1987). Electrophysiological studies have provided evidence for a presynaptic site of action on nociceptive afferents (Dickenson *et al.*, 1985; Henry, 1982). Baclofen inhibits Ca^{2+} currents in avian (Dunlap, 1984) and mammalian dorsal root ganglion neurons (Désarmenien *et al.*, 1984; Dolphin and Scott, 1986). Inhibition of Ca^{2+} currents by GABA and baclofen is blocked by pertussis toxin (Dolphin and Scott, 1987; Holz *et al.*, 1986), but is not affected by forskolin or intracellular cyclic AMP (Dolphin and Scott, 1987). No evidence was obtained to support the involvement of other second messengers such as inositol phosphate, protein kinase C or arachidonic acid in this action of baclofen, and it has been proposed that the effect on Ca^{2+} currents is due to a direct coupling of $GABA_B$ receptors to the Ca^{2+} channel via a G-protein (Dolphin *et al.*, 1989).

Inhibition of Ca^{2+} currents is considered to be a mechanism by which baclofen can inhibit neurotransmitter release in the peripheral and central nervous systems. GABA and baclofen inhibit release of substance P, a putative primary afferent sensory neurotransmitter, from cultured avian dorsal root ganglion neurons, and the action of GABA is pertussis toxin-sensitive (Holz *et al.*, 1989). However, baclofen and GABA do not inhibit release of substance P from adult mammalian preparations (Go and Yaksh, 1987; Pang and Vasko, 1986; Sawynok *et al.*, 1982). Excitatory amino acids also have been implicated in the transmission of nociceptive information (Aanonsen and Wilcox, 1987), and baclofen has been shown to decrease the release of excitatory amino acids from the spinal cord (Johnston *et al.*, 1980). A clear characterization of the sensory neurotransmitters affected by baclofen in the production of antinociception remains to be established.

Electrophysiological studies also have provided evidence for postsynaptic actions of baclofen within the spinal cord (Davies, 1981; Henry and Ben-Ari, 1976; Kangrga *et al.*, 1987). In behavioural studies, baclofen inhibits the biting, licking, scratching response induced by i.t. substance P (Hwang and Wilcox, 1989) but not that induced by i.t. excitatory amino acids (Aanonsen and Wilcox, 1989). These behavioural responses have been interpreted as postsynaptic activation of sensory pathways in the spinal cord, and inhibition of the response to substance P could contribute to antinociception. Previously, baclofen had been shown to inhibit depolarizing responses to substance P on spinal motoneurons (Otsuka and Yanagisawa, 1980; Saito *et al.*, 1975). Depolarization by substance P has been attributed to a decrease in K^+ conductance (Nowak and MacDonald, 1982), while postsynaptic effects of baclofen in the hippocampus have been attributed to an increase in K^+ conductance (Inoue *et al.*, 1985; Newberry and Nicoll, 1985). It has been noted that the characteristics of the K^+ conductance affected by baclofen and substance P are similar, and speculated that the same channel is affected by both agents (Ogata *et al.*, 1987). The increase in K^+ conductance by baclofen in the hippocampus and dorsal raphé involves a pertussis toxin-sensitive G-protein (Andrade *et al.*, 1986; Innis *et al.*, 1988). This effect also may occur via a direct coupling of a G-protein to an ion channel, as cyclic AMP and other second messengers do not appear to mediate this response to baclofen (Andrade *et al.*, 1986; Innis *et al.*, 1988). Within the spinal cord, G-proteins directly coupled to ion channels potentially are involved in the postsynaptic action of baclofen, as well as in presynaptic actions on dorsal root ganglion neurons, accounting for the pertussis toxin sensitivity of the spinal antinociceptive action of baclofen.

RECEPTORS INVOLVED IN SPINAL ANTINOCICEPTION BY BACLOFEN: STUDIES WITH ANTAGONISTS

The spinal antinociceptive effect of baclofen is stereoselective for the L-isomer (Sawynok and Dickson, 1985a; Smith, 1984; Wilson and Yaksh, 1978) and it has generally been assumed that activation of GABA$_B$ receptors is involved in this action. This hypothesis has been difficult to test directly until recently because of the lack of availability of specific receptor antagonists. However, in the guinea-pig ileum, an established GABA$_B$ preparation, antagonism of the action of baclofen has been reported with 3-aminopropylsulphonic acid (3-APS; Giotti *et al.*, 1983), δ-aminovaleric acid (DAVA; Kerr and Ong, 1984) and phaclofen (Kerr *et al.*, 1987). In the trigeminal nucleus, a functional homologue of the dorsal horn of the spinal cord, the D-isomer of baclofen was reported to antagonize electrophysiological effects of L-baclofen (Terrence *et al.*, 1983). These observations led us to examine antagonism of spinal analgesia by L-baclofen with these antagonists.

In initial studies, pretreatment with D-baclofen blocked the antinociceptive action of L-baclofen (Sawynok and Dickson, 1985a). However, effective doses of D-baclofen exhibited intrinsic antinociceptive activity which was probably due to contamination with the L-isomer as it too was blocked by pretreatment with D-baclofen (Sawynok and Dickson, 1985a). In a subsequent study using repurified D-baclofen, antagonism was observed in the absence of an intrinsic effect (Sawynok, 1986). D-Baclofen is a more potent antagonist of spinal antinociception than phaclofen or DAVA (Sawynok, 1986; Figure 22.3), while 3-APS lacks antagonist activity (Sawynok, 1986). Interestingly, D-baclofen and

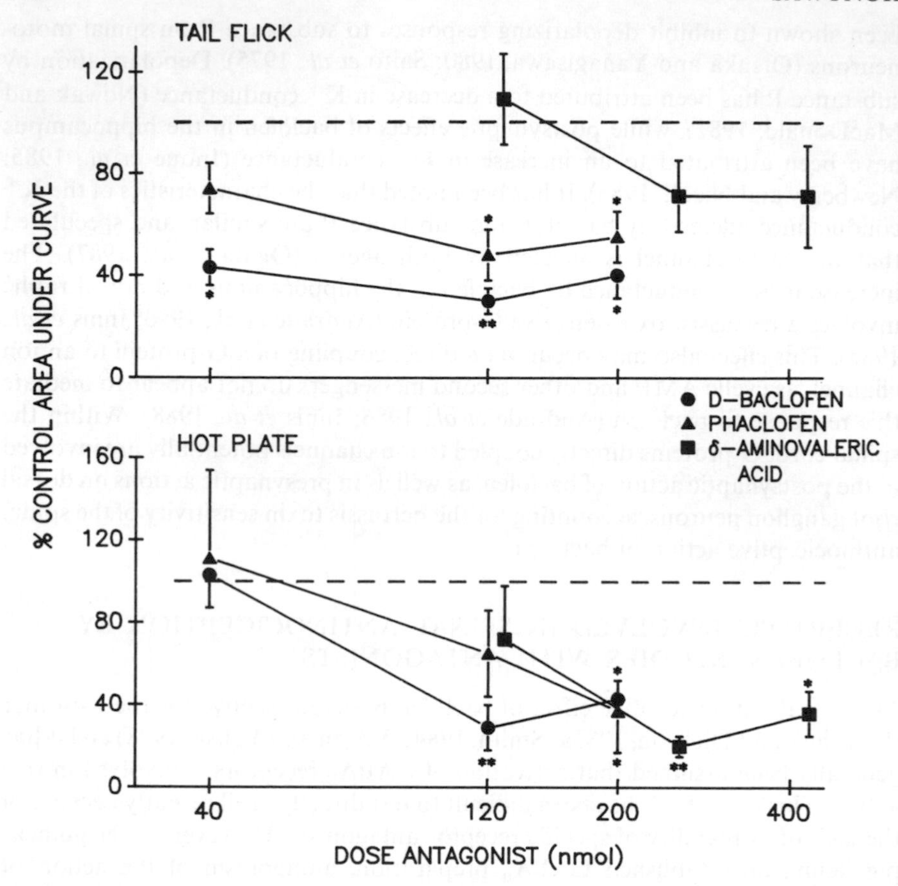

Figure 22.3. Effect of baclofen antagonists on the spinal antinociceptive action of L-baclofen in the tail-flick and hot-plate tests. Data are expressed as the mean percentage ± SEM of the control area under the time-response curve following intrathecal (i.t.) injection of 0.5 μg L-baclofen for determinations at 15-min intervals for 60 min. Control area under the curve values were 11.0–13.1 s in the tail-flick test and 37–67 s in the hot-plate test (n = 5–12 per group). All antagonists were injected i.t. 15 min prior to the L-baclofen, and were without intrinsic effect prior to the L-baclofen injection. *$P < 0.05$, **$P < 0.01$ compared to control group by Student's t-test analysis of raw data. Data are from unpublished experiments

phaclofen were more potent in the tail-flick test than in the hot-plate test, while the converse was the case with DAVA (Figure 22.3). Reversal of the action of L-baclofen by effective doses of these three antagonists could not be demonstrated (data not shown), which may reflect a weak affinity for receptors. D-Baclofen does not block the inhibitory effect of L-baclofen in the guinea-pig ileum, although DAVA and 3-APS are effective (Sawynok, 1986). Similarly, D-baclofen does not block the L-isomer in hippocampal slices (Haas *et al.*, 1985) or the cortex (Howe and Zieglgänsberger, 1986) although DAVA and phaclofen are antagonists at these sites (Karlsson *et al.*, 1988; Nakahiro *et al.*, 1985).

The profound difference in sensitivity to the antagonist action of D-baclofen has led to the proposal that spinal antinociception by baclofen may not be due to an interaction with classical $GABA_B$ receptors (Sawynok, 1986, 1987). It is unclear whether this receptor is a GABA receptor as i.t. GABA does not produce antinociception even in the presence of a GABA uptake inhibitor or a GABA transaminase inhibitor (Sawynok and Dickson, 1985a). In the hippocampus, the changes in K^+ conductance induced by GABA and baclofen exhibit quite different properties (Ogata *et al.*, 1987), suggesting hyperpolarization induced by baclofen may not be due to activation of a $GABA_B$ receptor. Other studies have provided evidence for interactions of baclofen with multiple recognition sites. Thus, pertussis toxin blocks the postsynaptic but not the presynaptic effects of baclofen in the hippocampus and dorsal raphé (Colmers and Williams, 1988; Dutar and Nicoll, 1988). In addition, phaclofen blocks the postsynaptic but not the presynaptic effect of baclofen (Dutar and Nicoll, 1988). These results suggest different actions of baclofen may be mediated by recognition sites with distinct pharmacological properties interacting with coupling mechanisms with a differential sensitivity to pertussis toxin and antagonists. It should be noted that the presynaptic action of baclofen in dorsal root ganglion neurons is pertussis toxin-sensitive (Dolphin and Scott, 1987). However, the effects of the various baclofen antagonists on dorsal root ganglion neurons have not been evaluated. The use of D-baclofen and phaclofen as antagonists and pertussis toxin in comparative studies may enable a clearer characterization of the receptors activated by baclofen in the spinal cord to produce antinociception, as well as in other parts of the central nervous system.

ACKNOWLEDGEMENTS

This work was supported by the Medical Research Council of Canada. I thank Charles Dickson and Allison Reid for skilled technical assistance throughout these studies.

REFERENCES

Aanonsen, L.M. and Wilcox, G.L. (1987) Nociceptive action of excitatory amino acids in the mouse: effects of spinally administered opioids, phencyclidine and sigma agonists. *J. Pharmacol. Exp. Ther.*, **243**, 9–19.

Aanonsen, L.M. and Wilcox, G.L. (1989) Muscimol, γ-aminobutyric $acid_A$ receptors and excitatory amino acids in the mouse spinal cord. *J. Pharmacol. Exp. Ther.*, **248**, 1034–1038.

Andrade, R., Malenka, R.C. and Nicoll, R.A. (1986) A G protein couples serotonin and $GABA_B$ receptors to the same channels in hippocampus. *Science*, **234**, 1261–1265.

Bartolini, A., Malmberg, P., Bartolini, R. and Giotti, A. (1982) Effect of antimigraine drugs on nonopioid analgesia. In: M. Critchley *et al.* (eds) *Advances in neurology*, vol. 33. New York: Raven Press, pp. 89–97.

Bowery, N.G. (1982) Baclofen: 10 years on. *Trends Pharmacol. Sci.*, **3**, 400–403.

Bowery, N.G., Hudson, A.L. and Price, G.W. (1987) $GABA_A$ and $GABA_B$ receptor site distribution in the rat central nervous system. *Neuroscience*, **20**, 365–383.

Colmers, W.F. and Williams, J.T. (1988) Pertussis toxin pretreatment discriminates between pre- and postsynaptic actions of baclofen in rat dorsal raphé nucleus in vitro. *Neurosci. Lett.*, **93**, 300–306.

Cutting, D.A. and Jordan, C.C. (1975) Alternative approaches to analgesia: baclofen as a model compound. *Br. J. Pharmacol.*, **54**, 171–179.

Dahlström, A. and Fuxe, K. (1964) Evidence for the existence of monoamine nerves in the central nervous system: I. Demonstration of monoamines in the cell bodies of brain stem neurons. *Acta Physiol. Scand.*, **64**, Suppl. 232, 1–55.

Dahlström, A. and Fuxe, K. (1965) Evidence for the existence of monoamine nerves in the central nervous system: II. Experimentally induced changes in intraneuronal amine levels of bulbo-spinal neuron systems. *Acta Physiol. Scand.*, **64**, Suppl. 247, 1–36.

Davies, J. (1981) Selective depression of synaptic excitation in cat spinal neurons by baclofen: an iontophoretic study. *Br. J. Pharmacol.*, **72**, 373–384.

Désarmenien, M., Felz, P., Occhipinti, G., Santangelo, F. and Schlichter, R. (1984) Coexistence of $GABA_A$ and $GABA_B$ receptors on Aδ and C primary afferents. *Br. J. Pharmacol.*, **81**, 327–333.

Dickenson, A.H., Brewer, C.M. and Hayes, N.A. (1985) Effects of topical baclofen on C fibre-evoked neuronal activity in the rat dorsal horn. *Neuroscience*, **14**, 557–562.

Dolphin, A.C. and Scott, R.H. (1986) Inhibition of calcium currents in cultured rat dorsal root ganglion neurones by (−)-baclofen. *Br. J. Pharmacol.*, **88**, 213–220.

Dolphin, A.C. and Scott, R.H. (1987) Calcium channel currents and their inhibition by (−)-baclofen in rat sensory neurones: modulation by guanine nucleotides. *J. Physiol.*, **386**, 1–17.

Dolphin, A.C., McGuirk, S.M. and Scott, R.H. (1989) An investigation into the mechanisms of inhibition of calcium channel currents in cultured sensory neurones of the rat by guanine nucleotide analogues and (−)-baclofen. *Br. J. Pharmacol.*, **97**, 263–273.

Dunlap, K. (1984) Functional and pharmacological differences between two types of GABA receptor on embryonic chick sensory neurons. *Neurosci. Lett.*, **47**, 265–270.

Dutar, P. and Nicoll, R.A. (1988) Pre- and postsynaptic $GABA_B$ receptors in the hippocampus have different pharmacological properties. *Neuron*, **1**, 585–591.

Foong, F.-W. and Satoh, M. (1983) Analgesic potencies of non-narcotic, narcotic and anesthetic drugs as determined by the bradykinin-induced biting-like responses in rats. *Jap. J. Pharmacol.*, **33**, 933–938.

Frye, G.D., Baumeister, A.D., Crotty, K., Newman, K.D. and Kotrla, K.J. (1986) Evaluation of the role of antinociception in self-injurious behavior following intranigral injection of muscimol. *Neuropharmacology*, **25**, 717–726.

Giotti, A., Luzzi, S., Spagnesi, A. and Ziletti, L. (1983) Homotaurine: a $GABA_B$ antagonist in guinea-pig ileum. *Br. J. Pharmacol.*, **79**, 855–862.

Go, V.L.W. and Yaksh, T.L. (1987) Release of substance P from the cat spinal cord. *J. Physiol.*, **391**, 141–167.

Guiliani, A., Evangelista, S., Borsini, F. and Meli, A. (1988) Intracerebroventricular phaclofen antagonizes baclofen antinociceptive activity in hot plate test with mice. *Eur. J. Pharmacol.*, **154**, 225–226.

Haas, H.L., Greene, R.W. and Olpe, H.-R. (1985) Stereoselectivity of L-baclofen in hippocampal slices of the rat. *Neurosci. Lett.*, **55**, 1–4.

Hammond, D.L. and Drower, E.J. (1984) Effects on intrathecally administered THIP, baclofen and muscimol on nociceptive threshold. *Eur. J. Pharmacol.*, **103**, 121-125.
Henry, J.L. (1982) Effects on intravenously administered enantiomers of baclofen on functionally identified units in lumbar dorsal horn of the spinal cat. *Neuropharmacology*, **21**, 1073-1083.
Henry, J.L. and Ben-Ari, Y. (1976) Actions of the *p*-chlorophenyl derivative of GABA, Lioresal, on nociceptive and non-nociceptive units in the spinal cord of the cat. *Brain Res.*, **117**, 540-544.
Hill, D.R. (1985) $GABA_B$ receptor modulation of adenylate cyclase activity in rat brain slices. *Br. J. Pharmacol.*, **84**, 249-257.
Hill, R.C., Maurer, R., Buescher, H.H. and Roemer, D. (1981) Analgesic properties of the GABA-mimetic THIP. *Eur. J. Pharmacol.*, **69**, 221-224.
Hoehn, K., Reid, A. and Sawynok, J. (1988) Pertussis toxin inhibits antinociception produced by intrathecal injection of morphine, noradrenaline and baclofen. *Eur. J. Pharmacol.*, **146**, 65-72.
Holz, G.G., Rane, S.G. and Dunlap, K. (1986) GTP-binding proteins mediate transmitter inhibition of voltage-dependent calcium channels. *Nature*, **319**, 670-672.
Holz, G.G., Kream, R.M., Spiegel, A. and Dunlap, K. (1989) G proteins couple α-adrenergic and $GABA_B$ receptors to inhibition of peptide secretion from peripheral sensory neurons. *J. Neurosci.*, **9**, 657-666.
Howe, J.R. and Zieglgänsberger, W. (1986) D-Baclofen does not antagonize the actions of L-baclofen on rat neocortical neurons in vitro. *Neurosci. Lett.*, **72**, 99-104.
Hwang, A.S. and Wilcox, G.L. (1989) Baclofen γ-aminobutyric $acid_B$ receptors and substance P in the mouse spinal cord. *J. Pharmacol. Exp. Ther.*, **248**, 1026-1033.
Innis, R.B., Nestler, E.J. and Aghajanian, G.K. (1988) Evidence for G protein mediation of serotonin- and $GABA_B$-induced hyperpolarization of rat dorsal raphé neurons. *Brain Res.*, **459**, 27-36.
Inoue, M., Matsuo, T. and Ogata, N. (1985) Characterization of pre- and postsynaptic actions of (−)-baclofen in the guinea-pig hippocampus in vitro. *Br. J. Pharmacol.*, **84**, 843-851.
Johnston, G.A.R., Hailstone, M.H. and Freeman, C.G. (1980) Baclofen: stereoselective inhibition of excitant amino acid release. *J. Pharm. Pharmacol.*, **32**, 230-231.
Kangrga, I., Randic, M. and Jeftinija, S. (1987) Adenosine and (−)-baclofen have a neuromodulatory role in the rat spinal dorsal horn. *Soc. Neurosci. Abstr.*, **13**, 1134.
Karbon, E.W. and Enna, S.J. (1985) Characterization of the relationship between γ-aminobutyric acid B agonists and transmitter-coupled cyclic nucleotide-generating systems in rat brain. *Mol. Pharmacol.*, **27**, 53-59.
Karlsson, G., Pozza, M. and Olpe, H.-R. (1988) Phaclofen: a $GABA_B$ blocker reduces long-duration inhibition in the neocortex. *Eur. J. Pharmacol.*, **148**, 485-486.
Kendall, D.A., Browner, M. and Enna, S.J. (1982) Comparison of the antinociceptive effect of γ-aminobutyric acid (GABA) agonists: evidence for a cholinergic involvement. *J. Pharmacol. Exp. Ther.*, **220**, 482-487.
Kerr, D.I.B. and Ong, J. (1984) Evidence that ethylenediamine acts in the isolated ileum of the guinea-pig by releasing endogenous GABA. *Br. J. Pharmacol.*, **83**, 169-177.
Kerr, D.I.B., Ong, J., Prager, R.H., Gynther, B.G. and Curtis, D.R. (1987) Phaclofen: a peripheral and central baclofen antagonist. *Brain Res.*, **405**, 150-154.
Levy, R.A. and Proudfit, H.K. (1977) The analgesic action of baclofen β-(4-*p*-chlorophenyl)-γ-aminobutyric acid. *J. Pharmacol. Exp. Ther.*, **202**, 437-445.
Levy, R.A. and Proudfit, H.K. (1979) Analgesia produced by microinjection of baclofen and morphine at brain stem sites. *Eur. J. Pharmacol.*, **57**, 43-55.
Liebman, J.M. and Pastor, G. (1980) Antinociceptive effects of baclofen and muscimol upon intraventricular administration. *Eur. J. Pharmacol.*, **61**, 225-230.

Lim, C.R., Garant, D.G. and Gale, K. (1985) GABA agonist induced analgesia elicited from the lateral preoptic area in the rat. *Eur. J. Pharmacol.*, **107**, 91–94.

Nakahiro, M., Saito, K., Yamada, I. and Yoshida, H. (1985) Antagonistic effect of δ-aminovaleric acid on bicuculline-insensitive γ-aminobutyric $acid_B$ ($GABA_B$) sites in the rat's brain. *Neurosci. Lett.*, **57**, 263–266.

Newberry, N.R. and Nicoll, R.A. (1985) Comparison of the action of baclofen with γ-aminobutyric acid on rat hippocampal pyramidal cells in vitro. *J. Physiol.*, **360**, 161–185.

Nicholson, D., Reid, A. and Sawynok, J. (submitted for publication) Effects of forskolin and phosphodiesterase inhibitors on the spinal antinociceptive action of morphine.

Nowak, L.M. and MacDonald, R.M. (1982) Substance P: ionic basis for depolarizing responses of mouse spinal cord neurons in cell culture. *J. Neurosci.*, **2**, 1119–1128.

Ogata, N., Inoue, M. and Matsuo, T. (1987) Contrasting properties of K^+ conductances induced by baclofen and γ-aminobutyric acid in slices of the guinea-pig hippocampus. *Synapse*, **1**, 62–69.

Otsuka, M. and Yanagisawa, M. (1980) The effects of substance P and baclofen on motoneurons of isolated spinal cord of the newborn rat. *J. Exp. Biol.*, **89**, 201–214.

Pang, I.-H. and Vasko, M.R. (1986) Morphine and noradrenaline but not 5-hydroxytryptamine and γ-aminobutyric acid inhibit the potassium-stimulated release of substance P from rat spinal cord slices. *Brain. Res.*, **376**, 268–279.

Price, G.W., Wilkin, G.P., Turnbull, M.J. and Bowery, N.G. (1984) Are baclofen-sensitive $GABA_B$ receptors present on primary afferent terminals of the spinal cord? *Nature*, **307**, 376–380.

Price, G.W., Kelly, J.S. and Bowery, N.G. (1987) The location of $GABA_B$ receptor binding sites in mammalian spinal cord. *Synapse*, **1**, 530–538.

Proudfit, H.K. and Levy, R.A. (1978) Delimitation of the neuronal substrates necessary for the analgesic action of baclofen and morphine. *Eur. J. Pharmacol.*, **47**, 159–166.

Rosenthal, W., Hescheler, J., Trautwein, W. and Schultz, G. (1988) Control of voltage-dependent Ca^{2+} channels by G protein-coupled receptors. *FASEB J.*, **2**, 2784–2790.

Saelens, J.K., Bernard, P.S. and Wilson, D.E. (1980) Baclofen as an analgesic. *Brain Res. Bull.*, **5**, Suppl. 2, 133–142.

Saito, K., Konishi, S. and Otsuka, M. (1975) Antagonism between Lioresal and substance P in rat spinal cord. *Brain Res.*, **97**, 177–180.

Sawynok, J. (1983) Monoamines as mediators of the antinociceptive effect of baclofen. *Naunyn-Schmiedeberg's Arch. Pharmacol.*, **323**, 54–57.

Sawynok, J. (1986) Baclofen activates two distinct receptors in the rat spinal cord and guinea pig ileum. *Neuropharmacology*, **25**, 581–586.

Sawynok, J. (1987) GABAergic mechanisms of analgesia: an update. *Pharmacol. Biochem. Behav.*, **26**, 463–474.

Sawynok, J. (1989) The role of ascending and descending noradrenergic and serotonergic pathways in opioid and non-opioid antinociception as revealed by lesion studies. *Can. J. Physiol. Pharmacol.*, **67**, 975–988.

Sawynok, J. and Dickson, C. (1985a) D-Baclofen is an antagonist at baclofen receptors mediating antinociception in the spinal cord. *Pharmacology*, **31**, 248–259.

Sawynok, J. and Dickson, D. (1985b) Evidence for the involvement of descending noradrenergic pathways in the antinociceptive effect of baclofen. *Brain Res.*, **335**, 89–97.

Sawynok, J. and LaBella, F.S. (1982) On the involvement of GABA in the analgesia produced by baclofen, muscimol and morphine. *Neuropharmacology*, **21**, 397–404.

Sawynok, J. and Reid, A. (1986a) Role of ascending and descending noradrenergic pathways in the antinociceptive effect of baclofen and clonidine. *Brain Res.*, **386**, 341–350.

Sawynok, J. and Reid, A. (1986b) Clonidine reverses baclofen-induced increase in noradrenaline turnover in rat brain. *Neurochem. Res.*, **11**, 723–731.
Sawynok, J. and Reid, A. (1988) Role of ascending and descending serotonergic pathways in the antinociceptive effect of baclofen. *Naunyn-Schmiedeberg's Arch. Pharmacol.*, **337**, 359–365.
Sawynok, J. and Reid, A. (1990) Supersensitivity to intrathecal 5-hydroxytryptamine, but not noradrenaline, following various methods of depleting spinal cord 5-hydroxytryptamine levels. *Naunyn-Schmiedeberg's Arch. Pharmacol.*, **342**, 1–8.
Sawynok, J., Kato, N., Havlicek, V. and LaBella, F.S. (1982) Lack of effect of baclofen on substance P and somatostatin release from rat spinal cord in vitro. *Naunyn-Schmiedeberg's Arch. Pharmacol.*, **319**, 78–81.
Scherer, R.W., Ferkany, J.W. and Enna, S.J. (1988) Evidence for pharmacologically distinct subsets of GABA$_B$ receptors. *Brain Res. Bull.*, **21**, 439–443.
Sivam, S.P. and Ho, I.K. (1983) GABAergic drugs, morphine and morphine tolerance: a study in relation to nociception and gastrointestinal transit in mice. *Neuropharmacology*, **22**, 767–774.
Smith, D.F. (1984) Stereoselectivity of spinal neurotransmission: effects of baclofen enantiomers on tail flick reflex in rats. *J. Neural Trans.*, **60**, 63–67.
Sweeney, M.I., White, T.D. and Sawynok, J. (1987) Involvement of adenosine in the spinal antinociceptive effects of morphine and noradrenaline. *J. Pharmacol. Exp. Ther.*, **243**, 657–665.
Tamayo, L., Rifo, J. and Contreras, E. (1988) Influence of adrenergic and cholinergic mechanisms in baclofen induced analgesia. *Gen. Pharmacol.*, **19**, 87–89.
Terrence, C.F., Sax, M., Fromm, G.H. and Yoo, C.A. (1983) Effect of baclofen enantiomorphs on the spinal trigeminal nucleus and steric similarities of carbamazepine. *Pharmacology*, **27**, 85–94.
Vaught, J.L., Pelley, K., Costa, L.G., Setler, P. and Enna, S.J. (1985) A comparison of the antinociceptive responses to the GABA-receptor agonists THIP and baclofen. *Neuropharmacology*, **24**, 211–216.
Watling, K.J. and Bristow, D.R. (1986) GABA$_B$ receptor-mediated enhancement of vasoactive intestinal peptide-stimulated cyclic AMP production in slices of rat cerebral cortex. *J. Neurochem.*, **46**, 1756–1762.
Wilson, P.R. and Yaksh, T.L. (1978) Baclofen is antinociceptive in the spinal intrathecal space of animals. *Eur. J. Pharmacol.*, **51**, 323–330.
Wojcik, W.L. and Neff, N.H. (1984) γ-Aminobutyric acid B-receptors are negatively coupled to adenylate cyclase in brain, and in the cerebellum these receptors may be associated with granule cells. *Mol. Pharmacol.*, **25**, 24–28.
Xu, J. and Wojcik, W.L. (1986) Gamma-aminobutyric acid B-receptor-mediated inhibition of adenylate cyclase in cultured cerebellar granule cells: blockade by islet-activating protein. *J. Pharmacol. Exp. Ther.*, **239**, 568–573.
Yaksh, T.L. and Reddy, S.V.R. (1981) Studies in the primate on the analgesic effects associated with intrathecal actions of opiates, α-adrenergic agonists and baclofen. *Anesthesiology*, **54**, 451–467.
Yeung, J.C. and Rudy, T.A. (1980) Multiplicative interaction between narcotic agonisms expressed at spinal and supraspinal sites of antinociceptive action as revealed by concurrent intrathecal and intracerebroventricular injections of morphine. *J. Pharmacol. Exp. Ther.*, **215**, 633–642.
Zorn, S.H. and Enna, S.J. (1985) The effect of mouse spinal cord transection on the antinociceptive response to the γ-aminobutyric acid agonists THIP (4,5,6,7-tetrahydroisoxazolo(5,4-*C*)pyradine-3-ol) and baclofen. *Brain Res.*, **338**, 380–383.

CLOSING REMARKS

Closing Remarks

B. GÄHWILER

Summing up or concluding meetings is always an extremely risky job. It is a balancing act because each participant is bound to view the impact or importance of his or her contribution in a very individual way. I have therefore opted for a very unbalanced but personal view of the meeting and I am going to discuss the various points which were most important to me. I have to apologize to those whose results I will not be mentioning.

A title such as the '1st International $GABA_B$ Symposium' raises certain expectations and mine were as follows. Firstly, I had hoped to get to know all about $GABA_B$ receptors; that is to say their location, the existence of subtypes, their development and isolation, and cloning of the receptors. Secondly, I was curious to hear more about the pharmacology of new specific agonists and antagonists of $GABA_B$ receptors. Thirdly, I had hoped to get as much information as possible about the physiological role played by $GABA_B$ receptors, the conductances involved, the ion channels, single-channel data etc. Fourthly, I wondered whether there were close links between $GABA_B$ receptors and diseases, where the agonists or antagonists of the $GABA_B$ receptor would be useful therapeutic agents.

Let's start with the receptors. A number of papers have been presented at this meeting showing that $GABA_B$ receptors are localized peripherally as well as in the central nervous system (CNS), but there is no need to go into detail since Norman Bowery has summarized these data very beautifully in his introductory lecture (see Bowery *et al.*, Chapter 1). Some evidence was presented for the existence of subtypes of $GABA_B$ receptors. We are used to the fact that researchers carrying out binding studies usually rather quickly postulate the existence of receptor subtypes to explain complex binding behaviour, but this time electrophysiologists have joined them rather rapidly and have hypothesized the existence of at least two types of $GABA_B$ receptors which are preferentially localized at pre- and postsynaptic sites. When Roger Nicoll gave his beautiful talk on the first day, everything appeared to be crystal clear, and I was actually quite convinced that there were two types of receptors. He showed that pre- and postsynaptic sites can be distinguished on the basis of their responsiveness to antagonists such as phaclofen (see Nicoll and Dutar, Chapter 11), and he got supporting evidence from Dr Gallagher's work showing that pertussis toxin preferentially blocks the postsynaptic actions of baclofen. Before the meeting started, the sensitivity to pertussis toxin presented one of the hard

$GABA_B$ Receptors in Mammalian Function. Edited by N.G. Bowery, H. Bittiger and H.-R. Olpe

facts I had in my head for the existence of receptor subtypes, but during the course of the meeting there were several reports showing that release of excitatory as well as inhibitory transmitters are both phaclofen-sensitive and pertussis toxin-sensitive. Taking all the data together that I have seen at this meeting, I tentatively conclude that subtypes may exist, but I do not think that we have had any hard data at the moment which really prove the existence of two different $GABA_B$ receptors.

There may, nevertheless, be more than a single receptor, as shown in the interesting observation of the mismatch discussed by Dr Misgeld between $GABA_B$ and baclofen in the hippocampus. In fact it appears that baclofen has actions which are not shared by GABA and this would point to a third type of receptor, not a $GABA_B$ but a baclofen receptor. I think we have to take this seriously, although if there is only one report on such a receptor one can take it philosophically and say, 'Okay, we'll wait, how long will it take?', but knowing Dr Misgeld I think he is going to provide additional arguments that will occupy a number of laboratories in the future.

Dr Schwartzkroin presented very nice data on the development of receptors in the immature hippocampus (see Schwartzkroin, Chapter 13). It is intriguing that there appears to be a closer interaction between $GABA_A$ and $GABA_B$ receptors during early development. There was also an interesting report by Professor Kuriyama dealing with the partial purification of $GABA_B$ sites in the cerebral cortex (see Kuriyama *et al.*, Chapter 10), but there appear to be only a few groups working on the molecular identification of $GABA_B$ receptors. I don't know why molecular biologists have avoided this area of $GABA_B$ receptors, but I guess that in the future the molecular approach will be used much more often than now. Dr Roberts even speculated that at the next GABA meeting 30–40% of the work will at least include the molecular approach, and I can only agree that this will most likely be the case.

What about the pharmacology of the receptor? I think that reports about the development of new GABA agonists and antagonists represent one of the highlights of this symposium. Breakthroughs in this area were urgently needed and this holds true for both agonists and antagonists which are specific and which cross the blood–brain barrier. Most substitutions have yielded agonists with lower potency than baclofen, that is with the exception of 3-aminopropylphosphonic acid (3-APA) and, of course, all the substances which are in the drawers of chemists who refused to show their data at this meeting. It is particularly interesting that at least some progress has been made with agonists but more significant progress has been achieved with the antagonists. The new $GABA_B$ receptor blocker CGP 35348 appears to fulfil most of the criteria for a selective and powerful $GABA_B$ receptor antagonist, and it is expected that such a tool will be extremely valuable for functional studies both in situ and in isolated preparations.

What about physiology? $GABA_B$ receptors, as pointed out before, are found in the CNS as well as in periphery. A series of papers described the action of baclofen in the spinal cord and various brain areas including the septum, substantia nigra, thalamus and, of course, the hippocampus, one of the favourite brain areas studied by physiologists. In the hippocampus, baclofen hyperpolarizes pyramidal cells as well as presumed interneurons (in the fascia dentata), and $GABA_B$ receptors are clearly involved in mediating slow inhibition. Both effects are blocked by $GABA_B$ antagonists.

Dr Malouf presented very interesting data on the role of the late inhibition in preventing afterdischarge bursts and was able to show that the $GABA_B$ antagonist phaclofen blocked the slow inhibitory postsynaptic potentials (IPSPs; see Malouf *et al.*, Abstract section, this volume). I hope the newly developed antagonists will be instrumental in helping to clarify this point. Several people in this audience, and again I point to Drs Misgeld and Gallagher, have raised the question, 'To what degree are the late IPSPs mediated by the activation of $GABA_B$ receptors?' Since several transmitters exert similar actions, the response will most likely be tissue-specific. I expect that in some tissues the late IPSP is mediated entirely by the activation of $GABA_B$ receptors, but it may be very different in other tissues, or may be mediated by a mixture of transmitters as pointed out by Dr Gallagher.

What about ionic mechanisms, conductances and ion channels? In the past there has been some controversy about whether the effects of GABA at $GABA_B$ sites are due to an increase in potassium conductance or a decrease in calcium conductance. Data presented at this meeting again emphasized the importance of both of these ions. Dr Dolphin described the current situation and showed that as far as calcium is concerned, the situation may be even more complicated, since low concentrations of baclofen have been shown to increase the calcium T-currents, whereas the higher concentrations inhibit calcium currents (see Dolphin *et al.*, Chapter 15). The possible interaction of calcium channels with $GABA_B$ agonists was also implicated by the binding studies of Dr Thalmann, who showed that dihydropyridine agonists increased and antagonists decreased baclofen binding (see Thalmann and Al-Dahan, Abstract section, this volume).

It is quite unfortunate that no patch-clamp data are available at the single-channel level and this holds true for both the calcium conductance and the potassium conductance. I am of course aware of the problems that are involved. Firstly, there are second messengers which complicate the story considerably and, secondly, nature has put the potassium conductances on the dendrites of very complicated cells, but I think the problems can be solved and I hope that progress will be reported in the near future.

We heard several contributions linking $GABA_B$ with both the adenlyl cyclase system and the phosphotidylinositol system. I have tried very hard to fully understand the kind of second messengers which are involved but, as the list of proposed messengers becomes ever longer, I am reminded of one of my teacher's famous sayings: 'If you know something exactly, then you can say it in one sentence. If you don't, then you have to write books about it', and as far as intracellular messengers for $GABA_B$ actions are concerned I am quite sure that we still have to write books, at least for the present.

What kind of therapeutic applications can be expected from drugs which interact centrally with $GABA_B$ receptors? We are of course all aware of the traditional indication for baclofen, i.e. spasticity. The most difficult question is related to the possible involvement of $GABA_B$ receptors with other disease states. The problem is not inheritant to $GABA_B$ agonists; similar difficulties were encountered with $GABA_A$ substances. We have witnessed, however, a lively debate about the possible involvement of $GABA_B$ in depression. In my opinion, progress in this area will most likely be rather fast in the future, given the availability of selective pharmacological tools, and there is certainly hope that some of the compounds presented and introduced at this meeting will be used as experimental tools for neurobiologists to elucidate functional aspects of $GABA_B$ receptor actions. To find new indications it is absolutely necessary that these compounds be tested in clinical trials, and I hope that the pharmaceutical companies which are involved have, despite an unknown outcome, enough patience as well as resources to do so.

Finally, as Chairman of the last session I take extreme pleasure in expressing the sincere thanks of all participants to the organizers and in particular to Norman Bowery. I think he has done an absolutely superb job. All the organizers have to be congratulated for setting up a very interesting, diverse and balanced programme, and we have had the privilege of participating in a perfectly organized and successful meeting, in a spectacularly beautiful setting. The organizers have succeeded in creating a superb atmosphere which has encouraged very exciting and challenging discussions. We have noted that this conference was called the '1st International $GABA_B$ Symposium'. Since 'B' may also stand for Bowery, we all hope that Norman Bowery is going to organize an equally successful 2nd Symposium.

ABSTRACTS

Abstracts: Contents

FREE COMMUNICATIONS

$GABA_B$ Mechanisms in Human Jejunum In Vitro 399
G. Gentilini, S. Franchi-Micheli, D. Pantalone, C. Cortesini and L. Zilletti

3-Aminopropylphosphinic Acid, Its $GABA_B$ Agonist Activity and the Effects of Phaclofen on In-vitro Pharmacological Preparations 399
J. M. Hills, W. Howson and M. M. Larkin

Prejunctional $GABA_B$ Inhibition of Neurally Induced Airway Constriction in Guinea-pigs 400
R. W. Chapman, G. Danko, C. Rizzo, P. J. Mauser, R. W. Egan and W. Kreutner

[^{3}H](−)-Baclofen Labels Specific Binding Sites in Rat Kidney Cortex 401
S. L. Erdö, A. Michler and J. R. Wolff

Interaction of D-Baclofen and L-Baclofen 401
G. H. Fromm, M. Nakata and T. Shibuya

$GABA_B$ Receptor Ligands Modulate Neuronal Outgrowth Through a Mechanism Independent of Inositol 1,4,5-Trisphosphate Production 402
A. Michler and S. L. Erdö

GABA Receptors and Female Sexual Behavior in the Rat 403
P. Soria and A. Agmo

$GABA_B$ Receptors and Sociosexual Behavior in the Rat 404
R. Paredes and A. Agmo

Hippocampal Damage Induced by Tetanus Toxin in Rats is Accompanied by a Decrease in $GABA_A$ but not $GABA_B$ Binding Sites 405
G. Bagetta, G. Nistico and N. G. Bowery

The Role of $GABA_B$ Inhibitory Postsynaptic Potentials in Thalamocortical Processing 405
V. Crunelli, S. Lightowler, I. Soltesz, M. Haby, D. Jassik-Gerschenfeld and N. Leresche

Pharmacological Actions of Baclofen and $GABA_B$ Antagonists in Rats: Effects on Muscle Tone and Spinal Reflex Transmission 406
U. Wüllner, T. Klockgether, M. Schwarz and K.-H. Sontag

Mismatch Between $GABA_B$ and Baclofen Effects on K^+ Conductance of Hippocampal Neurons 407
U. Misgeld, W. Müller and M. Bijak

Comparative Electrophysiological Investigations of the $GABA_B$ Blockers CGP 35348 and Phaclofen in the Hippocampus and the Spinal Cord In Vitro 407
G. Karlsson, M. F. Pozza, H.-R. Olpe and F. Brugger

Pertussis Toxin Blocks Post- but not Presynaptic Actions of Baclofen in the Rat Dorsolateral Septal Nucleus 408
J. P. Gallagher, K. D. Phelan, M. J. Twery and H. Hasuo

$GABA_B$ Receptor-mediated Hyperpolarization of Presumed Dopamine-containing Rat Substantia Nigra Neurons 409
M. G. Lacey and R. A. North

Involvement of Presynaptic $GABA_B$ Receptors in Paired-pulse Depression of GABAergic Inhibitory Postsynaptic Potentials in Rat Hippocampal Slices 409
C. H. Davies and G. L. Collingridge

Baclofen Facilitates the Induction of Long-term Potentiation 410
D. D. Mott, C. M. Ferrari, D. V. Lewis, W. A. Wilson and H. S. Swartzwelder

$GABA_B$ Receptors and Experimental Models of Epilepsy 411
H.-R. Olpe, G. Karlsson, M. Schmutz, K. Klebs and H. Bittiger

POSTERS

3-Heteroaromatic Aminobutyric Acids: Potent and Selective Ligands for the $GABA_B$ Receptor 415
P. Berthelot, C. Vaccher, N. Flouquet, M. Debaert, M. Luyckx and C. Brunet

Hydroxylated Analogues of GABA and δ-Aminovaleric Acid as $GABA_B$ Receptor Ligands 416
A. Hedegaard, B. Frølund, C. Herdeis, H. Hjeds and P. Krogsgaard-Larsen

Structural and Chemical Modifications of Baclofen, a Potent $GABA_B$ Agonist 417
A. Mann, A. Schöenfelder, S. Lecoz and C. G. Wermuth

Ethanol Antagonizes the Baclofen-induced Modification of [^{3}H]Noradrenaline and [^{3}H]GABA Release in Mouse Cortical Slices 417
M. Daoust, C. Vadon, J. P. Henry and H. Ollat

Functional Coupling of the $GABA_B$ Receptor with Phosphatidylinositol Turnover in the Rat Cerebral Cortex 418
Y. Ohmori and K. Kuriyama

Pharmacological Characterization of the GABA Receptor-mediating Inhibition of Substance P Release from Capsaicin-sensitive Neurons in Rat Trachea 419
N. J. Ray, A. J. Jones and P. Keen

Calcium Channel Ligand Binding Sites Affect $GABA_B$ Receptors 420
R. H. Thalmann and M. I. Al-Dahan

$GABA_B$ Autoreceptors Modulate [^{3}H]GABA Release in Rat Cerebral Cortical Slices 421
G. Bonanno, E. Fedele and M. Raiteri

Functional Coupling of the $GABA_B$ Receptor with the Adenylyl Cyclase System in the Brain 421
J.-I. Taguchi, M. Nishikawa and K. Kuriyama

An Investigation of the Relative Potencies of a Number of $GABA_B$ Antagonists 422
J. M. Hills, W. Howson, M. M. Larkin and C. S. Li

$GABA_B$ Receptors in Bovine Adrenal Medulla: Their Implication in the Regulation of Catecholamine Secretion 423
M. J. Oset-Gasque, E. Castro and M. P. González

GABA Receptors and Motor Functions 424
A. Agmo and R. Paredes

Pharmacology of GABA-evoked Responses in the Septohippocampal Neurons of the Rat: Implication of a Pertussis Toxin-sensitive G-Protein 425
P. Dutar, O. Rascol and Y. Lamour

Effects of Phaclofen on Visually Evoked Responses in the Cat Striate Cortex 425
U. Baumfalk and K. Albus

Use of the Neonatal Rat Spinal Cord in Studies on the $GABA_B$ Receptor 426
D. E. Jane, P. C. K. Pook, D. C. Sunter and J. C. Watkins

The $GABA_B$ Receptor Antagonist, Phaclofen, Reverses Stimulus-induced Disinhibition in the Dentate Gyrus 427
D. V. Lewis and D. D. Mott

Biphasic Phaclofen-sensitive Actions of Baclofen on Excitatory Synaptic Transmission in the Frog Optic Tectum In Vitro 427
L. Sivilotti, G. Mazda and A. Nistri

$GABA_B$-mediated Inhibitory Postsynaptic Potentials Prevent Afterdischarge Bursts in Cultured Hippocampal Slices 428
A. T. Malouf, C. A. Robbins and P. A. Schwartzkroin

Kainate and Quisqualate Block the $GABA_B$ Response of CA3 Hippocampal Pyramidal Cells 429
C. Rovira, M. Gho and Y. Ben-Ari

Effects of $GABA_A$ and $GABA_B$ Blockers on Paired-pulse Inhibition in the Hippocampus CA1 Region 429
M. F. Pozza, G. Karlsson, F. Brugger and H.-R. Olpe

Blockade of the Late Inhibitory Postsynaptic Potential in Rat CA1 Hippocampal Neurons by 2-Hydroxysaclofen 430
N. A. Lambert, N. L Harrison and T. J. Teyler

GABA Receptors in the Nervous System of an Insect, *Periplaneta americana* 431
C. Malecot, B. Hue, S. D. Buckingham and D. B. Sattelle

Free Communications

GABA$_B$ Mechanisms in Human Jejunum In Vitro

G. GENTILINI, S. FRANCHI-MICHELI*, D. PANTALONE*, C. CORTESINI and L. ZILLETTI

*Department of Preclinical and Clinical Pharmacology 'Mario Aiazzi Mancini' Viale G.B. Morgagni 65 and *Clinica Chirurgica III, Viale G.B. Morgagni 85, University of Florence, Florence, Italy*

Evidence of GABA$_B$ mechanisms in human jejunum in vitro is reported here. Longitudinal strips of human jejunum obtained from surgical intervention were set up in an organ bath and their motility was recorded. The effects of different drugs were evaluated using the 'motility index' (frequency × amplitude of contractions). Preparations generally exhibited spontaneous motility which could be abolished by 3×10^{-7} M hyoscine. The administration of GABA (3×10^{-6}–10^{-4} M) caused a dose-dependent reduction in motility index (median effective dose, $ED_{50} = 10^{-5}$ M). This effect was mimicked by (−)-baclofen, while 3×10^{-5} M homotaurine (3-aminopropylsulphonic acid, 3-APS) and 3×10^{-5} M muscimol had no significant effect. The effect of GABA was prevented by 3×10^{-7} M tetrodotoxin but not affected by 10^{-5} M picrotoxin. It was reduced by the weak GABA$_B$ antagonist 3×10^{-3} M 5-aminovaleric acid; 3×10^{-3} M 3-APS caused a parallel shift to the right in the dose–response curve for GABA with an EC_{50} ratio of 0.9 logarithmic units. The effect of GABA was not influenced by 3×10^{-5} M hexamethonium, 10^{-5} M phentolamine plus 10^{-5} M propranolol, 10^{-5} M naloxone or by 10^{-4} M ATP desensitization.

3-Aminopropylphosphinic Acid, Its GABA$_B$ Agonist Activity and the Effects of Phaclofen on In-vitro Pharmacological Preparations

J. M. HILLS, W. HOWSON* and M. M. LARKIN

*Departments of Pharmacology and *Medicinal Chemistry, Smith Kline & French Research Ltd., Welwyn, UK*

A novel structural analogue of GABA, 3-aminopropylphosphinic acid (3-APA), is reported to have high binding affinity for the GABA$_B$ receptor (Dingwall *et al.*, 1987) and to show functional GABA$_B$ agonist activity in the guinea-pig ileum and rat anococcygeus muscle (Hills *et al.*, in press). We have studied the agonist activity of 3-APA and the effects of phaclofen in the rat vas deferens (electrically stimulated) and guinea-pig colon.

	$IC_{50} \pm$ SEM (μM)	
	Rat vas deferens	Guinea-pig colon
3-APA	0.79 ± 0.23 ($n = 8$)	0.27 ± 0.1 ($n = 4$)
Baclofen	6.4 ± 0.9 ($n = 8$)	6.4 ± 2.8 ($n = 4$)

IC_{50}, half-maximal inhibitory concentration.

These results indicate that 3-APA is 8 and 24 times more potent than baclofen in the vas deferens and guinea-pig colon respectively. However, phaclofen (500 μM) was an effective antagonist only in the vas deferens but not in the colon.

REFERENCES

Dingwall, J. G., Ehrenfreund, J., Hall, R. G. and Jack, J. (1987) *Phosph. Sulf.*, **30**, 571–575.
Hills, J. M., Dingsdale, R. A., Parsons, M. E., Dolle, R. E. and Howson, W. (in press) *Br. J. Pharmacol.*

Prejunctional $GABA_B$ Inhibition of Neurally Induced Airway Constriction in Guinea-pigs

R. W. CHAPMAN, G. DANKO, C. RIZZO, P. J. MAUSER, R. W. EGAN and W. KREUTNER

Department of Allergy and Inflammation, Schering-Plough Corp., Bloomfield, NJ, USA

GABA is implicated as an inhibitory neurotransmitter in peripheral tissues including lung. In this study, we examined the potential inhibitory effects of GABA on neuronally induced airway contractions in guinea-pigs. In vitro, electrical field stimulation of isolated guinea-pig tracheal ring caused cholinergic contractions that were partially (40%) inhibited by GABA (3–100 μM), whereas GABA did not inhibit contractions due to exogenous acetylcholine. Baclofen, a $GABA_B$ agonist, mimicked the effect of GABA, whereas muscimol, a $GABA_A$ agonist, was inactive. The effects of baclofen were inhibited by the $GABA_B$ antagonist, 3-aminopropylphosphonic acid (3-APPA), but were unaffected by the $GABA_A$ antagonist, (+)-bicuculline. In vivo, electrical vagal nerve stimulation (5 V, 20 Hz, 0.5 ms, 5 S) caused cholinergic bronchoconstrictions in guinea-pigs that were inhibited by GABA (1–10 mg/kg, i.v.). In contrast, GABA did not inhibit the bronchospasm due to exogenous methacholine. The inhibitory effects of GABA on the vagally mediated bronchospasm were duplicated by baclofen (1–10 mg/kg, i.v.), and inhibited by 5 mg/kg 3-APPA i.v., but were unchanged by compounds acting at $GABA_A$ receptors. These results demonstrate that GABA inhibits neurally mediated airway contraction by acting at prejunctional $GABA_B$ receptors.

$[^3H](-)$-Baclofen Labels Specific Binding Sites in Rat Kidney Cortex

S. L. ERDÖ, A. MICHLER and J. R. WOLFF

Department of Anatomy, University of Göttingen, Göttingen, FRG

Knowledge of $GABA_B$ receptors outside the mammalian central nervous system is mostly derived from pharmacological (mainly release) and electrophysiological studies. However, there have been few binding studies, or autoradiographic experiments, other than the present, designed to characterize $GABA_B$ receptors in peripheral organs. In this study, the occurrence and distribution of specific baclofen binding sites were examined in cryostat sections of the rat kidney. Sections (7 μm) were cut at -15 °C, mounted on gelatine-coated slices, and incubated at 20 °C for 45 min in the presence of 20 nM $[^3H](-)$-baclofen (NEN, 46.5 Ci/mmol). Control sections were incubated in the additional presence of 0.1 mM unlabelled $(\pm)$-baclofen or 0.5 mM GABA. The incubation medium was a Tris–HCl buffer (50 mM, pH 7.4) containing 2.5 mM $CaCl_2$. The slides were subsequently rinsed, dried and exposed on Ultrofilm for 6–7 weeks at 4 °C. To quantify bound radioactivity, other sections were removed from the slides and put into liquid-scintillation vials. High-affinity baclofen binding, showing the features of $GABA_B$ receptors, was demonstrated. Autoradiographic examination of the distribution of binding sites revealed a predominant occurrence of silver grains in the renal cortex, suggesting a role for $GABA_B$ receptors in the modulation of ionic transport processes in the rat kidney.

ACKNOWLEDGEMENT

Supported by DFG grants Wo 279/6-2 and 6-5.

Interaction of D-Baclofen and L-Baclofen

G. H. FROMM, M. NAKATA and T. SHIBUYA

Department of Neurology, University of Pittsburgh, Pittsburgh, PA, USA

D-Baclofen has been reported to antagonize the action of L-baclofen when administered systemically in both experimental animals and humans, but several in-vitro studies did not find any antagonistic effect. We have therefore investigated the effect of the iontophoretic application of D-baclofen and L-baclofen on single neurons in the trigeminal nucleus of Sprague–Dawley rats anesthetized with halothane.

The iontophoresis of 10–20 nA L-baclofen depressed the response of trigeminal nucleus neurons to electrical stimulation of the face similar to the effect previously observed after systemic administration of L-baclofen. The concomitant iontophoresis of 10–20 nA D-baclofen blocked the action of iontophoretically applied L-baclofen. However, the iontophoresis of 30–60 nA D-baclofen had the opposite effect, facilitating the action of L-baclofen. Application of 200–300 nA D-baclofen depressed the neuron response, similar to the action of low doses of L-balcofen. We also found that 30–40 nA L-baclofen had a much stronger effect than previously obtained with systemic administration, and D-baclofen was not able to block it.

Our observations suggest that D-baclofen is a partial agonist of L-baclofen, and emphasize the importance of testing drugs at therapeutic concentrations in an appropriate model in order to predict their therapeutic effect.

$GABA_B$ Receptor Ligands Modulate Neuronal Outgrowth Through a Mechanism Independent of Inositol 1,4,5-Trisphosphate Production

A. MICHLER and S. L. ERDÖ

Department of Anatomy, University of Göttingen, Göttingen, FRG

The trophic effect of GABA on neuronal growth is thought to be mediated via specific receptors, which may be GABA-gated chloride channels ($GABA_A$), or may be coupled to a second messenger ($GABA_B$), such as inositol 1,4,5-trisphosphate (IP_3). In the present study, we attempted to investigate whether the trophic effects of GABA on cultured neurons may be mediated through the IP system. Primary cultures from embryonic (7-day-old) chick tectum were cultivated in a serum-free, defined NI medium at a density of 2×10^{-6} cells/8 cm^2, for 4 days, and then exposed (30 min) to various ligands of the GABA receptors. The production of IP_3 was examined by a competitive ^3H-binding assay kit from Amersham (TRK.1000). GABA and its specific agonists, THIP (4,5,6,7-tetrahydroisooxazolo[5,4-*c*]pyridin-3-ol) and baclofen (up to 100 μM), failed to influence the formation of IP_3. The $GABA_B$ receptor antagonist, phaclofen (10–1000 μM), but not 2-hydroxysaclofen (up to 100 μM), caused a dose-dependent stimulation. This effect could be evoked by phaclofen concentrations lower than those blocking $GABA_B$ receptors, and could not be antagonized by 100 μM baclofen. These findings indicate that, in cultured neurons, phaclofen stimulates the production of IP_3 by a mechanism not related to $GABA_B$ receptors, and thus is an antagonist of $GABA_B$ receptors less specific than 2-hydroxysaclofen. Although calcium and/or potassium channels seem to be involved in the GABAergic effects on neuronal growth, there is no direct evidence for these channels being identical with the $GABA_B$ receptor-coupled ones.

GABA Receptors and Female Sexual Behavior in the Rat

P. SORIA and A. AGMO

Department of Psychology, Universidad Anahuac, Mexico City, Mexico

In order to observe stimulatory and inhibitory effects of GABA on the lordotic response in female rats, several GABAergic agents were administered to ovariectomized animals treated with estradiol benzoate alone or with different doses of this substance plus progesterone. Our results indicate that balcofen, a $GABA_B$ receptor agonist, inhibits lordosis behavior in doses from 5 mg/kg. THIP (4,5,6,7-tetrahydroisooxazolo[5,4-*c*]-pyridin-3-ol), a $GABA_A$ receptor agonist, also inhibits this behavior in doses from 20 mg/kg. The administration of the GABA transaminase inhibitor, γ-acetylen-GABA (GAG), as well as the $GABA_A$ agonist, 3-aminopropylsulfonic acid, and the $GABA_A$ receptor antagonist, bicuculline, had no effect.

Bicuculline also failed to block the effects of THIP. These data suggest that the $GABA_B$ receptor may play an important role in the control of lordosis behavior, while the $GABA_A$ receptor may be of little importance.

Other experiments indicate that, in a situation in which the female can control sexual interaction with the male, precopulatory behaviors are also affected by the administration of baclofen in doses from 5 mg/kg. Under these conditions, the frequency of behaviors such as sniffing, genital exploration and approach to the male decreased, as did female attractivity. The last was reflected in the reduced number of mounts received by the drug-treated females. With this same drugs and dose, no effects were found upon behaviors related to motor activity such as rearing and resting. This leads us to think that the $GABA_B$ receptor may also be involved in the control of proceptive behavior.

On the other hand, the administration of GAG in doses from 100 mg/kg inhibited precopulatory behavior as well as motor activity but did not affect the lordosis response. This leads us to conclude that GAG has a non-specific effect and supports the hypothesis that stimulation of the $GABA_B$ receptor, rather than a generalized increase in GABA levels, is responsible for GABAergic control of female sexual behavior.

$GABA_B$ Receptors and Sociosexual Behavior in the Rat

R. PAREDES and A. AGMO

Department of Psychology, Universidad Anahuac, Mexico City, Mexico

The enhancement of GABAergic neurotransmission inhibits sexual behavior in the male rat. Neither the effects of $GABA_A$ agonists nor those of GABA transaminase inhibitors (GABA-TIs) can be blocked by $GABA_A$ antagonists. Baclofen also inhibits sexual behavior, suggesting that the $GABA_B$ rather than the $GABA_A$ receptor is involved in the control of this behavior. Moreover, baclofen inhibits sexual behavior in castrated male rats in doses that have no effect on motor execution, while the doses needed to inhibit copulatory behavior with $GABA_A$ agonists or GABA-TIs have strong motor effects. The possibility that GABAergic drugs produced a specific inhibition of sexual behavior by affecting sociosexual interactions was investigated. Animals treated with γ-acetylen-GABA (GAG), sodium valproate or baclofen were observed with a receptive female or a castrated male to differentiate the effects of these drugs on sexual interactions from those on sociosexual interactions (castrated male). The frequency and duration of the following behaviors were registered: rearing, sniffing, self-grooming, grooming partner, genital exploration and pursuit.

Sodium valproate (400 mg/kg) and GAG (100 mg/kg) inhibited sexual behavior. However, no interaction was found between sex and treatment in the sociosexual interactions, suggesting a nonspecific inhibitory effect of these drugs upon sexual behavior. Baclofen (5 mg/kg) also inhibited sexual behavior and a significant interaction was found between sex and treatment with respect to sociosexual interactions. Moreover, the inhibitory effects of baclofen were restricted to behavioral items related to sexual interaction, primarily those constituting precopulatory behaviors. It appears that baclofen has a specific inhibitory effect on behaviors associated with the initiation of copulatory activity. These results support previous observations suggesting that the $GABA_B$ receptor (not the $GABA_A$) is involved in the control of sexual behavior.

Hippocampal Damage Induced by Tetanus Toxin in Rats is Accompanied by a Decrease in $GABA_A$ but not $GABA_A$ Binding Sites

G. BAGETTA, G. NISTICO and N. G. BOWERY

Department of Pharmacology, School of Pharmacology, London, UK and Institute of Pharmacology, Faculty of Medicine and Surgery, Catanzaro, Italy

Pyramidal neurons in the CA1 hippocampal area receive both excitatory and inhibitory innervation, the overactivity of the former being responsible for neuronal damage produced in the same area by transient cerebral ischaemia in the rat. Neuronal damage in the hippocampus may also occur under experimental conditions in which an impairment in inhibitory mechanisms has been produced. We tested this hypothesis using tetanus toxin, a GABA inhibitory-transmission blocking agent. Unilateral injection of tetanus toxin (500–2000 MLDs; $n = 3$ rats per dose) into the rat CA1 hippocampal area produced, in comparison to the contralateral area, a dose- and time-dependent neuronal loss. The neuronal loss produced by 1000 MLDs tetanus toxin was accompanied by a significant reduction of $GABA_A$ but not of $GABA_B$ binding sites in the pyramidal cell layer, as determined by autoradiographic analysis. In conclusion, this may indicate that $GABA_B$ sites in the pyramidal cell layer are restricted to presynaptic sites.

The Role of $GABA_B$ Inhibitory Postsynaptic Potentials in Thalamocortical Processing

V. CRUNELLI, S. LIGHTOWLER, I. SOLTESZ, M. HABY*, D. JASSIK-GERSCHENFELD* and N. LERESCHE

*Department of Pharmacology, St George's Medical School, London, UK and *Department of Neuroscience of Vision, Université P & M Curie, Paris, France*

The $GABA_B$ inhibitory postsynaptic potential (IPSP) evoked in thalamocortical projection cells of the dorsal lateral geniculate nucleus (LGN) by stimulation of the optic tract is:

(1) late (35 ms), long-lasting (90–350 ms) and K^+-dependent;
(2) reversibly blocked by the $GABA_B$ antagonist phaclofen;
(3) weakly effective against strong repetitive firing;
(4) reversibly increased in amplitude by bicuculline;
(5) present (together with $GABA_A$ IPSPs) in both X and Y cells, indicating that it is not peculiar to any of the parallel visual channels;

(6) generated by thalamic interneurons, indicating that the late, long-lasting inhibition in the thalamus does not originate exclusively from the GABA cells of the perigeniculate/reticularis nucleus;
(7) able to evoke a low-threshold Ca^{2+} potential that, in turn, can bring thalamic cells to firing threshold (Figure 1). This property suggests a role of $GABA_B$ IPSPs in the generation of thalamic 4–6 Hz waves.

Interestingly, in cells of the ventral LGN (that do not project to the cortex but have functionally active $GABA_B$ receptors) no $GABA_B$ IPSP is evoked by optic-tract stimulation.

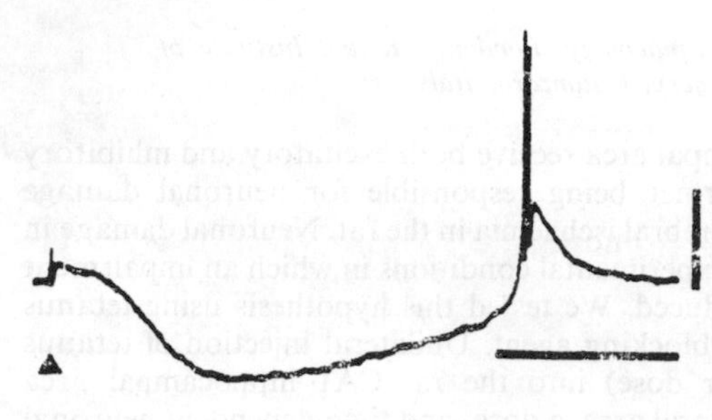

Figure 1. Stimulation of the optic tract (arrow) evokes an excitatory postsynaptic potential followed by a $GABA_B$ inhibitory postsynaptic potential which de-inactivates a low-threshold Ca^{2+} potential large enough to produce a burst of action potentials (calibration: 20 mV, 100 ms).

Pharmacological Actions of Baclofen and $GABA_B$ Antagonists in Rats: Effects on Muscle Tone and Spinal Reflex Transmission

U. WÜLLNER*, T. KLOCKGETHER†, M. SCHWRZ‡ and K.-H. SONTAG*

**Max-Planck-Institute for Experimental Medicine, Hermann-Rein-Strasse 3, Göttingen, FRG, †Department of Neurology, University of Tübingen, Tübingen, FRG, ‡Department of Neurology, Alfried Krupp Hospital, Essen, FRG*

Baclofen is used clinically to reduce spasticity. We therefore investigated the action of the presumed $GABA_B$ antagonists phaclofen and δ-aminovaleric acid (DAVA) on muscle relaxant and reflex suppressant properties of baclofen.

(1) Intrathecal (i.t.) injection of baclofen suppressed the monosynaptic H- and polysynaptic flexor. Phaclofen or DAVA antagonized the action of baclofen but did not affect the action of a $GABA_A$ agonist, muscimol. Intrathecal phaclofen alone had no effect on spinal reflex transmission and long-lasting inhibition.
(2) The effect of i.t. baclofen and DAVA was studied on increased muscle tone in genetically spastic Wistar rats. DAVA antagonized the muscle relaxant action of baclofen but not of muscimol.
(3) Stimulation of $GABA_B$ receptors in the ventromedial thalamic nucleus by local injections of baclofen induced catalepsy and limb rigidity. The action of baclofen was antagonized by DAVA.

Mismatch Between $GABA_B$ and Baclofen Effects on K^+ Conductance of Hippocampal Neurons

U. MISGELD, W. MÜLLER aand M. BIJAK

Department of Neurophysiology, Max-Planck-Institute for Psychiatry, Planegg-Martinsried, FRG

(−)-Baclofen (0.1–1 μM) disinhibits dentate granule cells of the guinea-pig hippocampus in vitro by a hyperpolarization of inhibitory neurons. Granule cells are less sensitive to (−)-baclofen than are hilar neurons and CA3 pyramidal cells. Such a difference is not known in GABA effects. In CA3 neurons carbachol (0.1–0.3 μM), through an M_1 receptor-mediated effect, reduces the K^+ conductance increase which follows the application of (−)-baclofen. This effect is pronounced if (−)-baclofen is applied in concentrations (<0.3 μM) which do not activate a maximal K^+ current. Carbachol did not, however, reduce the K^+ conductance increase which follows GABA application. Our findings are not easily accommodated by the assumption that baclofen specifically replaces GABA as an agonist at $GABA_B$ receptor sites activating K^+ conductance of hippocampal neurons.

Comparative Electrophysiological Investigations of the $GABA_B$ Blockers CGP 35348 and Phaclofen in the Hippocampus and the Spinal Cord In Vitro

G. KARLSSON, M. F. POZZA, H.-R. OLPE and F. BRUGGER

Research and Development Department, Pharmaceuticals Division, CIBA-Geigy, Basel, Switzerland

We compared the new $GABA_B$ receptor blocker CGP 35348 and phaclofen in the hippocampal slice (CA1 neurons) and the hemisected spinal cord preparation of the rat. In the hippocampus, at concentrations of 30–100 μM, CGP 35348 abolished both the membrane hyperpolarization induced by baclofen (10 μM) and the late inhibitory postsynaptic potential (IPSP). Marginal effects were seen with 10 μM CGP 35348. Phaclofen at 300 μM was equipotent to 10 μM CGP 35348 in reducing the late IPSP but had no or very weak effect on baclofen-induced hyperpolarization. At concentrations of 0.5–1 mM, phaclofen abolished the action of baclofen as well as the late IPSP. CGP 35348 (30–100 μM) and phaclofen (1 mM), to a lesser extent, induced weak transient membrane depolarizations (2–4 mV). Baclofen at 3 μM almost completely depressed the monosynaptic reflex and the spontaneous activity of the isolated spinal cord preparation of the rat. CGP 35348 (100 μM) and phaclofen (1 mM) clearly reduced the inhibitory effects of baclofen on the monosynaptic reflex (by 80% and 35% respectively) and the spontaneous activity. We thus demonstrated that CGP 35348 is a blocker of $GABA_B$ receptors and approximately 20–30 times more potent than phaclofen.

Pertussis Toxin Blocks Post- but not Presynaptic Actions of Baclofen in the Rat Dorsolateral Septal Nucleus

J. P. GALLAGHER, K. D. PHELAN, M. J. TWERY and H. HASUO

Department of Pharmacology, University of Texas Medical Branch, Galveston, TX, USA

Our laboratory has demonstrated that application of GABA or baclofen activates $GABA_B$ receptors and an associated K^+ conductance on dorsolateral septal nucleus (DLSN) neurons, while orthodromically induced GABA release results in a $GABA_B$-mediated late hyperpolarizing potential (LHP) (Gallagher *et al.*, 1984; Stevens *et al.*, 1985, 1987). These $GABA_B$ responses are blocked competitively by phaclofen (Hasuo and Gallagher, 1988).

Several laboratories have demonstrated that the central nervous system (CNS) postsynaptic $GABA_B$-receptor/K^+-channel complex is coupled via a pertussis toxin-sensitive G-protein. Since $GABA_B$ receptors have been localized presynaptically within the CNS (Bowery *et al.*, 1980), we initiated studies to determine whether a similar G-protein may be required for coupling both pre- and postsynaptic $GABA_B$ receptors to their effectors in the DLSN. In brain slices taken from rats 3–5 days after intracerebroventricular injection of pertussis toxin, the postsynaptic effect of baclofen and $GABA_B$-mediated LHPs were absent. However, evoked excitatory postsynaptic potentials were depressed, as in controls, by baclofen. The inability of pertussis toxin to block this presynaptic action of baclofen while preventing its postsynaptic hyperpolarizing action suggests a difference in the pertussis toxin sensitivity of the G-protein or a distinct coupling mechanism for the $GABA_B$-receptor/K^+-channel complex at pre- vs postsynaptic sites. Our results in the DLSN together with data collected in the hippocampus (Dutar and Nicoll, 1988) are consistent with the existence of more than one type of $GABA_B$-receptor/complex.

$GABA_B$ Receptor-mediated Hyperpolarization of Presumed Dopamine-containing Rat Substantia Nigra Neurons

M. G. LACEY* and R. A. NORTH

*Oregon Health Sciences University, Oregon, USA, *Present address: Smith Kline & French, Welwyn, UK*

Baclofen hyperpolarizes presumed dopamine-containing rat substantia nigra neurons in vitro, apparently due to activation of a G-protein-linked potassium conductance increase (Lacey *et al.*, 1988). The possibility that cyclic AMP might be involved in this response was examined. Application by superfusion of forskolin (1–10 μM), the activator of adenyl cyclase, was without effect on the hyperpolarizations caused by a submaximal concentration (3 μM) of baclofen (five cells). The rightward shift of the baclofen concentration–effect curve caused by phaclofen (1 mM) indicated an equilibrium dissociation constant for this antagonist of around 200 μM, further implicating $GABA_B$ receptors in this effect.

REFERENCE

Lacey, M. G., Mercuri and North, R. A. (1988) *J. Physiol.*, **401**, 437–453.

Involvement of Presynaptic $GABA_B$ Receptors in Paired-pulse Depression of GABAergic Inhibitory Postsynaptic Potentials in Rat Hippocampal Slices

C. H. DAVIES and G. L. COLLINGRIDGE

Department of Pharmacology, University of Bristol, Bristol, UK

Monosynaptic biphasic inhibitory postsynaptic potentials (IPSPs) in CA1 pyramidal neurons exhibit paired-pulse depression (Collingridge *et al.*, 1988). An explanation of this phenomenon is that GABA acts on presynaptic receptors to inhibit its own release (Barker *et al.*, 1988). We have tested this proposal.

Paired-pulse depression of the biphasic IPSP occurred for interstimuli intervals ranging between 10 and 5000 ms with an interval for maximal depression of 100–125 ms. At intervals of less than 10 ms, IPSPs summated.

(−)-Baclofen (0.5 μM) depressed the IPSP by 50% and reduced the extent of paired-pulse depression by 30%. 2-Hydroxysaclofen (1 mM) abolished the late phase and markedly depressed the early phase (50%) of the IPSP. In addition, it depressed paired-pulse depression of IPSPs.

The results support a feedback autoinhibition role for GABA mediated through presynaptic $GABA_B$ receptors as the underlying mechanism of paired-pulse depression of GABAergic IPSPs in the hippocampus.

REFERENCES

Collingridge, G. *et al.* (1988) *Neurol. Neurobiol.*, **46**, 171–178.
Harrison, N. L. *et al.* (1988) *Neurosci. Lett.*, **85**, 105–109.

Baclofen Facilitates the Induction of Long-term Potentiation

D. D. MOTT, C. M. FERRARI, D. V. LEWIS, W. A. WILSON and H. S. SWARTZWELDER

Departments of Pharmacology, Neurobiology, Pediatrics (Neurology), Medicine (Neurology) and Psychology, Duke University and VA Medical Centers, Durham, NC, USA

The induction of long-term potentiation (LTP) in the hippocampus is modulated by synaptic inhibition. We have previously reported a disinhibitory action of baclofen in the dentate gyrus. At low perforant path (PP) stimulus intensities, baclofen depressed dentate granule cell firing, but at high stimulus intensities it caused multiple population spikes. Using rat hippocampal slices, we now report that baclofen facilitates the induction of the LTP in the dentate gyrus.

In 6/6 slices PP stimulus trains (3 trains, 10 pulses at 100 Hz, every 20 min) produced no significant LTP. A single additional PP stimulus train in baclofen (10 μM) caused marked LTP of the PP-evoked population-spike amplitude (200%). Pre-exposure of the slice to MK-801 blocked the induction of LTP.

These results suggest that the disinhibitory action of baclofen facilitates the induction of LTP by unmasking *N*-methyl-D-aspartate receptors.

GABA$_B$ Receptors and Experimental Models of Epilepsy

H.-R. OLPE, G. KARLSSON, M. SCHMUTZ, K. KLEBS and H. BITTIGER

Research and Development Department, Pharmaceuticals Division, CIBA-Geigy, Basel, Switzerland

The role of GABA$_B$ receptors in epilepsy remains controversial. In man and animals both pro- and anticonvulsive effects have been observed with the GABA$_B$ agonist baclofen. New experimental possibilities to address this question are provided by the new GABA$_B$ receptor blocker CGP 35348, which in electrophysiological tests is about 20–30 times more potent than phaclofen and which penetrates the blood–brain barrier. CGP 35348 had no effects in various convulsion tests induced chemically or electrically in rodents. CGP 35348 did not accelerate epileptic-like discharges of hippocampal slice exposed to penicillin. Furthermore, multiple population spikes in hippocampal slices kept in low Mg^{2+} were not affected by the drug. In contrast to bicuculline, CGP 35348 did not induce multiple population spikes in slices perfused with regular medium. The effects of baclofen and CGP 35348 on kindling development will be described. In conclusion, GABA$_A$ receptors appear to be considerably more important for the regulation of epileptic phenomena than do GABA$_B$ receptors.

Posters

3-Heteroaromatic Aminobutyric Acids: Potent and Selective Ligands for the $GABA_B$ Receptor

P. BERTHELOT*, C. VACCHER*, N. FLOUQUET*, M. DEBAERT*, M. LUYCKX† and C. BRUNET†

Laboratoires de Pharmacie Chimique et de Pharmacodynamie†, Faculté de Pharmacie, Lille, France*

Within the central and peripheral nervous system, the inhibitory neurotransmitter GABA has been shown to act through at least two distinctly different receptor sites, $GABA_A$ and $GABA_B$. Until now, *β-p*-chlorophenyl-GABA (baclofen) was the only selective agonist for the $GABA_B$ receptor. We recently described the synthesis of 3-(benzo[*b*]furan-2-yl)-GABA, a new selective ligand for $GABA_B$ sites (Berthelot *et al.*, 1987).

In the course of our work and in an attempt to elucidate the structural requirements for access to the $GABA_B$ receptor, we undertook the synthesis and binding studies of new 3-(2- or 3-heteroaromatic)-GABA. These compounds were prepared from aldehyde according to Wittig's procedure.

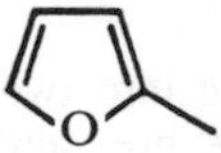

5a 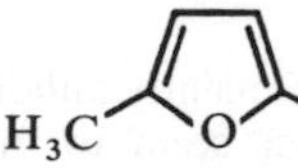 5b 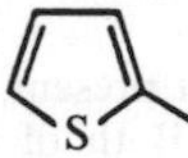 5c 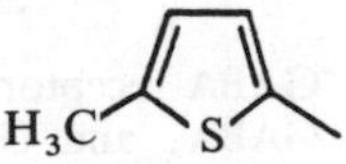 5d

R–CH–CH_2COOH with CH_2NH_2

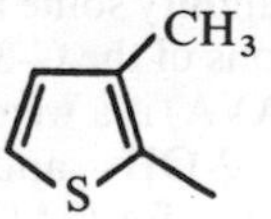

5e 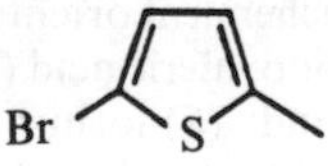 5f 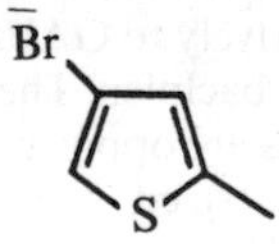 5g

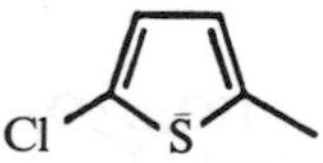

5h 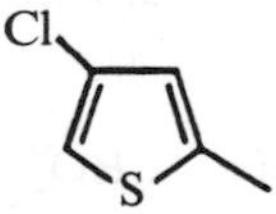 5i 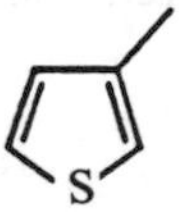 5j

These *racemic compounds*, especially 5d, 5f and 5h, are potent and *specific ligands* for the $GABA_B$ receptor, with a better half-maximal inhibitory concentration (IC_{50}) than achieved by us previously. The IC_{50} values in the presence of (*R*)-(−)[^{3}H]baclofen (0.33 μM) were as follows: 5d, 1.34; 5f, 1.86; 5h, 0.61.

In addition to previously known structural requirements for access to the $GABA_B$ receptor (Berthelot *et al.*, 1987), these results show that the substituent in the 'para' position on the aromatic ring should be of lipophilic nature in order to avoid excessive steric hindrance.

REFERENCE

Berthelot, P. *et al.* (1987) *J. Med. Chem.*, **30**, 743.

Hydroxylated Analogues of GABA and δ-Aminovaleric Acid as $GABA_B$ Receptor Ligands

A. HEDEGAARD, B. FRØLUND, C. HERDEIS*, H. HJEDS and P. KROGSGAARD-LARSEN

*Department of Organic Chemistry, Royal Danish School of Pharmacy, Copenhagen, Denmark and *Institut für Pharmazie und Lebensmittelchemie der Universität Würzburg, Würzburg, FRG*

GABA receptors are, at present, most conveniently subdivided into two main classes, $GABA_A$ and $GABA_B$. Both of these classes seem to include pre- and postsynaptic receptors as well as autoreceptors. There is an urgent need for selective agonists and antagonists for each of these subtypes of $GABA_B$ receptor as experimental tools, and such compounds have therapeutic interest.

(*R*)-(−)-Baclofen is a potent agonist for $GABA_B$ receptors. (*R*)-(−)-3-OH-GABA binds effectively to $GABA_B$ receptor sites with an affinity some four times lower than that of (*R*)-(−)-baclofen. The stereochemical orientations of the C-3 substituents of these two compounds are opposite. δ-Aminovaleric acid (DAVA) is a weak $GABA_B$ antagonist. We have synthesized the (*R*)- and (*S*)-forms of 2-OH- and 4-OH-DAVA and of 2-OH-6-NH_2-hexanoic acid and tested their affinities for $GABA_A$ and $GABA_B$ synaptic mechanisms. The stereoselectivity of the $GABA_B$ receptors with respect to the enantiomers of 4-OH-DAVA is more pronounced than with respect to those of 2-OH-DAVA, and both (*R*)- and (*S*)-2-OH-6-NH_2-hexanoic acid are inactive. In collaboration with Drs D.I.B. Kerr and J. Ong the pharmacology of these compounds is being studied in more detail.

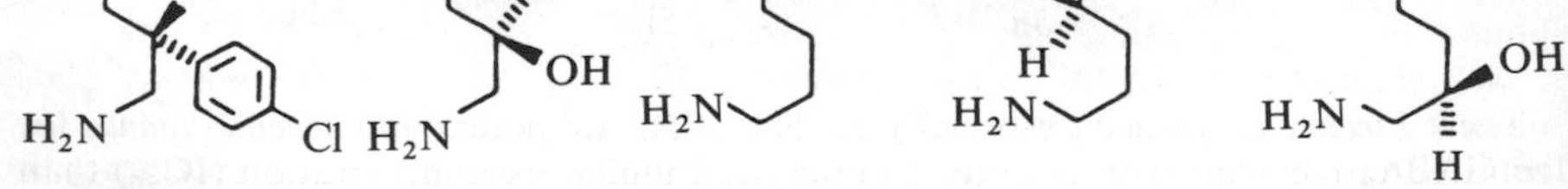

Structural and Chemical Modifications of Baclofen, a Potent $GABA_B$ Agonist

A. MANN, A. SCHÖENFELDER, S. LECOZ and C. G. WERMUTH

Centre de Neurochimie du CNRS, Département de Pharmacochimie Moléculaire, Strasbourg, France

Baclofen is a potent and unique ligand for the $GABA_B$ receptor. To study the potentiality of $GABA_B$ agonists or antagonists as pharmacological tools, more potent compounds are required. Taking a pharmacological approach using rigidification and manipulation of functions, we designed and synthesized a set of baclofen analogues in order to obtain more details about the $GABA_B$ pharmacophore. Therefore we locked the amino and acid functions of baclofen in defined space regions via incorporation in tetrahydronaphthalene, piperidine or indanyle rings; we transformed the amino functions in amidines (pyridazines) or imidazoles. The compounds were subjected to binding experiments, but none showed any significant affinity compared to baclofen. To account for this observed inactivity, several explanations are proposed, based on conformational analysis and solid-state structures.

Ethanol Antagonizes the Baclofen-induced Modification of [^{3}H]Noradrenaline and [^{3}H]GABA Release in Mouse Cortical Slices

M. DAOUST, C. VADON, J. P. HENRY and H. OLLAT*

*Pharmacochimie, UER Médecine-Pharmacie, St Etienne Rouvray, and *ANPP, Paris, France*

We previously showed that baclofen, but not $GABA_A$ receptor agonists, was able to reduce ethanol intake in rats. Moreover, we also demonstrated that noradrenaline and GABA were co-involved in the modulation of ethanol dependence.

The objective of the present work was to demonstrate that both noradrenaline and $GABA_B$ receptors were associated with ethanol-induced dysfunctioning in membrane-bound enzymes and membrane-mediated processes.

We studied [^{3}H]GABA and [^{3}H]noradrenaline release in mouse cortical slices obtained from alcoholized and non-alcoholized mice. Regulation of the release of the two transmitters by baclofen, or by drugs that decrease ethanol intake (acamprosate, desipramine), was investigated, using the superfusion model of Raiteri. Membrane depolarization was achieved using KCl, 12 mM for [^{3}H]GABA and 40 mM for

[^{3}H]noradrenaline. Stimulation values S_2/S_1 and net release were analysed. In vitro, baclofen (100 μM) decreased GABA release and in alcoholized mice this effect disappeared. Baclofen enhances the release of noradrenaline in vitro and this effect was prevented in alcoholized mice.

Interaction between these mediators and their ability to modulate ethanol dependence are discussed.

Functional Coupling of the $GABA_B$ Receptor with Phosphatidylinositol Turnover in the Rat Cerebral Cortex

Y. OHMORI and K. KURIYAMA

Department of Pharmacology, Kyoto Prefectural University of Medicine, Kyoto, Japan

It has been well established that GABA receptors in the brain are divided into $GABA_A$ and $GABA_B$ types. Although it has been reported that the $GABA_B$ receptor is coupled with the calcium ion channel and adenylyl cyclase system via an inhibitory GTP-binding protein, little is known on the functional coupling of $GABA_B$ receptors with other transmembrane biosignalling systems. In this study, the coupling of the $GABA_B$ receptor with phosphatidylinositol (PI) turnover was examined using rat cerebral cortical slices. Furthermore, the effect of phaclofen, which is considered to be a $GABA_B$ antagonist, was also studied. (−)-Baclofen, a $GABA_B$ agonist, significantly inhibited the accumulation of inositol-1-phosphate and inositol-1,4,5-trisphosphate, and this inhibition was counteracted by the addition of phaclofen. Furthermore, the inhibitory effect of (−)-baclofen was eliminated by treatment with islet-activating protein (IAP). These results suggest that the cerebral $GABA_B$ receptor may be coupled negatively with PI turnover, and phaclofen may exert an antagonistic effect on the $GABA_B$ receptor-mediated inhibition of PI turnover. The present results also suggest that this negative coupling may be modulated by the action of an IAP-sensitive GTP-binding protein.

Pharmacological Characterization of the GABA Receptor-mediating Inhibition of Substance P Release from Capsaicin-sensitive Neurons in Rat Trachea

N. J. RAY, A. J. JONES and P. KEEN

Department of Pharmacology, University of Bristol, The Medical School, Bristol, UK

Sensory neuropeptides, including substance P, have been proposed as mediators of non-adrenergic, non-cholinergic (NANC) neurogenic effects in airways. GABA and baclofen have been demonstrated to inhibit NANC bronchoconstriction in guinea-pig in vivo (Belvisi et al., 1988). We have developed a multisuperfusion system to study the control of the release of substance P-like (SP-LI) immunoreactivity and have investigated the effects of GABA receptor agonists and antagonists on K^+-stimulated SP-LI release from capsaicin-sensitive neurons in rat trachea in vitro. Eight or ten spirally cut tracheae were mounted in parallel oxygenated glass chambers and superfused with Krebs' solution (pH 7.4 37 °C, gassed with 95% O_2/5% CO_2) containing phosphoramidon (1 μM), captopril (100 μM), and bacitracin (20 mg/l) at 1 ml/min. Four fractions of 5 min from each trachea were simultaneously collected in vials containing a final concentration of 0.1% trifluoracetic acid. Fractions were then concentrated on Sep-Pak C_{18} cartridges, lyophilized and reconstituted for radioimmunoassay. Drugs were than added to the superfusion fluid reservoir for 4 min during the third fraction. Changes in SP-LI release were calculated as the difference in fractional release constants between the mean of fractions 1 and 2 and the mean of fractions 3 and 4 based on the recovery of exogenous substance P determined as $62.2 \pm 3.2\%$. Capsaicin, administered neonatally (50 mg/kg s.c.), caused a $93.2 \pm 6.4\%$ reduction in tracheal SP-LI content, suggesting that the great majority of tracheal SP-LI is neuronal in origin. A concentration of 60 mM K^+ significantly increased SP-LI release above spontaneous release values. Removal of calcium ions resulted in total abolition of 60 mM K^+-stimulated release ($P < 0.001$). GABA (1–100 μM) and ($\pm$)-baclofen (1–100 μM) did not affect spontaneous release but, when added 1 min prior to and during exposure to 60 mM K^+, produced dose-related inhibition of SP-LI release with maximum decreases of $77.7 \pm 18.8\%$ ($P < 0.001$) and $52.9 \pm 20.2\%$ ($P < 0.05$) respectively. The GABA agonist 3-aminopropylsulphonic acid (3-APS; 10–100 μM) was ineffective. The effect of ($\pm$)-baclofen was completely abolished by phaclofen (100 μM; $P < 0.001$), when present throughout the experiment, but was not sensitive to bicuculline (1 μM; $P > 0.05$) under identical conditions. $GABA_B$ receptor agonists may have therapeutic potential in respiratory disease, notably asthma.

REFERENCE

Belvisi *et al.* (1988) *Br. J. Pharmacol.*, **95**, 776P.

Calcium Channel Ligand Binding Sites Affect $GABA_B$ Receptors

R. H. THALMANN and M. I. AL-DAHAN

Baylor College of Medicine, Houston, TX, USA

A filtration assay was used to examine the binding of the $GABA_B$ agonist [^{3}H]baclofen to well-washed, permeabilized synaptic membranes prepared from adult rat cerebrum (Al-Dahan and Thalmann, in press).

This assay revealed a single binding site (equilibrium dissociation constant, K_D = 50 nM), binding to which was completely and specifically inhibited by an antagonist of $GABA_B$ responses, saclofen: half-maximal inhibitory concentration (IC_{50}) for inhibition of baclofen binding was 7 μM, but >1 mM for inhibition of the binding of the $GABA_A$ agonist [^{3}H]muscimol or the muscarinic antagonist [^{3}H]NMS.

The dihydropyridine calcium channel antagonist nifedipine inhibited baclofen binding (IC_{50} = 16 nM) by reducing the apparent receptor density (B_{max}) from 935 to 638 fmol/mg, while K_D was unaffected. The dihydropyridine calcium channel agonist BAY K 8644 stimulated baclofen binding up to 35% (median effective concentration, EC_{50} = 115 nM). The stereospecificity of this stimulatory effect was shown using the dihydropyridine 202-791: the calcium channel agonist (+)-202-791 stimulated binding up to 30% (EC_{50} = 8 nM), while (−)-202-791 was inactive. D-600, a calcium channel antagonist acting at a different site than the dihydropyridines, also reduced baclofen binding (IC_{50} = 27 nM).

Thus calcium channel ligand binding sites must communicate with $GABA_B$ receptors, perhaps reflecting the physiological direction of coupling from receptor to calcium channel.

REFERENCE

Al-Dahan, M.I. and Thalmann, R.H. (in press) *J. Neurochem.*, **53**.

$GABA_B$ Autoreceptors Modulate [^{3}H]GABA Release in Rat Cerebral Cortex Slices

G. BONANNO, E. FEDELE and M. RAITERI

Istituto di Farmacologia e Farmacaognosia, Università degli Studi di Genova, Genova, Italy

The release of [^{3}H]GABA and its modulation by autoreceptors were studied in rat cerebral cortical slices. Slices were prelabelled with [^{3}H]GABA in the presence of the glial GABA uptake inhibitor β-alanine, superfused with a medium containing amino-oxyacetic acid, β-alanine and the neuronal GABA uptake blocker SKF 89976A and stimulated electrically. The electrically evoked overflow of [^{3}H]GABA was tetrodotoxin- and Ca^{2+}-dependent. Exogenous GABA decreased in a concentration-dependent manner the release of [^{3}H]GABA. The $GABA_B$ receptor agonist (−)-baclofen, but not the $GABA_A$ receptor agonist muscimol, mimicked GABA. The GABA-induced inhibition of [^{3}H]GABA release was sensitive to the $GABA_B$ receptor antagonist phaclofen, which by itself increased the [^{3}H]GABA overflow, but not to the $GABA_A$ receptor antagonists bicuculline and SR 95531. The results show that $GABA_B$ autoreceptors are present in rat cerebral cortex.

Functional Coupling of the $GABA_B$ Receptor with the Adenylyl Cyclase System in the Brain

J.-I. TAGUCHI, M. NISHIKAWA and K. KURIYAMA

Department of Pharmacology, Kyoto Prefectural University of Medicine, Kyoto, Japan

It is well known that GABA receptors are divided into $GABA_A$ and $GABA_B$ types on the basis of their pharmacological properties. Although our previous studies indicated that the $GABA_A$ receptor was coupled functionally as well as structurally with the benzodiazepine receptor and chloride channel, the biochemical and pharmacological characteristics of the $GABA_B$ receptor have not been clearly defined. In this study, we have investigated the effect of GABA and (−)-baclofen, $GABA_B$ receptor agonists, on the cyclic AMP formation mediated by adenylyl cyclase in rat crude synaptic membrane (P_2) fraction. We have also studied the antagonistic effect of phaclofen, which was considered electrophysiologically to be a $GABA_B$ receptor antagonist, on the $GABA_B$ agonist-induced inhibition of cerebral adenylyl cyclase activity. GABA and (−)-baclofen significantly inhibited basal and forskolin-stimulated adenylyl cyclase activities in a dose-dependent manner, and phaclofen completely abolished the inhibitory effect of GABA and (−)-baclofen on forskolin-stimulated adenylyl cyclase activity. Furthermore, treatment of the fractions with islet-activating protein (IAP) eliminated the inhibitory effect of $GABA_B$ receptor agonists on forskolin-stimulated adenylyl cyclase activity. These results suggest that the $GABA_B$ receptor may be functionally coupled with the adenylyl cyclase system via an IAP-sensitive GTP-binding protein such as G_i or G_0. The present results also suggest that phaclofen may exhibit an antagonistic effect on $GABA_B$ receptor-mediated inhibition of cerebral adenylyl cyclase.

An Investigation of the Relative Potencies of a Number of $GABA_B$ Antagonists

J. M. HILLS, W. HOWSON*, M. M. LARKIN and C. S. LI*

*Departments of Pharmacology and *Medicinal Chemistry, Smith Kline & French, Welwyn, UK*

Recent publications have suggested that phaclofen and structurally related analogues have improved antagonist activity over other compounds (Kerr *et al.*, 1988, 1989). We have examined the antagonist potency of these componds in the electrically stimulated rat vas deferens. Each antagonist has been studied using at least three different concentrations ($n > 4$ animals for each concentration) in the range 10–1000 μM. pA_2 values ($P < 0.05$) have been derived from full Schild analysis where possible.

Antagonist	pA_2 (slope)
4-Aminobutylphosphonic acid (4-ABPA)	4.62 (0.98)
δ-Aminovaleric acid (DAVA)	4.42 (0.86)
Phaclofen	4.25 (1.01)
Saclofen	4.7*
2-Hydroxysaclofen	4.7*

*pA_2 values were estimated from a single concentration of antagonist due to non-linear Schild plots.

Our results suggest that in this functional assay no great advance in $GABA_B$ antagonist potency can be demonstrated.

REFERENCES

Kerr, D.I.B., Ong, J., Johnston, G.A.R., Abbenate, J. and Prager, R.H. (1988) *Neurosci. Lett.*, **92**, 92–96.

Kerr, D.I.B., Ong, J., Johnston, G.A.R. and Prager, R.H. (1989) *Brain Res.*, **480**, 312–316.

GABA$_B$ Receptors in Bovine Adrenal Medulla: Their Implication in the Regulation of Catecholamine Secretion

M. J. OSET-GASQUE, E. CASTRO and M. P. GONZÁLEZ

Instituto di Bioquímica, Facultad de Farmacia, Universidad Complutense, Madrid, Spain

The [^{3}H]GABA binding to bovine adrenal medulla membranes shows two components: one of low affinity (equilibrium dissociation constant, $K_D = 139 \pm 22$ nM; maximum binding capacity, $B_{max} = 3.2 \pm 0.4$ pmol/mg protein) and another of high affinity ($K_D = 41 \pm 6$ nM; $B_{max} = 0.35 \pm 0.26$ pmol/mg protein). On the basis of pharmacology and sensitivity to Ca^{2+} ions in the external medium, these sites show the characteristics of GABA$_A$ and GABA$_B$ receptor sites, respectively, since the binding of GABA to the site of low affinity is blockable by muscimol and insensitive to Ca^{2+} ions and the binding to the high-affinity site is blockable by baclofen and enhanced by increasing Ca^{+2} in the medium.

Both GABA binding sites are implicated in the modulation of catecholamine (CA) secretion in bovine chromaffin cells, but their effects are opposite. So, while GABA$_A$ agonists increase both basal and nicotine-evoked CA secretion, GABA$_B$ agonists (baclofen, 3-aminopropylphosphinic acid (3-APA) and GABA, when added together with bicuculline) inhibit the CA secretion evoked by nicotine and high KCl. The inhibitory effect of GABA$_B$ agonists on nicotine-evoked CA secretion is strongly dependent on the doses of nicotine used to stimulate cells. So, maximum responses (95% of inhibition) are achieved at doses of 1 μM, while only 30% inhibition is shown at 10 μM nicotine.

In respect to second messenger systems possibly implicated in this neuromodulatory effect, preliminary results indicate that:

1. GABA$_B$ agonists increase basal and forskolin-stimulated cyclic AMP levels, which indicates a possible coupling of GABA$_B$ receptors with adenylyl cyclase systems.
2. GTP analogues inhibit both the increase in basal and the decrease in nicotine-evoked release of CA. This could be due to a decrease in the affinity of the binding of the agonists to GABA$_B$ sites which would implicate the participation of regulatory G-proteins in these neuromodulatory effects.
3. Activation of protein kinase C by pretreatment of chromaffin cells with phorbol esters inhibits the neuromodulatory effects of GABA$_B$ receptors on basal CA secretion, an action possibly due to a reduction in the increase in cytosolic free Ca^{2+} mediating the exocytotic response of these agonists.

GABA Receptors and Motor Functions

A. AGMO and R. PAREDES

Department of Psychology, Universidad Anahuac, Mexico City, Mexico

Several GABAergic agents, such as receptor agonists or transaminase inhibitors, have been found to reduce ambulatory activity and to impair motor execution. In order to determine the contributions of subtypes of the GABA receptor to these effects, specific agonists were administered to male rats. The $GABA_A$ agonists THIP (4,5,6,7-tetrahydro-isooxazolo[5,4-*c*]pyridin-3-ol) and 3-aminopropylsulfonic acid (3-APS) reduced ambulatory activity, and THIP also impaired motor execution, as evaluated by a treadmill test. However, the actions of THIP could not be blocked by bicuculline or by picrotoxin. Since THIP does not bind to the $GABA_B$ receptor, it is possible that this drug acts at an additional GABA receptor, unrelated to $GABA_A$ or $GABA_B$ sites. 3-APS, when infused into the lateral cerebral ventricle, had no effects on motor functions, nor could the actions obtained after systemic administration be blocked by bicuculline or picrotoxin. Moreover, the motor effects of the GABA-transaminase inhibitor γ-acetylen-GABA or of the GABA reuptake inhibitor SKF 100330A cannot be blocked by bicuculline or by picrotoxin. This makes it unlikely that the $GABA_A$ receptor is important for motor effects of GABAergic agents. Baclofen has inhibitory actions on both ambulatory activity and motor execution when administered systemically or intracerebroventricularly. The effects are produced by the (*R*)-enantiomer, since racemic baclofen is half as potent and (*S*)-baclofen completely inactive. This coincides with data from receptor binding studies, and may suggest that the $GABA_B$ receptor participates in the control of motor functions. Interestingly, the doses required for reducing ambulatory activity are about half those required for impairing motor execution, while these two behaviors are affected by the same dose of THIP. Purposeful movement, such as running through a straight alley to receive a reward, is not affected by baclofen even in doses that impair motor execution. This seems to suggest that activation of the $GABA_B$ receptor reduces spontaneous activity but does not impede the motor aspects of goal-directed behaviors.

Pharmacology of GABA-evoked Responses in the Septohippocampal Neurons of the Rat: Implication of a Pertussis Toxin-sensitive G-Protein

P. DUTAR, O. RASCOL and Y. LAMOUR

Physiopharmacologie du Système Nerveux, INSERM U161, Paris, France

The effects of $GABA_A$ and $GABA_B$ agents have been studied in septohippocampal neurons (SHNs) in rats anaesthetized with urethane, using extracellular recordings and microiontophoresis. We studied the possible involvement of GABA in the rhythmical bursting activity (RBA) of the SHNs. GABA ($n = 43$), nipecotic acid ($n = 15$) and bicuculline (which blocked the GABA-induced inhibition; $n = 11$) had no effect on the RBA. Baclofen had no consistent effect on the RBA ($n = 37$); phaclofen ($n = 21$) induced a partial antagonism of the baclofen response in 9 out of 12 SHNs but did not affect the RBA. Responses to GABA and baclofen were next compared between control ($n = 10$) and pertussis toxin (PTX) -pretreated rats (2 μg injected intracerebroventricularly; $n = 12$). GABA inhibited virtually all neurons tested in control ($n = 43$) as well as PTX-treated rats ($n = 54$). In contrast, baclofen, which inhibited the activity of the SHNs in control rats (97%, $n = 61$), had either no effect in PTX-treated rats (59%, $n = 83$), or even an excitatory effect (14%). It inhibited only 26% of the SHNs. The frequency of the RBA was significantly decreased after PTX pretreatment. In conclusion, the response mediated by $GABA_B$ receptors in the medial septum depends on a PTX-sensitive G-protein, while the $GABA_A$ responses do not. The $GABA_A$ and $GABA_B$ agents are unable to disrupt the RBA, suggesting that it is not under control of a GABAergic input.

Effects of Phaclofen on Visually Evoked Responses in the Cat Striate Cortex

U. BAUMFALK and K. ALBUS

Abteilung für Neurobiologie, MPI für biophysikalische Chemie, Göttingen, FRG

The effects of the $GABA_B$ antagonist phaclofen on visually evoked responses (VER) of single neurons and on the baclofen-mediated inhibition of VER were investigated in anesthetized and paralysed cats by extracellular recording and iontophoretic drug application. Phaclofen (15–300 nA) facilitated, suppressed or did not change VER. Phaclofen-mediated inhibition was seen especially in simple and unimodal cells. In a few neurons VER were suppressed by lower ejection currents (40–100 nA) of phaclofen and became normal or even facilitated with higher iontophoretic currents. There seems to be a correlation between the phaclofen-mediated effects and the strength of the baclofen-induced inhibition: if the baclofen inhibition was strong ('low dose–high response relationship'), phaclofen facilitated or inhibited mostly without affecting the spontaneous activity. But when baclofen had to be applied in high doses and/or for a long time ('high dose–weak response relationship'), phaclofen normally had no effect on VER. The differential effects of phaclofen on single striate neurons might indicate differences in location and/or functional importance of receptor sites mediating baclofen effects. In 17 of 21 neurons tested so far phaclofen reversibly antagonized the baclofen-induced inhibition of VER. This antagonistic effect was seen irrespective of the effect of phaclofen alone. Phaclofen had no effect on GABA-induced inhibition tested in six neurons where it completely removed the baclofen-induced inhibition. Other putative $GABA_B$ antagonists were either much less effective than phaclofen (δ-aminovaleric acid (DAVA), Ba^{2+}) or interfered also with GABA-induced inhibition (Ba^{2+}). Our findings confirm that phaclofen is an effective and selective, but rather weak, antagonist of baclofen in the central nervous system of mammals. Preliminary findings indicate that in spite of this antagonistic property phaclofen does not change the directional selectivity of striate cortical cells. This suggests that inhibitory mechanisms mediated solely by the $GABA_B$ receptor are less important for functional properties of single neurons than are mechanisms mediated via the $GABA_A$ receptor. It is therefore supposed that the main task of $GABA_B$ receptors in striate cortex is to regulate the general excitability of the neurons.

Use of the Neonatal Rat Spinal Cord in Studies on the $GABA_B$ Receptor

D. E. JANE, P. C. K. POOK, D. C. SUNTER and J. C. WATKINS

Department of Pharmacology, School of Medical Sciences, Bristol, UK

A major feature of the action of the $GABA_B$ agonist baclofen is to suppress monosynaptic excitation in the spinal cord. In the hemisected isolated spinal cord preparation of the immature rat, this is manifested as an abolition or diminution of the amplitude of the fastest component of the dorsal root-evoked ventral root potential (DR-VRP) in an Mg^{2+}/AP5-containing medium which abolishes slower components of the DR-VRP mediated by *N*-methyl-D-aspartate (NMDA) receptors. A similar reduction in the fast component can be produced both by L-AP4, presumably via a presynaptic mechanism, and by non-NMDA receptor antagonists such as CNQX which block the postsynaptic action of the transmitter released from primary afferent terminals. Phaclofen (0.3–1 mM) reverses the action of baclofen (0.5–5 μM), while having little or no action on L-AP4 (0.5–5 μM) or CNQX (1–10 μM) -induced depressions of the fast component. Hydroxy-saclofen has an action which is similar to but approximately five times more potent than that of phaclofen. The actions of other baclofen and phaclofen analogues on this preparation will be discussed.

Note: L-AP4, L-2-amino-4-phosphonobutanoic acid; AP5, 2-amino-5-phosphonopentanoic acid; CNQX, 6-cyano-7-nitroquinoxaline-2,3-dione.

The $GABA_B$ Receptor Antagonist, Phaclofen, Reverses Stimulus-induced Disinhibition in the Dentate Gyrus

D. V. LEWIS and D. D. MOTT

Departments of Pediatries (Neurology), Neurobiology and Pharmacology, Duke University Medical Center, Durham, NC, USA

Recent studies have suggested the involvement of $GABA_B$ receptors in stimulus-induced disinhibition (SID). Using the hippocampal slice preparation we now report pharmacological evidence that supports this hypothesis.

Recurrent inhibition in the dentate gyrus was induced by stimulating the mossy fibers in the stratum lucidum of CA3b 5 ms prior to perforant path stimulation. This recurrent inhibition was significantly suppressed by a single conditioning pulse delivered to any one of several sites including the stratum lucidum, alveus or perforant path. SID was maximal 200–400 ms after the conditioning pulse and was completely reversed by 2000 ms. Application of baclofen (1 mM) significantly antagonized SID and this effect was reversible upon drug wash-out.

The results suggest that $GABA_B$ receptors are involved in SID in the dentate gyrus.

Biphasic Phaclofen-sensitive Actions of Baclofen on Excitatory Synaptic Transmission in the Frog Optic Tectum In Vitro

L. SIVILOTTI, G. MAZDA and A. NISTRI

Department of Pharmacology, St Bartholomew's Hospital Medical College, London, UK

In the optic tectum of the frog, optic nerve-evoked synaptic potentials are paradoxically enhanced by GABA or muscimol through a bicuculline-insensitive action (Nistri and Sivilotti, 1985). The aim of the present work was to establish the effect of compounds known to act selectively on $GABA_B$ receptors such as the agonist baclofen and the competitive antagonist phaclofen. The U_1 and U_2 synaptic components of the optic nerve-evoked field potential were recorded with a microelectrode from the superficial layers of the frog optic tectum maintained at 7 °C in vitro. Superfusion with 0.1–20 μM (−)-baclofen produced a slow onset, sustained depression of field potential amplitude. This response was dose-dependent, and reached a maximum with 5–10 μM baclofen (corresponding to 60% reduction in both the U_1 and U_2 waves). In 12 out of 15 preparations the depression was preceded by a transient enhancement of the U_1 and U_2 amplitude: this effect was apparently not dose-dependent (range 21–47% and 25–106% for the U_1 and U_2 waves, respectively). In a small minority of preparations baclofen evoked either depressions or enhancements. (+)-Baclofen was inactive at 1 μM. Phaclofen (0.5 mM) reversibly increased the U_2 wave amplitude by 58% and strongly reduced both the enhancing and depressant responses to 5 μM baclofen. Picrotoxin (50 μM) had no effect on either phase of the baclofen response, whereas it blocked the enhancing action of GABA or muscimol on this preparation. These results suggest that in the frog optic tectum the predominant action of baclofen was a reduction in excitatory synaptic transmission. This phenomenon might be partly due to depression of transmitter release from sensory afferents in a fashion similar to the action of baclofen on spinal cord synaptic transmission.

REFERENCE

Nistri, A. and Sivilotti, L. (1985) *Br. J. Pharmacol.*, **83**, 917–922.

$GABA_B$-mediated Inhibitory Postsynaptic Potentials Prevent Afterdischarge Bursts in Cultured Hippocampal Slices

A. T. MALOUF*, C. A. ROBBINS* and P. A. SCHWARTZKROIN*†

*Departments of *Neurological Surgery and †Physiology and Biophysics, University of Washington, Seattle, WA, USA*

The role of fast $GABA_A$-mediated inhibitory postsynaptic potentials (IPSPs) in the maintenance of normal, non-epileptic, activity in the hippocampus is well documented. The function of the slow $GABA_B$-mediated IPSP is not known, primarily due to the lack of an antagonist. We have applied the new $GABA_B$ antagonist, phaclofen, to hippocampal slice cultures to assess the role of the slow IPSP in the maintenance of normal hippocampal physiology. Slice cultures combine the advantage of long-term dissociated cultures and, in addition, maintain the proper organotypic structure and synaptic connections of the brain. We observed both spontaneous and stimulus-evoked triphasic PSPs in CA3 pyramidal neurons (excitatory PSP, EPSP, followed sequentially by fast and slow IPSPs), similar to the PSPs observed in acute slices following mossy-fiber stimulation. Our studies verify that phaclofen (1 mM) reversibly blocks the slow IPSP ($n = 14$). Following the loss of the slow IPSP, we observed a gradual increase in excitability, and the appearance of repetitive afterdischarge bursts. Slow IPSPs returned, and the frequency and duration of afterdischarges declined when phaclofen was washed out. Thus, in hippocampal slice cultures, slow IPSPs appear to prevent the formation of afterdischarge bursts.

Kainate and Quisqualate Block the $GABA_B$ Response of CA3 Hippocampal Pyramidal Cells

C. ROVIRA, M. GHO and Y. BEN-ARI

INSERM U29, Paris, France

It has been shown (Kehl and McLennan, 1985) that kainate (an excitatory amino-acid analog) reduces the slow $GABA_B$-mediated inhibitory postsynaptic potential (IPSP). We have used the voltage-clamp technique to study the effects of kainic and quisqualic acids on the slow inhibitory postsynaptic current (IPSC) recorded from CA3 hippocampal pyramidal neurons in vitro. Kainate (200–250 nM) induced an inward current. The reversal potential was unchanged. (−)-Baclofen (10 μM) and 5-hydroxytryptamine (30 μM), which activate a K^+ conductance controlled by a G-protein similar to the slow IPSC, were also blocked by kainate after a delay to the same extent as found with the slow IPSC. However, the postsynaptic $GABA_A$ response was not reduced. Similar effects were produced by quisqualate (5–10 μM). It is concluded that kainate and quisqualate block the postsynaptic $GABA_B$ responses.

REFERENCE

Kehl and McLennan (1985) *Exp. Brain Res.*, **60**, 299.

Effects of $GABA_A$ and $GABA_B$ Blockers on Paired-pulse Inhibition in the Hippocampus CA1 Region

M. F. POZZA, G. KARLSSON, F. BRUGGER and H.-R. OLPE

Research and Development Department, Pharmaceuticals Division, CIBA-Geigy, Basel, Switzerland

We have studied the effects of bicuculline methiodide, phaclofen, CGP 35348, a novel $GABA_B$ receptor antagonist, and L-baclofen on orthodromic paired-pulse inhibition in the hippocampal slice. An interpulse interval of 20 ms was used.

Bicuculline (1 μM) and L-baclofen (1 μM) reduced paired-pulse inhibition by $65 \pm 23\%$ (mean ± SD, $n = 8$) and $66 \pm 5\%$ ($n = 4$) respectively. In this paired-pulse paradigm, phaclofen (1 mM) was a weak but selective $GABA_B$ receptor antagonist since it attenuated the disinhibiting effect of L-baclofen (1 μM) on the inhibition. CGP 35348 also blocked the action of L-baclofen but was about 10 times more potent than phaclofen. Phaclofen had no effect against the action of the benzodiazepine midazolam which, in contrast to L-baclofen, strongly enhanced the paired-pulse inhibition. Phaclofen (1 mM) alone did not alter the inhibition, whereas CGP 35348 (100–300 μM) enhanced orthodromic, homosynaptic paired-pulse inhibition.

We describe opposite effects of the $GABA_B$ receptor blocker CGP 35348 and the $GABA_B$ agonist L-baclofen on paired-pulse inhibition in the hippocampus. We tentatively conclude that these actions occur at the level of presynaptic $GABA_B$ receptors located on tonically active GABAergic interneurons. The potentiation of GABAergic transmission by CGP 35348 conceivably results from blockade of presynaptic $GABA_B$ receptors located on GABAergic interneurons.

Blockade of the Late Inhibitory Postsynaptic Potential in Rat CA1 Hippocampal Neurons by 2-Hydroxysaclofen

N. A. LAMBERT, N. L. HARRISON* and T. J. TEYLER

*Department of Neurobiology, Northeastern Ohio Universities College of Medicine, Rootstown, OH, and *Department of Anesthesia and Critical Care, The University of Chicago, Chicago, IL, USA*

The baclofen analog 2-hydroxysaclofen (2-OH-S) is an antagonist at $GABA_B$ receptors in the guinea-pig ileum and cat spinal cord. We have studied the actions of 2-OH-S at postsynaptic $GABA_B$ receptors on rat CA1 pyramidal neurons, using intracellular recording in hippocampal slices. Stimulation of the Schaffer collateral/commissural fibers evoked a monosynaptic excitatory postsynaptic potential (EPSP), followed by an early (fast) inhibitory PSP (IPSP) and a late (slow) IPSP. In all neurons tested, 2-OH-S (50–200 μM) decreased the amplitude of the late IPSP by greater than 50% when added to the perfusion medium for 10 min. The EPSP and early IPSP were not altered by 2-OH-S. Complete or partial recovery of the late IPSP was obtained after washing for 20 min. A concentration of 20 μM 2-OH-S had no detectable effect. 2-OH-S decreased the amplitude of the late IPSP without changing its reversal potential and had no consistent effects on either resting membrane potential or input resistance.

At a concentration of 10–30 μM, baclofen induced a large (10 mV) hyperpolarization from the resting membrane potential, which was accompanied by a decrease in membrane input resistance. This effect of baclofen was blocked by 2-OH-S at concentrations that also blocked the late IPSP. These experiments indicate that 2-OH-S is a potent antagonist at postsynaptic $GABA_B$ receptors on rat CA1 hippocampal neurons, and should be useful for studying the role of postsynaptic $GABA_B$ receptors in inhibitory synaptic transmission in the brain.

ACKNOWLEDGEMENT

2-OH-S was a gift from Drs D. Kerr, J. Ong and R. Prager, Adelaide, Australia.

GABA Receptors in the Nervous System of an Insect, *Periplaneta americana*

C. MALECOT*, B. HUE*, S. D. BUCKINGHAM† and D. B. SATTELLE†

**Laboratoire de Neurophysiologie, Faculté de Médicine, Angers, France and †Department of Zoology, University of Cambridge, Cambridge, UK*

Motor neurons, sensory neurons and interneurons of the cockroach *Periplaneta americana* have been used to investigate the existence in the insect nervous system of receptors of the $GABA_A$ and $GABA_B$ subtypes, well known from studies on the mammalian brain. Experiments on a sensory neuron, giant interneuron 2 (GI 2) synapse reveal the presence of presynaptic GABA receptors but baclofen (1.0×10^{-4} M), the mammalian $GABA_B$ receptor agonist, fails to modify synaptic responses to GABA. Baclofen at this concentration is also ineffective in blocking the responses to GABA in the cell bodies of GI 2 and the fast coxal depressor motor neuron (D_f). Nevertheless, both these neuronal cell bodies possess a bicuculline-insensitive, GABA-operated Cl^- channel that is blocked by picrotoxin. A receptor/channel molecule of this type is also present in the postsynaptic membrane at inhibitory synapses onto GI 2. This receptor exhibits several differences from the mammalian $GABA_A$ receptor. For example, the pharmacological profile of the benzodiazepine binding site on the insect receptor resembles more closely that of the mammalian peripheral benzodiazepine binding site than the benzodiazepine binding site linked to the mammalian $GABA_A$ receptor.

Preliminary cell-attached, patch-clamp recordings have been obtained from GABA-operated Cl^- channels of dissociated cockroach neurons. Brief openings occurring in bursts are observed and two conductances can be detected (5.7 pS and 13.8 pS). Preliminary analysis of single-channel data reveals two open states and three closed states. Picrotoxin applied via perfusion of the patch pipette blocked GABA-operated Cl^- channels.

Bicuculline, though ineffective in blocking synaptic and cell-body (extrasynaptic) GABA receptors of GI 2, is an extremely effective antagonist of an insect α-bungarotoxin-sensitive nicotinic acetylcholine receptor present in GI 2 and motor neuron D_f, indicating that the antagonist specificity observed for mammalian transmitter-operated receptors/channels is not necessarily shared by the corresponding molecules of insects. The insect GABA-operated Cl^- channel is a major target for cyclodiene and cyclohexane insecticides. Pharmacological differences between GABA receptors of insects and vertebrates could conceivably be exploited in the design of novel, safer, insect control agents.

Index

4-ABPA *see* 4-Aminobutylphosphonic acid
Abstracts, 393–412
ACPPA, structure (*table*), 36
Adenosine, depressant effect, 349
Adenylyl cyclase
  assay of activity, 185
  forskolin-stimulated activity, effect of GABA and baclofen (*illus.*), 188
  functional coupling with $GABA_B$ receptors (*abstract*), 421
  $GABA_B$-mediated inhibition, Ca^{2+} influx and transmitter release, 144–154
    baclofen-mediated, 148–150
    discussion, 155–156
    GABA-mediated effects, 148–150
    pertussin sensitivity, 150–152
Adrenal medulla, regulation of catecholamine secretion (*abstract*), 423
β-Adrenoceptors
  autoradiography, 321
  defined by (−)-isoprenaline, Scatchard plots (*illus.*), 326
  iodopindolol binding (*illus.*), 324–325
  modulation by antidepressants, 322–327
Affinity column chromatography, 184, 190
Affinity gel chromatography, 187–188
Affinity gel, development for $GABA_B$ receptors, 183–193
4-Aminobutylphosphinic acid, structure (*table*), 36
4-Aminobutylphosphonic acid, structure (*table*), 36
cis-Aminocrotonic acid, structure (*table*), 32
Aminophylline, in antinociceptive action of baclofen (*illus.*), 375
3-Aminopropylphosphinic acid
  action on baclofen binding to $GABA_B$ receptors, 212
  decrease of amplitude of IPSPs (*illus.*), 212, 213
  $GABA_B$ agonist activity (*abstract*), 399
  structure (*illus.*), 13, 36
3-Aminopropylphosphonic acid
  analogue of GABA, 33
  effect on baclofen-induced augmentation of isoproterenol-stimulated cAMP accumulation, 135
  effect on isoproterenol-stimulated cyclic AMP accumulation (*illus.*), 134
  structure (*illus.*), 13, 36, 132
3-Aminopropylsulphonic acid
  antagonism of action of baclofen, 377–379
  characterization, 37–38
  in noradrenaline-induced inositol phosphate formation, 174
  structure (*illus.*), 38
2-OH-3-Aminopropylsulphonic acid, structure (*illus.*), 38
δ-Aminovaleric acid
  antagonism of action of baclofen, 377–379
  hydroxy-substituted analogues, 40–41
  (*R*) and (*S*) forms, as $GABA_B$ receptor ligands (*abstract*), 416
  structure (*table*), 32
Amitriptyline
  effect on biochemical indicators of GABA terminal function, 298
  enhanced dopamine release, autoradiography (*illus.*), 342
  GABA binding to $GABA_B$ receptors, 300
    autoradiographs, 320, 326
      (*illus.*), 322
    (*illus.*), 338
    Scatchard plots, 328–329
  Scatchard plots (*illus.*), 323, 330
  spiperone binding, autoradiography (*illus.*), 343
Anaesthesia, GABA effects, 102–103
Analgesia
  GABA analogues, effects, 102–103
  and $GABA_B$ receptors, 369–379

Antidepressants
activity on GABA synapses, 297–301
chronic administration, 319–332
effect on biochemical indicators of GABA terminal function (*illus.*), 298
and $GABA_B$ site upregulation, 297–305
prenatal administration,
neurochemical effects on dopamine receptors, 337–342
neurochemical effects on $GABA_B$ receptors, 335–346
repeated administration,
autoradiographic analysis, 319–332
$GABA_B$ receptor function (*table*), 304
reversal of loss of $GABA_B$ receptor-binding sites, 20–21
see also Depressiom
Antinociceptive Index, defined, 375
3-APA *see* 3-Aminopropylphosphinic acid
3-APPA *see* 3-Aminopropylphosphonic acid
3-APS *see* 3-Aminopropylphosphonic acid
Autoradiographic analysis
$GABA_B$ receptors, 319–332
methods, 320–327
summary, 327–332

Baclofen
(+)-baclofen
effects
[^{3}H]GABA overflow, different frequencies (*illus.*), 72
K^+-evoked overflow of [^{3}H]GABA, rat cortical synaptosomes (*illus.*), 84
the K^+-evoked release of [^{3}H]GABA (*illus.*), 91, 94
resting tone, distal colon (*illus.*), 113
inhibition of histamine-induced inositol phosphate formation, 168
(−)-baclofen,
bronchospasm, effects (*illus.*), 108
concentration–inhibition curves, electrically evoked overflow of [^{3}H]GABA (*illus.*), 88
defining labelling of $GABA_B$ receptors, 320
dose–response curve, human jejunum (*illus.*), 114
effects,
[^{3}H]GABA overflow, different frequencies (*illus.*), 72
forskolin-stimulated activity adenylyl cyclase (*illus.*), 188
K^+-evoked overflow of [^{3}H]GABA, rat cortical synaptosomes (*illus.*) , 84
the K^+-evoked release of [^{3}H]GABA (*illus.*), 91, 94
resting tone, distal colon (*illus.*), 113
ileum twitch response, dose-response curves (*illus.*), 106
inhibition of histamine-induced inositol phosphate formation, 168
inhibition of release of endogenous GABA, rat cortical synaptosomes (*illus.*), 86
analgesia studies, 369–379
analgesic effects, 102
antiepileptic action, 279–280
antinociceptive action, 370–371
receptors involved, 377–379
B-baclofen, structure (*illus.*), 39
blocking inward rectifying K^+ conductance, 279
cell culture electrophysiology, discussion, 218
cellular actions, neocortical neurons, 240–241
central and peripheral effects (*table*), 128
depression of population EPSPs (*illus.*), 214
effect of bath application on burst discharge, hippocampal region (*illus.*), 230, 231
effects,
antagonism by CGP 35348, 49–55
EPSPs, 249
field stimulation-induced vasoconstriction of artery (*illus.*), 105
IPSPs, 249–250
synaptic components, neocortical neurons, 241–254
synaptic transmission, 277–279
enantiomers, effects on release of endogenous GABA (*table*), 71
in epilepsyincidence of seizures (*table*), 361
experimental models of epilepsy, 351–362
results 352–354, (*table*), 358
inhibition of forskolin-stimulated cyclic AMP accumulation (*illus.*), 131
interaction of D-baclofen and L-baclofen (*abstract*), 401
noradrenaline and GABA release in cortical slices, antagonism by ethanol (*abstract*), 417
phaclofen sensitivity to excitatory synaptic transmission (*abstract*), 427
pharmacological action on muscle tone and spinal reflex transmission (*abstract*), 406

phosphonoanalogues as antagonists, 33–37
postsynaptic action, 197–199, 249
presynaptic inhibitory action (*illus.*), 202, 210–212
stereoselective effect on noradrenaline-stimulated cyclic AMP (*illus.*), 130
structural and chemical modifications (*abstract*), 417
structure (*illus.*), 13, 36, 132
Baclofen analogues
aryl and heterocyclic substitution at the β-carbon, 38–40
β-carbon substitution, structure (*illus.*), 39
phosphonic, structure (*illus.*), 36
sulphonic, structure (*illus.*), 38
Baclofen-epoxy-activated Sepharose-B
preparation, 184
structure (*illus.*), 188
Baclofen-mediated responses, effect of cyclic AMP analogue, 152–153
BC-3-APPA, structure (*illus.*), 39
BF-3-Aminosulphonic acid, structure (*illus.*), 38
Bicuculline
effect of GABA receptor antagonists on electrically-evoked overflow of [^{3}H]GABA (*illus.*), 89
effects,
GABA-induced inhibition of [^{3}H]GABA release (*illus.*), 85
intra- and extracellular recordings, epileptiform activity of hippocampus (*illus.*), 234
the K^+-evoked release of [^{3}H]GABA (*table*), 94
enantiomers, effects on release of endogenous GABA (*table*), 71
experimental models of epilepsy, results, 352–354
induction of epileptiform activity, 350, 357
and repeated administration of antidepressants, 313
sensitivity, action of GABA (*illus.*), 22
structure (*table*), 32
BP-3-APPA, structure (*illus.*), 39
BPG (β-phenyl-GABA), structure (*illus.*), 39
Bronchospasm, baclofen effects (*illus.*), 108
2-Butyl-GABA
effect on baclofen-induced augmentation of isoproterenol-stimulated cyclic AMP accumulation (*illus.*), 136
structure (*illus.*), 132
Ca^{2+}
in GABA modulation,
noradrenaline-induced inositol phosphate formation (*table*), 178
responses to histamine and noradrenaline, 176–178
Ca^{2+} channel currents
G-protein linked, 251
G-protein-mediated inhibition, voltage dependence, 266–268
$GABA_B$ inhibition, 262–264
$GABA_B$-mediated inhibition, 259–269
inward and outward, effect of baclofen (*illus.*), 266
neurotransmitter modulation and presynaptic inhibition, 251
mechanism, 265–266
Ca^{2+} channel ligand binding sites, effect on $GABA_B$ receptors (*abstract*), 420
Ca^{2+}-dependence, effect on the [K^+-evoked release of [^{3}H]GABA, human cortical synaptosomes (*table*), 93
CACA *see cis*-Aminocrotonic acid
Carbachol-induced inositol phosphate formation, 168–176
effect of GABA (*illus.*), 169
Catecholamine secretion, adrenal medulla, regulation of $GABA_B$ receptors (*abstract*), 423
CGP 20712A, binding to β-adrenoceptors (*illus.*), 324–325
CGP 35348
activity, 18, 19, 21
antagonism of baclofen-potentiated noradrenaline-stimulated adenylyl cyclase (*illus.*), 50
antagonism of effects of baclofen, 49–55
release of [^{3}H]GABA (*illus.*), 76
release of endogenous GABA (*illus.*), 77
rotarod test (*illus.*), 55
blocking depressant responses (*illus.*), 53
chemistry, 47–49
crossing blood–brain barrier, 350
effects,
alone, 56–58
behavioural, 58
enhancement of release of [^{3}H]GABA (*illus.*), 75
in epilepsy, 350–355
hyperpolarization of membrane potential (*illus.*), 51

CGP 35348 (*cont.*)
inhibition of binding,
GABA$_B$ and GABA$_B$ receptors (*table*), 48
receptor assays (*table*), 49
interaction with L-baclofen, EEG of freely moving rat (*illus.*), 54
reduction of IPSPs (*illus.*), 57
structure (*illus.*), 13, 48
summary and conclusions, 58–59
Chlorimipramine, chronic prenatal exposure, 336–337
β-p-Chlorophenyl-GABA *see* Baclofen
Chromatography
affinity column chromatography, GABA$_B$ receptors, 183–193
gel filtration column chromatography, 184
Circulation, effects of GABA$_B$, 102–106
Citalopram
effect on biochemical indicators of GABA terminal function, 298
enhancing GABA$_B$ binding, 300
Cortical synaptosomes
release-regulating GABA$_B$ receptors,
human, 92–95
rat, 82–86

DAVA *see* *δ*-Aminovaleric acid
2-Decyl GABA
effect on baclofen-induced augmentation of isoproterenol-stimulated cyclic AMP accumulation (*illus.*), 136
lack of effect on baclofen-induced forskolin-stimulated cyclic AMP accumulation (*illus.*), 137
structure (*illus.*), 132
Dentate gyrus, reversal of stimulus-induced disinhibition, by phaclofen (*abstract*), 427
Depression
animal models and GABA$_B$ receptors, 301–303
approaches to treatment, 335–337
involvement of GABA, 313–316
see also Antidepressants
Desipramine
in autoradiography of GABA$_B$ receptors, 320, 326 (*illus.*), 322, 328–329
comparison of effects on GABA$_B$ receptors and *β*-adrenoceptors (*table*), 332
effect on biochemical indicators of GABA terminal function, 298
effects of repeated administration, GABA$_B$ receptor function (*table*), 305
enhancing GABA$_B$ binding, 300
Scatchard plots (*illus.*), 323, 330
Desmethylimipramime
effects on 5-hydroxytryptamine receptors and GABA$_B$ binding (*table*), 311
low affinity component in GABA$_B$ receptors, 313
Detergents, GABA$_B$ receptor solubilization (*table*), 186
5,7-Dehydroxytryptamine, analgesia studies, 371–373
DMI *see* Desmethylimipramime
DOC-Na, in chromatographic separation of GABA$_B$ receptors, 185–191
Dopamine receptors, effects of prenatal administration of antidepressants, 337–342
Dopamine system, ontogeny, 336–337
Dorsal root ganglion neurons
effect of baclofen (*illus.*), 260, 263, 264, 265
GABA$_B$-mediated inhibition of Ca^{2+} currents, 259–269
lack of effect of baclofen when pretreated with pertussis toxin (*illus.*), 261, 262

EDA *see* Ethylenediamine
Epilepsy, temporal lobe, 279
Epileptiform activity
induction, 350
postsynaptic GABA$_B$ receptors, 230–234
suppression by baclofen, 279–280
EPSPs, effects of baclofen, 249
Ethanol, antagonism of baclofen-induced modification of noradrenaline and GABA release in cortical slices (*abstract*), 417
Ethylenediamine, release of endogenous GABA from ileum, 108

Fengabine
effect on biochemical indicators of GABA terminal function, 298
enhancing GABA$_B$ binding, 300
Fluoxetine
effect on biochemical indicators of GABA terminal function, 298
enhancing GABA$_B$ binding, 300
Forskolin
in antinociceptive action of baclofen (*illus.*), 375
cyclic AMP accumulation, inhibition by baclofen (*illus.*), 131

stimulation of activity of adenylyl cyclase, effect of GABA, 188
stimulation of cAMP,
baclofen-induced,
effect of 2-decyl GABA, 137
effect of 2-OH-saclofen, 134
baclofen-inhibited, 131

G-proteins
action of pertussis toxin, 8-9
in antinociceptive action of baclofen, 375-377
pertussis toxin-sensitive,
involvement in inhibition of Ca^{2+} currents, 259-262
and pharmacology of $GABA_B$ receptors (*abstract*), 425
receptor autoradiography, 15-17
GABA
[^{3}H]GABA, release,
as function of stimulation frequency (*illus.*), 67
K^+-stimulated (*illus.*), 69
modulation by $GABA_B$ receptors, cortical slices (*abstract*), 421
analogues,
3-aminopropylphosphinic acid binding affinity (*abstract*), 399
β-carbon substituted, 38-40
hydroxy-substituted analogues, 40-41
hydroxylated analogues (*abstract*), 416
in noradrenaline-induced inositol phosphate formation (*table*), 174
phosphonic analogues, structure (*illus.*), 36
phosphonoanalogues as antagonists, 33-37
sulphonic analogues, 37-38
see also Bicuculline; Picrotoxin
bicuculline sensitivity (*illus.*), 22
concentration-dependent inhibition of the K^+-evoked release of [^{3}H]GABA (*illus.*), 90
concentration-inhibition curves, electrically evoked overflow of [^{3}H]GABA (*illus.*), 88
conformation for attachment at $GABA_A$ receptors, 30-31 and depression, 313-316
2-decyl *see* 2-Decyl GABA
dose-response curve, human jejunum (*illus.*), 114
effects,
field stimulation-induced vasoconstriction of artery (*illus.*), 105
forskolin-stimulated activity adenylyl cyclase (*illus.*), 188
resting tone, distal colon (*illus.*), 113
3-heteroaromatic GABA, synthesis and binding (*abstract*), 415
hippocampal cells, action on (*illus.*), 198
inhibition of K^+-evoked overflow of [^{3}H]GABA, rat cortical synaptosomes (*illus.*), 83
ionic mechanisms, 284-285
multiphasic membrane effect, 280-284
postsynaptic action, 200
release in cortical slices, baclofen-induced modifications, antagonism by ethanol (*abstract*), 417
release as function of stimulation frequency (*illus.*), 67
structure (*illus.*), 32, 132
uptake inhibitors, effect on [^{3}H]GABA release (*table*), 87
uptake system, potentiation by nipecotic acid, 175-176
GABA receptors, antagonists, effect on electrically-evoked overflow of [^{3}H]GABA (*illus.*), 89
GABA-mediated IPSPs, prevention of afterdischarge bursts (*abstract*), 428
$GABA_A$ agonists, in noradrenaline-induced inositol phosphate formation, 174
$GABA_A$ effects, ileum, 109
$GABA_A$ receptors
contrasted with $GABA_B$ receptors (*table*), 8
hippocampal cell culture electrophysiology, 223
influence of Ca^{2+}, binding sites (*table*), 7
$GABA_B$
effects,
at intestinal level, 108-114
on circulation, 102-106
on respiration, 106-108
on urogenital system, 114-116
effects, peripheral,
review, 101-117
summary, 116-117
(*table*), 117
$GABA_B$ antagonists, investigation of relative potencies (*abstract*), 422
$GABA_B$ receptor-mediating inhibition of substance P, capsaicin-sensitive rat neurons (*abstract*), 419

$GABA_B$ receptors
activation, 8–9
associated actions (*table*), 17
effects, 17–21
adrenal medulla, bovine, regulation of catecholamine secretion (*abstract*), 423
affinity gel, development, 183–193
agonists,
and antagonists, 29–42
structures (*illus.*), 13
and analgesia, 369–379
antagonists,
baclofen, 37–38
effects within CNS, 19–21
structures (*illus.*), 13
sulphonic analogues of GABA, 37–38
antidepressant drug actions, 309–313
autoradiography, 10–12
analysis, following chronic antidepressant drug administration, 319–332
and G-proteins, 15–17
binding,
assay, 185
effect of GTP analogues (*table*), 187
biochemical aspects, 125–160
in brain, confirmation of existence, 22
characteristics, 8–9
contrasted with $GABA_A$ receptors (*table*), 8
control of socio-sexual behaviour in rats (*abstract*), 404
effects,
calcium channel ligand binding sites (*abstract*), 420
desmethylimipramime (*table*), 311
multiple effects, 143–144
zimelidine (*table*), 311
electrophysiological aspects, 9–10, 195–289
and epilepsy, experimental models, 349–362
(*abstract*), 411
evidence for existence of $GABA_B$ receptor subpopulations in CNS, 127–138
functional coupling,
with adenylyl cyclase (*abstract*) 421
with phosphatidylinositol turnover (*abstract*), 418
heterogeneity, 12–15
high and low affinity, 312
historical aspects, 3–8
hyperpolarization of dopamine-containing rat substantia nigra neurons (*abstract*), 409
influence of Ca^{2+}, binding sites (*table*), 7
inhibition,
adenylyl cyclase, 141–143
Ca^{2+} channel currents, pharmacology, 268–269
cyclic AMP accumulation, 141–156
involvement in release of GABA, 65–78
absence of effects of muscimol, bicuculline, 69–70
effect of baclofen, 70–73
effect of $GABA_B$ antagonists, 73–77
frequency of stimulation, 67–68
K^+ concentration, 68
methodology, 66–67
summary and conclusions, 77–78
model, original hypothesis (*illus.*), 4
modulation of [^{3}H]GABA release, cortical slices (*abstract*), 421
and motor functions (*abstract*), 424
multiplicity,
evidence for, 128–137
(*table*), 129
neocortical IPSPs, frequency dependence, 239–254
neonatal rat spinal cord (*abstract*), 426
partial agonists or antagonists, 41–42
peripheral autonomic sympathetic nerve terminals, 103–117
peripheral organs, specific binding sites (*abstract*), 401
Periplaneta americana (*abstract*), 410
pharmacology, implication of a pertussis toxin-sensitive G-protein (*abstract*), 405
postmortem, rat and man, 314
postsynaptic $GABA_B$ receptors, 197–200
control of excitability, 229–230
epileptiform activity, 230–234
pre- and postsynaptic location, 10–12
prejunctional inhibition of neurally induced airway constriction (*abstract*), 400
presynaptic $GABA_B$ receptors, 201–203, 207–219, 224–229
properties, 127–128
rat spinal cord, neonatal (*abstract*), 426
release-regulating, 81–95
human cortical synaptosomes, 92–95
rat cortical slices, 86–90
rat cortical synaptosomes, 82–86
rat substantia nigra synaptosomes, 90–92
and slow IPSPs, 287–289
solubilization and partial purification, 183–193

materials and methods, 183-185
results, 185-190
discussion, 191-193
solubilized, effect of Ca^{2+} channel blockers (*illus.*), 187
specific racemic ligands, 3-heteroaromatic GABA (*abstract*), 415
summary and closing remarks, 387-390
and transmitter release, 63-65
$GABA_B$-mediated inhibition, Ca^{2+} currents, 259-269
$GABA_B$-mediated responses, effect of cyclic AMP analogue, 152-153
GABAergic dysfunction, affective disorders, 336-337
GABAergic inhibitory neurons, presynaptic $GABA_B$ receptors, 210-212
GABOB *see* β-hydroxy-GABA
Gel filtration column chromatography, 184, 188-190
Glutamate release, inhibition, $GABA_B$ receptors (*illus.*), 268

Haloperidol, displacement of dopamine binding sites, 337
3-Heteroaromatic GABA, synthesis and binding (*abstract*), 395
Hippocampus
CA1 region, $GABA_A$ and $GABA_B$ blockers, effect on paired pulse inhibition (*abstract*), 430
CA3 region, $GABA_B$ response blocked by kainite and quisqualate (*abstract*), 429
cell culture electrophysiology, 208
field potentials, 208-209
intracellular recordings, 209
effects of CGP 35348 and phaclofen (*abstract*), 407
$GABA_B$ receptors, development and control of excitability, 223-238
$GABA_B$-mediated IPSPs prevent afterdischarge bursts (*abstract*), 428
hyperpolarization by baclofen,
$GABA_B$ receptor sites (*abstract*), 407
(*illus.*), 274
ionic mechanism, 274-276
physiology and pharmacology, 276-277
postsynaptic action, 277-279
long-term potentiation, facilitated by baclofen (*abstract*), 410
multiphasic membrane effects, baclofen, 280-284
GABA, 280-284
paired-pulse depression of GABAergic IPSPs (*abstract*), 409
postsynaptic $GABA_B$ receptors, 197-200, 224-225
control of excitability, 229-230
epileptiform activity, 230-234
discussion, 234-235
presynaptic $GABA_B$ receptors, 201-203, 207-219, 225-228
control of excitability, 228-229
in immature tissue, 226-228
tetanus toxin-induced damage, and presynaptic $GABA_B$ receptors (*abstract*), 405
Histamine-induced inositol phosphate formation, 164-168
effect of baclofen and other GABA mimetics, 165-168
effect of GABA (*illus.*), 169
GABA analogues (*table*), 179
GABA inhibition, 178-180
inhibition by muscimol (*illus.*), 168
inhibition by stereospecific baclofen (*illus.*), 167
time-course (*illus.*), 166
Homotaurine
effect on field stimulation-induced vasoconstriction of artery (*illus.*), 105
ileum twitch response, dose-response curves (*illus.*), 106
5-HT_2 *see* 5-Hydroxytryptamine receptors
β-Hydroxy-GABA, 40-41
cis-4-Hydroxynipecotic acid *see* Nipecotic acid
2-Hydroxysaclofen
blockade of late IPSPs in CA1 hippocampus region (*abstract*), 410
blocking of baclofen,
population EPSPs (*illus.*), 215, 216
presynaptic effects (*illus.*), 217
blocking of IPSPs at postsynaptic $GABA_B$ receptors, 213-214
effects,
baclofen-induced augmentation of isoproterenol-stimulated cyclic AMP accumulation, 133
(*illus.*), 136
baclofen-induced forskolin-stimulated cyclic AMP accumulation (*illus.*), 134
structure (*illus.*), 13, 38, 132
5-Hydroxytryptamine, analgesia studies, 371-373

5-Hydroxytryptamine receptors
effect of desmethylimipramime (*table*), 311
effect of zimelidine (*table*), 311
ontogeny, 336–337

ICI 118551, binding to β-adrenoceptors (*illus.*), 324–325
Imipramine, effects of repeated administration, $GABA_B$ receptor function (*table*), 305
Inositol 1,4,5-trisphosphate, inositol phospholipid metabolism, 161–180
Inositol phosphates, formation, histamine-induced, 164–165
ratios, effect of GABA (*illus.*), 173
Inositol phospholipid metabolism
GABA-induced phosphate formation, 163–164
pathways (*illus.*), 162
transmembrane signalling mechanism, 161–180
Interpeduncular nucleus, presynaptic location of $GABA_B$ receptors, 10–12
Intestine, effects of $GABA_B$, 108–114
Iodopindolol binding
Scatchard plots (*illus.*), 326
to β-adrenoceptors, effect of antidepressants (*table*), 331
Iprindole
chronic prenatal exposure, 336–337
enhanced dopamine release, autoradiography (*illus.*), 342
GABA binding to $GABA_B$ receptors (*illus.*), 343
spiperone binding, autoradiography (*illus.*), 343
IPSPs
early/late, 242
effects of baclofen (*illus.*), 246–247
fast, ionic mechanisms, 285–287
$GABA_B$ evoked, lateral geniculate nucleus (*abstract*), 405
neocortical neurons, $GABA_B$ receptor-mediated frequency dependence, 239–254
properties (*table*), 243
role of late component in $GABA_B$ receptors, 228–234
slow, $GABA_B$ receptors, 287–289
Isoguvacine, $GABA_A$ agonist, 174
(−)-Isoprenaline, binding to β-adrenoceptors (*illus.*), 324–325

Jejunum human, 'motility index' (*abstract*), 399

K^+ conductance
activation causing slow IPSPs, mammalian brain, 288–289
baclofen-induced hyperpolarization, 273–274
ionic mechanism, 274–276
pharmacology, 276
physiology, 276
$GABA_B$ receptor-mediated, 273–289
Kainic acid, experimental models of epilepsy, 357
Kainite, and quisqualate, blocking of $GABA_B$ response (*abstract*), 429

Leaky-condenser model, K^+ conductance (*illus.*), 275
Lung, $GABA_B$ effects, 106–108

Methoxamine, partial agonist, noradrenaline-induced inositol phosphate formation, 170
Mg^{2+} levels
experimental models of epilepsy,
methods, 350–352
results, 352–354
(*table*), 357
Mianserin, chronic prenatal exposure, 336–337
Morphine, enhancement of locomotor activity, 337
Motility index, human jejunum (*abstract*), 399
Motor functions, and $GABA_B$ receptors (*abstract*), 424
Muscimol
concentration–inhibition curves, electrically evoked overflow of $[^3H]$GABA (*illus.*), 88
effects,
field stimulation-induced vasoconstriction of artery (*illus.*), 105
K^+-evoked overflow of $[^3H]$GABA, rat cortical synaptosomes (*illus.*), 84
K^+-evoked release of $[^3H]$GABA (*illus.*), 91, 94
enantiomers, effects on release of endogenous GABA (*table*), 71
inhibition of histamine-induced inositol phosphate formation, 168
structure (*table*), 32

Neocortical neurons, IPSPs, $GABA_B$ receptor-mediated frequency dependence, 239–254
Neurotensin receptor binding sites, 11

Nifedipine
  and Ca^{2+}, in GABA modulation, 177
  effect on baclofen-mediated responses (*illus.*), 153–154
Nipecotic acid
  effects on IPSPs, neocortical neurons, 244–245
  potentiation, GABA uptake system, 175–176
  prolongation of hippocampal IPSPs, 201
Nomifensine
  chronic prenatal exposure, 336–337
  high affinity component in $GABA_B$ receptors, 312
Noradrenaline
  analgesia studies, 371–373
  and GABA release in cortical slices, baclofen-induced modifications, antagonism by ethanol (*abstract*), 417
  ontogeny of noradrenaline system, 336–337
  tritiated, GABA-induced depression of release (*illus.*), 5
Noradrenaline-induced inositol phosphate formation, 168–176
  EC_{50}, 171
  effect of GABA (*illus.*), 169
  GABA potentiation, concentration-effect curves (*illus.*), 172
  (*illus.*), 170

8-OH-DPAT, population EPSPs (*illus.*), 214

Paroxetine
  autoradiographs (*illus.*), 322, 328–329
  in autoradiography of $GABA_B$ receptors, 320, 326
  Scatchard plots (*illus.*), 323, 330
Penicillin, induction of epileptiform activity, 350, 357
PENS, structure (*illus.*), 38
Periplaneta americana, $GABA_B$ receptors (*abstract*), 431
Pertussis toxin
  in antinociceptive action of baclofen (*illus.*), 374
  blocking postsynaptic baclofen in rat dorsolateral septal nucleus (*abstract*), 408
  G-proteins,
    and pharmacology of $GABA_B$ receptors (*abstract*), 425
    involvement in inhibition of Ca^{2+} currents, 259–262
    action on G-proteins, 8–9
  effects,
    GABA receptor-binding *in vitro* (*illus.*), 16
    GABA receptor-binding *in vivo* (*table*), 15
Phaclofen
  antagonism of GABA-induced inhibition of [^{3}H]GABA release, rat cortical synaptosomes (*table*), 85
  antagonism of inhibitory effect of (−)-baclofen on release of [^{3}H]GABA (*illus.*), 76
  effects,
    GABA receptor antagonists on electrically-evoked overflow of [^{3}H]GABA (*illus.*), 89
    visually evoked responses (*abstract*), 425
    [^{3}H]GABA overflow, different frequencies (*illus.*), 74
  enhancement of release of [^{3}H]GABA (*illus.*), 75
  experimental models of epilepsy, 355
  reversal of stimulus-induced disinhibition, dentate gyrus (*abstract*), 427
  selective antagonism of $GABA_B$ responses (*illus.*), 199
  sensitivity to baclofen action on excitatory synaptic transmission (*abstract*), 427
  structure (*illus.*), 13, 36
  synthesis, 33
Phenylephrine, partial agonist, noradrenaline-induced inositol phosphate formation, 170
β-Phenyl-GABA, structure (*illus.*), 39
Phosphatidylinositol-4,5-bisphosphate, in inositol phospholipid metabolism, 162
Phosphatidylinositol, turnover, functional coupling with $GABA_B$ receptors (*abstract*), 418
Phosphoinositidase C *see* Phospholipase C
Phospholipase C, PI pathway, 161
Phosphono-baclofen *see* Phaclofen
PIC *see* Phospholipase C
Picrotoxin
  effect on GABA-induced inhibition of [^{3}H]GABA release (*illus.*), 85
  experimental models of epilepsy, 351–355
Postsynaptic inhibitory potentials, 200–201
Protein kinase C, inositol phospholipid metabolism (*illus.*), 162
PTX *see* Pertussis toxin
Pyramidal cells, spatial responsiveness to ionophoretically applied GABA (*illus.*), 284

Quisqualate, blocking of $GABA_B$ response (*abstract*), 429
Respiration, effects of $GABA_B$, 106–108
Ro 201724, in antinociceptive action of baclofen (*illus.*), 375
Rotarod test (*illus.*), 55

Saclofen
- characterization, 37–38
- structure (*illus.*), 13, 38
- *see also* 2-Hydroxysaclofen

Sepharose-B
- baclofen-epoxy-activated, preparation, 184
- structure (*illus.*), 188

Serotonin, activation of K^{2+} channel by baclofen, 197, 199
Sexual behaviour, GABAergic control in female rats (*abstract*), 403
SL 75102
- activity at $GABA_B$ receptors, 31
- structure (*table*), 32

Sodium deoxycholate, in chromatographic separation of $GABA_B$ receptors, 185–191
Spiperone
- binding,
  - autoradiography, 338
  - mean values (*table*), 339
- displacement of dopamine binding sites, 337

SR 95531
- effects,
  - GABA receptor antagonists on electrically evoked overflow of [^{3}H]GABA (*illus.*), 89
  - GABA-induced inhibition of [^{3}H]GABA release (*illus.*), 85
  - the K^+-evoked release of [^{3}H]GABA (*table*), 94

SR 95 series, structure (*illus.*), 32
Substance P
- blocking inward rectifying K^+ conductance, 278–279
- $GABA_B$ receptor-mediating inhibition, capsaicin-sensitive rat neurons (*abstract*), 419
- neuronal release of GABA, 110–111
- selective blocking by baclofen (*illus.*), 278

Substance P-cholinergic-GABA processing mechanisms, 111
Substantia nigra synaptosomes, release-regulating $GABA_B$ receptors, rat, 90–92
Suicide
- demographic details (*table*), 315
- $GABA_B$ receptors (*tables*), 315, 316
- and GABA, 313–316
- *see also* Depression

THIP (4,5,6,7-Tetrahydroisooxazolo[5,4-*c*] pyridin-3-ol)
- in CGP 35348 studies, 52–53
- and baclofen, experimental models of epilepsy, 354
- $GABA_A$ agonist, depressant effect, 349

Transmembrane signalling, inositol phospholipid metabolism, 161–180

Urogenital system, effects of $GABA_B$, 114–116

VDCC, effect of nifedipine, 153–154
Viloxazine
- effect on biochemical indicators of GABA terminal function, 298
- enhancing $GABA_B$ binding, 300

Visually evoked responses, effect of phaclofen (*abstract*), 425

Z-5, structure (*table*), 32
Zimelidine
- effects on 5-hydroxytryptamine receptors and $GABA_B$ binding (*table*), 311
- low affinity component in $GABA_B$ receptors, 313

Index compiled by June Morrison